Applied and Computational Measurable Dynamics

Mathematical Modeling and Computation

About the Series

The SIAM series on Mathematical Modeling and Computation draws attention to the wide range of important problems in the physical and life sciences and engineering that are addressed by mathematical modeling and computation; promotes the interdisciplinary culture required to meet these large-scale challenges; and encourages the education of the next generation of applied and computational mathematicians, physical and life scientists, and engineers.

The books cover analytical and computational techniques, describe significant mathematical developments, and introduce modern scientific and engineering applications. The series will publish lecture notes and texts for advanced undergraduate- or graduate-level courses in physical applied mathematics, biomathematics, and mathematical modeling, and volumes of interest to a wide segment of the community of applied mathematicians, computational scientists, and engineers.

Appropriate subject areas for future books in the series include fluids, dynamical systems and chaos, mathematical biology, neuroscience, mathematical physiology, epidemiology, morphogenesis, biomedical engineering, reaction-diffusion in chemistry, nonlinear science, interfacial problems, solidification, combustion, transport theory, solid mechanics, nonlinear vibrations, electromagnetic theory, nonlinear optics, wave propagation, coherent structures, scattering theory, earth science, solid-state physics, and plasma physics.

Erik M. Bollt and Naratip Santitissadeekorn, *Applied and Computational Measurable Dynamics*

Daniela Calvetti and Erkki Somersalo, *Computational Mathematical Modeling: An Integrated Approach Across Scales*

Jianke Yang, *Nonlinear Waves in Integrable and Nonintegrable Systems*

A. J. Roberts, *Elementary Calculus of Financial Mathematics*

James D. Meiss, *Differential Dynamical Systems*

E. van Groesen and Jaap Molenaar, *Continuum Modeling in the Physical Sciences*

Gerda de Vries, Thomas Hillen, Mark Lewis, Johannes Müller, and Birgitt Schönfisch, *A Course in Mathematical Biology: Quantitative Modeling with Mathematical and Computational Methods*

Ivan Markovsky, Jan C. Willems, Sabine Van Huffel, and Bart De Moor, *Exact and Approximate Modeling of Linear Systems: A Behavioral Approach*

R. M. M. Mattheij, S. W. Rienstra, and J. H. M. ten Thije Boonkkamp, *Partial Differential Equations: Modeling, Analysis, Computation*

Johnny T. Ottesen, Mette S. Olufsen, and Jesper K. Larsen, *Applied Mathematical Models in Human Physiology*

Ingemar Kaj, *Stochastic Modeling in Broadband Communications Systems*

Peter Salamon, Paolo Sibani, and Richard Frost, *Facts, Conjectures, and Improvements for Simulated Annealing*

Lyn C. Thomas, David B. Edelman, and Jonathan N. Crook, *Credit Scoring and Its Applications*

Frank Natterer and Frank Wübbeling, *Mathematical Methods in Image Reconstruction*

Per Christian Hansen, *Rank-Deficient and Discrete Ill-Posed Problems: Numerical Aspects of Linear Inversion*

Michael Griebel, Thomas Dornseifer, and Tilman Neunhoeffer, *Numerical Simulation in Fluid Dynamics: A Practical Introduction*

Khosrow Chadan, David Colton, Lassi Päivärinta, and William Rundell, *An Introduction to Inverse Scattering and Inverse Spectral Problems*

Charles K. Chui, *Wavelets: A Mathematical Tool for Signal Analysis*

Applied and Computational Measurable Dynamics

Erik M. Bollt
Clarkson University
Potsdam, New York

Naratip Santitissadeekorn
University of North Carolina
Chapel Hill, North Carolina

Society for Industrial and Applied Mathematics
Philadelphia

10 9 8 7 6 5 4 3 2 1

Library of Congress Cataloging-in-Publication Data

Bollt, Erik M., author.
Applied and computational measurable dynamics / Erik M. Bollt, Clarkson University, Potsdam, New York, Naratip Santitissadeekorn, University of North Carolina, Chapel Hill, North Carolina.
pages cm. – (Mathematical modeling and computation)
Includes bibliographical references and index.
ISBN 978-1-611972-63-4
1. Dynamics. 2. Graph theory. 3. Ergodic theory. 4. Dynamics–Data processing.
I. Santitissadeekorn, Naratip, author. II. Title.
QA845.B65 2013
515'.39–dc23

2013027536

I would like to thank my boys,
Keith, Scott, and Adam,
and especially my wife Elizabeth,
who have so patiently let me do "my thing"
on this work since forever, and
who have been the heart in my life.

Erik Bollt

I gratefully dedicate this book to my mother,
Prapai Pupattanakul, and my wife,
Thanomlak Angklomklieo,
for their tremendous support.

Naratip Santitissadeekorn

Contents

Preface

Measurable dynamics has traditionally referred to ergodic theory, which is in some sense a sister topic to dynamical systems and chaos theory. However, the topic has until recently been a highly theoretical mathematical topic which is generally less obvious to those practitioners in applied areas, who may not find obvious links to practical, real-world problems. During the past decade, facilitated by the advent of high-speed computers, it has become practical to represent the notion of a transfer operator discretely but to high resolution thanks to rapidly developing algorithms and new numerical methods designed for the purpose. An early book on this general topic is *Cell-to-Cell Mapping: A Method of Global Analysis for Nonlinear Systems* [167] from 1987.[1] A tremendous amount of progress and sophistication has come to the empirical perspective since then.

Rather than discussing the behaviors of complex dynamical systems in terms of following the fate of single trajectories, it is now possible to empirically discuss global questions in terms of evolution of density. Now complementary to the traditional geometric methods of dynamical systems transport study, particularly by stable and unstable manifold structure and bifurcation analysis, we can analyze transport activity and evolution by matrix representation of the Frobenius–Perron transfer operator. While the traditional methods allow for an analytic approach, when they work, the new and fast-developing computational tools discussed here allow for detailed analysis of real-world problems that are simply beyond the reach of traditional methods. Here we will draw connections between the new methods of transport analysis based on transfer operators and the more traditional methods. The goal of this book is not to become a presentation of the general topic of dynamical systems, as there are already several excellent textbooks that achieve this goal in a manner better than we can hope. We will bring together several areas, as we will draw connections between topological dynamics, symbolic dynamics, and information theory to show that they are also highly relevant to the Ulam–Galerkin representations. In these parts of the discussion, we will compare and contrast notions from topological dynamics to measurable dynamics, the latter being the first topic of this book. That is, if measurable dynamics means a discussion of a dynamical system in consideration of how much, how big, and other notions that require measure structure to discuss transport rates, topological dynamics can be considered as a parallel topic of study that asks similar questions in the absence of a measure that begets scale. As such, the mechanism and geometry of transport are more the focus. Therefore, including a discussion of topological dynamics in our primary discussion here on measurable dynamics should be considered complementary.

[1] Recent terminology has come to call these "set oriented" methods.

There are several excellent previous related texts on mathematical aspects of transfer operators which we wish to recommend as possible supplements. In particular, Lasota and Mackay [198] give a highly regarded discussion of the theoretical perspective of Frobenius–Perron operators in dynamical systems, whose material we overlap in as far as we need these elements for the computational discussion here. Boyarsky and Gora [50] also give a sharp presentation of an ensembles density perspective in dynamical systems, but more specialized for one-dimensional maps, and some of the material and proofs therein are difficult to find elsewhere. Of course the book by Baladi [11] is important in that it gives a thoroughly rigorous presentation of transfer operators, including a unique perspective. We recommend highly the book by Zhou and Ding, [324], which covers a great deal of theoretical information complementary to the work discussed in this book, including Ulam's method and piecewise constant approximations of invariant density, piecewise linear Markov models, and especially analysis of convergence. Also an in-depth study can be found concerning connections of the theory of Frobenius–Perron operators and the adjoint Koopman operator, as well as useful background in measure theory and functional analysis. The book by McCauley [215] includes a useful perspective regarding what is becoming a modern perspective on computational insight into behaviors of dynamical systems, especially experimentally observed dynamical systems. That is, finite realizations of chaotic data can give a great deal of insight. This is a major theme which we also develop here toward the perspective that a finite time sample of a dynamical system is not just an estimate of the long time behavior, as suggested perhaps by the traditional perspective, but in fact finite time samples are most useful in their own right toward understanding finite time behavior of a dynamical system. After all, any practical, real-world observation of a dynamical system can be argued to exist only during a time window which cannot possibly be infinite in duration.

There are many excellent textbooks on the general theory of dynamical systems, clearly including Robinson [268], Guckenheimer and Holmes [146], Devaney [95], Alligood, Sauer, and Yorke [2], Strogatz [301], Perko [251], Meiss [218], Ott [244], Arnold [4], Wiggins [316], and Melo and van Strein [89], to name a few. Each of these has been very popular and successful, and each is particularly strong in special aspects of dynamical systems as well as broad presentation. We cannot and should not hope to repeat these works in this presentation, but we do give what we hope is enough background of the general dynamical systems theory in order that this work can be somewhat self-contained for the nonspecialist. Therefore, there is some overlap with other texts insofar as background information on the general theory is given, and we encourage the reader to investigate further in some of the other cited texts for more depth and other perspectives. More to the point of the central theme of this textbook, the review article by Dellnitz and Junge [87] and then later the Ph.D. thesis by Padberg [247] (advised by Dellnitz) both give excellent presentations of a more computationally based perspective of measurable dynamical systems in common with the present text, and we highly recommend them. A summary of the German school's approach to the empirical study of dynamical systems can be found in [112], and [82]. Also, we recommend the review by Froyland [121]. Finally, we highly recommend the book by Hsu [167], and see also [166], which is an early and less often cited work in the current literature, as we rarely see "cell-to-cell mappings" cited lately. While lacking the transfer oriented formalism behind the analysis, this cell-to-cell mapping paradigm is clearly a precursor to the computational methods which are now commonly called set oriented methods. Also, we include a discussion and contrast to the early ideas by Ulam [307] called the Ulam method. Here we hope to give a useful broad presentation in a manner that includes some necessary background to allow a sophisticated but otherwise not specialized student or researcher to dive into this topic.

Acknowledgments. Erik Bollt would like to thank the several students and colleagues for discussions and cooperation that have greatly influenced the evolution of his perspective on these topics over several years, and who have made this work so much more enjoyable as a shared activity. He would also like to thank the National Science Foundation and the Office of Naval Research and the Army Research Office, who have supported several aspects of this work over the recent decade.

Chapter 1

Dynamical Systems, Ensembles, and Transfer Operators

1.1 Ergodic Preamble

In this chapter, we present the heuristic arguments leading to the Frobenius–Perron operator, which we will restate with more mathematical rigor in the next chapter. This chapter is meant to serve as a motivating preamble, leading to the technical language in the next chapter. As such, this material is a quick start guide so that the more detailed discussion can be followed with more motivation. It also provides enough background so that the techniques in subsequent chapters can be understood without necessarily reading all of the mathematical theory in the middle chapters.

In terms of practical application, the field of measurable dynamics has been hidden in a forest of formal language of pure mathematics that may seem impenetrable to the applied scientist. This language may be quite necessary for mathematical proof of the methods in the field of ergodic theory. However, the proofs often require restricting the range of problems quite dramatically, whereas the utility may extend quite further. In reality, the basic tools one needs to begin practice of measurable dynamics by transfer operator methods are surprisingly simple, while still allowing useful studies of transport mechanisms in a wide array of real-world dynamical systems. It is our primary goal to bring out the simplicity of the field for practitioners. We will attempt to highlight the language necessary to speak properly in terms necessary to prove convergence, invariance, steady state, and several of the other issues rooted in the language of ergodic theory. But above all, we wish to leave a spine of simple techniques available to practitioners from outside the field of mathematics. We hope this book will be useful to those experimentalists with real questions coming from real data, and to any students interested in such issues.

Our discussion here may be described as a contrast between the Lagrangian perspective of following orbits of single initial conditions and the Eulerian perspective associated with the corresponding dynamical system of the transfer operator which describes the evolution of measurable ensembles of initial conditions while focusing at a location. This leads to issucs traditionally affiliated with ergodic theory, a field which has important practical implications in the applied problems of transport study that are of interest here. Thus wc hope the reader will agree that both perspectives allow important information to be derived from a dynamical system. In particular, the transfer operator approach will allow us to discuss

- exploring global dynamics and characterization of the global attractors,
- estimating invariant manifolds,
- partitioning the phase space into invariant regions, almost invariant regions, and coherent sets,
- rates of transport between these partitioned regions,
- decay of correlation,
- associated information theoretic descriptions.

As we will discuss throughout this book, the question of transport can be boiled down to a question of walks in graphs, stochastic matrices, Markov chains, graph partitioning questions, and matrix analysis, together with Galerkin's methods for discussing the approximation. We leave this section with a picture in Fig. 1.1, which in some sense highlights so many of the techniques in the book. We will refer back to this figure often throughout this text. For now, we just note that the figure is an approximation of the action on the phase space of a Henon mapping as the action of a directed graph. The Henon mapping,

$$\begin{aligned} x_{n+1} &= y_{n+1} - a x_n^2, \\ y_{n+1} &= b x_n, \end{aligned} \tag{1.1}$$

for parameter values $a = 1.4$, $b = 0.3$, is frequently used as a research tool and as a pedagogical example of a smooth chaotic mapping in the plane. It is a diffeomorphism that highlights many issues of chaos and chaotic attractors in more than one dimension. Such mappings are not only interesting in their own right, but they also offer a step toward understanding differential equations by Poincaré section mappings.

1.2 The Ensemble Perspective

The dynamical systems point of view is generally Lagrangian, meaning that we focus on following the fate of trajectories corresponding to the evolution of a single initial condition. Such is the perspective of an ODE, Eq. (2.1), as well as a map, Eq. (2.7). Here we contrast the Lagrangian perspective of following single initial conditions to the Eulerian perspective rooted in following measurable ensembles of initial conditions, based on the associated dynamical system of transfer operators and leading to ergodic theory. We are most interested here in the transfer operator approach in that it may shed light on certain applied problems to which we have already alluded and we will detail.

Example 1.1 (following initial conditions, the logistic map). The logistic map,

$$x_{n+1} = L(x_n) = 4x_n(1 - x_n), \tag{1.2}$$

is a model of resource limited growth in a population system. The logistic map is an extremely popular model of chaos, partly for pedagogical reasons of simplicity of analysis,

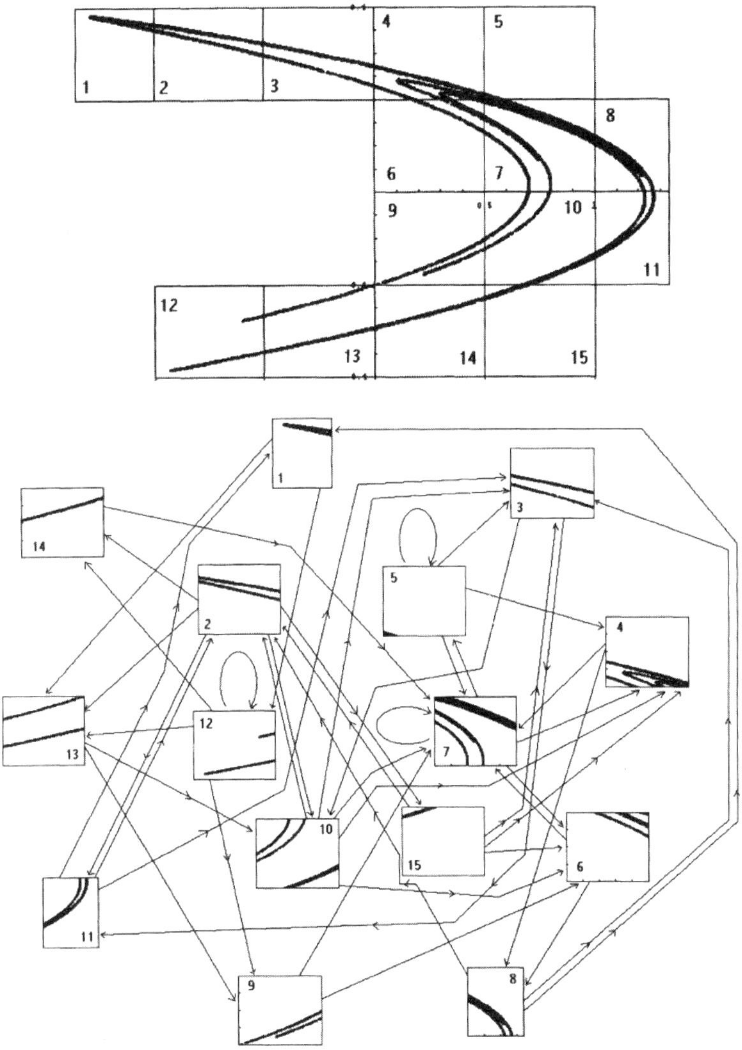

Figure 1.1. ***Approximate action of a dynamical system on density looks like a directed graph: Ulam's method is a form of Galerkin's method.*** *In a sense, this could be the mantra of this book. (Above) we see an attractor of the Henon map, Eq.* (1.1) *partitioned by an arbitrary graph, with the grid laid out according to a natural order of the plane phase space. (Below) The action of the dynamical system which moves (ensembles of) initial conditions is better represented as a directed graph. The action shown here is faithful (match the numbered boxes) but approximate, since a Markov partition was not used, and a refinement would apparently be beneficial.* [27]

and partly for historical reasons. In Fig. 1.2 we see the mapping and the time series it produces for a specific initial condition, $x_0 = 0.4$, where a time series is simply the function of the output values with respect to time. An **orbit** *is a sequence starting at a single initial*

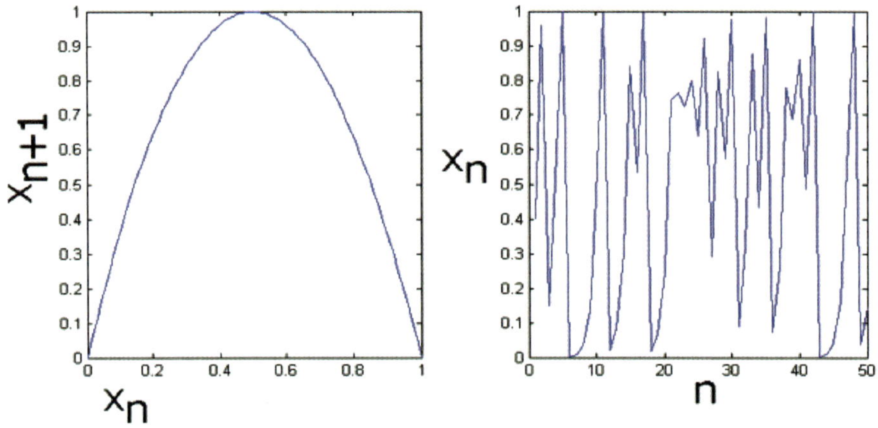

Figure 1.2. *The logistic map (left) produces a time series (right) shown for a given initial condition* $x_0 = 0.4$.

condition,

$$\{x_0, x_1, x_2, x_3 \dots\} = \{x_0, L(x_0), L^2(x_0), L^3(x_0), \dots\}, \tag{1.3}$$

where

$$x_i = L^i(x_0) \equiv L \circ L \circ \cdots \circ L(x_0), \text{ which denotes the } i\text{th-composition.} \tag{1.4}$$

In this case, the orbit from $x_0 = 0.4$ is the sequence $\{x_0, x_1, x_2, \dots\} = \{0.4, 0.96, 0.1536, \dots\}$. The time series perspective illustrates the trajectory of a single initial condition in time, which as an orbit runs "forever" and we are simply inspecting a finite segment. In this perspective, we ask how a single initial state evolves. Perhaps there is a limit point? Perhaps there is a stable periodic orbit? Perhaps the orbit is unbounded? Or perhaps the orbit is chaotic?

At this stage it is useful to give some definition of a **dynamical system**. A mathematically detailed definition of a dynamical system is given in Chapter 2 in Definitions 2.1–2.3. Said plainly for now, a dynamical system is

1. a state space (phase space), usually a manifold, together with

2. a notion of time, and

3. an evolution rule (often a continuous evolution rule) that brings forward states to new states as time progresses.

Generally, dynamical systems can be considered of two general types, continuous time as in flows (or semiflows) usually from differential equations, and discrete time mappings. For instance, the mapping $x_{n+1} = L(x_n)$ in Example 1.1, Eq. (1.2), is a discrete time map, $L : [0,1] \to [0,1]$. In this case, (1) the state space is the unit interval, $[0,1]$, (2) time is taken to be the iteration number and it is discrete, and (3) the mapping $L(x_n) = 4x_n(1 - x_n)$ is the evolution rule which assigns new values to the old values. The phrase dynamical system is usually reserved to mean that the evolution rule is deterministic, meaning the same input will always yield the same output in a function mapping–type relationship, whereas the phrase **stochastic dynamical system** can be used to denote those systems with some kind of randomness in the behavior. Each of these will be discussed in subsequent chapters.

Another perspective we pursue will be to ask what happens to the evolution of many different initial conditions, the so-called **ensemble of initial conditions**. To illustrate this idea, we provide the following example.

Example 1.2 (following an ensemble of initial conditions in the logistic map). Imagine that instead of following one initial condition, we choose N initial conditions, $\{x_0^i\}_{i=1,\ldots,N}$ (let $N = 10^6$, a million, for the sake of specificity). Choosing those initial conditions by a random number generator, approximating uniform, $U(0,1)$, we follow each one. Now it would not be reasonable to plot a time series for all million states. The corresponding plot to Fig. 1.2 (right) would be too busy; we would only see a solid band. Instead, we accumulate the information as a histogram, an empirical representation of the probability density function. A histogram of N uniformly chosen initial conditions is shown in Fig. 1.4 (left). Iterating each one of the initial conditions under the logistic map yields $\{x_1^i\}_{i=1,\ldots,N} = \{L(x_0^i)\}_{i=1,\ldots,N}$ and so forth, through each iteration. Due to the very large number of data points, we can only reasonably view the data statistically, as histograms, the profile of each evolving upon each successive iteration, as shown in the successive panels of Fig. 1.3.

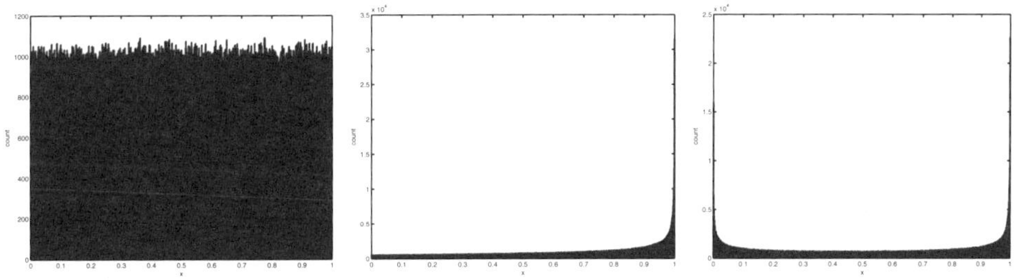

Figure 1.3. *Histograms depicting the evolution of many ($N = 10^6$) initial conditions under the logistic map. (Left) Initially, $\{x_0^i\}_{i=1,\ldots,N}$ are chosen uniformly by $U(0,1)$ in this experiment. (Middle) After one iterate, each initial condition x_0^i moves to its iterate, $x_1^i = L(x_0^i)$, and the full histogram, $\{x_1^i\}_{i=1,\ldots,N}$ is shown. (Right) The histogram of the second iterate, $\{x_1^i\}_{i=1,\ldots,N}$ is shown.*

There are central tenants of ergodic theory to be found in this example. The property of ergodic is defined explicitly in Sections 3.5 and 3.5.1, but a main tenant is highlighted by the Birkhoff theorem describing coincidence of time averages and space averages (see Eq. (1.5)). Two major questions that may be asked of this example are

1. Will the profile of the histogram settle down to some form, or will it change forever?

2. Does the initial condition play a role, and, in particular, how does the specific initial condition play a role in the answer to question 1?

It is not always true that a dynamical system will have a long-term steady state distribution, as approximated by the histogram; the specific dynamical system is relevant, and for many

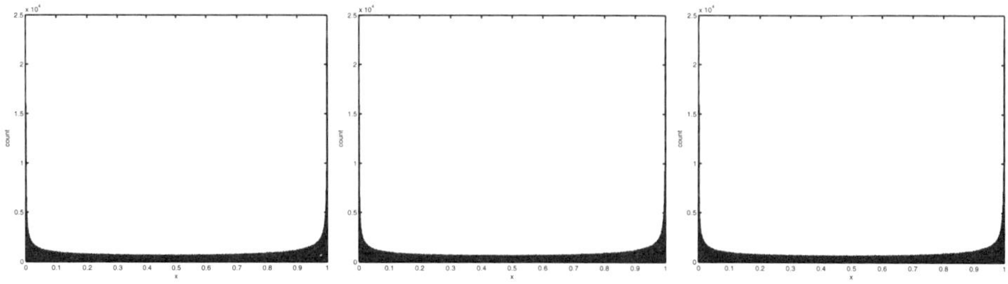

Figure 1.4. *Following Fig. 1.3, histograms of* $\{x_{10}^i\}_{i=1,\ldots,N}$ *(left) and* $\{x_{25}^i\}_{i=1,\ldots,N}$ *(middle) are shown. Arguably, a limit is apparent in the profile of the histogram. (Right) The histogram of the orbit of a single initial condition gives apparently the same long-term limit density. This is empirical evidence suggesting the ergodic hypothesis.*

dynamical systems, but not all, the initial condition may be relevant. When there is a steady state distribution, loosely said, we will discuss the issue of natural measure, which is a sort of stable ergodic invariant measure [171]. By invariant measure, we mean that the ensemble of orbits may each move individually, but in such a way that their distribution nonetheless remains the same.

More generally, there is the notion of an **invariant measure** (see Definition 3.4), where invariant measure and ergodic invariant measure are discussed further in Sections 3.5 and 3.5.1. A transformation which has an invariant measure μ need not be ergodic, which is another way of saying it favors just part of the phase space, or it may even be supported on just part of the phase space. By contrast, the density[2] as illustrated here by the histogram shown covers the whole of $[0,1]$, suggesting at least by empirical inspection[3] that the invariant density is absolutely continuous.[4]

Perhaps the greatest application of an ergodic invariant measure follows **Birkhoff's ergodic theorem**. Stated roughly, with respect to an ergodic T-invariant measure μ on a measurable space $(X, \mathcal{A})$, where $\mathcal{A}$ is the sigma algebra of measurable sets, time averages and spatial averages may be exchanged,

$$\lim_{n\to\infty} \frac{1}{n} \sum_{i=1}^{n} f \circ T^i(x_0) = \int_X f(x)\,d\mu(x), \tag{1.5}$$

for μ-almost every initial condition. This is evidenced in that a long orbit segment of a single initial condition $\{x_j\}_{j=1}^{10^6}$ yields essentially the same result as the long-term ensemble, as seen in Fig. 1.3.

[2]We will often speak of measure and density interchangeably. In fact they are dual. This is best understood when there is a Radon–Nikodým derivative [190], in the case of a positive absolutely continuous measure-μ, $d\mu(x) = g(x)dx$. g is the density function when it exists, which expression denotes, $\mu(B) = \int_B d\mu(x) = \int_B g(x)dx$. In the case the measurable functions are cells of a histogram's partition, this is descriptive of the histogram. In the case of continuous functions g this reminds us of the fundamental theorem of calculus.

[3]The result does in fact hold by arguments that the invariant measure is absolutely continuous to Lebesgue measure, which will not be presented here [50].

[4]A positive measure $\mu(x)$ is called absolutely continuous when it has a Radon–Nikodým derivative preimage [190] to Lebesgue measure $d\mu(x) = g(x)dx$.

Example 1.3 (Birkhoff ergodic theorem and histograms). The statement that a histogram reveals the invariant measure for almost all initial conditions can be sharpened by choosing the measurable function f in Eq. (1.5), as the characteristic (indicator) functions:

$$f(x) = \chi_{B_i}(x) = \begin{pmatrix} 1 & \text{if} & x \in B_i \\ 0 & & \text{else} \end{pmatrix}. \tag{1.6}$$

A histogram in these terms is an occupancy count of a data set "sprinkled" in a topological partition, $X = \cup_i \overline{(B_i)}$. Then in these terms, considering how many points of a sample orbit $\{x_j\}_{j=1}^n$ occupy a cell B_i as part of building a histogram, in Eq. (1.5)

$$\sum_{j=1}^{n} f \circ T^j(x) = \sum_{j=1}^{n} \chi_{B_i}(x_j). \tag{1.7}$$

Also, Eq. (1.5) promises that we will almost never choose a bad initial condition but still converge toward the same occupancy for cell B_j. Likewise, repeating for each cell in the partition produces a histogram such as in Fig. 1.3.

Example 1.4 (what can Birkhoff's ergodic theorem say about Lyapunov exponents?). In Chapter 8, we will discuss finite time Lyapunov exponents (FTLEs), which in brief are related to derivatives averaged along *finite* orbit segments but multiplicatively, and how the results vary depending on where in the phase space the initial condition is chosen, the time length of the orbit segment, and how this information relates to transport. This is in dramatic contrast to the traditional definition of Lyapunov exponents, which are almost the same quantity, but averaged along an infinite orbit. In other words, if we choose a measuring function

$$f(x) = \ln|T'(x)|,^{5} \tag{1.8}$$

then μ-almost every initial condition again will give the same result. Perhaps this "usual" way of thinking of orbits as infinitely long, and Lyapunov exponents as limit averages with the Birkhoff theorem stating that almost every starting point is the same, prevented the discovery of the brilliant-for-its-simplicity but powerful idea of FTLEs, which are intrinsically spatially dependent due to the finite time aspect.

To state more clearly the question of how important the initial condition is, in brief the answer is almost not at all. In this sense, almost every initial condition stated in the measure theoretic sense is "typical." This means that with probability one we will choose an initial condition which will behave as the ergodic case. To put the statement of this rarity in perspective, in the same sense we may say that if we choose a number randomly from the unit interval, with probability zero the number will be rational, and with probability one the number will be irrational. Of course this does not mean it is impossible to choose a rational number, just that the Lebesgue measure of the rationals is zero. By contrast, the situation is opposite when performing the random selection on a computer. The number will always be rational because (1) the random number generator is just a model of the uniform random

[5] We are specializing to a one-dimensional setting so that we do not need to discuss issues related to the Jacobian derivative matrices and diagonalizability at this early part of the book. In this case the Birkhoff theorem describes the Lyapunov exponents as discussed here. The more general scenario in more than one dimension requires Oseledets multiplicative ergodic theorem to handle products along orbits, as discussed in Section 8.2, in contrast to the one-dimensional scenario in Section 8.1.

variable, as must be all algorithms descriptive of a pseudorandom number generator [310], and (2) the computer can only represent rational numbers, and, in fact, a finite number of those. Nonetheless, when selecting a pseudorandom number, it will be ergodic-typical in the sense above for a "typical" dynamical system.

1.3 Evolution of Ensembles

Perhaps a paradoxical fact, but one that is central to the analysis of this book, is that while a chaotic dynamical system[6] may be nonlinear, causing particular difficulty in predicting the fate of individual orbits, the evolution of density is an associated *linear*[7] dynamical system which turns out to be especially straightforward to predict. That is, the dynamical system

$$f : X \to X \tag{1.9}$$

moves initial conditions, whereas there is an associated linear dynamical system,

$$P_f : L^1(X) \to L^1(X), \tag{1.10}$$

which is descriptive of the evolution of densities of ensembles of initial conditions. The operator, P_f, is called the Frobenius–Perron operator. Initially, we will specialize for simplicity of presentation to the logistic map as follows. The general theory will be saved for Chapter 2.

The evolution of density follows a principle of mass conservation: ensembles of initial conditions evolve forward in time, and no individual orbits are lost. In terms of densities, it must be assumed that the transformation is nonsingular in that this will guarantee that densities map to densities.[8] If there are N initial conditions $\{x_{0,i}\}_{i=1}^N$, then in general they may be distributed in X according to some initial distribution $\rho_0(x)$, for which we write $x_0 \sim \rho_0(x)$. The question of evolution of density is as follows. After one iteration by f, each $x_{0,i}$ moves to $x_{1,i} = f(x_{0,i})$ for each i. Generally, if we investigate the distribution of the points in their new positions, we must allow that the distribution of them may be different than their initial configuration. If the actual new configuration of $\{x_{1,i}\}_{i=1}^N$ distributes according to some new density $\rho_1(x)$, then the problem becomes one of finding $\rho_1(x)$ given $\rho_0(x)$. Likewise, we can look for the orbit of distributions,

$$\{\rho_0(x), \rho_1(x), \rho_2(x), \ldots\}. \tag{1.11}$$

From the principle conservation of initial conditions follows a **discrete continuity equation**,

$$\int_B \rho_1(x)dx = \int_{f^{-1}(B)} \rho_0(x)dx \ \forall B \in \mathcal{A}, \tag{1.12}$$

from which will follow the dynamical system,

$$\begin{aligned} P_f &: L^1(X) \to L^1(X), \\ \rho_0(x) &\mapsto \rho_1(x) = P_f[\rho_0](x), \end{aligned} \tag{1.13}$$

[6]A dynamical system is defined to be *chaotic* if it displays sensitive dependence to initial conditions and a dense orbit [95, 12, 314, 213]. Or, according to [2], an *orbit is chaotic* if it is bounded, not asymptotically periodic, and has a positive Lyapunov exponent.

[7]A dynamical system $T : X \to X$ is *linear* if for any $x_1, x_2 \in X$, and $a, b \in \mathbb{R}$, $T(ax_1 + bx_2) = aT(x_1) + bT(x_2)$; otherwise the dynamical system is nonlinear.

[8]More precisely, we wish that absolutely continuous densities map to absolutely continuous densities.

and the assignment of a new density by the operator P_f is interpreted at each point x. This continuity equation may be interpreted as follows. Formally, over a measure space $(X, \mathcal{A}, \mu)$, $B \in \mathcal{A}$ is any one of the measurable subsets. For simplicity of discussion, we may interpret the B's to be any one or collection of the cells used in describing the histograms such as shown in Figs. 1.3 or 1.8. Then $\rho_0(x)$ is an initial density descriptive of an ensemble, such as the approximation depicted in Fig. 1.4 (left). When asking where each of the initial conditions goes under the action of f, we are better suited to ask where orbits distributed by ρ_1 came from after one iteration of f. The preimage version Eq. (1.12) and

$$\int_{f(B)} \rho_1(x')dx' = \int_B \rho_0(x')dx' \tag{1.14}$$

would give the same result as Eq. (1.13) if the mapping were piecewise smooth and one-one, but many examples, including the logistic map, are not one-one, as shown in Fig. 1.5.

The continuity equation Eq. (1.12) is well stated for any measure space, $(X, \mathcal{A}, \mu)$, but assuming that X is an interval $X = [a,b]$ and μ is Lebesgue measure, then we may write

$$\int_a^x P_f \rho(x')dx' = \int_{f^{-1}([a,x])} \rho(x')dx' \ \forall x \in [a,b], \tag{1.15}$$

thus representing those B's which are intervals $[a,x]$. For a noninvertible f, f^{-1} denotes the union of all the preimages. Differentiating both sides of the equation, and assuming differentiability, then the fundamental theorem of calculus gives

$$P_f \rho(x) = \frac{d}{dx} \int_{f^{-1}([a,x])} \rho(x')dx' \ \forall x \in [a,b]. \tag{1.16}$$

Further assuming that f is invertible and differentiable allows application of the fundamental theorem of calculus and the chain rule to the right-hand side of the equation:

$$P_f \rho(x) = \frac{d}{dx} \int_{f^{-1}(a)}^{f^{-1}(x)} \rho(x')dx' = \rho(f^{-1}(x)) \frac{d}{dx}(f^{-1}(x)) = \frac{\rho(f^{-1}(x))}{|f'(f^{-1}(x))|} \ \forall x \in [a,b]. \tag{1.17}$$

However, since generally f may not be invertible, the integral derivation is applied over each preimage, resulting in a commonly presented form of the Frobenius–Perron operator for deterministic evolution of density in maps:

$$P_f[\rho](x) = \sum_{y:x=f(y)} \frac{\rho(y)}{|f'(y)|}. \tag{1.18}$$

The nature of this expression is a functional equation for the unknown density function $\rho(x)$. Questions of the existence and uniqueness of solutions are related to the fundamental unique ergodicity question in ergodic theory. That is, can one find a special distribution function that should be stated as a centrally important principle in the theory of Frobenius–Perron operators?

An invariant density is a fixed "point" of the Frobenius–Perron operator: Functionally, this can be stated as

$$\rho^*(x) = P_f[\rho^*](x). \tag{1.19}$$

One note is that we say *an* invariant density rather than *the* invariant density as there can be many and even infinitely many. However, usually we are interested in the "dominant" invariant measure, or other information related to dominant behaviors such as almost-invariant sets. Further, is this fixed density (globally) stable? This is the critical question: will general (most or all?) ensemble distributions settle to some unique density profile? $\rho_i(x) \to \rho^*(x)$ as $n \to \infty$? In the following example, we offer a geometric interpretation of the form of the Frobenius–Perron operator and its relationship to unique ergodicity. More on this principle can be found discussed in Section 3.4.

Example 1.5 (Frobenius–Perron operator of the logistic map). The (usually two) preimage(s) of the logistic map in (1.2) at each point may be written as

$$L_{\pm}^{-1}(x) = \frac{1 \pm \sqrt{1-x}}{2}. \tag{1.20}$$

Therefore, the Frobenius–Perron operator in (1.18) specializes to

$$P_f[\rho](x) = \frac{1}{4\sqrt{1-x}}\left(\rho\left(\frac{1-\sqrt{1-x}}{2}\right) + \rho\left(\frac{1+\sqrt{1-x}}{2}\right)\right). \tag{1.21}$$

This functional equation can be interpreted pictorially, as in Fig. 1.5; the collective ensemble at cell B comes from those initial conditions at the two preimages $f_{\pm}^{-1}(B)$ shown. The preimage of the set may as shown in the cobweb diagram, scaled roughly as the inverse of the derivative at the preimages. Roughly, the scaling occurs almost as if we were watching ray optics, where the preimages $f^{-1}(B)$ focus on B through mirrors by the action of

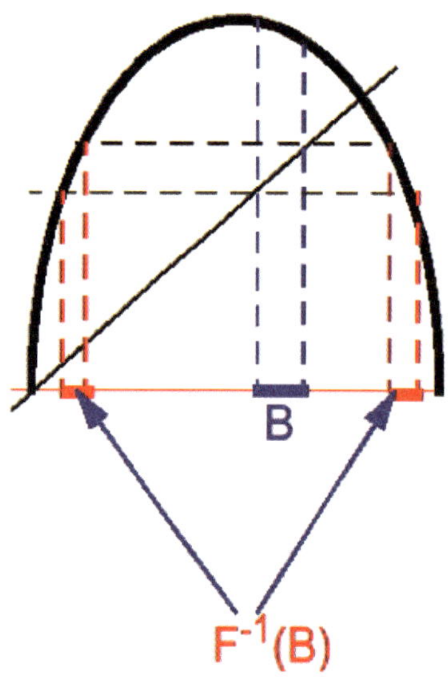

Figure 1.5. *Cobweb of density. Note how infinitesimal density segments of B grow or shrink inversely proportionally to the derivative at the preimage, as prescribed by Eq.* (1.18).

the map, placed at B by the map, but scaled as if focused by the inverse of the derivative because the ensemble of initial conditions at $f^{-1}(B)$ shuttles into B.

It is a simple matter of substitution, and therefore application of trigonometric identities, to check that the function

$$\rho(x) = \frac{1}{\pi\sqrt{x(1-x)}} \tag{1.22}$$

is a fixed point of the operator in (1.21). This guess and check method is a valid way for validating an invariant density. However, how does one make the guess? Comparison of the experimental numerical histograms validate this density, comparing Eq. (1.22) to Figs. 1.3 and 1.4. By comparison to a simpler system, where the invariant density is easy to guess, the invariant density of this logistic map is straightforward to derive.

Example 1.6 (Frobenius–Perron operator of the tent map). The tent map serves as a simple example to derive invariant density and for comparison to the logistic map.

$$x_{n+1} = T(x_n) = 2\left(1 - 2\left|x - \frac{1}{2}\right|\right) \tag{1.23}$$

may also be viewed as a dynamical system on the unit interval, $T : [0,1] \to [0,1]$, shown in Fig. 1.6 (left), and a "typical" time series is shown in Fig. 1.6 (middle). Repeating the experiment of the evolution of an ensemble of initial conditions as was done for the logistic map, Figs. 1.3 and 1.4, yields the histogram in Fig. 1.6 (right). Apparently from the empirical experiment, the uniform density, $U(0,1)$, is invariant. This is straightforward to validate analytically by checking that the Frobenius–Perron operator, Eq. (1.18), specializes to

$$P_T[\nu](x) = \frac{1}{4}\left(\nu\left(\frac{x}{4}\right) + \nu\left(1 - \frac{x}{4}\right)\right). \tag{1.24}$$

Further, invariance (1.19) has a solution:

$$\nu(x) = 1. \tag{1.25}$$

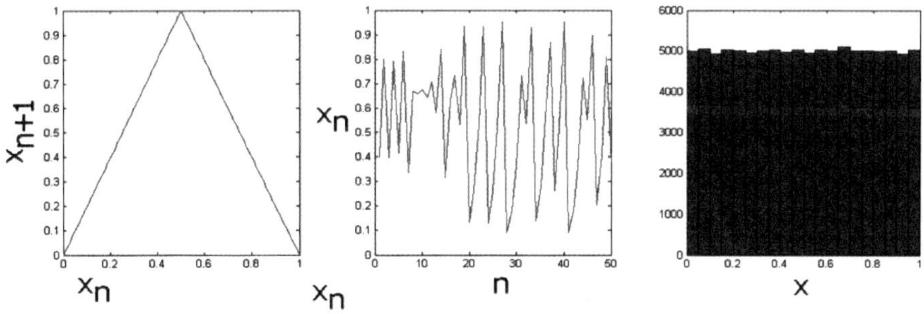

Figure 1.6. *The tent map (left),* (3.56), *a sample time series (middle), and a histogram of a sample ensemble (right). This figure mirrors Fig.* 1.3 *shown for the logistic map. Apparently here, the tent map suggests an invariant density which is uniform,* $U(0,1)$.

The well-known change of variables between the dynamics of the fully developed chaotic[9] tent map (slope $a = 2$) and the fully developed logistic map ($r = 4$) is through the change of variables

$$h(x) = \frac{1}{2}(1 - \cos(\pi x)), \tag{1.26}$$

which is formally an example of a *conjugacy* in dynamical systems.

The fundamental equivalence relationship in the field of dynamical system, comparing two dynamical systems, $g_1 : X \to X$ and $g_2 : Y \to Y$, is a conjugacy.

Definition 1.1 (conjugacy). Two dynamical systems,

$$g_1 : X \to X \text{ and } g_2 : Y \to Y, \tag{1.27}$$

are a conjugate if there exists a function (a change of variables)

$$h : X \to Y, \tag{1.28}$$

such that h commutes (a pointwise functional requirement),

$$h \circ g_1(x) = g_2 \circ h(x), \tag{1.29}$$

often written as a commuting diagram,

$$\begin{array}{ccc} X & \xrightarrow{g_1} & X \\ h\downarrow & & \downarrow h \\ Y & \xrightarrow{g_2} & Y \end{array}, \tag{1.30}$$

and h is a *homeomorphism* between the two spaces X and Y. The function h is a homeomorphism if

- h is one-one,
- h is onto,
- h is continuous,
- h^{-1} is continuous.

Change of variables is a basic method in mathematical sciences since it is fair game to change from a coordinate system where the problem may be hard (say, Cartesian coordinates) to a coordinate system (say, spherical coordinates) where the problem may be easier in some sense, the goal often being to decouple variables. The most basic requirement is that in the new coordinate system, solutions are neither created nor destroyed, as the above definition allows. The principle behind defining a good coordinate transformation to be

[9]**Fully developed chaos** as used here refers to the fact that as the parameter (a or r for the tent map or logistic map, for example) is varied, the corresponding symbol dynamics becomes complete in the sense that the corresponding symbol dynamics is a fullshift, meaning the corresponding grammar has no restrictions. The symbol dynamics theory will be discussed in detail in Chapter 6. A different definition can be found in [215], which differs from our use largely by including the notion that the chaotic set should densely fill the interval.

a homeomorphism is that the two dynamical systems should take place in topologically equivalent phase spaces. Further, solutions should correspond to solutions with the same behavior in a continuous manner, and this is further covered by the pointwise commuting principle $h \circ g_1(x) = g_2 \circ h(x)$. By contrast, for example, without requiring that h is one-one, two solutions may come from one, and so forth.

Returning to comparing the logistic and the tent maps, it can be checked that the function

$$h(x) = \frac{1}{2}(1 - \cos(\pi x)), \tag{1.31}$$

shown in Fig. 1.7 (upper right), is a conjugacy between g_1 as the logistic map of Eq. (1.2), (with the parameter value 4) and g_2 as the tent map (3.56) (with parameter value 2), with $X = Y = [0,1]$.

A graphical way to represent that commuter function (a function simply satisfying Eq. (1.30) whether or not that function may be a homeomorphism [292]) is by what we call a **quadweb diagram**, as illustrated in Fig. 1.7. A quadweb is a direct and pointwise graphical representation of the commuting diagram. In [292], we discuss further how representing the commuting equation even when two systems may not be conjugate (and therefore the commuter function is not a homeomorphism) has interesting relevance to relating dynamical systems. Here we will simply note that the quadweb illustrates that a conjugacy is a

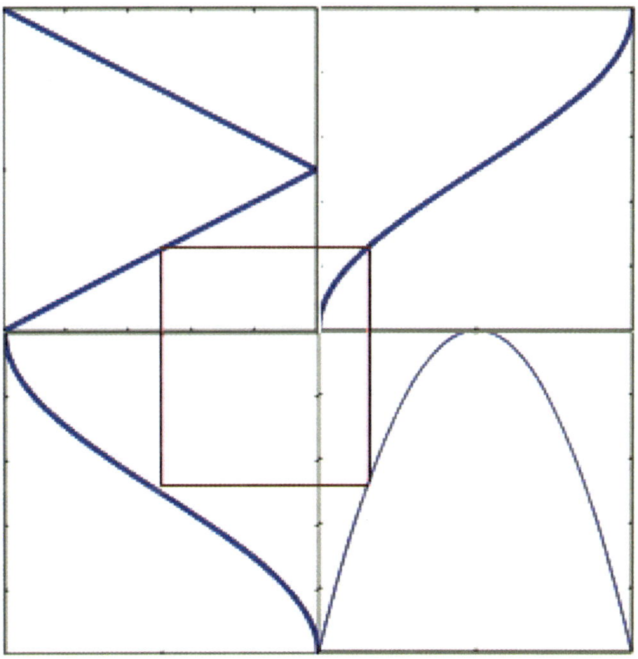

Figure 1.7. *A quadweb is a graphical way to pointwise represent the commuting diagram* (1.30). *Further, when h is a homeomorphism, then the two maps compared are conjugate. Shown here is the conjugacy* $h(x) = \frac{1}{2}(1 - \cos(\pi x))$, *changing variables between the full tent map and the full logistic map, Eq.* (1.31).

pointwise relationship. Of course, we named a quadweb as such since it is a direct play on the better-known term "cobweb" diagram. When further h is a homeomorphism, then the two maps compared are conjugate.

Inspecting Eq. (1.31), we see that the function is not simply continuous, it is differentiable. As it turns out, this most popular example of a conjugacy is atypical in the sense that it is stronger than required. In fact, it is a diffeomorphism.

Definition 1.2 (diffeomorphism). A diffeomorphism is a homeomorphism h which is bi-differentiable (meaning h and h^{-1} are differentiable), and when stated that two dynamical systems are diffeomorphic, there is a conjugacy which is bi-differentiable.

Conjugacy is an equivalence relationship with many conserved quantities between dynamical systems, including notably topological entropy. Diffeomorphism is a stronger equivalence relationship which conserves quantities such as metric entropy and Lyapunov exponents. Interestingly, despite the atypical nature of diffeomorphism, in the sense of genericity implying that most systems if conjugate have nondifferentiable conjugacies, the sole explicit example used for introduction in most textbooks is a diffeomorphism, Eq. (1.31). A nondifferentiable conjugacy of two maps in the interval will be a Lebesgue singular function [292], meaning it will be differentiable almost everywhere, but wherever it is differentiable, the derivative is zero. Nonetheless the function is monotone nondecreasing in order to be one-one. These are topologically exotic in the sense that they are a bit more like a devil's staircase function [271, 328] than they are like a cosine function.

Most relevant for our problem here is the comparison between invariant densities of the logistic map and the tent map, for which we require the differentiability of the conjugacy. Thus h must further be a diffeomorphism to execute the change of density. We require the infinitesimal comparison[10]

$$\rho(x)dx = \nu(y)dy, \tag{1.32}$$

from which follows

$$\rho(x) = \frac{dy}{dx} = \frac{1}{\pi\sqrt{x(1-x)}}. \tag{1.33}$$

This result is in fact the fixed density already noted in Eq. (1.22), which agrees with Figs. 1.3 and 1.4.

Finally, we illustrate the ensemble perspective of invariant density for an example of a mapping whose phase space is more than an interval—the Henon mapping from Eq. (1.1). This is a diffeomorphism of the plane,

$$H : \mathbb{R}^2 \to \mathbb{R}^2. \tag{1.34}$$

As such, a density is a positive function over the phase space,

$$\rho : \mathbb{R}^2 \to \mathbb{R}^+. \tag{1.35}$$

In Fig. 1.1 we illustrated both the chaotic attractor as well as the action of this mapping, which is approximately a directed graph. The resulting invariant density, of a long time

[10]This is equation in the simplest problems is called "u-change of variables" in elementary calculus books, but is a form of the Radon–Nikodým derivative theorem in more general settings [190].

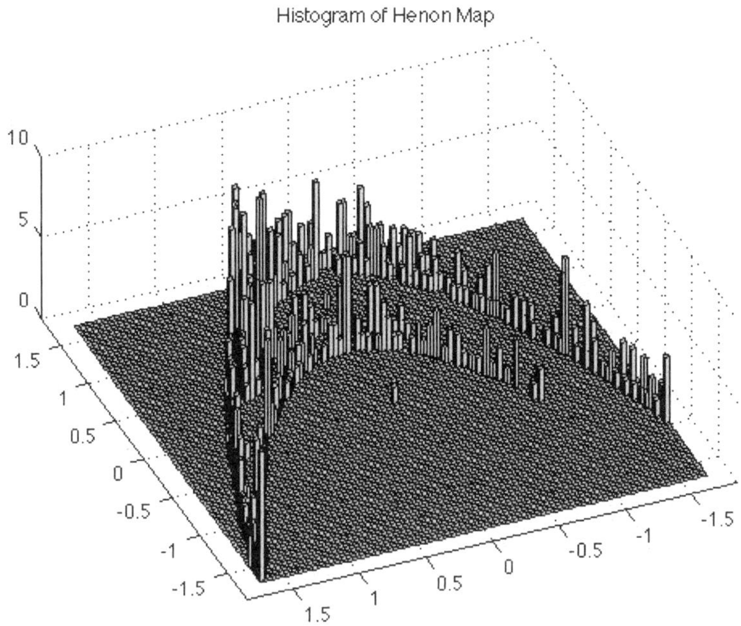

Figure 1.8. *Henon map histogram approximating the invariant density. Notice the irregular nature typical of the densities of such chaotic attractors, which are often suspected of not being absolutely continuous.*

settling an ensemble of initial conditions, or alternatively of a long time behavior of one typical orbit, is illustrated in Fig. 1.8. As we will describe further in the next chapter, this invariant density derived here by a histogram of a long orbit may also be found as the dominant eigenvector of transition matrix of the graph shown in Fig. 1.1; this is the Ulam conjecture [307].

1.4 Various Useful Representations and Invariant Density of a Differential Equation

An extremely popular differential equation considered often and early in the presentation of chaos in nonlinear differential equations, and historically central in the development of the theory, is the **Duffing equation**:

$$\ddot{x} + a\dot{x} - x + x^3 = b \sin \omega t. \tag{1.36}$$

This equation in its most basic physical realization describes the situation of a massless ball bearing rolling in a double-welled potential,

$$P(x) = -x^2 + \frac{1}{4}x^4, \tag{1.37}$$

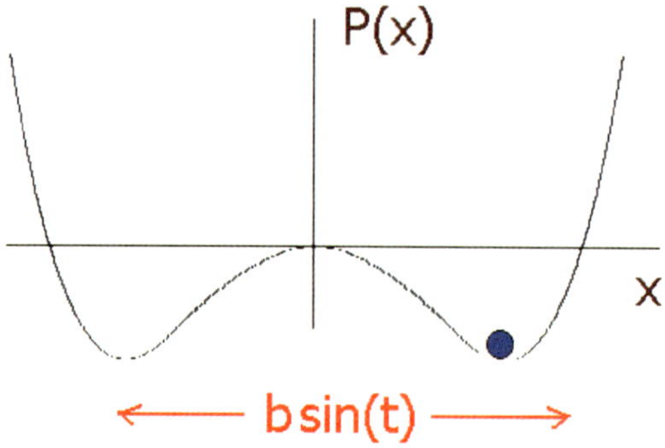

Figure 1.9. *Duffing double-well potential, Eq.* (1.37)*, corresponding to the Duffing oscillator ($a = 0$ case). Unforced, the gradient flow can be illustrated as a massless ball bearing in the double well as shown. Further forcing with a sinusoidal term can be thought of as a ball bearing moving in the well, but the floor is oscillating, causing the ball to sometimes jump from one of the two wells to the other.*

which is then sinusoidally forced, as depicted in Fig. 1.9.[11] This is a standard differential equation in the pedagogy of dynamical systems. We use this problem as an example to present the various presentations in representing the dynamics of a flow, including

- time series, Fig. 1.10,
- phase portrait, Fig. 1.11,
- Poincaré map, Figs. 1.12 and 1.13,
- attractor, also seen in Fig. 1.12,
- invariant density, Fig. 1.14.

Written in a convenient form as a system of first-order equations, with the substitution

$$y \equiv \dot{x}, \tag{1.38}$$

gives a **nonautonomous**[12] two-dimensional equation,

$$\begin{aligned} \dot{x} &= y, \\ \dot{y} &= -ay - x - x^3 + b\cos\omega t. \end{aligned} \tag{1.39}$$

As a **time series** of measured position $x(t)$ and velocity $y(t)$ of these equations, with $a = 0.02$, $b = 3$, and $\omega = 1$, we observe a signature chaotic oscillation as seen in Fig. 1.10. This time series of an apparently erratic oscillation nonetheless comes from the

[11]The gradient system case, where the autonomous part can be written $-\frac{\partial P}{\partial x}$, occurs when the viscous friction part is zero, $a = 0$.

[12]An autonomous differential equation can be written $\dot{x} = F(x)$ without explicitly including t in the right-hand side of the equation, and otherwise the differential equation is nonautonomous when it must be written $\dot{x} = f(x,t)$.

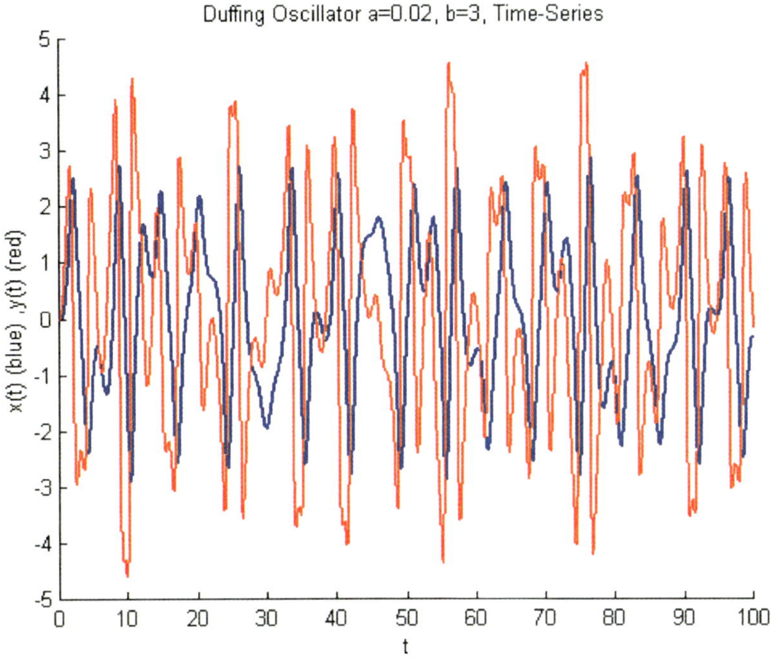

Figure 1.10. *A Duffing oscillator can give rise to a chaotic time series, shown here for both $x(t)$ and $y(t)$ solutions from Eq.* (1.39)*, with $a = 0.02$, $b = 3$, and $\omega = 1$.*

deterministic evolution of the ODE (1.36). This is simply a plot of the variables x or y as a function of time t.

A **phase portrait** in **phase space**, however, suppresses the time variable. Instead, the t serves as a parameter which for representation of a solution curve in parametric form $(x(t), y(t)) \in \mathbb{R}^2$ is seen in Fig. 1.11.

Augmenting with an extra time variable, $\tau(t) = t$, from which $\frac{d\tau}{dt} = \dot{\tau} = 1$ gives the **autonomous** three-dimensional equations of this flow:

$$\begin{aligned} \dot{x} &= y, \\ \dot{y} &= -ay - x - x^3 + b \sin \omega\tau, \\ \dot{\tau} &= 1. \end{aligned} \tag{1.40}$$

This form of the dynamical system allows us to represent solutions in a phase space, $(x(t), y(t), \tau(t)) \in \mathbb{R}^3$, for each t. In this representation, the time variable is not suppressed as we view the solution curves, $(x(t), y(t), \tau(t))$. Thus generally one can represent a nonautonomous differential equation as an autonomous differential equation by embedding in larger phase space.

A convenient way to study topological and measurable properties of the dynamical system presented by a flow is to produce a discrete time mapping by the **Poincaré section** method to produce a **Poincaré mapping**. That is, a codimension-1 "surface" is placed transverse to the flow so that (almost every) solution will pierce it, and then rather than

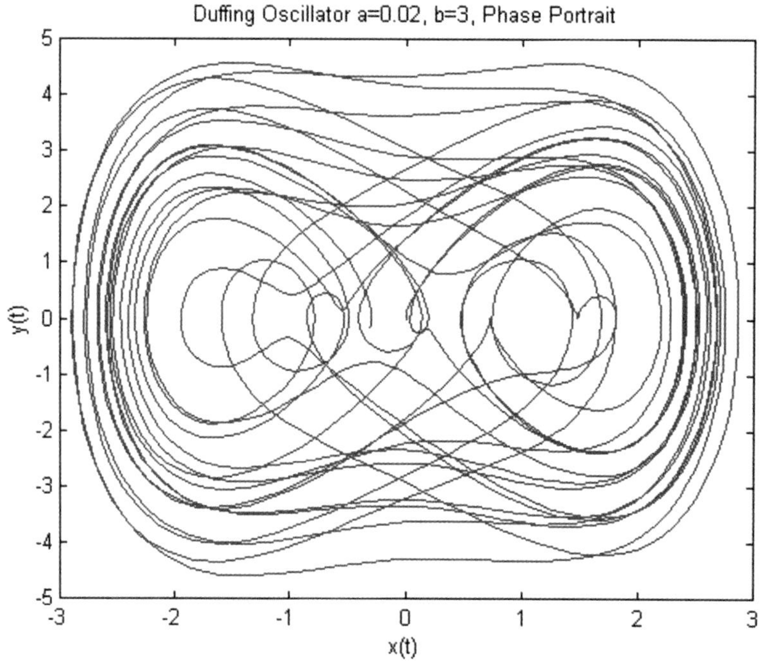

Figure 1.11. *In the Duffing equations* (1.39), *nonautonomous phase space is* $(x(t), y(t)) \in \mathbb{R}^2$, *with* $a = 0.02$, $b = 3$, *and* $\omega = 1$.

recording every point on the flow, it is sufficient to record the values at the instants of piercing.

In the case of the Duffing oscillator, a suitable Poincaré surface is a special case called a "stroboscopic" section, by $\omega\tau = 2\pi k$ for $k \in \mathbb{Z}$. The brilliance of Poincaré's trick allows the ordered discrete values $(x(t_k), y(t_k))$, $t_k = 2\frac{\pi k}{\omega}$, or rather we simply write (x_k, y_k), to represent the flow on its attractor. In this manner, Fig. 1.12 replaces Fig. 1.11, and in many ways this representation as a discrete time mapping,

$$(x_{k+1}, y_{k+1}) = F(x_k, y_k), \tag{1.41}$$

is easier to analyze, or at least there exists a great many new tools otherwise not available to the ODE perspective alone. For the sake of classification, when the right-hand side of the differential equation is in the form of an autonomous vector field, as we represented in the case of Eq. (1.40), we write specifically

$$\begin{aligned} G &: \mathbb{R}^3 \to \mathbb{R}^3, \\ G(x, y, \tau) &= \langle y, -ay - x - x^3 + b, 1 \rangle. \end{aligned} \tag{1.42}$$

Then simply let

$$z = (x, y, \tau) \text{ and } \dot{z} = G(z), \tag{1.43}$$

which is a general form for Eq. (1.40).

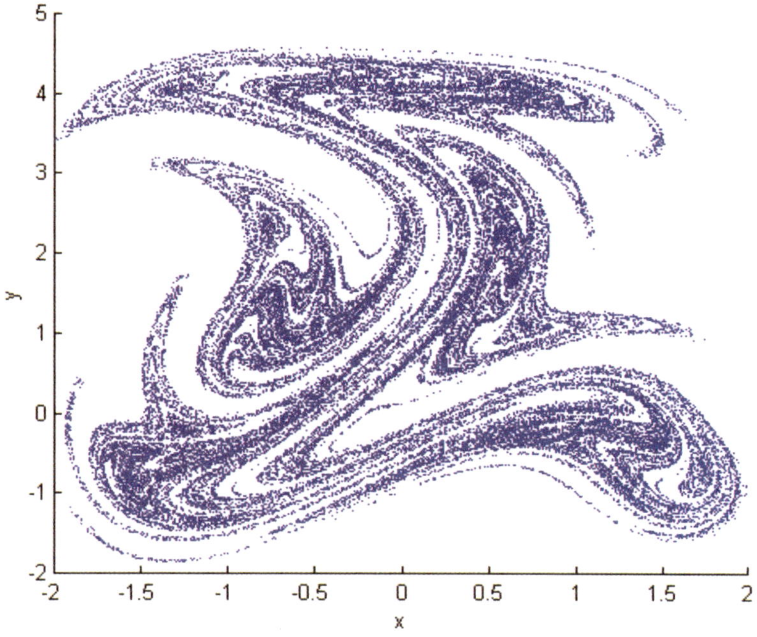

Figure 1.12. *A Poincaré-stroboscopic mapping representation of the Duffing oscillator. The discrete time mapping in* $\mathbb{R}^2$ *is derived by recording* (x,y) *each time that* $(x(t), y(t), \tau(t)) \in \Sigma$*; the Poincaré surface in this case,* $\Sigma = \{(x,y,\tau) : (x,\tau) \in \mathbb{R}^2, \tau \in \frac{2\pi k}{\omega}\}$ *as caricatured in Fig.* 1.13 (1.39)*, with* $a = 0.02$*,* $b = 3$*, and* $\omega = 1$*.* [30]

If the vector field G is Lipschitz,[13,14] then it is known that there is continuous dependence both with respect to initial conditions and with respect to parameters as proven through Gronwall's inequality [164, 251]. It follows that the Poincaré mapping F in Eq. (1.41) must be a continuous function in two dimensions, $F : \mathbb{R}^2 \to \mathbb{R}^2$, corresponding to a two-dimensional dynamical system in its own right. If, further, $G \in C^2(\mathbb{R}^3)$, then F is a diffeomorphism which brings with it a great many tools from the field of differentiable dynamical systems, such as transport study by stable and unstable manifold analysis.

[13] $G : \mathbb{R}^n \to \mathbb{R}^n$ is Lipschitz in a region $\Omega \subset \mathbb{R}^n$ if there exists a constant $L > 0$, $\|G(z) - G(\tilde{z})\| \leq L\|z - \tilde{z}\|$ for all $z, \tilde{z} \in \Omega$; the **Lipschitz** property can be considered as a form of stronger continuity (often called Lipschitz continuity) but not quite as strong as differentiability, which allows for the difference quotient limit $z \to \tilde{z}$ to maintain the constant L.

[14] Perhaps the most standard **existence and uniqueness** theorem used in ODE theory is the Picard–Lindelöf theorem: an initial value problem $\dot{z} = G(t,z)$, $z(t_0) = z_0$ has a unique solution $z(t)$ at least for time $t \in [t_0 - \epsilon, t_0 + \epsilon]$ for some time range $\epsilon > 0$ if G is Lipschitz in z and continuous in t in an open neighborhood containing $(t_0, z(t_0))$. The standard proof relies on Picard iteration of an integral form of the ODE, $z(t) = z(t_0) = \int_{t_0}^{t} G(s, z(s))ds$, which with the Lipschitz condition can be proven to converge in a Banach space by the contraction mapping theorem [251]. Existence and uniqueness is a critical starting condition to discuss an ODE as a dynamical system, meaning one initial condition does indeed lead to one outcome which continues (at least for awhile), and correspondingly often the analysis herein may be as a discrete time mapping by Poincaré section.

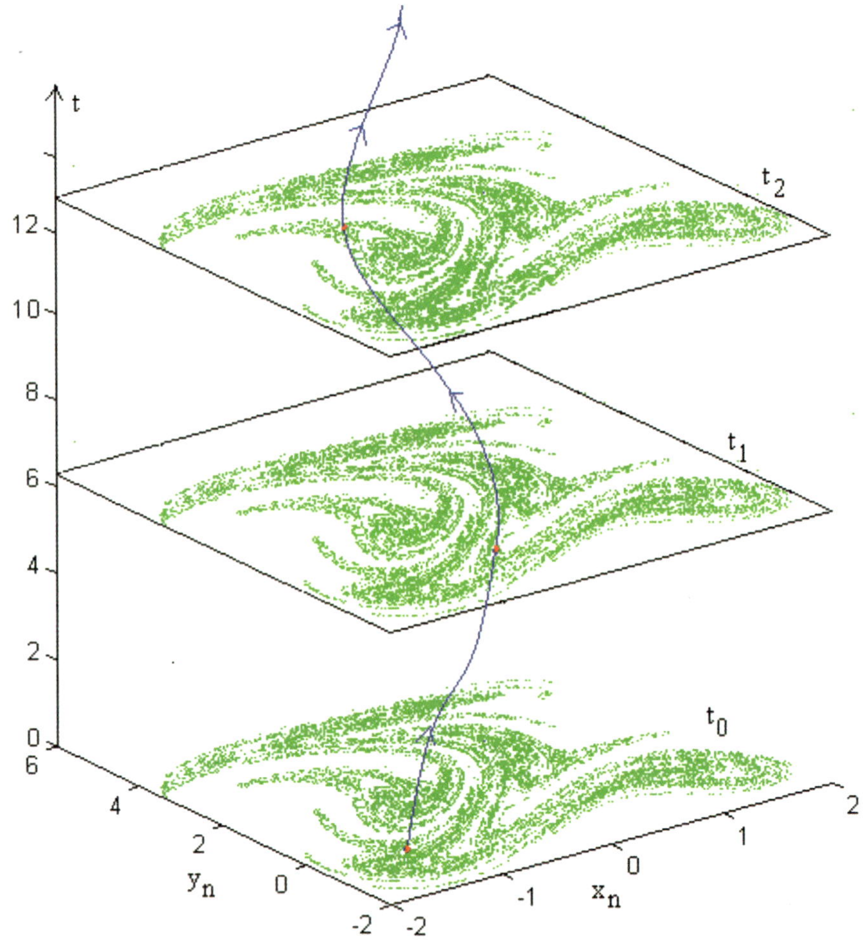

Figure 1.13. *The Poincaré mapping shown in Fig.* 1.12 *are surfaces* $\Sigma = \{(x, y, \tau) : (x, \tau) \in \mathbb{R}^2, \tau \in \frac{2\pi k}{\omega}\}$ *caricatured as the flow from Eq.* (1.39)*, with* $a = 0.02$*,* $b = 3$*, and* $\omega = 1$*, pierces the surfaces.*

In fact, the Duffing oscillator is an excellent example for presentation of the Poincaré mapping method. There exists a two-dimensional Duffing mapping—in this case a diffeomorphism. Such is common with differential equations arising from physical and especially mechanical problems. All this said, the common scenario is that we cannot explicitly represent the function $F : \mathbb{R}^2 \to \mathbb{R}^2$. In Fig. 1.12 we show the attractor corresponding to the Duffing oscillator on the left, and a caricature of the stroboscopic method whose flight produces F on the right. In practice, a computer is required for all examples we have experienced to numerically integrate chaotic differential equations, and thus further to estimate the mapping F and a finite number of sample points.

Just as in the case of the logistic map, where a histogram as in Figs. 1.3, 1.4, and 1.6 gives further information regarding the long-term fate of ensembles of initial con-

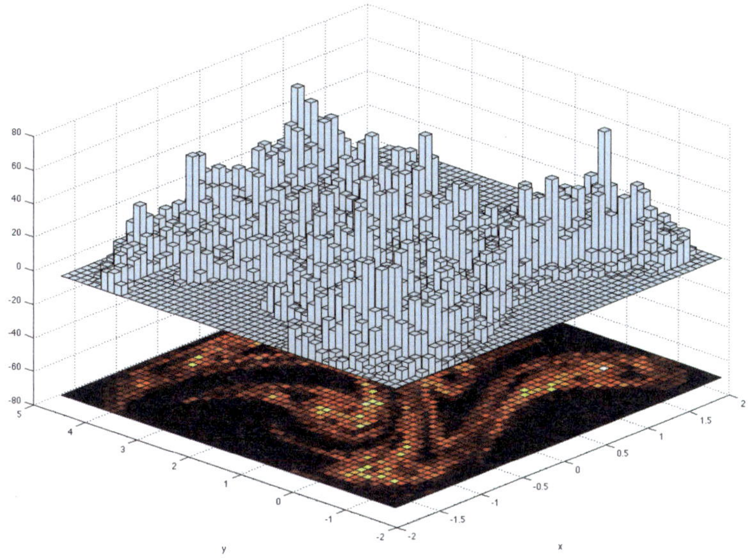

Figure 1.14. *Duffing density of the Poincaré-stroboscopic mapping method, estimated by simulation of a single initial condition evolved over* 100,000 *mapping periods, and density approximated by a histogram. The density is shown both as block heights (above) and as a color intensity map (below). Compare to the attractor shown in Fig.* 1.12.

ditions, we can make the same study in the case of differential equations by using the Poincaré mapping representation. The question is the same, but posed in terms of the Poincaré mapping. How do ensembles of initial conditions evolve under the discrete mapping, $(x_{k+1}, y_{k+1}) = F(x_k, y_k)$, as represented by a histogram over $\mathbb{R}^2$? See Fig. 1.14. The result of an experiment of a numerical simulation of one initial condition is expected to represent the same fate of many samples, for almost all initial conditions. That is true if one believes the system is ergodic, and thus follows the Birkhoff ergodic theorem (1.5). See also the discussion regarding natural measure near Eq. (3.78). The idea is that the same long-term averages sampled in the histogram boxes are almost always the same with respect to choosing initial conditions. Making these statements of ergodicity into mathematically rigorous statements turns out to be notoriously difficult even for the most famous chaotic attractors from the most favored differential equations from physics and mathematics. This mathematical intricacy is certainly beyond the scope of this book and we refer to Lai-Sang Young for a good starting point [323]. This is true despite the apparent ease with which we can simulate and seemingly confirm ergodicity of a map or differential equation through computer simulations. Related questions include existence of a natural measure, presence of uniform hyperbolicity, and representation by symbolic dynamics, to name a few.

In subsequent chapters, we will present the theory of transfer operator methods to interpret invariant density, mechanism, and almost invariant sets leading to steady states, almost steady states, and coherent structures partitioning the phase space. Further, we will show how the action of the mapping by a transfer operator may be approximated by a

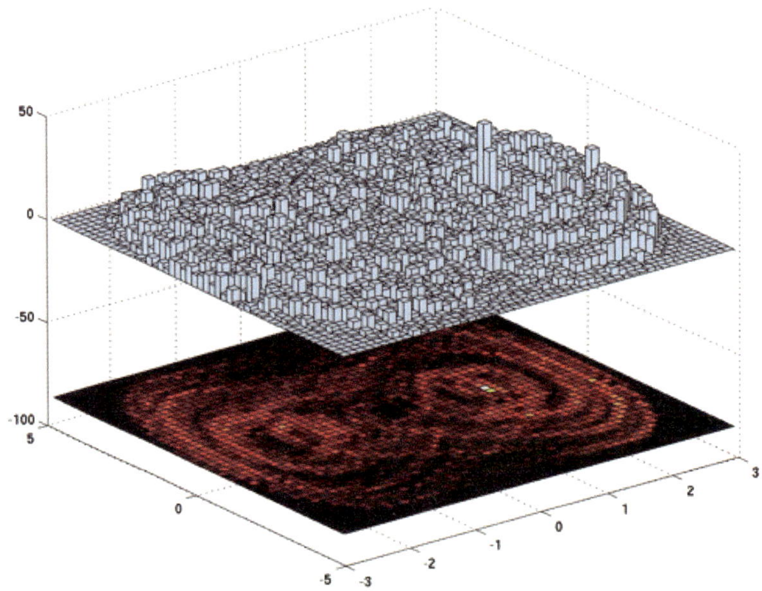

Figure 1.15. *The attractor of the Duffing oscillator flow in its phase space* $(x(t), y(t))$ *has relative density approximated by a histogram. Contrasts to the Poincaré mapping presentation of the same orbit segments are shown in Fig.* 1.14.

graph action generated by a stochastic matrix, through the now classic Ulam method, and how graph partitioning methods approximate and can be pulled back to present relevant structures in the phase space of the dynamical system.

Finally in this section for sake of contrast, we may consider the histogram resulting directly from following a single orbit of the flow as shown in the phase space but without resorting to the Poincaré mapping. That is, it is the approximation of relative density from the invariant measure of the attractor of the flow in the phase space. See Figure 1.15. This is in contrast to the density of the more commonly used and perhaps more useful Poincaré mapping as shown in Fig. 1.14. As was seen for the Henon map in Fig. 1.1, considering the action of the mapping on a discrete grid leads to a directed graph approximation of the action of the map. We will see that this action becomes a discrete approximation of the Frobenius–Perron operator, and as such it will serve as a useful computational tool. The convergence with respect to refinement we call the Ulam–Galerkin method and will be discussed in subsequent chapters. Also a major topic of this book will be the many algorithmic uses for this presentation as a method for transport analysis. There are a great number of computational methods that we will see become available when considering these directed graph structures. We will be discussing these methods, as well as the corresponding questions of convergence and representation, in subsequent chapters.

A major strength of this computational perspective for global analysis is the possibility to analyze systems known empirically only through data. As a case study and an important application [36], consider the spreading of oil following the 2010 Gulf of Mexico

Deepwater Horizon oil spill disaster. On April 20, 2010, an oil well cap explosion below the Deepwater Horizon, an offshore oil rig in the Gulf of Mexico, started the worst human-caused submarine oil spill ever. Though a historic tragedy for the marine ecosystem, the unprecedented monitoring of the spill in real time by satellites and increased modeling of the natural oceanic flows has provided a wealth of data, allowing analysis of the flow dynamics governing the spread of the oil. In [36] we studied two computational analyses describing the mixing, mass transport, and flow dynamics related to oil dispersion in the Gulf of Mexico over the first 100 days of the spill. Transfer operator methods were used to determine the spatial partitioning of regions of homogeneous dynamics into almost-invariant sets, and FTLEs were used to compute pseudobarriers to the mixing of the oil between these regions. The two methods give complementary results, which we will give in subsequent chapters. As we will present from several different perspectives, these data make a useful presentation for generating a comprehensive description of the oil flow dynamics over time, and for discussing the utility of many of the methods described herein.

Basic questions in oceanic systems concern large-scale and local flow dynamics which naturally partition the seascape into distinct regions. Following the initial explosion beneath the Deepwater Horizon drilling rig on April 20, 2010, oil continued to spill into the Gulf of Mexico from the resulting fissure in the well head on the sea floor. Spill rates have been estimated at 53,000 barrels per day by the time the leak was controlled by the "cap" fix three months later. It is estimated that approximately 4.9 million barrels, or 185 million gallons, of crude oil flowed into the Gulf of Mexico, making it the largest ever submarine oil spill. The regional damage to marine ecology was extensive, but impacts were seen on much larger scales as well, as some oil seeped into the Gulf Stream, which transported the oil around Florida and into the Atlantic Ocean. Initially, the amount of oil that would disperse into the Atlantic was overestimated, because a prominent dynamical structure arose in the Gulf early in the summer preventing oil from entering the Gulf Stream. The importance of computational tools for analyzing the transport mechanisms governing the advective spread of the oil may therefore be considered self-evident in this problem. Fig. 1.16. shows a satellite image of the Gulf of Mexico off the coast of Louisiana on May 24, 2010 just over a month after the initial explosion. The oil is clearly visible in white in the center of the image, and the spread of the oil can already be seen. During the early days of the spill the Gulf Stream was draining oil out of the Gulf and, eventually, into the Atlantic. This spread was substantially tempered later in the summer, due to the development of a natural eddy in the central Gulf of Mexico, which acted as a barrier to transport.

The form of the data is an empirical nonautonomous vector field $f(x,t)$, $x \in \mathbb{R}^2$, here derived from an ocean modeling source called the HYCOM model [173]. One time shot from May 24, 2010 is shown in Fig. 1.17. Toward transfer operator methods, in Fig. 1.18 we illustrate time evolution of several rectangle boxes suggesting the Ulam–Galerkin method to come in Chapter 4, and analogous to what was already shown in Figs. 1.1 and 1.19. In practice a finer grid covering would be used rather than this coarse covering which is used for illustrative purposes. The kind of partition result we may expect using these directed graph representations of the Frobenius–Perron transfer operator can be seen in Fig. 1.20. Discussion of almost-invariant sets, coherent sets, and issues related to transport and measure based partitions in dynamical systems from Markov models are discussed in detail in Chapter 5. For now we can say that the prime direction leading to this computational avenue is asking three simple questions:

Figure 1.16. *Satellite view of the Gulf of Mexico near Louisiana during the oil spill disaster, May* 24, 2010. *The oil slick spread is clearly visible and large. The image, taken by NASA's Terra satellite, is in the public domain.*

- Where does the product (oil) go, at least in relatively short time scales?
- Where does the product not go, at least in relatively short time scales?
- Are there regions which stay together and barriers to transport between regions?

These are questions of transport and of partition of the space relative to which transport can be discussed. Also related to partition is the boundary of partition for which there is a complementary method that has become useful. The theory of FTLEs will be discussed in Chapter 8, along with highlighting interpretations as barriers to transport and shortcomings for such interpretations. See, for example, an FTLE computation for the Fig. 1.17 Gulf of Mexico data in Fig. 1.21. In the following chapters these computations and supporting theory will be discussed.

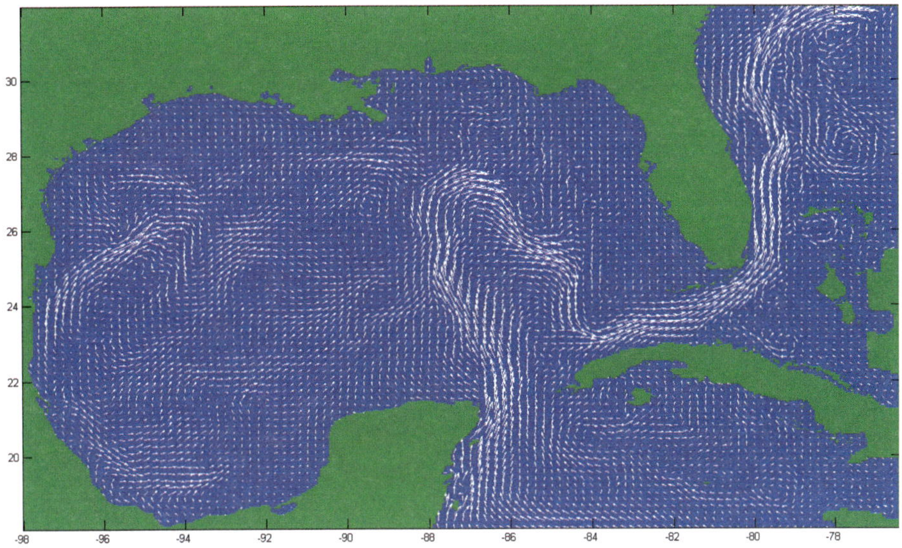

Figure 1.17. *Vector field describing surface flow in the Gulf of Mexico on May* 24, 2010, *computed using the HYCOM model* [173]. *Note the coherence of the Gulf Stream at this time. Oil spilling from south of Louisiana could flow directly into the Gulf Stream and out toward the Atlantic. Horizontal and vertical units are degrees longitude (negative indicates west longitude) and degrees latitude (positive indicates north latitude), respectively.*

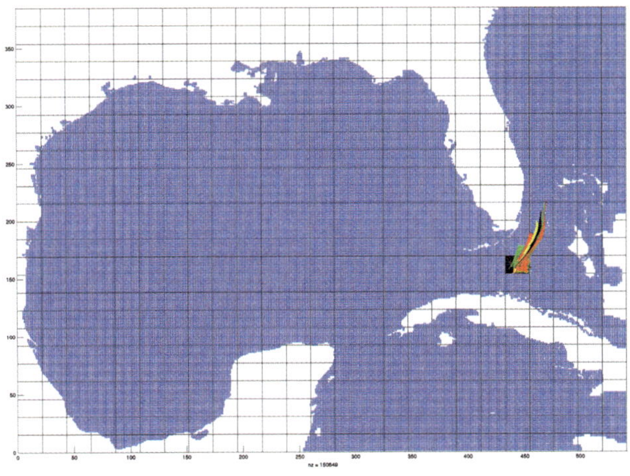

Figure 1.18. *Evolution of rectangles of initial conditions illustrates the action of the Frobenius–Perron transfer operator in the Gulf as estimated on a coarse grid by the Ulam–Galerkin method. Further discussion of such methods can be found in Chapter* 4. *Compare to Figs.* 1.1 *and* 1.19.

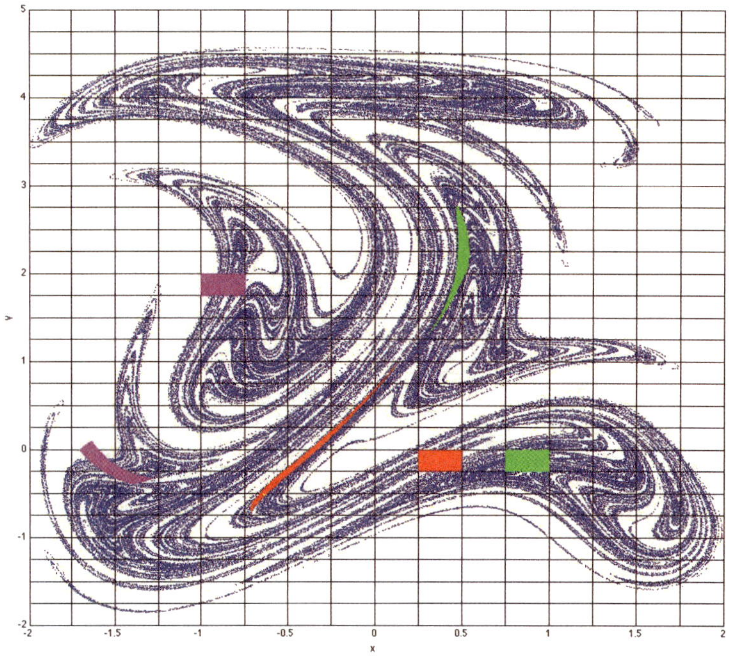

Figure 1.19. *Toward the Ulam–Galerkin method in the Duffing oscillator in a Poincaré mapping representation as explained in Chapter* 4. *Covering the attractor with rectangles, three are highlighted by colors magenta, red, and green. Under the Poincaré mapping in Eq.* (1.41), *F (rectangle) yields the distorted images of the rectangles shown. Each rectangle is mapped correspondingly to the same colored regions. Considering the relative measures of how much of each of these rectangles maps across other rectangles leads to a discrete approximation of the Frobenius–Perron operator, akin to the graph presentation of the Henon map's directed graph shown in Fig.* 1.1. *For generality, compare this figure to a similar presentation in the Gulf of Mexico in Fig.* 4.3 *allowing global analysis of a practical system known only through data.*

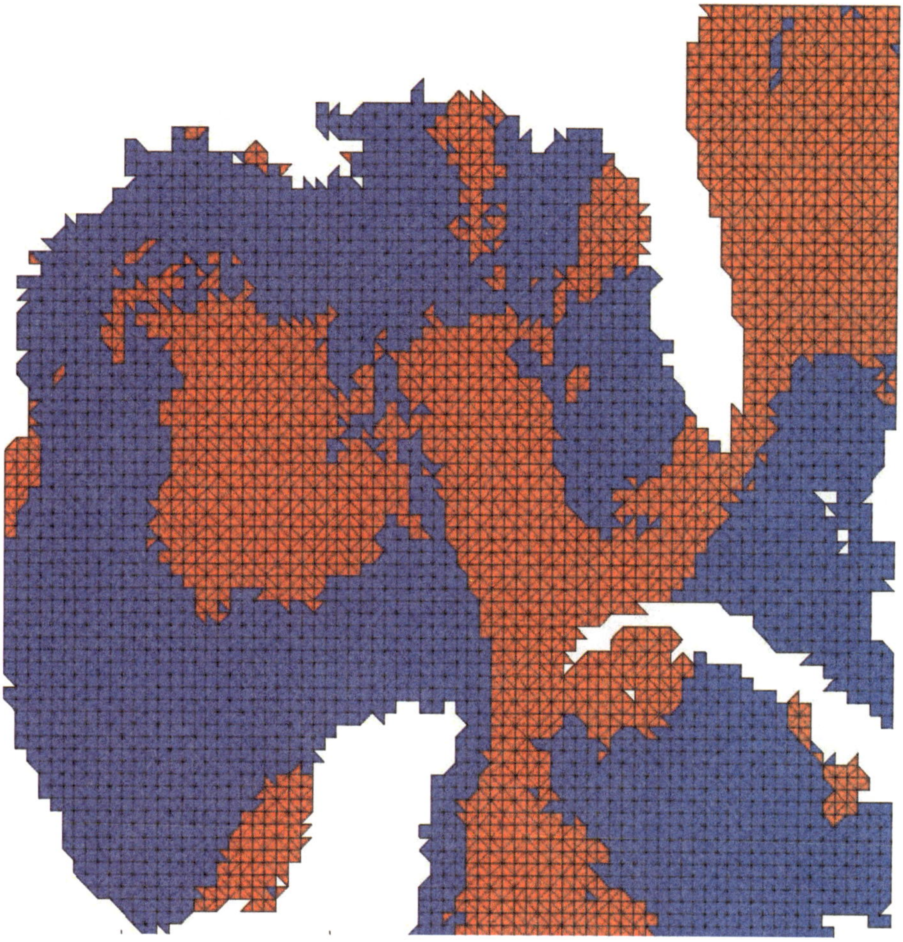

Figure 1.20. *Partition of the Gulf of Mexico using the transfer operator approach to be discussed in Chapter 5. Regions in red correspond to coherent sets, i.e., areas into and out of which little transport occurs.* [36]

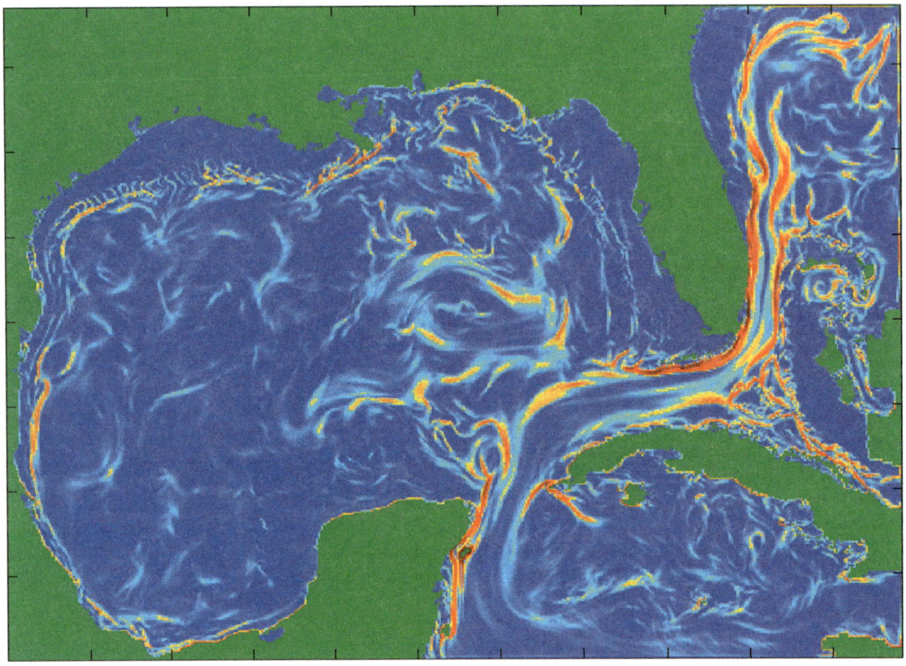

Figure 1.21. *Finite time Lyapunov exponents in the Gulf of Mexico help with understanding of transport mechanisms, as discussed in Chapter* 8, *including both interpretations and limitations. Roughly stated, the redder regions represent slow to almost no transport across these ridges.* [36]

Chapter 2

Dynamical Systems Terminology and Definitions

Some standard dynamical systems terminology and concepts will be useful, and we review here those elements that will be used in what follows. Some general and popular references for the following materials include [146, 268, 316, 218]. The material in this chapter uses a bit more formal presentation than the quick start guide of the previous chapter. But only necessary background will be given to support the central topic of the book, in an attempt to not overly repeat several of the many excellent general dynamical systems textbooks.

In its most general form, dynamical systems can be described as the study of group action on manifolds, and perhaps including differentiable structure. Throughout the majority of this work, we will not require this most general perspective, but some initial description can be helpful. Two major threads in the field of dynamical systems involve the study of

- topological properties,
- measurable properties,

of either groups or semigroups [218, 268] therein. Discussion of topological properties will be addressed in Chapter 6 regarding symbolic dynamics, and information theoretic aspects are addressed in Chapter 9. Discussion of measurable properties is more closely allied with the specific theme here, which involves transport mechanism of the fate of ensembles of initial conditions. In a basic sense, these two perspectives are closely related, as can be roughly understood by inspecting the Henon map example depicted in Fig.1.1; the figure depicts the action of the map on the phase space as approximated by a directed graph. In **topological dynamics**, we are not concerned with the relative scale of the sets being mapped. Therefore, the approximations by directed graphs have no weights. Thus follows unweighted graphs and adjacency matrices which generate them. On the other hand, **measurable dynamics** is concerned with relative weights of the sets, and so the directed graph approximation must be a weighted graph, with the weights along the edges describing either probability or relative ratios of transitions. Correspondingly, the graphs are generated by stochastic matrices rather than adjacency matrices. As we will see, both of these perspectives have their place in the algorithmic study of applied measurable dynamical systems.

2.1 The Form of a Dynamical System

Here, we shall write

$$\dot{x} = f(x,t),\ x(t_0) = x_0,\ x(t) \in M,\ t \in \mathbb{R}, \tag{2.1}$$

to denote a nonautonomous continuous time dynamical system, an ODE with initial condition. We shall assume sufficient regularity to allow existence and uniqueness [251] to suggest a dynamical system. By this, we mean that (semi)group action leads to a (semi)dynamical system with solutions that are (semi)flows.

Definition 2.1 (flow [218, 251]). A flow is a two-parameter family of differentiable maps on a manifold:[15]

1. Two-parameter mapping: $x(t;t_0,x_0) : \mathbb{R} \times \mathbb{R} \times M \to M$ which we interpret as mapping phase space M, through the time parameters t_0 and t, i.e., $x_0 \mapsto x(t;t_0,x_0)$.
2. Identity property: for each "initial" $x_0 \in M$ at the initial time t_0, the identity evolution is $x(t_0;t_0,x_0) = x_0$.
3. Group addition property: For each time t and s in $\mathbb{R}$, $x(s;t,x(t;t_0,x_0)) = x(t+s;t_0,x_0)$.
4. The function $x(\cdot;\cdot,\cdot)$ is differentiable with respect to all arguments.

Definition 2.2 (semiflow). A semiflow is identical to that of a flow, weakened only by the lack of time reversibility. That is, property 3 is changed so that the parameters t and s must come from the positive reals, $\Re^+$.

See also Definition 3.1. Whereas a flow is a group isomorphic to addition on the reals, a semiflow is not reversible. Hence group action is now a semigroup since we cannot expect the property that each element has an inverse. The concatenation of an element and its inverse is expressed by

$$x(-t+t_0;t,x(t;t_0,x_0)) = x(-t+t_0+t;t_0,x_0) = x(t_0;t_0,x_0) = x_0, \tag{2.2}$$

requiring use of a negative time $-t$, meaning prehistory.

The concept of semiflows arises naturally in certain physical systems, such as a heat equation $u_t = ku_{xx}$ to cite a simple PDE example, or the leaky bucket problem to cite an ODE example. Such are problems that "forget" their history, and the forgetting process we will eventually discuss alternatively as dissipation or a friction.

Example 2.1 (the leaky bucket—no reversibility [301]). The initial value problem

$$\dot{x} = -\sqrt{|x|}, \quad x(0) = x_0, \tag{2.3}$$

describes the height of a column of water in a bucket, which leaks out due to a hole at the bottom. We assume that the rate of water loss due to the leak depends on the pressure above

[15] If $f(x,t) \equiv f(x)$, the ODE is called autonomous, and the flow associated with an autonomous equation is reduced to the one-parameter family of maps, i.e., $x_0 \mapsto x(t;x_0)$.

it, which in turn is proportional to the volume of the column of water above the hole. It can be shown by substitution that this problem when $x_0 = 0$ allows solutions of the form

$$x(t) = \begin{bmatrix} \frac{1}{4}(t-c)^2, & t < c, \\ 0, & t \geq c, \end{bmatrix} \tag{2.4}$$

for any c, as well as the constant zero solution, $x(t) = 0$. Analytically, it should be noted that the right-hand side of Eq. (2.3) fails to be Lipschitz at $t = 0$, and hence it is not Lipschitz in any open set containing the initial condition. Therefore, the usual Picard-uniqueness theorem [251, 190] fails to hold, suggesting that nonuniqueness is at least a possibility; as demonstrated by multiple solutions, nonuniqueness does in fact occur. Nonuniqueness in this example quite simply corresponds to a physically natural observation, that an empty bucket cannot "remember" when it used to be full, or even if it was ever full.

The function on the right-hand side of Eq. (2.1),

$$f : M \times \mathbb{R} \to M, \tag{2.5}$$

denotes a vector field in the phase space manifold, M. However, the same ansatz form can be taken to denote a wider class of problems, even semidynamical systems (see the formal definition in Section 3.1) including certain PDEs such as reaction diffusion equations when M is taken to be a Banach space [269]. In such a case, through Galerkin's method when M is a Hilbert space, the PDE corresponds to an infinite set of ODEs describing energy oscillating between the time-varying "Fourier" modes. While the idea is straightforward for our purposes, to make this statement rigorous it is necessary to properly understand regularity and convergence issues by methods between functional analysis and PDE theory [269, 64].

Here, we will be most interested in the ODE case. In particular, we contrast Eq. (2.1) with the autonomous case,

$$\dot{x} = f(x),\ x(t_0) = x_0 \in M, \tag{2.6}$$

where the right-hand side does not explicitly incorporate time. Note that a nonautonomous dynamical system can be written as an autonomous system in a phase space of one more dimension, by augmenting the phase space to incorporate the time.[16]

A flow incorporates continuous time, whereas maps are a widely studied class of dynamical systems descriptive of discrete time:

$$x_{n+1} = F(x_n),\ \text{given } x_0. \tag{2.7}$$

For convenience, we will denote both the mapping $x(t; x_0) : \mathbb{R} \times M \to M$ descriptive of a flow, and those of a mapping Eq. (2.7), by the notation, $\phi_t(x_0)$; the domain of the independent t variable will determine the kind of dynamical system.

Definition 2.3. Dynamical systems including maps and semiflows can be classified within the language of the flow in Definition 2.1:

- If $\phi_t(\cdot)$ denotes a **flow**, then require the time domain to be $\mathbb{R}$.
- If $\phi_t(\cdot)$ denotes a **semiflow**, then require the time domain to be $\mathbb{R}^+$.

[16]Given Eq. (2.1), let $\tau = t$ and hence $\dot{\tau} = d\tau/dt = 1$, and $\dot{x} = f(x,t)$ is autonomous in $M \times \mathbb{R}$.

- If $\phi_t(\cdot)$ denotes an **invertible discrete time mapping**, then require the time domain to be $\mathbb{Z}$.
- If $\phi_t(\cdot)$ denotes a **noninvertible discrete time mapping**, then require the time to be $\mathbb{Z}^+$.

We will refer to dynamical systems which are either semiflows or noninvertible mappings together as **semidynamical systems** (see the formal definition in Section 3.1), acknowledging the semigroup nature.

Both discrete time and continuous time systems will be discussed here, as there are physically relevant systems which are naturally cast in each category. A stereotypical discrete time system is a model descriptive of compounding interest at the bank, where interest is awarded at each time epoch. On the other hand, discrete time sampling of continuous systems, as well as numerical methods to compute estimates of solutions, both take the form of discrete time systems. As seen in the previous chapter, a flow gives rise to a discrete time map through the method of Poincaré mapping as in Fig. 1.12.

2.2 Linearization

In this section, we summarize the information to be gained from the linearization of a map, which we will use to study and classify the local behavior "near" the fixed or periodic points of nonlinear differential equations. Loosely speaking, the Hartman–Grobman theorem [146, 251] shows when the local behavior near the hyperbolic fixed point is similar to that of the associated linearized system.

Consider maps $f : U \subset \mathbb{R}^k \to \mathbb{R}^n$, where U is an open subset of $\mathbb{R}^m$. The first partial derivative of f at a point p can be expressed in an $n \times k$ matrix form called a Jacobian derivative matrix,

$$Df_p = \left[\frac{\partial f_i}{\partial x_j}\right]. \tag{2.8}$$

We consider this matrix as a linear map from $\mathbb{R}^m$ to $\mathbb{R}^n$, that is, $Df_p \in L(\mathbb{R}^m, \mathbb{R}^n)$, and recall that $L(\mathbb{R}^m, \mathbb{R}^n)$ is isomorphic to $\mathbb{R}^{mn}$. With this in mind we may define the (Fréchet) derivative in the following way.

Definition 2.4. A map $f : U \subset \mathbb{R}^m \to \mathbb{R}^n$ is said to be **Fréchet differentiable** at $x \in \mathbb{R}^m$ if and only if there exists $Df_x \in L(\mathbb{R}^m, \mathbb{R}^n)$, called the **Fréchet derivative** of f at x, for which

$$f(x+h) = f(x) + Df_x h + o(\|h\|) \quad \text{as } \|h\| \to 0. \tag{2.9}$$

The derivative is called *continuous* provided the map $Df : U \to L(\mathbb{R}^m, \mathbb{R}^n)$ is continuous with respect to the Euclidean norm on the domain and the operator norm on $L(\mathbb{R}^m, \mathbb{R}^n)$. If the first partial derivatives at all points in U exist and are continuous, the derivative is also continuous and the map f is called *continuously differentiable* or $f \in C^1$. Higher-order derivatives can be defined recursively. For example,

$$Df(x+k) = Df(x) + D^2 f_x k + o(\|k\|) \tag{2.10}$$

defines the second derivative, $D^2 f_x \in L(\mathbb{R}^m, L(\mathbb{R}^m, \mathbb{R}^n))$. We again note that $L(\mathbb{R}^m, \mathbb{R}^n)$ is isomorphic to bilinear maps in $L(\mathbb{R}^m \times \mathbb{R}^m, \mathbb{R}^n)$. The mapping $(h,k) \mapsto (D^2 f_x k)h$ is bounded and bilinear from $\mathbb{R}^m \times \mathbb{R}^n$ into $\mathbb{R}^n$. If $v, w \in \mathbb{R}^m$ are expressed in terms of the standard basis as $v = \sum_i v_i \mathbf{e}^i$ and $w = \sum_i w_i \mathbf{e}^i$, then we have

$$(D^2 f_x v)w = \sum_{i,j} \left(\frac{\partial^2 f}{\partial x_i \partial x_j} \right)_x v_i w_j. \tag{2.11}$$

Since the mixed cross derivatives are equal,[17] $D^2 f_x$ is a symmetric bilinear form. Generally, if all the derivatives of order $1 \le j \le r$ exist and are continuous, then f is said to be r-continuously differentiable, or $f \in C^r$. If $f : \mathbb{R}^m \to \mathbb{R}^m$ is C^r, Df_x is a linear isomorphism at each point $x \in \mathbb{R}^k$, and f is one-one and onto, then f is called a C^r *diffeomorphism.*

Now, consider a dynamical system as the flow from an ODE,

$$\dot{x} = f(x,t), \tag{2.12}$$

and assume that $f : M \to M$ is at least C^2 in the domain M of x and C^1 in time, t. Then it is possible to linearize a dynamical system about a point $\tilde{x} \in M$. Suppose an initial point $\tilde{x}$ is advected to $x(t) = \tilde{x} + \delta x(t)$ after time t by a vector field. Then the linearization of $\dot{x} = f(x,t)$ is

$$\frac{d\delta x}{dt} = Df_{\tilde{x}} \cdot \delta x + o(\|\delta x\|). \tag{2.13}$$

Note that $Df_{\tilde{x}(t)}$ is generally time-dependent along the trajectory $\tilde{x}(t)$, and it is also nonsingular at each time t by the existence and uniqueness of the solution. If $\gamma(t)$ is a trajectory of the vector field, we write the associated linearized system of $\dot{x} = f(x,t)$ as

$$\dot{\xi} = Df(\gamma(t),t) \cdot \xi. \tag{2.14}$$

Example 2.2 (the Duffing oscillator and associated variational equations). In practice, the variational Eqs. (2.14) together with the base orbit Eq. (2.12) should simply be integrated together as a single coupled system. To make this clear, consider the specific example of a Duffing equation in autonomous form, Eq. (1.40).[18] Also review Figs. 1.12 to 1.15, and 1.19. The Jacobian matrix of the vector valued function f is

$$Df = \begin{pmatrix} \frac{\partial f_1}{\partial x_1} & \frac{\partial f_1}{\partial x_2} & \frac{\partial f_1}{\partial x_3} \\ \frac{\partial f_2}{\partial x_1} & \frac{\partial f_2}{\partial x_2} & \frac{\partial f_2}{\partial x_3} \\ \frac{\partial f_3}{\partial x_1} & \frac{\partial f_3}{\partial x_2} & \frac{\partial f_3}{\partial x_3} \end{pmatrix} = \begin{pmatrix} 0 & 1 & 0 \\ -1-3x_1 & -a & -\omega b \sin \omega x_3 \\ 0 & 0 & 0 \end{pmatrix}. \tag{2.15}$$

We see that the variables x_1 and x_3, but not x_2 in this particular case, appear explicitly in the derivative matrix. These are evaluated along an orbit solution, $\gamma(t) = (x_1(t), x_2(t), x_3(t))$. Therefore, to solve the variational equation, Eq. (2.14), we must solve simultaneously

$$\dot{x} = f(x,t), \tag{2.16}$$

$$\dot{\xi} = Df(x(t),t) \cdot \xi. \tag{2.17}$$

Notice the forward-only coupling from Eq. (2.16) to Eq. (2.17). In this Duffing example,

$$\dot{x} = \begin{pmatrix} \dot{x_1} \\ \dot{x_2} \\ \dot{x_3} \end{pmatrix} = \begin{pmatrix} x_2 \\ -ax_2 - x_1 - x_1^3 + b\cos \omega x_3 \\ 1 \end{pmatrix}, \quad \begin{pmatrix} x_1(0) \\ x_2(0) \\ x_3(0) \end{pmatrix} = \begin{pmatrix} x_{1,0} \\ x_{2,0} \\ x_{3,0} \end{pmatrix}, \tag{2.18}$$

[17] Clauraut's theorem asserts symmetry of the partial derivatives with sufficient continuity [265].

[18] $\dot{x_1} = f_1(x_1,x_2,x_3) = x_2, \dot{x_2} = f_1(x_1,x_2,x_3) = -ax_2 - x_1 - x_1^3 + b\cos \omega x_3, \dot{x_3} = f_3(x_1,x_2,x_3) = 1.$

and

$$\dot{\xi} = \begin{pmatrix} \dot{\xi}_1 \\ \dot{\xi}_2 \\ \dot{\xi}_3 \end{pmatrix} = \begin{pmatrix} 0 & 1 & 0 \\ -1-3x_1 & -a & -\omega b \sin \omega x_3 \\ 0 & 0 & 0 \end{pmatrix} \cdot \begin{pmatrix} \xi_1 \\ \xi_2 \\ \xi_3 \end{pmatrix}. \tag{2.19}$$

An orthonormal set of initial conditions for Eq. (2.19) allows exploring the complete fundamental set of solutions. As such, it is convenient to choose

$$\begin{pmatrix} \xi_1(0) \\ \xi_2(0) \\ \xi_3(0) \end{pmatrix} = \begin{pmatrix} 1 \\ 0 \\ 0 \end{pmatrix}, \begin{pmatrix} 0 \\ 1 \\ 0 \end{pmatrix}, \text{ or } \begin{pmatrix} 0 \\ 0 \\ 1 \end{pmatrix}. \tag{2.20}$$

Counting the size of the problem, we see that $x(t) \in \mathbb{R}^n$, $n = 3$, for Eqs. (2.16) and (2.19). The variations $\xi(t) \in \mathbb{R}^n$ of the variational equations (2.17) and (2.19) are also $n = 3$ dimensional, but exploring a complete basis set (2.20) requires $n^2 = 3^2 = 9$ dimensions. Therefore a complete "tangent bundle" [265] which is the base flow together with the fundamental solution of variations requires $n + n^2 = 3 + 3^2 = 12$ dimensions. See Fig. 2.1. We emphasize that computationally these equations should in practice all be integrated simultaneously in one numerical integrator subroutine. In terms of the flow, $\Phi_T(x(0)) = x(t)$, the variational equations evolved for a basis set as discussed, produce a matrix $M = D\Phi_T|_{x(0)}$ which is often called a "monodromy" matrix [265]. Thus, treating the time-T evolution of the flow and the derivative matrix of variations together forms a discrete time mapping and its derivative matrix.

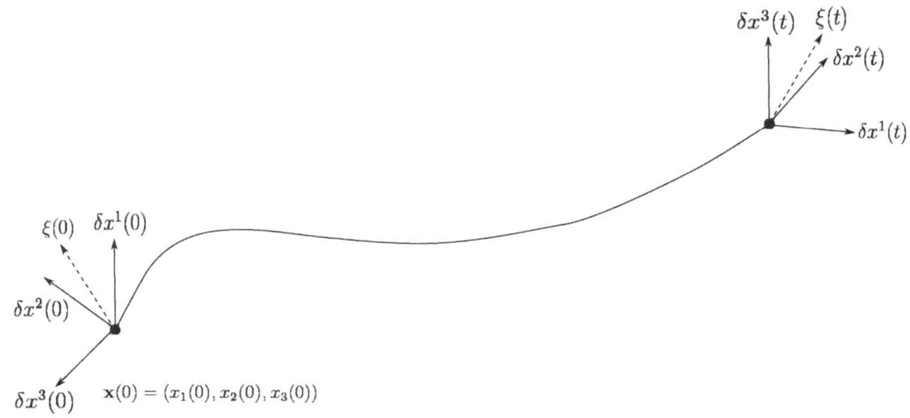

Figure 2.1. *The variational equations for an orthogonal set of initial conditions are evolved together with the base flow* $x(t)$. *See Eqs.* (2.16) *and* (2.17).

2.3 Hyperbolicity

Part of the importance of hyperbolicity in applied dynamical systems is that it allows significant simplifications for the analysis of complex and even chaotic dynamics. This section is intended to provide a sufficient background of hyperbolic dynamics for our discussion. We will limit ourselves to the notion of uniform hyperbolicity but note that a thorough treatment of hyperbolic dynamics and nonuniform hyperbolicity and its relation to ergodic theory can be found in Pollicott and Yuri [260]. In Chapter 8, we will see how the related finite time Lyapunov exponent (FTLE) theory has recently developed a central role in the empirical study of transport mechanism and partitioning the phase space of a complex system into dynamically related segments. The ideas behind that theory begin with an analogy to the linear theory. See Fig. 2.2 and compare it to the many figures in Chapter 8.

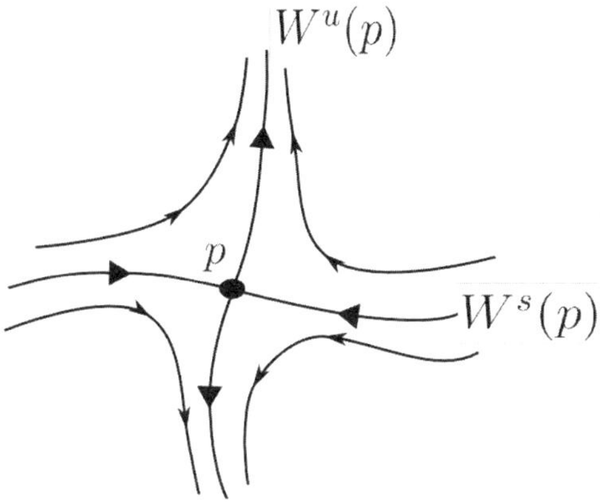

Figure 2.2. *Stable and unstable invariant manifolds of a fixed point p serve a critical role in organizing global behaviors such as invariant and almost-invariant sets.*

In this section, we are interested in the (discrete time) map $f: M \to M$, where M is a compact n-dimensional C^∞ Riemannian manifold. For our purposes here, it is sufficient to describe a Riemannian manifold as a differentiable topological space with an inner product on tangent spaces and which is locally isomorphic to a Euclidean metric space.

The Anosov diffeomorphism is a special prototypical example of hyperbolic behavior.

Definition 2.5 (Anosov diffeomorphism [316]). A diffeomorphism $f: M \to M$ is *Anosov* if there exist constants $c > 0, 0 < \lambda < 1$, and a decomposition of continuous tangent space,

$$T_x M = E_x^s \oplus E_x^u, \tag{2.21}$$

at each $x \in M$ satisfying the following properties:

(a) E_x^s and E_x^u are invariant with respect to the derivative Df_x:

$$\begin{aligned} Df_x E_x^s &= E_{f(x)}^s, \\ Df_x E_x^u &= E_{f(x)}^u. \end{aligned} \tag{2.22}$$

(b) The expansion rate under the forward and backward iterations are uniformly bounded:

$$\begin{aligned} \left\| Df_x^n v_s \right\| &= c\lambda^n \left\| v_s \right\| \qquad \text{for } v_s \in E_x^s, \\ \left\| Df_x^{-n} v_u \right\| &= c\lambda^n \left\| v_u \right\| \qquad \text{for } v_u \in E_x^u. \end{aligned} \tag{2.23}$$

Since the constants c and λ given in the above definition are the same for each point $x \in M$, Anosov systems are **uniformly hyperbolic** systems. By uniformly hyperbolic, it can be said roughly that at every point in the phase space, the hyperbolic splitting (2.22) and (2.23) holds.

We provide the following definitions.

Definition 2.6. A closed set $A \subset M$ is an **invariant** set with respect to a transformation f if $f(A) \subseteq A$.

Definition 2.7. If f is uniformly hyperbolic at each point $x \in A$, then A is called a uniformly **hyperbolic invariant** set.

Definition 2.8. If the invariant set is a single point, $A = \{\bar{x}\}$ (that is, $f(\bar{x}) = \bar{x}$), then $\bar{x}$ is called a **hyperbolic fixed point**.

Hyperbolic periodic points are similarly defined. Part of the importance of hyperbolicity is that the dynamics in a (small) neighborhood of the hyperbolic points exhibit expanding and contracting behavior in a manner which is homeomorphic to the associated linearized system; this statement is made precise by the Hartman–Grobman theorem [146, 268, 316].

A central concept in global analysis of dynamical systems in the presence of hyperbolicity is due to the presence of stable and unstable manifolds. Part of their importance is that these are codimension-one invariant sets, and as such they can serve to partition the phase space. Furthermore, stable and unstable manifolds tend to organize transport activity, as we will see developed in what follows. See Figs. 2.3 and 2.4. Their definition follows from the following existence theorems.

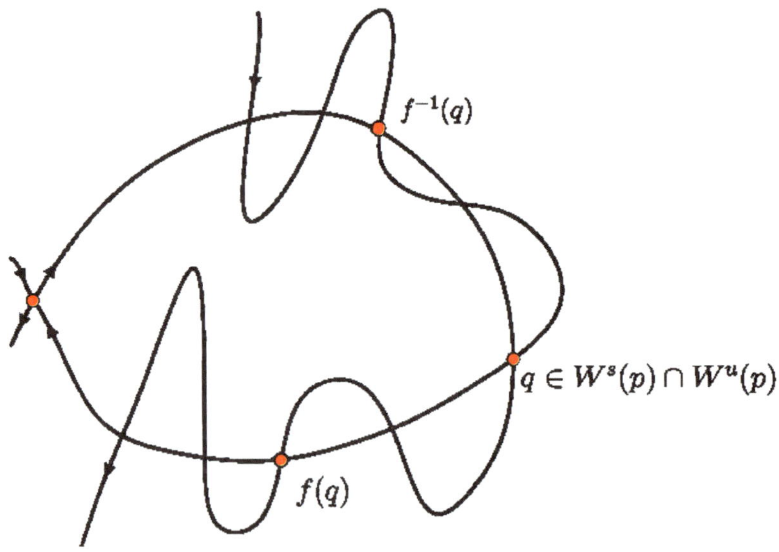

Figure 2.3. *An example of a homoclinic orbit of a hyperbolic fixed point p. Here q is the homoclinic point and the map f is assumed to be an orientation-preserving map, in which there must be at least one homoclinic point between q and $f(q)$. Note also that each of the infinitely many points $\{f^i(q)\}_{i=-\infty}^{\infty}$ on the orbit of q is homoclinic.*

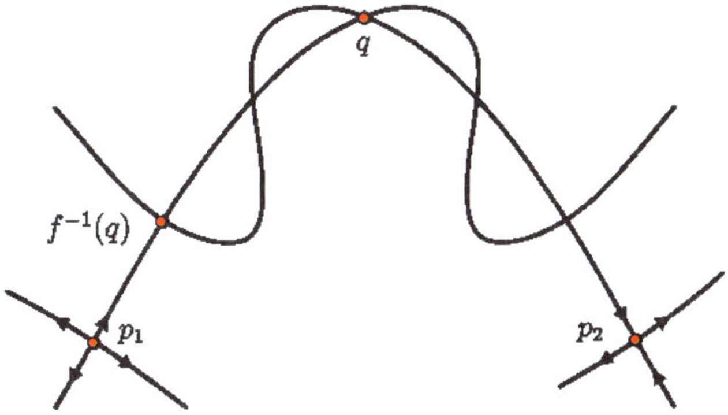

Figure 2.4. *An example of a heteroclinic orbit connecting two hyperbolic fixed points p_1 and p_2. Here q is the heteroclinic point and the map f is assumed to be an orientation-preserving map, as in Fig.* 2.3.

Definition 2.9 (local stable and unstable manifold [268]). Let $A \subset M$ be a hyperbolic invariant set, $x \in A$ and a neighborhood $N(x,\epsilon)$ for $\epsilon > 0$ sufficiently small. The local stable manifold, $W^s_\epsilon(x)$, and the local unstable manifold, $W^u_\epsilon(x)$, are given by

$$\begin{aligned} W^s_\epsilon(x) &= \{y \in N(x,\epsilon) | d(f^n(y),x) \to 0 \quad \text{as } n \to \infty\}, \\ W^u_\epsilon(x) &= \{y \in N(x,\epsilon) | d(f^{-n}(y),x) \to 0 \quad \text{as } n \to \infty\}. \end{aligned} \tag{2.24}$$

Note that the local unstable manifold can still be defined even if f is not invertible by abusing the notation $f^{-n}(x)$ as an iterated preimage of x.

Theorem 2.1 (stable manifold theorem [316]). Let $f : M \to M$, A be a hyperbolic invariant set with respect to f, and $x \in A$. Then there exist local stable and unstable manifolds $W^s_\epsilon(x)$ and $W^u_\epsilon(x)$ that satisfy the following properties:

(i) $W^s_\epsilon(x)$ and $W^u_\epsilon(x)$ are differentiable submanifolds and depend continuously on x.

(ii) Speaking of the tangent spaces, $TW^s_\epsilon(x) = E^s_x$ and $TW^u_\epsilon(x) = E^u_x$, i.e., $W^s_\epsilon(x)$ and $W^u_\epsilon(x)$ are of the same dimension and tangent to the subspace E^s_x and E^u_x, respectively.

The previous result can be extended to the entire domain M by the methods of the Hadamard–Perron manifold theorem (see Wiggins [316]). Therefore, we can discuss the local stable and unstable manifold at each point in the domain of an Anosov diffeomorphism.

Construction of the global stable and unstable manifolds follows by taking unions of backward and forward iterates of local stable and unstable manifolds, respectively. This is detailed in the following.

Definition 2.10 (global stable and unstable manifolds [268]). The global stable and unstable manifolds at each $x \in M$, $W^s(x)$ and $W^u(x)$, respectively, are given by

$$\begin{aligned} W^s(x) &\equiv \{y \in M | d(f^j(y), f^j(x)) \to 0 \quad \text{as } j \to \infty\}, \\ W^s(x) &= \bigcup_{n \geq 0} f^{-n}(W^s_\epsilon(f^n(x))) \quad \text{and} \\ W^u(x) &\equiv \{y \in M | d(f^{-j}(y), f^{-j}(x)) \to 0 \quad \text{as } j \to \infty\}, \\ W^u(x) &= \bigcup_{n \geq 0} f^n(W^u_\epsilon(f^{-n}(x))). \end{aligned} \tag{2.25}$$

It is clear from the above definition that the stable and unstable manifolds are invariant since they are a union of trajectories. Also, it is worth mentioning that a stable (unstable) manifold can cross neither itself nor other stable (unstable) manifolds of another point; otherwise, the crossing point must be iterated to two different points by definition, but this is not permitted by uniqueness of solutions. By contrast, an unstable manifold is allowed to intersect a stable manifold, which simply implies a coincidence in future outcomes from prior histories. Very interesting behavior can result as we discuss below.

The following definitions are some types of special orbits that are particularly important to transport and mixing theory, since they often form the transport barrier between regions of qualitatively different dynamics.

Definition 2.11 (homoclinic orbit [268]). Let p be a hyperbolic periodic point of period n for a diffeomorphism f, and let $\mathcal{O}(p)$ be the orbit of p. Let

$$\begin{aligned} W^{s,u}(\mathcal{O}(p)) &= \bigcup_{j=0}^{n-1} W^{s,u}(f^j(p)) \qquad \text{and} \\ \hat{W}^{s,u}(\mathcal{O}(p)) &= W^{s,u}(\mathcal{O}(p)) \setminus \mathcal{O}(p). \end{aligned} \tag{2.26}$$

A point $q \in \hat{W}^s(\mathcal{O}(p)) \cap \hat{W}^u(\mathcal{O}(p))$, if it exists, is called a *homoclinic point* for p. The orbit of q is then called the *homoclinic orbit* for p. See Figs. 2.3 and 2.4.

It follows from the above definition that a point in the homoclinic orbit will asymptotically approach the same hyperbolic periodic point p both in forward and backward time. The Smale–Birkhoff homoclinic theorem provides that the existence of a transverse homoclinic point induces a Smale horseshoe in the homoclinic tangle [268, 316, 315]. This important concept in the theory of transport mechanism will be detailed in Section 6.1.5.

Definition 2.12 (heteroclinic orbit [268]). Let $\{p_i\}, i = 1, \ldots, n$, be a collection of hyperbolic periodic orbits for a diffeomorphism f. A point $q \in \hat{W}^s(\mathcal{O}(p_i)) \cap \hat{W}^u(\mathcal{O}(p_j))$ for some $i \neq j$, if it exists, is called a *heteroclinic point*. Similarly, the orbit of q is called the *heteroclinic orbit*.

In contrast to a homoclinic orbit, a heteroclinic orbit asymptotically approaches a hyperbolic periodic orbit forward in time and a different periodic orbit backward in time; see Fig. 2.4.

2.4 Hyperbolicity: Nonautonomous Vector Fields

In the preceding section, the linearization of a fixed point or an invariant set was used to discuss the hyperbolicity property. However, for a general time-dependent system, we have to define the notion of the hyperbolicity for the time-dependent trajectory, for which the linearization along the trajectory does not necessarily give rise to a constant matrix associated with the linearized vector field. The standard notion to characterize the hyperbolicity of a time-dependent trajectory is that of the exponential dichotomy described below. We also summarize below the concept of stable and unstable manifolds of the hyperbolic trajectory and their significance and relation to the Lagrangian coherent structure (LCS), which will be discussed in Section 6.1.5.

As in the case of an autonomous system, we begin by defining an appropriate decomposition into stable and unstable subspaces along a trajectory of a nonautonomous system. This is a straightforward extension of the hyperbolicity of the autonomous vector field. First consider the linearized vector field of the form (2.14) and keep in mind that the coefficient matrix is now time-dependent.

Definition 2.13 (exponential dichotomy [316]). Consider a time-dependent linear differential equation

$$\dot{\xi} = A(t)\xi, \qquad \xi \in \mathbb{R}^n, \tag{2.27}$$

where $A(t) \in \mathbb{R}^{n\times n}$ is a time-dependent coefficient and continuous in time, $t \in \mathbb{R}$. Let $X(t) \in \mathbb{R}^{n\times n}$ be the solution matrix such that $\xi(t) = X(t)\xi(0)$ and $X(0) = I$. Then (2.27) is said to possess an *exponential dichotomy* if there exists a projection operator P, $P^2 = P$, and constants $K_1, K_2, \lambda_1, \lambda_2 > 0$ such that

$$\begin{aligned} \left\| X(t)PX^{-1}(t) \right\| &\le K_1 \exp(-\lambda_1(t-\tau)), \qquad t \ge \tau, \\ \left\| X(t)(I-P)X^{-1}(t) \right\| &\le K_2 \exp(\lambda_2(t-\tau)), \qquad t \le \tau. \end{aligned} \tag{2.28}$$

A generalization of a hyperbolic fixed point in autonomous systems is that of the hyperbolic trajectory of a time-dependent vector field, which can be defined via the exponential dichotomy.

Definition 2.14 (hyperbolic trajectory [316]). Let $\gamma(t)$ be a trajectory of the vector field $\dot{x} = f(x,t)$. Then $\gamma(t)$ is called a hyperbolic trajectory if the associated linearized system $\dot{\xi} = D_x(f(\gamma(t),t))\xi$ has an exponential dichotomy.

The geometry of a hyperbolic trajectory can be understood in the extended phase space:

$$\mathcal{E} \equiv \{(x,t) \in \mathbb{R}^n \times \mathbb{R}\}. \tag{2.29}$$

The nonautonomous vector field can then be viewed as an autonomous one:

$$\begin{aligned} \dot{x} &= f(x,t), \\ \dot{t} &= 1. \end{aligned} \tag{2.30}$$

In the extended phase space $\mathcal{E}$, we denote the hyperbolic trajectory by $\Gamma(t) = (\gamma(t),t)$ and define a *time slice* of the extended phase space $\mathcal{E}$ by

$$\Sigma_\tau \equiv \{(x,t) \in \mathcal{E} | t = \tau\}. \tag{2.31}$$

Then the condition (2.28) requires that there exists a projection onto a subspace of Σ_τ, called $E^s(\tau)$, so that a trajectory of an initial vector at time $t=0$ projected onto $E^s(\tau)$ by the associated linearized vector field (2.27) will have to decay to zero at an exponential rate, determined by λ_1 as $t \to \infty$. Similarly, the complementary projection $(I-P)$ onto a subspace of Σ_τ, called $E^u(\tau)$, exists so that the initial vector projected onto $E^u(\tau)$ will decay at an exponential rate, determined by λ_2 as $t \to -\infty$. Moreover, $\Sigma_\tau = E^s(\tau) \oplus E^u(\tau)$. Therefore, the exponential dichotomy guarantees the existence of a stable (unstable) subspace for which the initial conditions on these spaces asymptotically approach a hyperbolic trajectory at an exponential rate at a forward (backward) time; see Fig. 2.5. This is analogous to the autonomous vector field, where the initial conditions on the stable (unstable) manifold converge to some critical points, e.g., fixed points or periodic orbits, forward (backward) in time. However, the difference is that the stable (unstable) manifolds for an autonomous system are time-independent, whereas those of a nonautonomous case vary in time.

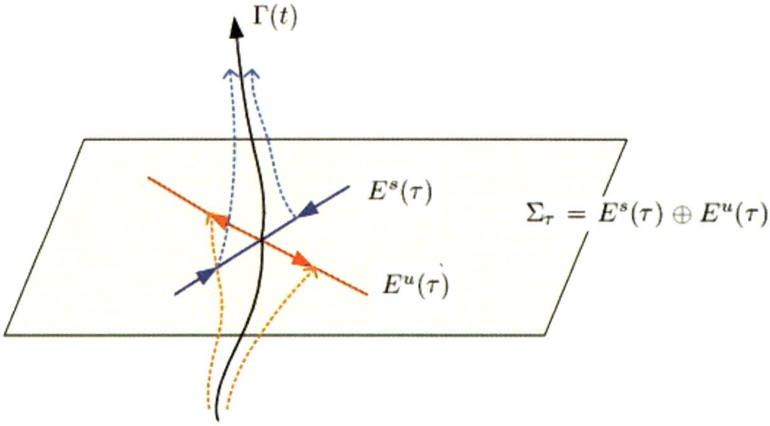

Figure 2.5. *The time slice $\Sigma_\tau = E^s(\tau) \oplus E^u(\tau)$. A trajectory of an initial segment projected onto $E^s(\tau)$ by the associated linearized vector field* (2.27) *approaches the hyperbolic trajectory $\Gamma(t)$ at an exponential rate as $t \to \infty$. Similarly, the initial vector projected onto $E^u(\tau)$ approaches $\Gamma(t)$ at an exponential rate as $t \to -\infty$.*

In particular, for a two-dimensional vector field, the stable (unstable) manifold in the autonomous case is an invariant curve, but the stable (unstable) manifold in the nonautonomous case becomes a time-varying curve or an invariant surface in the extended space $\mathcal{E}$. With the geometrical explanation of the exponential dichotomy in mind, we can now state the theorem that describes the existence of local stable and unstable manifolds of a hyperbolic trajectory.

Theorem 2.2 (local stable and unstable manifolds [316]; see similarly [268]). Let $D_\rho(\tau)\Sigma_\tau$ denote the ball of radius ρ centered at $\gamma(\tau)$ and define the tubular neighborhood of $\Gamma(t)$ in $\mathcal{E}$ by $\mathcal{N}_\rho(\Gamma(t)) \equiv \cup_{\tau\in\mathbb{R}}(D_\rho(\tau),\tau)$. There exist $(k+1)$-dimensional C^r manifold $W^s_{loc}(\Gamma(t)) \subset \mathcal{E}$, $(n-k+1)$-dimensional C^r manifold $W^u_{loc}(\Gamma(t)) \subset \mathcal{E}$, and ρ_0 sufficiently small such that for $\rho \in (0,\rho_0)$ the following hold:

(i) $W^s_{loc}(\Gamma(t))$, the local stable manifold of $\Gamma(t)$, is invariant under the forward-time evolution generated by (2.30); $W^u_{loc}(\Gamma(t))$, the local unstable manifold of $\Gamma(t)$, is invariant under the backward-time evolution generated by (2.30).

(ii) $W^s_{loc}(\Gamma(t))$ and $W^u_{loc}(\Gamma(t))$ intersect along $\Gamma(t)$, and the angle between the manifolds is bounded away from zero uniformly for all $t \in \mathbb{R}$.

(iii) Every trajectory on $W^u_{loc}(\Gamma(t))$ can be continued to the boundary of $\mathcal{N}_\rho(\Gamma(t))$ backward in time, and every trajectory on $W^u_{loc}(\Gamma(t))$ can be continued to the boundary of $\mathcal{N}_\rho(\Gamma(t))$ forward in time.

(iv) Trajectories starting on $W^s_{loc}(\Gamma(t))$ at time $t = \tau$ approach $\Gamma(t)$ at an exponential rate $e^{-\lambda'(t-\tau)}$ as $t \to \infty$, and trajectories starting on $W^u_{loc}(\Gamma(t))$ at time $t = \tau$ approach $\Gamma(t)$ at an exponential rate $e^{-\lambda'|t-\tau|}$ as $t \to \infty$, for some constant $\lambda' > 0$.

(v) Any trajectory in $\mathcal{N}_\rho(\Gamma(t))$ not on either $W^s_{loc}(\Gamma(t))$ or $W^u_{loc}(\Gamma(t))$ will leave $\mathcal{N}_\rho(\Gamma(t))$ both forward and backward in time.

The above theorem suggests a way to determine global stable and unstable manifolds of a hyperbolic trajectory [210, 211, 212]. In short, it allows us to extend an initial segment of a local unstable manifold, $W^u(\gamma(t_1))$, of a hyperbolic trajectory $\gamma(t_1)$ on the time slice Σ_{t_1} for some $t_1 < \tau$ by evolving this segment forward in time. Similarly, evolving the initial segment of a stable manifold, $W^s(\gamma(t_2))$, of a hyperbolic trajectory $\gamma(t_2)$ on the time slice Σ_{t_2} for some $t_2 > \tau$ backward in time yields a global stable manifold. Fig. 2.6 illustrates this concept.

One rigorous method to determine a meaningful hyperbolic trajectory, called *distinguished hyperbolic trajectory* (DHT), and its stable and unstable manifolds can be found in [210, 211, 212]. Since a dynamical system can possess infinitely many hyperbolic trajectories, e.g., all trajectories in the stable or unstable manifolds, the bounded hyperbolic trajectory with the special properties is used as the reference hyperbolic trajectory for "growing" the stable and unstable manifolds [210, 211]. It is demanded in [174] that a DHT remain in a bounded region for all time and that there exist a neighborhood $\mathcal{N}$ of the DHT such that all other hyperbolic trajectories within $\mathcal{N}$ have to leave $\mathcal{N}$ in a finite time, either forward or backward. If one attempts to grow the stable or unstable manifold from a hyperbolic trajectory different from the DHT, it is possible to observe a "drifting" phenomenon in the flow due to a slow expansion rate as demonstrated in a number of numerical examples in [212].

A rigorous method of determining the DHT in general is still an active research area. In most cases, DHT has to be selected with a careful observation and knowledge of the system. In another approach, Haller and Poje [156] determined a finite time, uniformly hyperbolic trajectory based on the local maxima of the time duration in which an initial point remains hyperbolic. The contour plot of such hyperbolicity time for each grid point of initial conditions reveals the uniformly hyperbolic invariant set, which can be used as a "seed" to gradually construct the corresponding global stable and unstable manifolds by a traditional technique, e.g., the straddling technique of You, Kostelich, and Yorke [321]. Nevertheless, this technique demands the deformation of the uniformly hyperbolic trajectory to be at a slower speed than the speed of individual particles.

In a recent approach, following the work in a series of papers by Haller [156, 153, 154], the search of DHT or hyperbolic invariant sets can be circumvented by computing

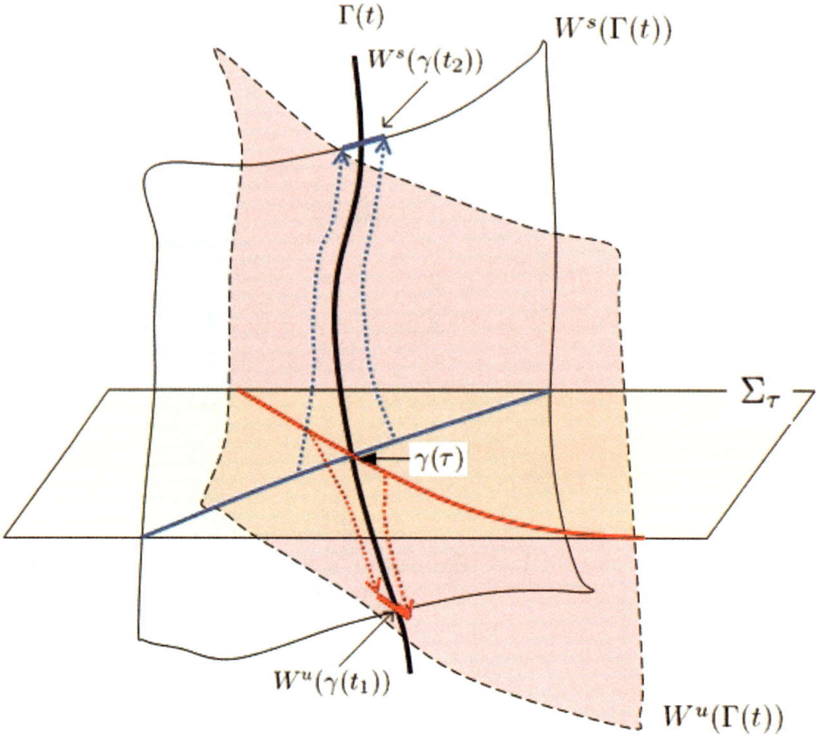

Figure 2.6. *The forward-time evolution of a segment of $W^u(\gamma(t_1))$ from the time slice Σ_{t_1} to the time slice Σ_τ yields the unstable manifold of $W^u(\gamma(\tau))$. Similarly, the backward-time evolution of a small segment of $W^s(\gamma(t_2))$ from the time slice Σ_{t_2} to the time slice Σ_τ yields the unstable manifold $W^s(\gamma(\tau))$.*

the FTLE field [287, 286], which directly captures the region of a large expansion rate; see Chapter 8. These regions correspond to the stable (unstable) manifold of the hyperbolic trajectory when computing forward (backward) in time. This technique is based on the fact that the stable (unstable) manifold of a hyperbolic trajectory is repelling (attracting) in the normal direction to the manifold, and hence initial points straddling the stable (unstable) manifold will eventually be separated from (attracted to) each other at some exponential rate. However, this approach gives us merely a scalar field indicating the expansion rate of the grid point in a domain of a nonautonomous system, whereas the preceding techniques yield the parameterized curve representing invariant manifolds. Nevertheless, the second-derivative ridge of the FTLE field can be used as a curve that represents the global stable (or unstable) manifold of a nonautonomous dynamical system by using a ridge detection method.

Chapter 3

Frobenius–Perron Operator and Infinitesimal Generator

This chapter is devoted to a review of some classical tools and techniques for studying the evolution of density under a continuous time process. Specifically, we concentrate on the concept of the Frobenius–Perron operator and its infinitesimal generator. The continuous time problem will also be cast in terms of the related discrete time problem. The discussion herein expands a more mathematical discussion of these transfer operators which we already introduced in Chapter 1.

Here we present the technical details of the Frobenius–Perron operator—the evolution operator point of view alluded to in Chapter 1. A good further review of this material is found in [198]. In principle, this chapter could be skipped if a more computational perspective is required.

3.1 Frobenius–Perron Operator

We define a continuous process in a topological Hausdorff space X by a family of mappings

$$S_t : X \to X, \qquad t \geq 0. \tag{3.1}$$

For example, a continuous time process generated by an autonomous d-dimensional system of differential equations

$$\frac{dx}{dt} = F(x) \tag{3.2}$$

or

$$\frac{dx_i}{dt} = F_i(x), \qquad i = 1,\dots,d, \tag{3.3}$$

where $x = (x_1,\dots,x_d)$ and $F : \mathbb{R}^d \to \mathbb{R}^d$ is sufficiently smooth to guarantee the existence and uniqueness of solutions. Then Eq. (3.2) defines a transformation $S_t(x^0) = x(t)$ where $x(t)$ is the solution of Eq. (3.2) at time t starting from the initial condition x^0 at time 0.

Definition 3.1. *A semidynamical system* $\{S_t\}_{t\geq 0}$ on X is a family of transformations S_t : $X \to X$, $t \in \mathbb{R}$, satisfying

(a) $S_0(x) = x$ for all $x \in X$,

(b) $S_t(S_{t'}(x)) = S_{t+t'}(x)$ for all $x \in X$ with $t, t' \in \mathbb{R}^+$, and

(c) the mapping $(x,t) \to S_t(x)$ from $X \times \mathbb{R}^+$ into X is continuous.

See also Definitions 2.1 and 2.2.

Remark 3.1. The reason we call $\{S_t\}_{t\geq 0}$ in the above definition a semidynamical system instead of a dynamical system is to allow for the possibility of lack of invertibility. By restricting $t \in \mathbb{R}^+$ a family of transformations that satisfies the above conditions possesses an Abelian semigroup properties and hence may be called a **semigroup of transformations**.

Now we revisit the density transfer operator known as the **Frobenius–Perron operator**, already discussed in the simplest case in Eqs. (1.10)–(1.18) of discrete time mappings of the interval. Now we present this important tool for studying a propagation of densities in the more general settings.

Let (X, Σ, μ) be a σ-finite measure space, where Σ denotes the σ-algebra of Borel sets. Assume that each transformation of a semidynamical system $\{S_t\}_{t\geq 0}$ is a nonsingular measurable transformation on (X, Σ, μ), that is,

$$\mu(S_t^{-1}(A)) = 0 \qquad \text{for each } A \in \Sigma \text{ such that } \mu(A) = 0. \tag{3.4}$$

Therefore, measure preserving transformations $\{S_t\}$ is necessarily nonsingular with respect to μ.[19] The **Frobenius–Perron operator**, $P_t : L^1(X) \to L^1(X)$ with respect to the transformation S_t, is defined [198] by the condition of conservation of mass

$$\int_{S_t^{-1}(A)} f(x)d\mu = \int_A P_t f(x) d\mu \qquad \text{for each } A \in \Sigma. \tag{3.5}$$

In what follows, we will consider only the action of P_t on the space $D(X, \Sigma, \mu)$ defined by

$$D \equiv D(X, \Sigma, \mu) = \{f \in L^1(X, \Sigma, \mu) : f \geq 0 \text{ and } \|f\| = 1\}. \tag{3.6}$$

So, D is a set of a probability density function (PDF) of $L^1(X)$. Therefore, $P_t f(x)$ is also a PDF, which is unique a.e.,[20] and depends on the transformations $\{S_t\}$ and the initial PDE $f(x)$. It is straightforward to show [198] that P_t satisfies the following properties:

$$\begin{aligned} &\text{(a) } P_t(\lambda_1 f_1 + \lambda_2 f_2) = \lambda_1 P_t f_1 + \lambda_2 P_t f_2 \qquad \forall f_1, f_2 \in L^1, \lambda_1, \lambda_2 \in \mathbb{R}; \\ &\text{(b) } P_t f \geq 0 \qquad \text{if } f \geq 0; \\ &\text{(c) } \int_X f(x) d\mu = \int_X P_t f(x) d\mu \qquad \forall f \in L^1. \end{aligned} \tag{3.7}$$

By using the above properties, one may prove that P_t satisfies properties (a) and (b) of the definition of a semidynamical system.

[19] Transformations $\{S_t\}$ are **measure preserving** with respect to μ if $\mu(S_t^{-1}(A)) = \mu(A)$ for all $A \in \Sigma$.

[20] The uniqueness can be established by applying the Radon–Nikodým theorem. For a given function $f \in L^1(X)$, we may define the left-hand side of (3.5) as a real measure $\mu_f(A)$. Since $\{S_t\}$ are nonsingular for every t, μ_f are absolutely continuous with respect to μ. By the Radon–Nikodým theorem, there exists a unique function (the so-called Radon–Nikodým derivative), denoted by $P_t f \in L^1(X)$, such that $\mu_f(A) = \int_A P_t f d\mu$ for every $A \in \Sigma$.

Consider in addition to the differential equation (3.2) an observable $K_t f$ defined by

$$K_t f(x) = f(S_t(x)), \tag{3.8}$$

where $f, K_t f \in L^\infty(X)$. Hence, the operator $K_t : L^\infty(X) \to L^\infty(X)$ as defined in Eq. (3.8), for every $t \geq 0$, can be interpreted as the operator that evolves an observable $f(S_t(x))$ of a semidynamical system $\{S_t\}_{t \geq 0}$ given by Eq. (3.2). The operator K_t is known as the **Koopman operator** associated with the transformation S_t. It is easy to check that $\{K_t\}_{t \geq 0}$ is a semigroup. An important mathematical relation between the Frobenius–Perron and Koopman operators is that they are adjoint. That is,

$$\langle P_t f, g\rangle = \langle f, K_t g\rangle \tag{3.9}$$

for all $f \in L^1(X), g \in L^\infty(X)$, and $t \geq 0$.[21]

Note that although the Frobenius–Perron operator preserves the L^1-norm $\|\cdot\|_{L^1}$ for $f \geq 0$ (recall the discrete continuity equation, Eq. (1.12)),

$$\|P_t f\|_{L^1} = \|f\|_{L^1}, \tag{3.10}$$

the Koopman operator, on the other hand, satisfies the inequality

$$\|K_t f\|_{L^\infty} \leq \|f\|_{L^\infty}. \tag{3.11}$$

In a subsequent section we will derive the infinitesimal operators of the Frobenius–Perron and Koopman operators. Therefore, we will concentrate on the semigroup of contracting linear operators defined in the following.

Definition 3.2. For $L = L^p$, $1 \leq p \leq 1$, a family $\{T_t\}_{t \geq 0}$ of operators, $T_t : L \to L$ is called a **semigroup of contracting operators** if T_t has the following properties:

(a) $T_t(\lambda_1 f_1 + \lambda_2 f_2) = \lambda_1 T_t f_1 + \lambda_2 T_t f_2$,

(b) $\|T_t f\|_L \leq \|f\|_L$,

(c) $T_0 f = f$, and

(d) $T_{t+t'} f = T_t(T_{t'} f)$,
for $f, f_1, f_2 \in L$, and $\lambda_1, \lambda_2 \in \mathbb{R}$.

The semigroup is called a **continuous semigroup** if it satisfies

$$\lim_{t \to t_0} \|T_t f - T_{t_0} f\|_L = 0 \qquad \text{for } f \in L, t_0 \geq 0. \tag{3.12}$$

[21]The notation $\langle q, r\rangle$, $q \in L^1(X), r \in L^\infty(X)$, denotes a bilinear form and here takes the form of integration. Often, to be explicit we emphasize the spaces from which the functions are drawn, $\langle q, r\rangle_{L^1(X) \times L^\infty(X)} = \int_X q(x) r(x) dx$.

3.2 Infinitesimal Operators

For a continuous semigroup of contractions $\{T_t\}_{t\geq 0}$ we define $D(A)$ by the set of all $f(x) \in L^p(\mathbb{X})$, $1 \leq p \leq \infty$, such that the limit

$$Af = \lim_{t\to 0} \frac{T_t f - f}{t} \tag{3.13}$$

exists in the sense of strong convergence. That is,

$$\lim_{t\to 0} \left\| Af - \frac{T_t f - f}{t} \right\|_{L^p} = 0. \tag{3.14}$$

The operator

$$A : D(A) \to L \text{ for } L \equiv L^p \tag{3.15}$$

is called the **infinitesimal generator**. Let

$$I(t) \equiv I(x,t) = T_t f(x) \text{ for fixed } f(x) \in D(A). \tag{3.16}$$

The function $I'(t) \equiv I'(t)(x) \in L$ is said to be the **strong derivative** of $I(t)$ if it satisfies the following condition:

$$\lim_{t\to 0} \left\| I'(t) - \frac{I(t) - f(x)}{t} \right\|_{L} = 0. \tag{3.17}$$

In this sense, $I'(t)$ describes the derivative of the ensemble of points with respect to time t. The following theorem demonstrates an important relationship between the infinitesimal generator and the strong derivative.

Theorem 3.1 (see [198]). Let $\{T_t\}_{t\geq 0}$ be a continuous semigroup of contractions and $A : D(A) \to L$ the corresponding infinitesimal generator. Then for each fixed $f \in D(A)$ and $t \geq 0$, the function $I(t) = T_f t$ has the properties

(a) $I(t) \in D(A)$,

(b) $I'(t)$ exists, and

(c) $I(t)$ satisfies

$$I'(t) = AI(t) \tag{3.18}$$

with the initial condition $I(0) = f(x)$.

Example 3.1. Consider the family of operators $\{T_t\}_{t\geq 0}$ defined by

$$T_t f = f(x - ct) \qquad \text{for } x \in \mathbb{R}, t \geq 0. \tag{3.19}$$

Under this operation a function $f(x)$ is translated in a positive direction of x by the length of ct. By using the "change of variable" formula we can see that the L^p-norm is preserved for $1 \leq p \leq \infty$. The conditions (a), (c), and (d) of Definition 3.2 straightforwardly follow from Eq. (3.19). Thus, $\{T_t\}$ is a semigroup of contracting operators. It is slightly more complicated to show that $\{T_t\}$ is also continuous; see [198]. Now assume that f is bounded and at least $C^1(\mathbb{R})$; then by the mean value theorem we have

$$\frac{f(x-ct)-f(x)}{t} = -cf'(x-\theta ct), \tag{3.20}$$

where $|\theta| \leq 1$. This implies that

$$Af = \lim_{t\to 0} \frac{T_t f - f}{t} = -cf' \tag{3.21}$$

and the limit is strong in L^p, $1 \leq p \leq \infty$, if f has compact support. Therefore, it follows from (3.18) that at each point in the (x,t)-plane $u(t,x)$ satisfies the partial differential equation

$$\frac{\partial u}{\partial t} + c\frac{\partial u}{\partial x} = 0 \qquad \text{with } u(0,x) = f(x), \tag{3.22}$$

where $u(t,x)$ is in $D(A)$ for each fixed $t \geq 0$.

Remark 3.2. This example offers some insight into the relationship between the semigroup of continuous operators, strong derivatives, and the corresponding partial differential equations. It is well known that the solution of Eq. (3.22) at time t is $T_t f$ as defined in Eq. (3.19). See [269] for a discussion of PDE theory as related to infinite-dimensional dynamical systems, and see [111] for the modern functional analysis formulation PDE theory.

Now consider a calculation of the infinitesimal generator of the semigroup of the Frobenius–Perron operators $\{P_t\}_{t\geq 0}$ as defined in Eq. (3.5) and the evolution of the time-dependent density function $I(x,t)$ under an action of the Frobenius–Perron operator. This will be done indirectly through the adjoint property of the Frobenius–Perron and Koopman operators.

It follows directly from the definition of the Koopman operator Eq. (3.8) that the infinitesimal of the Koopman operator denoted by A_K is

$$A_K g(x) = \lim_{t\to 0} \frac{g(S_t(x^0)) - g(x^0)}{t} = \lim_{t\to 0} \frac{g(x(t)) - g(x^0)}{t}. \tag{3.23}$$

If g is continuously differentiable with compact support, we can apply the mean value theorem to obtain

$$A_K g(x) = \lim_{t\to 0} \sum_{i=1}^{d} \frac{\partial g}{\partial x_i}(x(\theta t))x_i'(\theta t) = \sum_{i=1}^{d} \frac{\partial g}{\partial x_i} F_i(x), \tag{3.24}$$

where $0 < \theta < 1$. Combining (3.18) and (3.24) we conclude that the function

$$I(x,t) = K_t f(x) \tag{3.25}$$

satisfies the first-order PDE

$$\frac{\partial I}{\partial t} - \sum_{i=1}^{d} \frac{\partial I}{\partial x_i} F_i(x) = 0. \tag{3.26}$$

This leads to a derivation of the infinitesimal generator for the semigroup of Frobenius–Perron operators generated by the family $\{S_t\}_{t\geq 0}$ defined in Eq. (3.1).

Let $f \in D(A_{FP})$ and $g \in D(A_K)$, where A_{FP} and A_K denote the infinitesimal operators of the semigroups of the Frobenius–Perron and Koopman operators, respectively. Using the adjoint property of the two operators it can be shown that

$$\langle (P_t f - f)/t, g\rangle = \langle f, (K_t g - g)/t\rangle. \tag{3.27}$$

Taking the limit as $t \to 0$ we obtain

$$\langle A_{FP} f, g\rangle = \langle f, A_K g\rangle. \tag{3.28}$$

Provided that g and f are continuously differentiable and g has compact support, it follows that [198]

$$\langle A_{FP} f, g\rangle = \left\langle -\sum_{i=1}^{d} \frac{\partial f F_i}{\partial x_i}, g\right\rangle. \tag{3.29}$$

Hence, we conclude that

$$A_{FP} f = -\sum_{i=1}^{d} \frac{\partial f F_i}{\partial x_i}. \tag{3.30}$$

Again, using Eqs. (3.18) and (3.30) we conclude that the function

$$I(x,t) = P_t f(x) \tag{3.31}$$

satisfies the PDE (continuity equation)

$$\frac{\partial I}{\partial t} + \sum_{i=1}^{d} \frac{\partial I F_i}{\partial x_i} = 0 \tag{3.32}$$

or, symbolically,

$$\frac{\partial I}{\partial t} + F \cdot \nabla I + I \nabla \cdot F = 0. \tag{3.33}$$

Note that this equation is actually the same as the well-known continuity equation in fluid mechanics and many fields, but now it is a statement of conservation of density function of ensembles of trajectories.[22] In the case when F is a divergence-free vector field, i.e., $\nabla \cdot F = 0$, Eq. (3.32) corresponds to incompressible fluids such as water and it can be simplified to

$$\frac{dI}{dt} = \frac{\partial I}{\partial t} + F \cdot \nabla I = 0. \tag{3.34}$$

A comparison of Eqs. (3.32) and (3.34) to the classical optical flow problem was discussed in [275].

[22] Compare the continuous continuity equation from the infinitesimal generator of the Frobenius–Perron operator, Eq. (3.32), to the discrete time continuity equation noted in Eq. (1.12). Said simply, both essentially state that existence and uniqueness imply that orbits from all initial conditions are conserved, so histograms and hence densities of many initial conditions must change in time only to the movement or advection of the orbits.

Example 3.2. Now consider a Duffing oscillator in the domain $[0,1]\times[0,1]$ given by the following differential equation:

$$\begin{aligned}\frac{dx}{dt} &= 4y-2,\\ \frac{dy}{dt} &= 4x-2-8(2x-1)^3.\end{aligned} \tag{3.35}$$

According to Eq. (3.32), given the initial density $u(x,0)=f(x)$ the flow of the density$(u(x,t)$ $=P_t f(x))$ under the Duffing oscillator is given by

$$\frac{\partial u}{\partial t}+(4y-2)\frac{\partial u}{\partial x}+(4x-2-8(2x-1)^3)\frac{\partial u}{\partial y}=0. \tag{3.36}$$

A numerical simulation of this continuity equation of the density propagation by Eq. (3.36) along with the vector field of this ODE is shown in Fig. 3.1 using an initial density $u(x,0)$ illustrated in Fig. 3.1(a). Notice that the density is stretched and folded due to the hyperbolic structure of the system.

3.3 Frobenius–Perron Operator of Discrete Stochastic Systems

In more realistic situations, we consider a stochastic differential equation, in which case the evolution of densities is described by the well-known Fokker–Planck equation, which is a special case of the continuity equation as derived in Eq. (3.32). However, a study of a continuous time system with a continuous stochastic perturbation requires a great deal of preliminary concepts in stochastic differential equations and more advanced concepts in semigroup theory that are beyond the scope of this book. Nevertheless, the discrete time analogue of this problem is less involved and can be readily developed.

In what follows, we will consider a stochastic perturbation as a random variable.

Definition 3.3. A **random variable**,

$$X:\Omega\to\mathbb{R}, \tag{3.37}$$

is a measurable function[23] from a measure space $(\Omega,\mathcal{F},\mu)$ to a measurable space $(\mathbb{R},\mathcal{B}(\mathbb{R}))$, where $\mathcal{B}(\mathbb{R})$ denotes the Borel σ-algebra on $\mathbb{R}$. The realization of such a selection, $X(\omega)$, is often called a "random experiment."

We may interpret the random variable as a measurement device that returns a real number, or the random experiment in our language, for a given subset of Ω. Recall that for a measure space $(\Omega,\mathcal{F},\mu)$, a measure μ is called a probability measure if $\mu:\mathcal{F}\to[0,1]$ and $\mu(\Omega)=1$; hence a measure space $(\Omega,\mathcal{F},\mu)$ will also be referred to as a probability space. With a probabilistic viewpoint in mind, the random variable tells us that the probability to

[23]Given a measurable space $(\Omega,\mathcal{F})$ and a measurable space $(S,\mathcal{S})$, an "($\mathcal{F}$-)measurable function," or simply "measurable function," is a function $f:\Omega\to S$ such that $f^{-1}(A)\equiv\{\omega\in\Omega: f(\omega)\in A\}\in\mathcal{F}$ for every $A\in\mathcal{S}$.

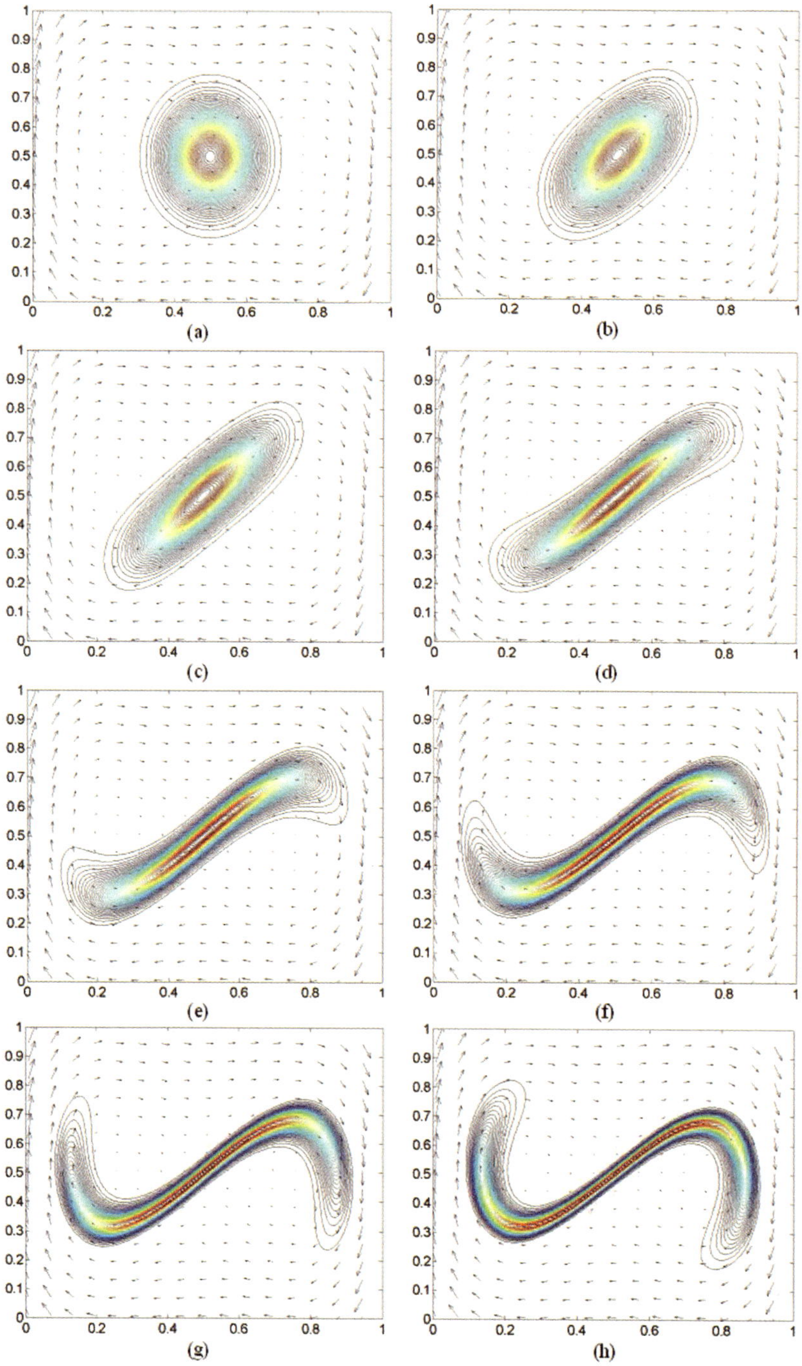

Figure 3.1. *A simulation of the flow of a density function according to the continuity equation* (3.36) *with the velocity field given by Eq.* (3.35)*. The time increment between each snapshot is* $t = 0.08$*. The range of high to low density is plotted from red to blue.*

observe a measurement outcome in some set $A \in \mathcal{B}(\mathbb{R})$ based on a probability measure μ is precisely $\mu(X^{-1}(A))$, which makes sense only if X is measurable.

We now extend the well-established formulation of the deterministic Frobenius–Perron operator in the preceding section to study the phase space transport of a discrete system with constantly applied stochastic perturbation. In particular, let $S,T : X \to X$ be (nonsingular) measurable functions acting on $X \subset \mathbb{R}^d$.

We consider a process with both **additive** and **multiplicative** stochastic perturbation defined by

$$x_{n+1} = \nu_n T(x_n) + S(x_n), \tag{3.38}$$

where ν_n are identically and independently distributed (i.i.d.) random variables each having the same density g. Note that if we set $S \equiv 0$, we would have a process with a multiplicative perturbation, whereas when $T \equiv 1$ we have a process with an additive stochastic perturbation. Suppose the density of x_n given by ρ_n. Such a system can be considered descriptive of both **parametric noise** and **additive noise** terms. We desire to show that the relationship of ρ_n and ρ_{n+1} in the above stochastic process analogous to Eq. (3.47) assumes that $S(x_n), T(x_n)$, and ν_n are independent [276].

Let $h : X \to X$ be an arbitrary, bounded, measurable function, and recall that the expectation of $h(x_{n+1})$ is given by

$$E[h(x_{n+1})] = \int_X h(x)\rho_{n+1}(x)dx. \tag{3.39}$$

Then, using Eq. (3.38), we also obtain

$$E[h(x_{n+1})] = \int_X \int_X h(zT(y)+S(y))\rho_n(y)g(z)dydz. \tag{3.40}$$

By a change of variable, it follows that

$$E[h(x_{n+1})] = \int_X \int_X h(x)\rho_n(y)g\big((x-S(y))T^{-1}(y)\big)|J|dxdy, \tag{3.41}$$

where $|J|$ is the Wronskian derivative which is equivalently stated as the determinant of the Jacobian derivative matrix of the transformation

$$x = zT(y) + S(y). \tag{3.42}$$

Since h was an arbitrary, bounded, measurable function, we can equate Eqs. (3.39) and (3.41) to conclude that

$$\rho_{n+1}(x) = \int_X \rho_n(y)g\big((x-S(y))T^{-1}(y)\big)|J|dy. \tag{3.43}$$

Based on the above expression, the (stochastic) Frobenius–Perron operator for this general form of a stochastic system with both parametric and additive terms may be defined by

$$P_\nu \rho(x) = \int_X \rho(y)g\big((x-S(y))T^{-1}(y)\big)|J|dy. \tag{3.44}$$

It is interesting to consider special cases. Specifically, for the case of the multiplicative perturbation, where $S(x) \equiv 0$, we have

$$P_\nu \rho(x) = \int_X \rho(y)g\big(xT^{-1}(y)\big)T^{-1}(y)dy. \tag{3.45}$$

Similarly, the stochastic Frobenius–Perron operator for the additive perturbation, where $T(x) \equiv 1$, is

$$P_\nu \rho(x) = \int_X \rho(y) g\big(x - S(y)\big) dy. \tag{3.46}$$

Finally, the process becomes deterministic when setting the density g to a delta function $\delta(x - S(y))$. The Frobenius–Perron operator associated with the map S can then be defined by [198]

$$P\rho(x) = \int_X \delta(x - S(y))\rho(y) dy. \tag{3.47}$$

Thus $P\rho(x)$ gives us a new probability density function.

3.4 Invariant Density Is a "Fixed Point" of the Frobenius–Perron Operator

In Eq. (1.19), we already observed that an invariant density is a solution of the Frobenius–Perron fixed point equation, which we repeat:

$$\rho^*(x) = P_f \rho^*(x). \tag{3.48}$$

This equation may be taken as defining the term **invariant density**. However, the invariant densities may exhibit singularities, which may be dense in a given subspace, and hence it is not absolutely continuous w.r.t. the Lebesgue measure. Therefore, in the situations when it is impossible to sensibly define an invariant density function, we will alternatively deal with the corresponding invariant measure, which can still be defined in general. Also, in such situations, the term invariant densities will be replaced by invariant measures instead. This requires a definition of invariant measure.

Definition 3.4. **Invariant measure** assumes a transformation $T : X \to X$ on a measure space (X, Σ, μ) and requires $\mu(T^{-1}(B)) = \mu(B)$ for each $B \in \Sigma$. That is, the "weight" ensembles of initial conditions of each B are the same before and after application of the transformation T. Here T^{-1} denotes "preimage" rather than inverse in the case the preimage may be multiply branched. In the case of a flow, a measure μ is a ϕ_t **invariant measure** if for every measurable set B, $\mu(\phi_t^{-1}(B)) = \mu(B)$.

Observe that Eq. (3.48) is a functional equation; solutions are functions, $\rho^*(x)$. Notice the plural use of functions. Generally, a unique solution is not expected. In particular, the following hold:

- If the dynamical system f has a fixed point x, $x = f(x)$, then there will be an invariant measure which is atomic (delta function) supported over this fixed point. $\rho^*(x) = \delta(x)$. See, for example, Fig. 3.2 (middle).
- If the dynamical system f has a periodic orbit, $\{x_1, x_2, \ldots, x_p\}$, then there will be an invariant measure which is atomic supported of this periodic orbit that is a sum of delta functions, $\rho^*(x) = \sum_{i=1}^p \delta(x_i)$.
- A chaotic dynamical system is characterized by having infinitely many periodic orbits [95], and from the above there follows infinitely many atomic invariant measures.

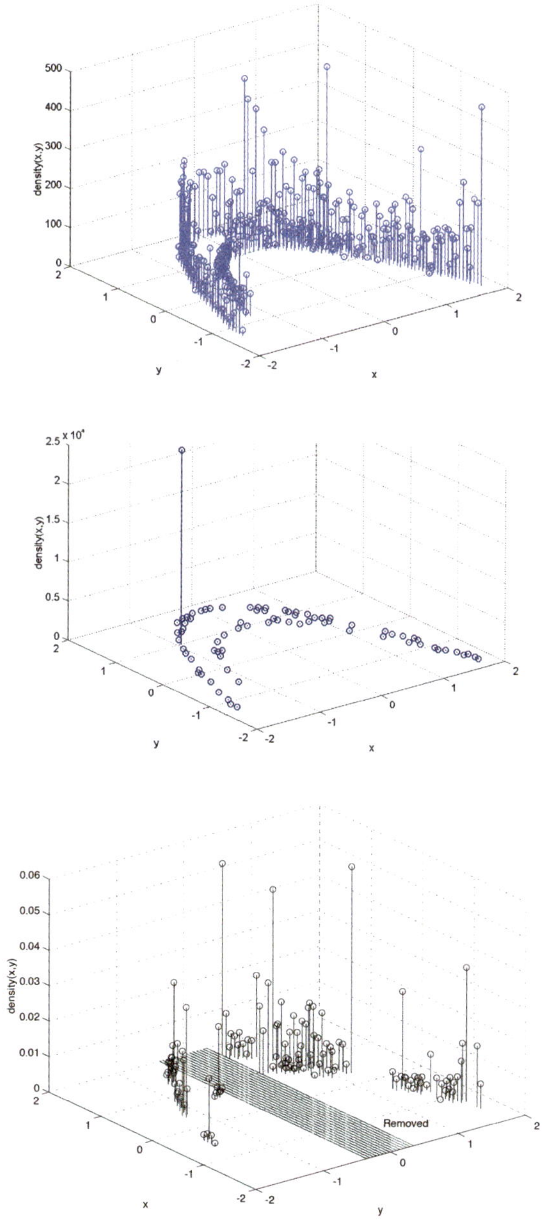

Figure 3.2. *Invariant densities of the Henon map. (Left) Density corresponding to the apparent natural measure. (Middle) Atomic invariant density supported over one of the fixed points. (Right) An invariant density supported over an unstable chaotic saddle which is here a Cantor set which avoids the strip labeled "Removed." Compare to Figs.* 7.29–7.32 *and the discussion of unstable chaotic saddles in Section* 7.4.3. [25]

- Linearity of the Frobenius–Perron operator allows convex combinations of any invariant densities to be invariant densities. That is, if

$$\rho_i^*(x) = P_f \rho_i^*(x) \quad \text{for } i = 1, \dots, q, \tag{3.49}$$

then

$$\rho^*(x) = \sum_{i=1}^{q} \alpha_i \rho_i^*(x) \tag{3.50}$$

is invariant, choosing $\sum_{i=1}^{q} \alpha_i = 1$ to enforce the convex combination statement. From the above, there are infinitely many convex combinations of the infinitely many periodic orbits.

- There can be other exotic invariant sets, such as unstable chaotic saddles which are generally Cantor sets. See the discussion of these sets in Section 7.4.3 and Definition 7.6. Each of these supports a typically atomic invariant measure (atomic if this saddle has Lebesgue measure zero). A picture of one for the Henon map can be found in Fig. 7.28. Illustrations of such atomic invariant measures supported over unstable chaotic saddles can be found in [25] and Fig. 3.2 (right).

- There can be invariant sets which are not Lebesgue measure zero, such as in Figs. 3.9 and 5.1, and each of these can support an invariant density, and likewise convex combinations of these will be invariant densities.

We are often interested in the natural invariant measure; see Eq. (3.78) for the definition of the natural measure. Numerical estimates popularly resort to Ulam's method [307, 103], as discussed in Chapter 4 in Sections 4.3.1–4.4. Roughly stated for now, the invariant measure may be estimated by computing the dominant eigenvector of the stochastic matrix estimate of the Frobenius–Perron operator.

3.5 Invariant Sets and Ergodic Measure

The topological dynamical feature of an invariant set and the measure theoretic concept of ergodicity are related in spirit, since invariance does have measurable consequences. An invariant set is a set that evolves (in)to itself under the dynamics. The general situation may not be one of invariance, but rather one such as shown in Fig. 3.3, where a set C is shown mapping across C. We sharpen the notion of invariance with the following definitions and examples.

Definition 3.5 (invariant set of a dynamical system). A set C is **invariant** with respect to a **dynamical system** ϕ_t if $\phi_t(C) = C$ for all $t \in \mathbb{R}$ in the case of a flow, or $t \in \mathbb{Z}$ in the case of a mapping. A set C is **positively invariant** if $\phi_t(C) \subset C$ for all $t > 0$.

However, invariance does not require that points be stationary. In the case of a semidynamical system, a slightly different definition is needed, since there may be multiple prehistories to each trajectory.

Definition 3.6 (invariant set of a semidynamical system). A set C is **invariant** with respect to a **semidynamical system** (see Definitions 2.1–2.3) ϕ_t if $\phi_t^{-1}(C) = C$ for all

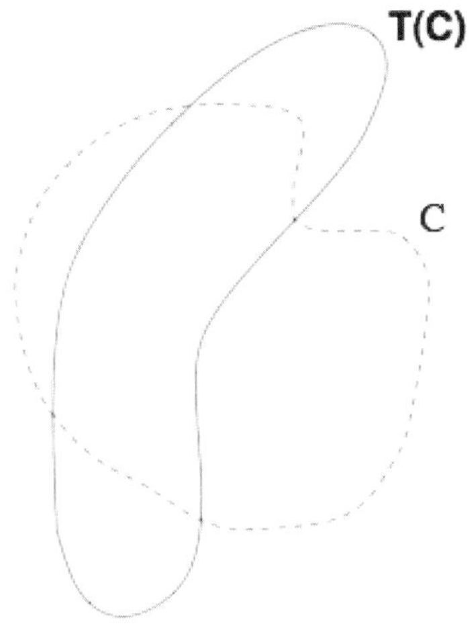

Figure 3.3. *A set C is shown mapping across C, since $C \cap T(C) \neq \emptyset$, but $C \neq F(C)$. C as shown is therefore not invariant with respect to T, and points in T more so are not stationary.*

$t \in \mathbb{R}$. Note that while a semiflow is not invertible, ϕ_t^{-1} denotes the preimage of a set, which may be multiply branched.[24]

Example 3.3 (an invariant set in a flow). The linear equation

$$\begin{aligned} \dot{x}_1 &= 2x_2, \\ \dot{x}_2 &= x_1 - x_2 \end{aligned} \tag{3.51}$$

has an invariant set which is the line $C = \{x_1, x_2 \in \mathbb{R} : x_1 + x_2 = 0\}$, and furthermore the origin $(x_1, x_2) = (0,0)$ is an invariant subset, but it is the only stationary point in the set C.

Example 3.4 (an invariant set in a map). The logistic map may be the most popular examples for pedagogical presentation in beginning texts, as a one-dimensional map which both presents chaotic oscillations and admits to many of the standard methods of analysis. Already introduced here, Eq. (1.2), the logistic map may be presented as a map of the real line,

$$\begin{aligned} f &: \mathbb{R} \to \mathbb{R}, \\ x &\mapsto f(x) = ax(1-x), \end{aligned} \tag{3.52}$$

but the map is more often presented in an iterating form, and as an initial value map,

$$x_{n+1} = f(x_n) = ax_n(1-x_n), x_0 \in \mathbb{R}. \tag{3.53}$$

When a is chosen to be $a = 4$, the logistic map can be proven [12, 95] to display fully developed chaos, in the sense of a fullshift symbolic dynamics. Details of this statement

[24]The logistic map $x_{n+1} = ax_n(1-x_n)$ is a quadratic function, and hence has up to two preimages at each point, $x_n = \frac{a \pm \sqrt{a^2-4}}{2a}$, following the usual quadratic formula.

will be given in Chapter 6. For the purposes of this example, for any value of a, in the most studied parameter range $a \in [0,4]$, the logistic map has an invariant set of $[0,1]$. That is, in the standard function statement,

$$f : \text{Domain} \to \text{Range}, \tag{3.54}$$

the range of the mapping is contained within the domain of the mapping which allows for repeated iteration, and in this case they are equal:

$$f : [0,1] \to [0,1]. \tag{3.55}$$

As for the rest of the domain, all initial conditions $x_0 \in [0,1]$ are in the basin of $-\infty$.

The same story is reflected in the tent map,

$$x_{n+1} = \left\{ \begin{array}{ccc} bx_n & \text{for} & x_n < 1/2 \\ b(1-x_n) & \text{for} & x_n \geq 1/2 \end{array} \right\}, \tag{3.56}$$

which, when $b = 2$ has fully developed chaos, also as a fullshift. This tent map is in fact conjugate (Definition 1.1) to the logistic map, $a = 4$; this map and its invariant sets are shown in Fig. 3.4 as a cobweb diagram suggesting the invariant set, $[0,1]$, and that $Basin(-\infty) = [\bar{0,1}]$.

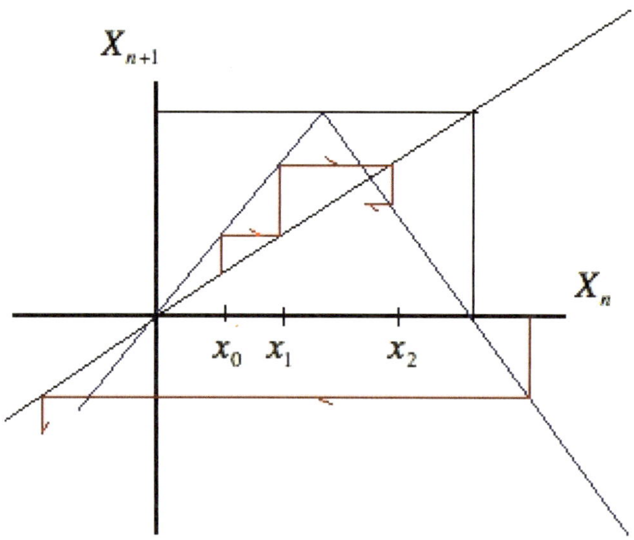

Figure 3.4. *A cobweb diagram of the tent map, Eq.* (3.56), *when* $b = 2$.

Example 3.5 (a stable invariant set in a flow, slow manifold). The following illustrative example found in [17] of a fast-slow system, also called a singularly perturbed system, illustrates a common scenario: strong dissipation often leads to a nontrivial invariant set and correspondingly two time scales,

$$\begin{aligned} \epsilon\dot{x} &= -x + \sin(y), \\ \dot{y} &= 1. \end{aligned} \tag{3.57}$$

The slow manifold here is a curve given in the singular limit. That is, choosing $\epsilon = 0$,

$$x = h_0(y) = \sin(y), \tag{3.58}$$

and the dynamics restricted to the slow manifold are solved to be

$$x(t) = \sin(y(t)) \quad \text{and} \quad y(t) = y_0 + t. \tag{3.59}$$

The so-called *associated system,* for change of time variable to *long time scales,*

$$s = \frac{t}{\epsilon}, \tag{3.60}$$

changes the system (3.57),

$$x' = -x + \sin(y), y = \text{ constant, where }' \equiv \frac{d}{ds}, \tag{3.61}$$

has solution

$$x(s) = (x_0 - \sin(y))e^{-s} + \sin(y), \tag{3.62}$$

from which we see all solutions must converge to the slow manifold $x = \sin(y)$, as $s \to \infty$. The direct solution of the system (3.57) can be found by the method of variation of parameters [110], written

$$x(t) = [x_0 - \bar{x}(0,\epsilon)]e^{-t/\epsilon} + \bar{x}(t,\epsilon), \quad y(t) = y_0 + t, \tag{3.63}$$

where

$$\bar{x}(t,\epsilon) = \frac{\sin(y_0 + t) - \epsilon\cos(y_0 + t)}{1 + \epsilon^2}. \tag{3.64}$$

For small t,

$$x(t) = [x_0 - \sin(y_0)]e^{-t/\epsilon} + \sin(y_0) + O(\epsilon) + O(t) \tag{3.65}$$

by approximation of

$$\bar{x}(t,\epsilon) = \sin(y_0) + O(\epsilon) + O(t), \tag{3.66}$$

which well approximates the associated solution (3.62). However, for larger time, there may be drift, but $e^{-t/\epsilon}$ decreases quickly. For times scales,

$$t = k\epsilon|\log\epsilon|, \tag{3.67}$$

the exponential term goes to zero faster than any power of ϵ, and solution (3.63) becomes close to

$$x(t) = \frac{\sin(y(t)) - \epsilon\cos(y(t))}{1 + \epsilon^2}, \quad y(t) = y_0 + t. \tag{3.68}$$

This solution suggests that solutions stay within $O(\epsilon)$ from the slow manifold (3.58), and this slow manifold approximation is good for time scales

$$t \gg \epsilon|\log\epsilon|. \tag{3.69}$$

This behavior is illustrated in Fig. 3.5, where we see that solutions remain near the slow manifold, since there is an invariant manifold $x = h_\epsilon(y)$ which further lies within $O(\epsilon)$ of

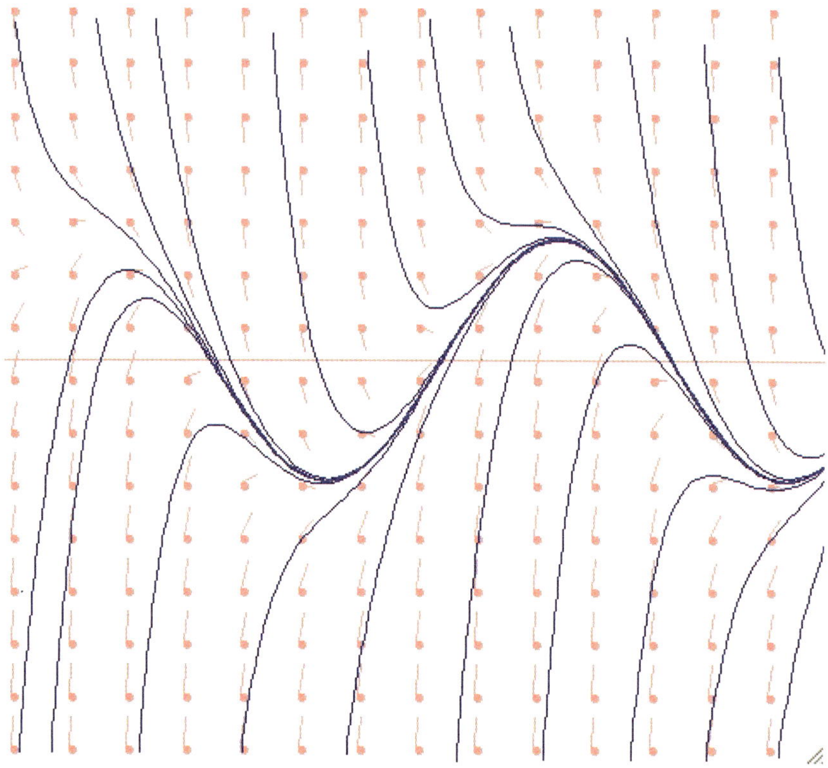

Figure 3.5. *A slopefield from a fast-slow system* (3.57) *together with several solutions for various initial conditions. Notice that the long time behavior is close to the slow manifold* $x = h_0(y) = \sin(y)$, *as can be seen explicitly by comparing Eqs.* (3.64) *and* (3.65).

the $\epsilon = 0$ slow manifold, $x = h_0(y) = \sin(y)$, (3.58). Furthermore, this invariant manifold is stable. The concepts seen here are illustrative of the sufficient conditions provided by the Tikhonov theorem [306], and reviewed in [17, 309, 92], the idea being a decomposed system with at least two time scales,

$$\begin{aligned} \epsilon \dot{x} &= f_1(x, y), \\ \dot{y} &= f_2(x, y), \end{aligned} \tag{3.70}$$

in which one can search for an invariant solution $x = h_\epsilon(y)$ near the slow manifold $x = h_0(y)$.

3.5.1 Ergodic Measure

In our "Ergodic Preamble" and "The Ensemble Perspective," Sections 1.1 and 1.2, we have described that considering many initial conditions leads naturally to considering the evolution of histograms of orbits. Ergodic theory may be thought of as a sister topic to topological dynamical systems, concerning itself not just with how initial conditions move, but also with how measured ensembles of initial conditions evolve. So again referring to Fig. 3.3, we understand a measure space which describes the relative "weight" of the ensemble in C, and how that measure moves under the dynamical system. As we have already alluded to in Sections 1.1 and 1.2, ergodic relates the evolution of measure under a dynamical system, by the Koopman operator (3.8), and this is dual to evolution of density by the Frobenius–Perron operator by Eq. (3.5). Therefore, in this section we will describe some of the simplest key terms carefully in the mathematical language of the rich and related field of ergodic theory.

Definition 3.7 (ergodic). If there exists a measure μ of a nonsingular[25] measure space (X, Σ, μ) and a nonsingular semidynamical system $\phi_t : X \to X$ satisfying the following properties, then the measure is called an **ergodic invariant measure** corresponding to the dynamical system, which is simply called **ergodic** since from this point on we will be discussing ergodicity mostly in a situation of using invariant measures to find average.[26] These properties are

- μ must be a ϕ_t invariant measure;[27]
- every measurable ϕ_t invariant set A measures either zero, $\mu(A) = 0$, or its complement measures zero, $\mu(X - A) = 0$.

An ergodic dynamical system is often associated with complicated behavior, and even with chaotic behavior. However, it is not strictly correct to associate the ideas, as these notions are separate concepts. As the following example demonstrates, the dynamics of and ergodic transformation can in fact be quite simple.

Example 3.6 (irrational rotation is ergodic). Perhaps the simplest example of an ergodic dynamical system is an irrational rotation on a circle. The **circle map** is defined as

$$\begin{aligned} f(x) : [0,1] &\to [0,1] \\ x &\mapsto f(x) = x + r \mod 1. \end{aligned} \tag{3.71}$$

It is easy to show the following behavior:

1. If r is rational, $r = p/q$, $p, q \in Z$, then every initial condition $x \in [0,1]$ corresponds to an at least q-periodic orbit.
2. If r is irrational, then there are no periodic orbits, and the Lebesgue measure is ergodic.

Notice that in the second case, while the transformation is ergodic, none of the chaotic properties is satisfied, and indeed it is hard to image a simpler transformation that is not trivial.

---

[25]We are referring to mutually **nonsingular measures**. A measure $\mu(\cdot)$ is called nonsingular with respect to another measure, usually Lebesgue $m(\cdot)$ measure if no other qualifier is mentioned, if there are no two disjoint sets A and B such that all subsets of A μ-measure 0, and all subsets of B m-measure zero, but $A \cap B - \emptyset$ and $A \cup B = X$.

[26]Without specifying with respect to which other measure it is ergodic, then the ergodic transformation with respect to Lebesgue measure is meant.

[27]The issue of existence of an invariant measure is a problem which requires a great deal of effort. Consider for example, the Krylov–Bogoliubov theorem—any continuous mapping of a metrizable compact space to itself has an invariant Borel measure [291].

Example 3.7 (the logistic map, $a = 4$ is ergodic). The much-studied Logistic map, Eq. (1.2), $f(x):[0,1]\to[0,1], x_{n+1} = ax_n(1-x_n)$, when $a=4$, can be shown to be ergodic since its invariant density,

$$\rho(x) = \frac{1}{\pi\sqrt{x(1-x)}}, \tag{3.72}$$

Eq. (1.22), generates the invariant measure

$$\mu(A) = \int_A \rho(x)dx \;\forall \text{ measurable } A \subset X. \tag{3.73}$$

This example is perhaps more stereotypical of the concept of ergodic since here the property of ergodic coincides with a complicated behavior which is even chaotic in this case. When $a < 4$ there are many specific transformations corresponding to what is believed to be both chaotic and also ergodic behavior, at least for those accumulation points in the Feigenbaum diagram.

Example 3.8 (an example where the invariant set is Lebesgue measure zero). Consider again the tent map $f(x):[0,1]\to[0,1], x_{n+1} = ax_n$ if $x_n < 1/2$ and $a - ax_n$ if $x \geq 1/2$. It is straightforward to show that if $a > 2$, then there is an interval of points of initial conditions $x_0 \in [1/a, 1-1/a]$ such that the orbit of those points leaves $[0,1]$ in one iteration. Considering the cobweb diagram in Fig. 3.4 and comparing a similar scenario shown in Fig. 5.2 exemplifies this statement. Likewise (up to two) preimages of this set $f^{-1}[1/a, 1-1/a]$ include a set of points which leave $[0,1]$ in two iterates. And $f^{-2}[1/a, 1-1/a]$ is up to four subintervals which leave $[0,1]$ in three iterates. Following this discussion indefinitely constructs a Cantor set which is the invariant set Λ of $[0,1]$. The measure of this invariant set is $m(\Lambda) = \lim_{n\to}(1-2/a)^n = 0$. This set Λ supports an ergodic measure μ which is not absolutely continuous to Lebesgue measure, and topologically the description of the dynamics on Λ is chaotic in the sense that it is conjugate to a shift map. Such sets are often called chaotic saddles [242, 28]. Perhaps less exotic, an atomic invariant measure exists for each periodic orbit, and these correspond to delta functions supported over the periodic points. There are infinitely many of these.

A central idea behind the importance of ergodic transformations is that *time averages exchange with space averages in the long time limit*. A historical origin of the ergodic hypothesis come from statistical physics and the study of the thermodynamics and states of gases.[28] See the Birkhoff ergodic theorem (1.5) and Examples 1.3 and 1.4.

The question of unique ergodicity is nontrivial. The following simple example has two invariant components, and it is therefore definitely not ergodic.

Example 3.9 (map with two invariant components). The map of the interval $f:[0,2]\to[0,2]$ shown in Fig. 3.6 has an invariant measure which is uniform, corresponding to the

[28]In statistical physics, it is said that in the long time study, passing to mean field, particles are said to spend time in microstates of the phase space with energy which is proportional to the volume, thus corresponding to equal probability. Boltzmann's ergodic hypothesis developed in the 1870s in his studies of statistical mechanics, initially intuitively by simply disregarding classical dynamics in studies of gas theory of many particles [42]. A statement of Boltzmann's ergodic hypothesis may be taken as, *Large systems of dynamically interacting particles display a state in the long run where time averages approximate ensemble equilibrium averages*, and he called such systems "ergode." Gibbs [195] later called such states the canonical ensemble in his development of what is nearly a modern classical thermodynamics theory. The Birkhoff ergodic theorem encompasses an essentially modern formulation of this idea [21].

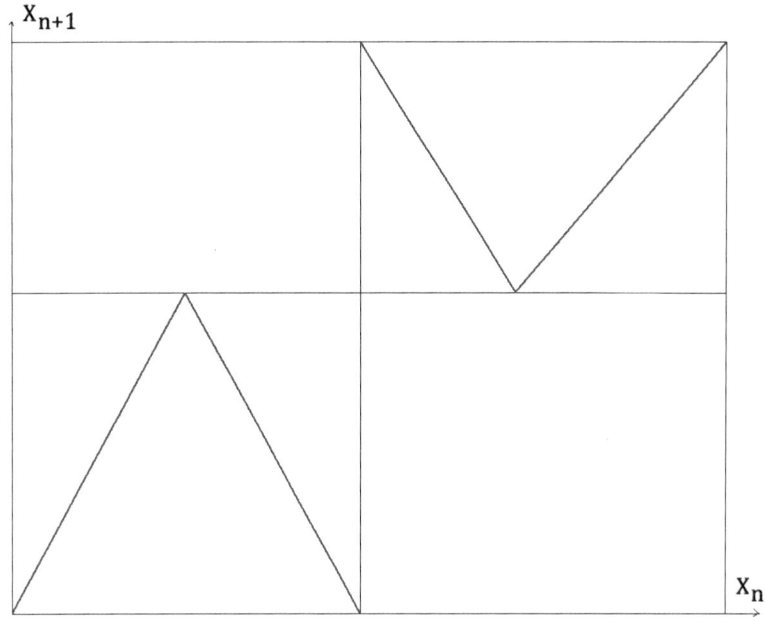

Figure 3.6. *A tent-like map with two invariant components in Example* 3.9. *Contrast to the example in Fig.* 5.2 *in which this system develops into one with a single invariant measure but with long transients, also known as weakly transitive in Section* 5.2. *Compare to Figs.* 5.1 *and* 5.2 *in Example* 5.1.

density $\rho(x) = 1/2$ on $[0,2]$. However, it also has two other nontrivial invariant measures generated by each of the following densities, respectively:

$$\rho_1(x) = \begin{pmatrix} 1 & \text{if} & 0 \leq x \leq 1 \\ 0 & \text{else} & \end{pmatrix}, \quad \rho_2(x) = \begin{pmatrix} 1 & \text{if} & 1 \leq x \leq 2 \\ 0 & \text{else} & \end{pmatrix}. \tag{3.74}$$

Since the invariant set $X = [0,2]$ measures with each of these invariant densities,

$$\mu_{\rho_1}(X) = \int_0^2 \rho_1(x)dx = \mu_{\rho_2}(X) = \int_0^2 \rho_2(x)dx = \frac{1}{2}, \tag{3.75}$$

this contradicts that these may be ergodic, since ergodic measures by definition must measure invariant sets to be either zero or one.

3.5.2 Almost-Invariant Sets

While invariance is purely a topological concept, with ergodic consequence when measure structure is assumed, the concept of *almost-invariance* becomes essentially a measurable concept, since it asks the question, relatively *how much* of a set may remain in place under the action of the dynamical system. Any question with the phrase "how much" in it requires a measure structure to discuss. To quantify almost-invariant sets we will formally define a notion of the almost-invariant set based on the Markov model (4.6).

Definition 3.8. For a dynamical system, with a flow $S_t : X \to X$, a Borel measurable set $A \subset X$ is called *μ-almost invariant* [87] if

$$\frac{\mu(A \cap S_t^{-1}(A))}{\mu(A)} \approx 1, \tag{3.76}$$

where μ is a probability measure and S_t^{-1} denotes the preimage when an inverse does not exist.

While we see that the judgment of almost-invariance depends on the flow period t and the choice of measure, that is, how we weight sets, there are two measures in particular that may be most sensible to choose in most discussions of almost-invariance. One choice of course is to let μ to be Lebesgue measure,

$$\mu(A) \equiv m(A) = \int_A dx, \tag{3.77}$$

when appropriate. In such case, simple Monte Carlo simulations often lead to useful results which are easy to interpret. Another sensible choice that is true to the behavior of the dynamical system is the so-called *natural measure*, which may be described as the invariant measure on the attractor. These words need some explanation.

Roughly speaking, the **natural measure** $\mu(B)$ of a measurable set B is high if the amount of time the orbit of Lebesgue almost-all points $x \in X$ spends in B is large. As such, $\mu(B)$=0 if no orbits enter B or revisit B after a certain number of iterations. When the limit

$$\mu(B) = \lim_{N\to\infty} \frac{\#\{F^n(x) \in B : 0 \leq n \leq N\}}{N} \tag{3.78}$$

exists[29] for a mapping F, we call it the natural measure of the set B. More precisely, it can be called a natural invariant measure as it is evident from its definition that if a natural measure exists, it must also be invariant, but we will refer to it just as a natural measure for short. This is also called a rain gauge measure [2] and, despite the notoriously difficult theoretical nuances involved to prove existence, the idea is quite simple to use in practice, but the result can be quite misleading in the scenario of very long transients. In practice, an initial point is selected uniformly at random. Then a trajectory through the space is computed by iterating the system some large number of times, and then estimating the limiting occupancy fraction behavior. In fact, this is also a reason that a natural measure is meant to be carried by an attractor where the fraction of iterates falling into a set in the attractor is the same for almost initial points (w.r.t. the Lebesgue measure) in the basin of attraction.

When the above measure is used to define the almost-invariant set, as expressed in Eq. (3.76), the interpretation is that if A is an almost-invariant set, then the probability that a point in A remains in A after one iteration is close to 1. Generally, the definition of the natural measure is related to the Birkhoff ergodic theorem highlighted in Eq. (1.5).

[29]We are avoiding here the detailed discussion as to when such measures do in fact exist, and for which kinds of systems there are attractors on which such measures exist. See [323, 171] for discussions regarding existence of SRB (Sinai–Ruelle–Bowen) measures and the beginning of a discussion of construction of "rain gauge" measures in [2], such as Eq. (3.78).

Remark 3.3. What if a system has several nontrivial invariant sets, and how does this relate to ergodicity? Recall that a measure is called invariant under S_t if $\mu(S_t^{-1}(A)) = \mu(A)$ for all measurable sets A. But an invariant measure is defined to be ergodic if $\mu(A) = 0$ or $\mu(A) = 1$ for all measurable sets A such that $S_t(A) = A$. This notion of ergodicity emphasizes indecomposibility of a space to be studied. In short, if a space comprises more than one invariant set of positive measure, then one could study them separately. Several descriptions of ergodicity and its relation to the notion of topological transitivity can be found in [303].

3.5.3 Almost-Invariant Sets as Approximated by Stochastic Matrices

The fact is that for many dynamical systems, the asymptotic notion of the natural measure in Eq. (3.78) can be difficult to compute for several reasons. First, it is possible that a dynamical system has a very long transient state, and consequently we need to compute a long sequence of iterations to observe eventual behavior of such a system. Therefore, we may not be able to observe the equilibrium distribution unless t in (3.78) is sufficiently large. Furthermore, in an extreme case, the round-off problem can prevent us from discovering the equilibrium distribution; for example, a sequence of iterations of the tent map (3.56) when $b = 2$ will send almost any initial point to negative infinity, as is easy to confirm by simulations. A way to circumvent these problems is to find the invariant measure based on the left eigenvector with the eigenvalue one from the matrix P defined in Eq. (4.6) and discussed further in the next chapter.

More precisely, the invariant measure of the transition matrix (4.7) is a good approximation of the natural measure of the Frobenius–Perron operator P_{S_t}.[30] Note that a deterministic chaotic dynamical system will have infinitely many invariant measures, one for each invariant set such as periodic orbits and convex combinations of these measures. However, the question is if there is only one natural measure. In general, there is no guarantee that the invariant density we discovered is the natural measure. In this regard, it was shown [121, 222] that the supports of the approximate invariant measures contain the support of the natural measure. So the consequence of such result is that at least we capture all the regions with the positive natural measure.

Now consider invariant density as found from the transition matrix Markov model approximating the action of the Frobenius–Perron operator. Let p be the left eigenvector of P with eigenvalue one. After normalizing p so that

$$\sum_{i=1}^{n} p_i = 1, \tag{3.79}$$

one may approximate the natural measure μ_N for $\{B_i\}_{i=1}^N$ by

$$\mu_N(B_i) = p_i. \tag{3.80}$$

[30] Approximation of the action of the Frobenius–Perron operator by a sequence of refinements and corresponding stochastic matrices is the subject of Chapter 4. In particular, the Ulam method [307] is the conjecture that this process can be used to compute the invariant density.

Then the approximate invariant measure of a measurable set B can be defined by

$$\mu_N(B) := \sum_{i=1}^{N} \mu_N(B_i) \frac{m(B_i \cap B)}{m(B_i)}. \tag{3.81}$$

Comparing Eq. (4.6) with Eq. (3.80) we see that it is just a restatement of the matrix times vector statement pA. Note that the quality of this approximation with respect to a refining topological partition $\{B_i\}_{i=1}^N$ depends upon the number and size of partition elements. Results concerning the convergence of such approximations can be found in [87].

By combining Eqs. (4.6) and (3.81), one can alternatively define almost invariance based on the estimated invariant density p and the transition matrix P, as in the following lemma [122].

Lemma 3.1 (see [122]). Given a box partition set $\{B_i\}_{i=1}^N$ with $\mu_N(B_i) = p_i$, a Borel measurable set $P = \bigcup_{i \in I} B_i$ for some set of box indices I is called **almost invariant** if

$$\frac{\sum\limits_{i,j \in I} p_i P_{ij}}{\sum\limits_{i \in I} p_i} \approx 1. \tag{3.82}$$

Proof. It is easy to see from (3.81) that $\mu_N(B_i) = p_i$ and hence we have

$$\mu_N(B) = \sum_{i \in I} p_i. \tag{3.83}$$

Now observe that for each fixed $i \in I$, we have

$$\begin{aligned} \mu_N(B_i \cap S_t^{-1}(B_j)) &= \sum_{j \in I} \frac{m(B_i \cap S_t^{-1}(B_j))}{m(B_i)} p_i \\ &= \sum_{j \in I} P_{ij} p_i. \end{aligned} \tag{3.84}$$

The desired result follows after summing the above equation over all i. □

3.6 Relation between the Frobenius–Perron and Koopman Operators

In this section we will show the connections between the Frobenius–Perron and Koopman operators beyond their adjointness, which was already developed. First, we will demonstrate that while the Frobenius–Perron operator acts as a "push forward" operator for a density function, the Koopman operator may be intuitively thought of as a "pull back" of a function. To make this description more precise, let us consider a semidynamical system $\{S_t\}_{t>0}$ on (X, Σ, μ) and some $A \subset \Sigma$ such that $S_t(A) \in \Sigma$. If we take $f(x) = 0$ for $x \in X - A$ and $g(x) = \chi_{X - S_t(A)}$, then we have $f \in L^1$ and $g \in L^\infty$. By the adjoint property of the two operator, we have

$$\begin{aligned} \int_X \chi_{X/S_t(A)} P_t f(x) d\mu &= \int_X f(x) \chi_{X/S_t(A)}(S_t f(x)) d\mu \\ &= \int_A \chi_{X/S_t(A)}(S_t f(x)) d\mu = 0. \end{aligned} \tag{3.85}$$

This means that the integrand on the left-hand side of the above equation must be zero almost everywhere, which happen if and only if

$$P_t f(x) = 0 \qquad \text{for } x \in X/S_t(A). \tag{3.86}$$

In other words, P_t "pushes forward" a density function supported on A to a function supported on $S_t(A)$.

Conversely, if we take $f(x) = 0$ for $x \in X - A$ again but consider it as $f \in L^\infty$, we have

$$K_t f(x) = f(S_t(x)) = 0 \qquad \text{if } S_t(x) \in X - A. \tag{3.87}$$

This means that whenever f is zero outside a set A, then $K_t f(x)$ is zero outside a set $S_t^{-1}(A)$. Therefore, K_t "pulls back" the function supported on A to a function supported on $S_t^{-1}(A)$.[31]

Now we contrast the dynamical description exhibited by P_t and K_t. Recall that the function $f(t,x) = P_t f(x)$ satisfies the continuity equation (3.32), which describes the evolution of a density function. It is well known that first-order PDEs of the form (3.32) can be solved by the method of characteristics. This method gives a solution of the density along the solution of the initial value problem (3.2). Let $x(t)$ denote a solution of (3.2) with $x(0) = x_0$ and define $\rho(t) = f(t,x(t))$ by parameterizing x by t. By applying the chain rule, the function $\rho(t)$ must satisfy

$$\begin{aligned} \frac{d\rho}{dt} &= \frac{\partial f(t,x(t))}{\partial t} + \sum_{i=1}^{d} \frac{\partial f(t,x(t))}{\partial x_i} \frac{dx_i}{dt} \\ &= \frac{\partial f(t,x(t))}{\partial t} + \nabla(f)\cdot F(x). \end{aligned} \tag{3.88}$$

Also, $f(t,x)$ obeys the continuity equation:

$$\frac{\partial f}{\partial t} = -\nabla\cdot(fF) = -\nabla(F)f - \nabla(f)\cdot F. \tag{3.89}$$

Comparing the above two equations suggests that

$$\frac{d\rho}{dt} = -\nabla(F)f. \tag{3.90}$$

This means that for a given $X(0) = x_0$ and $\rho(0) = f_0(x_0)$, the continuity equation can be solved pointwise by solving the initial value problem of $(d+1)$-dimensional ODE system

$$\begin{aligned} \frac{dx}{dt} &= F(x), \\ \frac{d\rho}{dt} &= -\nabla\cdot(F)f. \end{aligned} \tag{3.91}$$

[31]It is important to recall that the P_t acts on a density, which is defined in L^1, but K_t is defined on L^∞. It is not always legitimate to think of K_t as an operator that generally pushes a density backward in time. This can be easily noticed by comparing $K_t f(x) = f(S_t(x))$ with $P_t f(x) = f(S_t^{-1}(x))|D_x S_t^{-1}(x)|$. Clearly, the Jacobian term is not included in the definition of K_t and hence it can be regarded as an operator that pulls back a density only if the Jacobian equals to unity everywhere, that is, the semidynamical system is area-preserving $(\nabla\cdot(F) = 0)$, in which case the Koopman operator becomes isometric and it can be defined on L^P for $1 \le P \le \infty$.

The negative sign in the above equation should be intuitive, for if we start with an infinitesimal parcel of particles, as the parcel moves along with the flow its volume will be expanded, say $\nabla \cdot (F) > 0$, which results in a lower density parcel.

Furthermore, the right-hand side in the expression (3.88) can be expressed another way as

$$\frac{d\rho}{dt} = \frac{\partial f(t, x(t))}{\partial t} + A_K. \tag{3.92}$$

The quantity $\frac{d\rho}{dt}$ in the above expression is called the Lagrangian derivative (it is also called variously the material, the substantial, or the total derivative). It has a physical meaning of the time rate of change of some characteristics (represented by ρ) of a particular element of fluid (i.e., x_0 in the above context).[32] At this point, we may think of the Frobenius–Perron operator as a Eulerian PDF method, since its corresponding continuity equation has to be solved on a fixed coordinate. In contrast, the Lagrangian derivative derived from the Koopman operator by using a reverse time $-t$ to obtain $-A_k$ yields a Lagrangian PDF method, which has to be solved simultaneously with the flow for a specific particle x_0.

Finally, it is worth mentioning again that we introduce the concept of the Frobenius–Perron operator to utilize it to identify almost-invariant sets in subsequent chapters. In particular, it will be shown that the information embedded in the spectrum of the Frobenius–Perron operator is useful for our purpose. Nevertheless, it is not surprising that the Koopman operator, as an adjoint operator of the Frobenius–Perron operator, can also be used for the same purpose and it yields similar results. There has been extensive work on using the Koopman operator to identify invariant sets, which can be primarily referred to in Mezić [221, 220].

[32]If the flow is area-preserving, it is easy to see that $\frac{d\rho}{dt} = 0$. This can be interpreted to mean that the physical quantity described by ρ is conserved along the flow.

Chapter 4

Graph Theoretic Methods and Markov Models of Dynamical Transport

The topic of this chapter stems from the idea that Frobenius–Perron operators can be understood as if they were infinitely large stochastic matrices acting on an infinite-dimensional linear space. Furthermore, there are finite rank (corresponding to finite sized matrices) representations that can give excellent results. Such is the story of compact operators,[33] and this leads not only to better understanding of the operator, but most importantly also to computable methods for carrying forward a practical basis for numerics on digital computers.

We use what may be called the **Ulam–Galerkin method**—a specialized case of the Galerkin method [201]—to approximate the (perhaps stochastic) Frobenius–Perron operator (3.44). In this chapter, we flesh out the discussion mentioned earlier that the *approximate* action of dynamical system on density looks like a directed graph, and that Ulam's method is a form of Galerkin's method. To hopefully offer some insight, we again refer the reader to the caricature partitioning of the action of the Henon mapping in Fig. 1.1.

The approximation by **Galerkin's method** is based on the projection of the infinite-dimensional linear space $L^1(M)$ with basis functions,

$$\{\phi_i(x)_{i=1}^{\infty}\} \subset L^1(M), \tag{4.1}$$

onto a finite-dimensional linear subspace with a subset of the basis functions,

$$\Delta_N = \text{span}\{\phi_i(x)\}_{i=1}^{N}. \tag{4.2}$$

For the Galerkin method, the projection

$$\Pi_N : L^1(M) \to \Delta_N \tag{4.3}$$

maps an operator from the infinite-dimensional space to an operator of finite rank, an $N \times N$

[33] A compact operator is defined in the field of functional analysis in terms of having an atomic spectrum [8]. Compact operators are most easily described in a Hilbert space [191] (a complete inner product space) as these are the operators that are the closure of operators of finite rank. In other words, their action is "well approximated" by matrices, and approximated in the appropriate sense.

matrix, by using the inner product,[34]

$$P_{i,j} = \langle P_{F_\nu}(\phi_i), \phi_j \rangle = \int_M P_{F_\nu}(\phi_i(x))\phi_j(x)dx. \tag{4.4}$$

The advantage of such a projection is that the action of the Markov operator which is initially a transfer operator in infinite dimensions reduces approximately to a Markov matrix on a finite-dimensional vector space. Such is the usual goal of Galerkin's method in PDEs, and similarly it is used here in the setting of transfer operators.

Historically, Ulam's conjecture was proposed by S. Ulam [307] in a broad collection of interesting open problems from applied mathematics, including the problem of approximating Frobenius–Perron operators. His conjecture has been essentially proved as cited below, including both of the following elements:

Ulam's Conjecture (see [307]).

1. A finite rank approximation of the Frobenius–Perron operator by Eq. (4.6); and
2. the conjecture that the dominant eigenvector (corresponding to eigenvalue equal to 1 as is necessary for stochastic matrices) weakly approximates[35] the invariant distribution of the Frobenius–Perron operator.

Ulam did not write his conjecture in the formal language of a Galerkin projection, (4.1)–(4.4), but due to the equivalence to such, we will use the phrase **Ulam–Galerkin matrix** to refer to any projection of the Frobenius–Perron operator by an inner product as in Eq. (4.4), not necessarily including the infinite time limit part of the conjecture regarding steady state, item 2.

Ulam's method is often used to describe the process of using Ulam's conjecture, by developing what we call here the Ulam–Galerkin matrix, and then using the dominant eigenvector of this stochastic matrix to estimate the invariant density. Sometimes, however, the phrase Ulam's method is used to simply describe what we call here the developing of the Ulam–Galerkin matrix. Some discussion of computational aspects of the Ulam–Galerkin matrix and the Ulam method can be found in Appendix A. As we will see, the one-step action of the transfer of the operator is well approximated by Ulam–Galerkin matrices. The analysis to describe the approximation of the one-step action is much simpler than that of the infinite limit referred to in the Ulam method, and other issues such as decomposition of the space into almost-invariant sets is naturally approximated as well by the short time representation. Also, in a special case the approximation is in fact exact, as discussed in Section 4.4. This exact representation may occur when the *dynamical system is Markov.* Ulam's conjecture [307] has been proven in the special case of one-dimensional maps by methods of bounded variation [201]. In higher-dimensional dynamical systems, a rigorous footing of Ulam's conjecture is incomplete, except for special cases [49, 101, 102, 117, 119, 120], and it remains an active area of research. We should also point out recent

[34]Our use of the inner product structure requires the further assumption that the density functions are in the Hilbert space $L^2(M)$ rather than just the Banach space $L^1(M)$, which uses the embedding $L^2(M) \hookrightarrow L^1(M)$, provided M is of finite measure.

[35]Weak approximation by functions may be defined as convergence of the functions under the integral relative to test functions. That is, if $\{f_n\}_{n=1}^{\infty} \in L^1(M)$, it is defined that $f_n \overset{w}{\to} f^*$ if $\lim_{n\to\infty} \int_M |f^*(x) - f_n(x)| h(x)dx = 0$ for all $h \in L^\infty(M)$, which is referred to as the test function space.

developments in a nonuniformly expanding setting [236]. Nonetheless in practice it is easy and common to simply proceed to use the dominant eigenvector of the stochastic matrix and then to refer to it as an approximation of the invariant density. See Section 4.2, and compare to Fig. 1.1.

4.1 Finite-Rank Approximation of the Frobenius–Perron Operator

The quality of the Ulam–Galerkin approximation is discussed in several references, as is the convergence of Ulam's method [118, 117, 57, 101, 172, 49, 34]. It is straightforward to cast Ulam's method as Galerkin's method with a special choice of basis functions, as follows. For Ulam's method, the basis functions are chosen to be a family of characteristic functions:

$$\phi_i(x) = \chi_{B_i}(x) = 1 \text{ for } x \in B_i \text{ and zero otherwise.} \tag{4.5}$$

Generally, B_i is chosen to be a simple tiling of the region of interest in the phase space, meaning some region covering a stable invariant set such as an attractor. For convenience, B_i may be chosen as a simple covering of rectangle boxes, but we have also had success in using triangular tessellations using software packages often affiliated with PDE-based finite element methods technology. In the deterministic case, using the inner product (4.4) the matrix approximation of the Frobenius–Perron operator has the form

$$P_{i,j} = \frac{m(B_i \cap F^{-1}(B_j))}{m(B_i)}, \tag{4.6}$$

where m denotes the normalized Lebesgue measure on M and $\{B_i\}_{i=1}^N$ is a finite family of connected sets with nonempty and disjoint interiors that covers M. That is, $M = \cup_{i=1}^N B_i$, and indexed in terms of nested refinements [307]. These $P_{i,j}$ can be interpreted as the ratio of the fraction of the box B_i that will be mapped inside the box B_j after one iteration of a map to the measure of B_i.

Note that one may consider this matrix approximation of the Frobenius–Perron operator as a finite Markov chain, where the partition set $\{B_i\}_{i=1}^N$ represents a set of "states" and P_{ij} characterizes transition probabilities between states. It is well known that the matrix **P** in (4.6) is stochastic and has a left eigenvector with the eigenvalue one. Simply put, this eigenvector characterizes the equilibrium distribution of the Frobenius–Perron operator. In fact, it can be proven [50] that if the partition $\{B_i\}_{i=1}^N$ is a **Markov partition**, then the (unique) left eigenvector of the matrix **P** defines a good approximation of the equilibrium distribution, a statement that will be made precise in the next subsection. This leads to a straightforward way to understand the approximation theory for the generic non-Markov case, by approximation using Markov representations. Definition of the phrase Markov partition will be the subject of Section 4.2.

However, first we note a more readily computable experimental perspective to stand in for Eq. (4.6), essentially by a Monte Carlo sampling.

Remark 4.1. A key observation is that the kernel form of the operator in Eq. (3.44) allows us to generally approximate the action of the operator with test orbits as follows. If we only have a test orbit $\{x_j\}_{j=1}^N$, which is actually the main interest of this work, the Lebesgue measure can be approximated by

$$P_{i,j} \approx \frac{\#(\{x_k | x_k \in B_i \text{ and } F(x_k) \in B_j\})}{\#(\{x_k \in B_i\})}. \tag{4.7}$$

This statement can be made precise in terms of quality of the approximation and number of sample points by Monte Carlo sampling theory for integration of the inner product, Eq. (4.4). See Section A.1.1 and the demo in Appendix A for a MATLAB implementation of this orbit sampling method of developing an Ulam–Galerkin matrix.

See Fig. 4.1 for a graphical description of the situation of the approximations described in the previous paragraphs. Compare this to Fig. 4.2, which is developed using Eq. (4.7) in the case of a Henon map, and a rather course partition for the sake of illustration. Then recording the relative transition numbers according to Eq. (4.6) as estimated by Eq. (4.7) leads to stochastic matrices as presented by the transition matrix shown in Fig. 1.1. See Figs. 4.2 and 4.3 for illustrations of the set oriented methods reflected by the Ulam–Galerkin approach in realistic systems, the Henon map, and the flow of the Gulf, respectively.

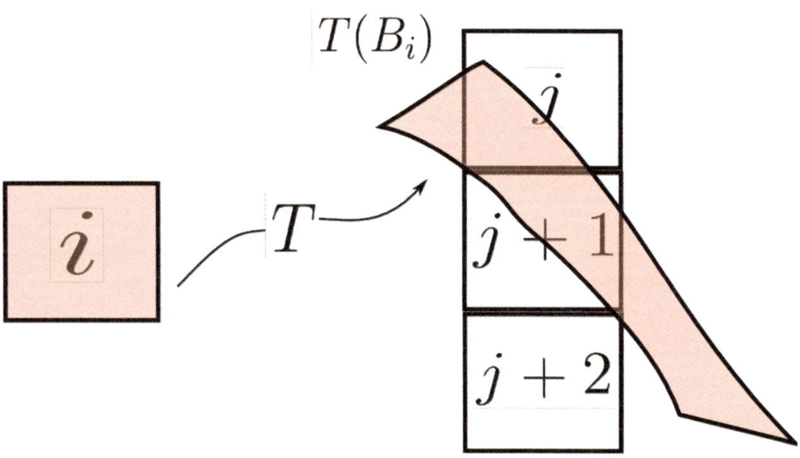

Figure 4.1. *An Ulam–Galerkin method approximation of the Frobenius–Perron operator is described by the action of a graph, as estimated by Eq.* (4.7). *Here we see a box i that maps roughly across j, $j+1$, and $j+2$. As such a graph G_A generated by matrix A would have an edge between vertex i and each of j, $j+1$, and $j+2$. The matrix A is formally described by the inner product, Eq.* (4.4), *where the basis functions are chosen to be characteristic functions $\chi_k(x)$ supported over each of the boxes in the covering, including j, $j+1$, $j+2$, and i. As shown, the $T(j)$ does not cover the i boxes in a way that allows a Markov partition, and thus the lost measure causes the finite rank to be only an approximation.*

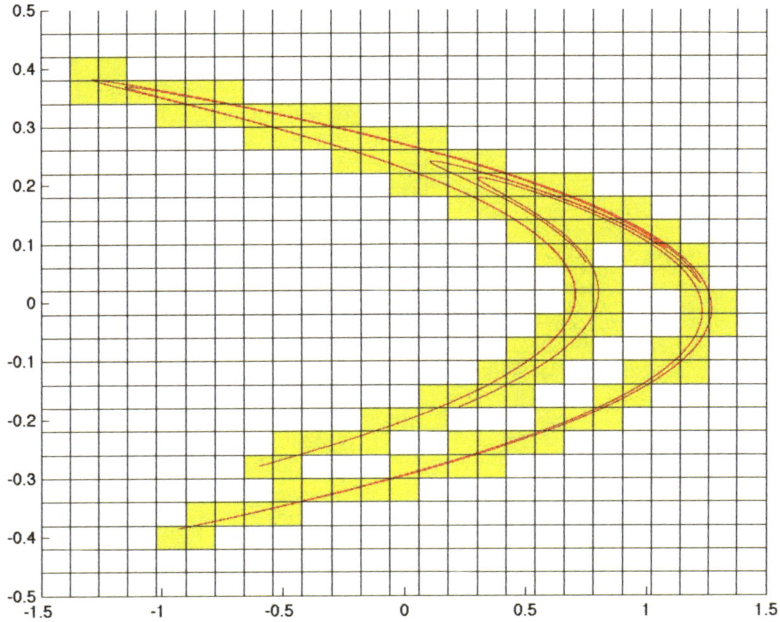

Figure 4.2. *The Ulam–Galerkin approximation of the Frobenius–Perron operator in its simplest terms is a "set oriented method" in that how points in sets map to other sets is recorded and used as an approximation of the action of the dynamical system on points. This figure shows a covering of boxes B_i which will be used to estimate the action of the Henon map dynamical system in those cells B_i in which the attractor is embedded (yellow). Compare this covering to the action of a Henon mapping so estimated and shown in Fig.* 1.1. *The underlying estimates of the stochastic matrix are summarized by the computations shown in Eqs.* (4.6) *and* (4.7). *Compare to Figs.* 4.1 *and* 4.3.

4.2 The Markov Partition: How It Relates to the Frobenius–Perron Operator

Generically, most dynamical systems are not Markov, meaning that they do not admit a Markov partition, and most partitions for such systems will not be Markov partitions. But when we do have such, then the corresponding Frobenius–Perron operator can be *exactly* represented by an operator of finite rank. We will give the necessary technical details and interpretations here. In the next section, we will show how this perspective of Markov partitions can be used to formulate a notion of approximation in the non-Markov case. Thus, as we will see, besides having an important role in a symbolic dynamics (from topological dynamics) of a dynamical system where the concept of a generating partition of the symbols is significantly simplified when there is a Markov partition, in measurable dynamics Markov partitions allow for a greatly simplified finite rank description of the Frobenius–Perron operator. Hence the computation of relevant statistical measures is greatly simplified.

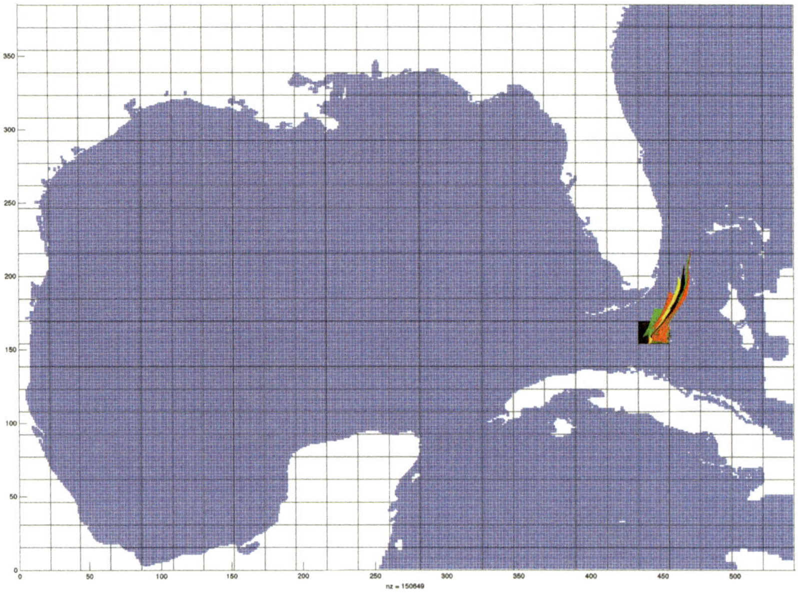

Figure 4.3. *The Ulam–Galerkin approximation of the Frobenius–Perron starts with the study of evolution of an ensemble of initial conditions from a single sell. Here a (rather large for artistic reasons) box B_i in a flow developed from an oceanographic model of the fluid flow in the Gulf reveals how a single square progressively becomes swept in the Gulf Stream. The useful time scale is one that reveals some nontrivial dynamic evolution, stretching across several image cells, but not so long that a loss of correlation may occur. Compare this image to a similar image developed for the Henon mapping, Figs.* 4.1 *and* 4.2, *and stochastic matrix estimates in Eqs.* (4.6) *and* (4.7). *The operator in its simplest terms is a "set oriented method" in that how points in sets map to other sets is recorded and this is used as an approximation of the action of the dynamical system on points. Again, compare the box covering as shown in Figs.* 1.1 *and* 4.2 *as in Eqs.* (4.6) *and* (4.7). *Compare to a similar presentation in the Duffing oscillator, Fig.* 1.16.

4.2.1 More Explicitly, Why a Markov Partition?

To simplify analysis of a dynamical system, we often study a topologically equivalent system using **symbol dynamics**, representing trajectories by infinite length sequences using a finite number of symbols. (An example of this idea is that we often write real numbers as sequences of *digits*, a finite collection of symbols.) Symbolic dynamics will be discussed in some detail in the next chapter.

To represent the state space of a dynamical system with a finite number of symbols, we must *partition* the space into a finite number of elements and assign a symbol to each one. In probability theory, the term "Markov" denotes a finite memory property. In other words the probability of each outcome conditioned on all previous history is equal to conditioning on only the current state; no previous history is necessary. The same idea has been adapted to dynamical systems theory to denote a partitioning of the state space so that all of the past information in the symbol sequence is contained in the current symbol, giving rise to the idea of a *Markov transformation*.

4.2.2 Markov Property of One-Dimensional Transformations

In the special but important case that a transformation of the interval is Markov, the symbol dynamic is simply presented as a finite directed graph. A Markov transformation in $\mathcal{R}^1$ is defined as follows [50].

Definition 4.1. Let $I = [c,d]$ and let $\tau : I \to I$. Let $\mathcal{P}$ be a partition of I given by the points $c = c_0 < c_1 < \cdots < c_p = d$. For $i = 1,\dots,p$, let $I_i = (c_{i-1}, c_i)$ and denote the restriction of τ to I_i by τ_i. If τ_i is a homeomorphism from I_i onto a union of intervals of $\mathcal{P}$, then τ is said to be *Markov*. The partition $\mathcal{P}$ is said to be a *Markov partition* with respect to the function τ.

Example 4.1 (one-dimensional example). Map 1, (Fig. 4.4(a)) is a Markov map with the associated partition $\{I_1, I_2, I_3, I_4\}$. The **symbol dynamics** are captured by the transition graph (Fig. 4.4(b)). Although map 2 (Fig. 4.4(c)) is piecewise linear and is logically partitioned by the same intervals as map 1, the partition is not Markov because interval I_2 does not map onto (in the mathematical sense) a union of any of the intervals of the partition. However, we are not able to say that map 2 is not Markov. There may be some other partition that satisfies the Markov condition [38].

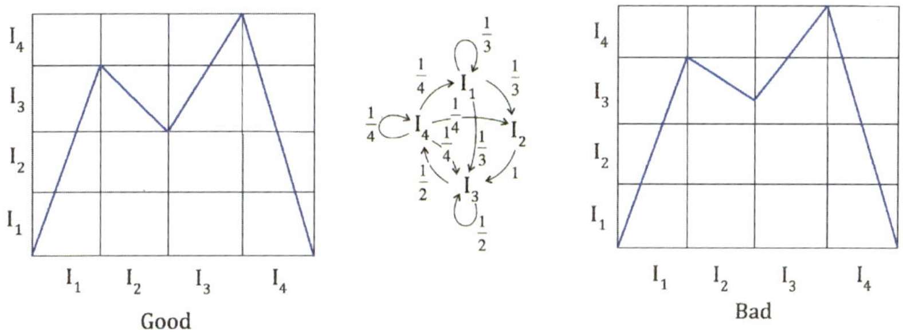

(a) (b) (c)

Figure 4.4. (a) *A Markov map with partition shown.* (b) *The transition graph for map a.* (c) *The partition is not Markov because the image of I_2 is not equal to a union of intervals of the partition.*

4.2.3 Markov Property in Higher Dimensions

Definition 4.2. A **topological partition** of a topological space (M, τ) is a finite collection $\mathcal{P} = \{P_1, P_2, \dots, P_r\}$ of disjoint open sets whose closures cover M in the sense that $M = \overline{P_1} \cup \cdots \cup \overline{P_r}$.

Definition 4.3. A **topological space** (M, τ) is a set M together with the set of subsets $\tau \subset 2^M$[36] that are defined to be open; as such τ must include the empty set $\emptyset$ and all of M, and τ must be closed under arbitrary unions and finite intersections [235].

Any topological partitioning of the state space will create **symbol dynamics** for the map. In the special case where the partition is *Markov*, the **symbol dynamics** capture the essential dynamics of the original system.

[36] 2^M denotes the "power-set" of M, meaning it is the set of all subsets.

Definition 4.4. Given a metric space M and a map $f : M \to M$, a **Markov partition** of M is a topological partition of M into rectangles $\{R_1, \dots, R_m\}$ such that whenever $x \in R_i$ and $f(x) \in R_j$, then $f[W^u(x) \cap R_i] \supset W_u[f(x)] \cap R_j$ and $f[W^s(x) \cap R_i] \subset W_s[f(x)] \cap R_j$ [46, 44].

In simplified terms, this definition says that whenever an image rectangle intersects a partition element, the image must stretch completely across that element in the expanding directions and must be inside that partition element in the contracting direction (see Fig. 4.5) [38].

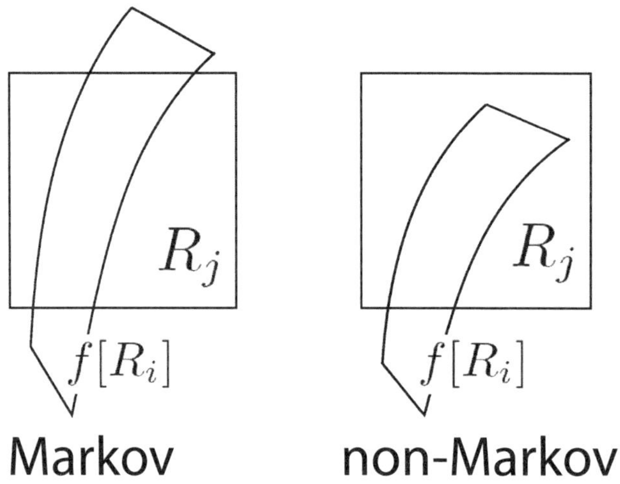

Figure 4.5. *In the unstable (expanding) direction, the image rectangle must stretch completely across any of the partition rectangles that it intersects.*

4.2.4 Generating Partition

It is important to use a "good" partition so that the resulting symbolic dynamics of orbits through the partition well represents the dynamical system. If the partition is Markov, then goodness is most easily ensured. However, a broader notion, called *generating partition*, may be necessary to capture the dynamics. A *Markov partition* is *generating*, but the converse is not generally true. See [272, 39] for a thorough discussion of the role of partitions in representing dynamical systems.

Definition 4.5. Given a topological space (M, τ) (Definition 4.3) and a topological partition, $\mathcal{P} = \{P_1, P_2, \dots, P_r\}$ is a topological **generating partition** for a mapping $T : M \to M$ if

$$\tau = \vee_{i=0}^{\infty} T^{-i} \mathcal{P} \tag{4.8}$$

(or require $\tau = \vee_{i=-\infty}^{\infty} T^{-i} \mathcal{P}$, for invertible dynamical systems).

As usual, T^{-i} denotes the ith preimage (possibly with many branches), but it is the ith composition of the inverse map if the map is invertible. This definition is in terms of the **join** of partitions, which is defined recursively.

Definition 4.6. The **join** of two partitions, $\mathcal{P}$ and $\mathcal{P}'$, is defined as

$$\mathcal{P} \vee \mathcal{P}' = \{P_k \cap P'_l : 0 \le k \le |\mathcal{P}| - 1, 0 \le l \le |\mathcal{P}'| - 1\}. \tag{4.9}$$

Thus terms such as $T^{-i}\mathcal{P}$ in the definition of the generating partition are joined with other iterates of the original partition, $T^{-j}\mathcal{P}$. The idea of a generating partition is that this process of joining many iterates of open sets creates collections of open sets. Proceeding infinitely in this manner creates infinitely many open sets, and the question is whether all of the open sets in the topology are generated.

If, furthermore, a measure space is assumed $(M, \mathcal{A}, \mu)$, where $\mathcal{A}$ is defined as the Borel sigma algebra of sets which are μ-measurable, then the question a generating partition becomes as follows.

Definition 4.7. Given a measure space $(M, \mathcal{A}, \mu)$, then a topological partition $\mathcal{P} = \{P_1, P_2, \ldots, P_r\}$ of measurable sets is a measurable **generating partition** for a mapping $T : M \to M$ if

$$\mathcal{A} = \vee_{i=0}^{\infty} T^{-i}\mathcal{P} \tag{4.10}$$

(or require $\tau = \vee_{i=-\infty}^{\infty} T^{-i}\mathcal{P}$, for invertible dynamical systems). We require that all the measurable sets are generated.

Example 4.2 (two-dimensional example—toral automorphism). The **cat map**, defined by

$$\mathbf{x} = (\mathbf{A}\mathbf{x}) \bmod 1, \tag{4.11}$$

$$\text{where } \mathbf{A} = \begin{bmatrix} 2 & 1 \\ 1 & 1 \end{bmatrix}, \tag{4.12}$$

yields a map from the unit square onto itself. This map is said to be on the toral space $\mathbf{T}^2$ because the mod 1 operation causes the coordinate $1 + z$ to be equivalent to z. A *Markov partition* for this map is shown in Fig. 4.6. The *cat map* is part of a larger class of functions called toral Anosov diffeomorphisms, and [268] provides a detailed description of how to construct *Markov partitions* for this class of maps [38].

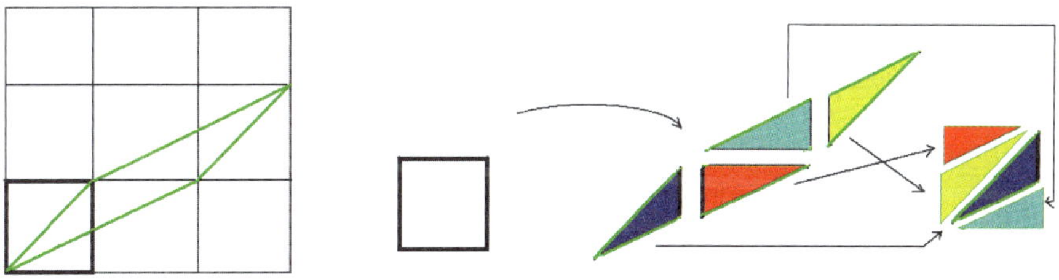

(a) (b) (c)

Figure 4.6. *The* ***cat map*** *is a toral automorphism.* (a) *The operation of the linear map on the unit square.* (b) *Under the mod operation, the image is exactly the unit square.* (c) *Tessellation by rectangles R_1 and R_2 forms an infinite partition on $\boldsymbol{R}^2$. However, since the map is defined on the toral space $\boldsymbol{T}^2$, only two rectangles are required to cover the space. The filled gray boxes illustrate that R_1 and R_2 are mapped completely across a union of rectangles.*

Example 4.3 (generating partition of the Henon map). Consider again the Henon map, $T(x,y) = (1-ax^2+y, bx)$, $(a,b) = (1.4, 0.3)$, the prototypical diffeomorphism of the plane with the famous strange attractor. See Fig. 6.31 in which a piecewise linear curve C produced as connecting tangencies of stable and unstable manifolds, according to the well-regarded conjecture [71, 144], produces what is apparently a generating partition but not a Markov partition. See also the generating partition discussion of the Ikeda map (9.87) in Example 9.3 shown in Fig. 9.8.

4.3 The Approximate Action of Dynamical System on Density Looks Like a Directed Graph: Ulam's Method Is a Form of Galerkin's Method

The title of this section says that the approximate action of dynamical system on a density looks like a directed graph: Ulam's method is a form of Galerkin's method. This is a perspective in which we already discussed the theory of Galerkin's method in Section 4.1. In fact, as stated above, when the dynamical system is Markov and using the Markov partition and the corresponding basis functions which are supported over the elements of that Markov partition, the action of the dynamical system is *exactly* represented by a directed graph. In this case, the inner product form (4.4) becomes exactly (4.6), resulting in a stochastic matrix A whose action and steady state are both discussed by the Frobenius–Perron theorem. In this section, we will summarize these statements more carefully, but first we motivate with the following examples.

Example 4.4 (finite rank transfer operator of a one-dimensional transformation). The map shown in Fig. 4.4(a) was already discussed to be a Markov map in the interval, with a Markov partition $\{I_1, I_2, I_3, I_4\}$ as shown, and according to the definition of Markov partition shown in Fig. 4.4.

In this piecewise linear case, it is easy to directly integrate the Galerkin method (4.4) integrals when choosing the basis functions to be one of each of the four characteristic functions,[37] $\{\chi_{I_1}(x), \chi_{I_2}(x), \chi_{I_3}(x), \chi_{I_4}(x)\}$. Let

$$\phi_i(x) = \chi_{I_i}(x), \quad i = 1,2,3,4. \tag{4.13}$$

For the sake of example, we will explicitly write one such integral here. From the drawing, writing the function in Fig. 4.4(a) explicitly, for simplicity assuming the uniform partition shown,

$$I_i = [i-1, i], \quad i = 1,2,3,4, \tag{4.14}$$

and that

$$\cup_{i=1}^4 \overline{I_i} = [0,4], \tag{4.15}$$

then $F : [0,4] \to [0,4]$ may be written

$$F(x) = \left\{ \begin{array}{ccc} 3x & \text{if} & x < 1 \\ -x+4 & \text{if} & 1 \leq x < 2 \\ 2x-2 & \text{if} & 2 \leq x < 3 \\ -4x+16 & \text{else} & x \geq 3 \end{array} \right\}. \tag{4.16}$$

[37] A characteristic function, also called an indicator function, takes on the value 1 inside the argument set, $\chi_A(x) : M \to \{0,1\}$, $\chi_A(x) = if(x \in A, 1, 0)$.

Now we may write and evaluate the inner product (4.4) to derive each of the elements of the 4×4 matrix A which describes the relative movement of ensemble density with each iteration of the map. Substitution into Eq. (4.4) gives

$$P_{i,j} = \langle P_F(\phi_i), \phi_j \rangle = \int_{x=0}^{4} P_{F_v}(\chi_i(x))\chi_j(x)dx, \quad i,j = 1,2,3,4. \tag{4.17}$$

Continuing to produce A requires $16 = 4^2$ such integrations for each $\chi_i(x), \chi_j(x)$ basis function pairing, but in this simplified case where we have piecewise linear functions, we will illustrate with just one pairing, $i = 1$ and $j = 2$, to produce $P_{1,2}$ using the basis functions $\chi_1(x)$ and $\chi_2(x)$ in the Galerkin method's inner product computation,

$$\begin{aligned} P_{1,2} &= \int_{x=0}^{4} P_{F_v}(\chi_1(x))\chi_2(x)dx, \\ &= \int_{x=1}^{2} P_{F_v}(\chi_1(x))dx, \end{aligned} \tag{4.18}$$

by considering the nonzero support of $\chi_2(x)$. Evaluating the Frobenius–Perron image of the initial density $\chi_1(x)$ for the integration by referring to the definition and specializing to this example,

$$P_F[\chi_1(x)] = \sum_{y \in F^{-1}(x)} \frac{\chi_1(x)}{|\det D_y F|} = \frac{1}{3}\chi_1(x). \tag{4.19}$$

Again we recall the region of nonzero support required to be considered in evaluating the integral in Eq. (4.18). In this region, colored as shown in Fig. 4.7, each x has exactly one preimage, still in I_1, and thus, $|\det D_y F| = 3$ there. Hence continuing,

$$P_{1,2} = \int_{x=1}^{2} \frac{1}{3}\chi_1(x)dx = \frac{1}{3}. \tag{4.20}$$

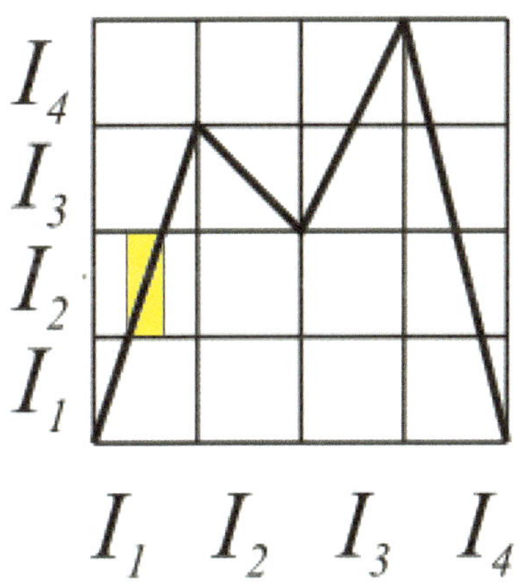

Figure 4.7. *The Markov map shown in Fig.* 4.4(a) *is again shown here, but the yellow region here is shaded to highlight the set which is active when computing the integrations necessary to produce the matrix element* $P_{1,2}$ *as described in* (4.13) *to* (4.20).

Similarly repeating this computation in kind for each i, j pairing results in the rest of the transition matrix:

$$A = \begin{bmatrix} 1/3 & 1/3 & 1/3 & 0 \\ 0 & 0 & 1 & 0 \\ 0 & 0 & 1/2 & 1/2 \\ 1/4 & 1/4 & 1/4 & 1/4 \end{bmatrix}. \tag{4.21}$$

As a side note, observe that when computing $P_F[\chi_i(x)]$, there is a potential for multiple preimages, but then considering the inner product $\langle P_F(\phi_i), \phi_j \rangle$ means that we will always select exactly one of those branches since the other function in the inner product, $\chi_j(x)$, selects just one branch and no more. This follows from the fact that we have specialized to characteristic functions supported over the elements of a Markov partition; thus we implicitly rely on the monotonicity which follows the required homeomorphism for each leg.

Thus the transfer operator becomes an operator of finite rank in terms of Galerkin's method with finitely many terms. Furthermore, the approximation is exact when a Markov partition and associated basis functions are used. The resulting A may be interpreted as a transfer matrix, for which it is easy to check that all row sums are 1,

$$\sum_j P_{i,j} = 1 \quad \forall j. \tag{4.22}$$

As such, it is a stochastic matrix. Correspondingly, this generates a weighted directed graph for a representing Markov process, as shown in Fig. 4.4(b).

It is general that matrices produced in this way have positive elements, and they will be stochastic matrices when the Markov partition property exists and is used.[38] As such, the theory of Frobenius–Perron matrices[39] applies and specializes to stochastic matrices. Some elements of this theory will be reviewed in the next section, but we will illustrate a main idea continuing with the example of Fig. 4.4(a). That is, the dominant eigenvalue equals 1, and uniquely if the dynamical system is ergodic or correspondingly the resulting matrix is irreducible. The dominant eigenvector corresponds to the steady state, and therefore associates to the steady state measure.

Example 4.5 (steady state of a finite rank transfer operator). Continuing with Example 4.4 from Fig. 4.4(a), we compute the steady state distribution. The steady state distribution of the Markov process, Fig. 4.4(b), can be shown from the Frobenius–Perron theory of positive matrices (see Section 4.3.3) to equal the eigenvector of the transition matrix corresponding to the largest eigenvalue, which is 1. This eigenvector of matrix (4.21) is

$$v = \frac{1}{\sqrt{226}} \begin{bmatrix} 3 \\ 3 \\ 12 \\ 8 \end{bmatrix}. \tag{4.23}$$

[38]Without using the existing Markov partition, or if one does not exist, then a stochastic matrix does not result, but it may be "almost" stochastic in that row sums are almost 1, reflecting some leak of measure in the representation, while not reflecting the reality of the dynamical system.

[39]Contrast that phrase to "Frobenius–Perron operators" already discussed here. While the names of these mathematicians associated with the operator are found ordered both ways in the literature, we will take here a common convention that FP denotes the operators, and PF denotes the finite rank situation and the related linear algebra.

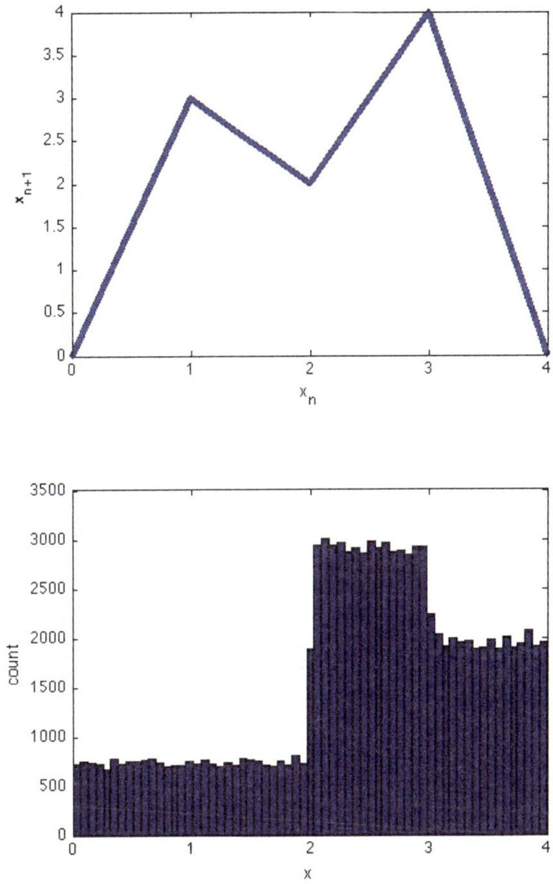

Figure 4.8. *Steady state of the system in Fig.* 4.4(a,b) *and Fig.* 4.7 *and defined by Eq.* (4.16) *was noted to be Markov, and as such the corresponding stochastic matrix from the Ulam–Galerkin method gives the dominant eigenvector Eq.* (4.23). *(Top) The map shown again to help match the regions of the partition relevant for the states in the density. Note that each piecewise linear section of the map matches piecewise constant sections of the invariant density. (Bottom) The histogram for a* 10^5 *iterate test orbit agrees quite well with the density predicted by the eigenvector* (4.23).

Now this eigenvector represents the steady state distribution of ensembles of initial conditions over long times, where each partition element of the Markov partition should see a relative number of points land in them, according to the relative proportions of this vector. To punctuate this point, we illustrate the histogram from a long orbit of 10^5 iterates in Fig. 4.8. Note that by showing the histogram of a long orbit, rather than the histogram produced by an ensemble of many initial conditions, we have again appealed to the notion that time averages and spatial averages can be exchanged. Such correspondence is the premise of the Birkhoff ergodic theorem, which is valid since this example is an ergodic system. The resulting matrix is irreducible and the system is Markov. Thus the finite rank representation exactly describes the steady state measure.

4.3.1 Graph Theory, Reducible Matrices

A broader theme of this book is the analysis of global transport, which can be defined in terms of a partition to discuss how ensembles may propagate from one element of the partition to others. The discrete counterpart of this discussion for a directed graph also corresponds to notions of partitions of a graph into subgraphs, as discussed here.

In this section we briefly review some basic concepts of reducible matrices and corresponding relationships to the graph induced by the matrices due to discrete approximations of a Frobenius–Perron operator. In particular, we introduce the question of how to discover a permutation of the transition matrix (4.7) that reveals the basin structure of a dynamical system revealed to us in the form of a test orbit segment $\{x_j\}_{j=1}^N$. See Fig. 4.9.

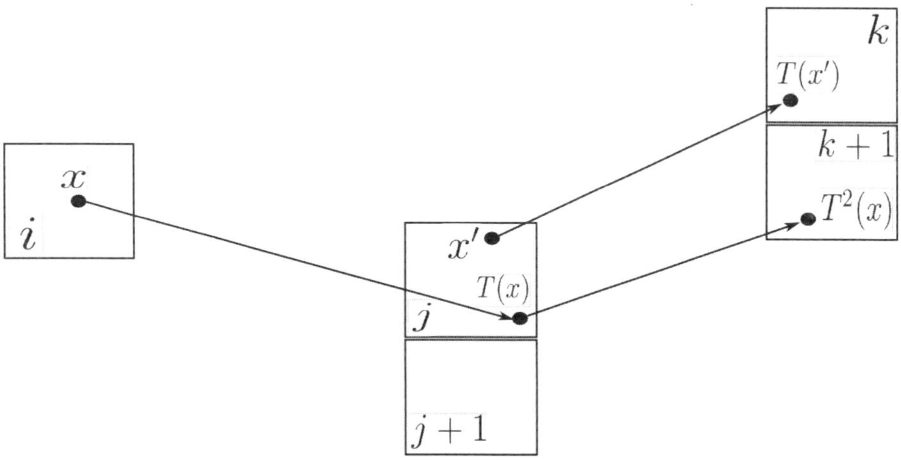

Figure 4.9. *A path in the graph corresponding to the transfer matrix produced from orbits in a dynamical system may not imply an orbit in the dynamical system passing through the same corresponding boxes, but rather an epsilon-chain pseudo-orbit. Here the graph has an orbit from boxes labeled $i \to j \to k$, but $x \in i$, $T(x) \in j$, and perhaps $T^2(x) \in k+1$, but perhaps there is an x' such that $T(x') \in k$.*

A graph G_A associated to a matrix $\mathbf{A}$ consists of a set of vertices V and a set of edges $E \subset V \times V$. The entry P_{ij} is nonzero when there exists an edge $(i, j) \in E$ that connects the vertices i and j.

As we will describe, borrowing a term from the recent activity in complex networks, it is also useful to consider structure within a graph. First we have components, and then generalizing that we have communities. If a disjoint collection $\{S_i\}_{i=1}^k$ of subsets $S_i \subset V$ consists of vertices such that there is a higher density of edges within each S_i than between them, then we say roughly that $\{S_i\}_{i=1}^k$ forms a *community structure* for G_A [240, 241].

For the discussion of this section, it is only necessary to consider the sign of the entries of a transition matrix A, or perhaps of a stochastic matrix A. We shall define

$$B = \text{sign}(A), \tag{4.24}$$

which when A is a nonnegative matrix, $B_{i,j} = 0$ or 1, and comparing to the directed graph

$G_A = (E, V)$ generated by A implies that there is an edge from j to i,

$$B_{i,j} = \left(\begin{array}{cc} 1 & \text{if } j \to i \\ 0 & \text{else} \end{array} \right). \tag{4.25}$$

Such is called an **adjacency matrix** when it encodes just possible transitions without weights. That is, a one-step path is on an edge from a vertex labeled j to vertex labeled i.

When the graph represents the action of the Frobenius–Perron operator from a dynamical system, a one-step path means that there exists a direct transfer from mode $\phi_j(x)$ to $\phi_i(x)$ when those modes are characteristic functions. This in turn implies existence of at least a one-step orbit of the dynamical system between the cells when basis functions are characteristic functions. If an n-step path in the graph is possible, this does not imply that an n-step orbit through the corresponding cells covering the phase space of the dynamical system is possible. Instead, the correct statement is that an epsilon-chain pseudo-orbit[40] is possible, but perhaps not a true orbit segment.

An essential question then is if a given **path**[41] in the graph is possible, which has significance to the distribution of measure in the corresponding Markov chain? A few issues relevant to transition matrices are captured by the following language.

Definition 4.8. A square matrix $A_{n \times n}$ is called **primitive** if there exists a (time) k such that $[A_{n \times n}]^k_{i,j} > 0$ for all i, j.

Remark 4.2. A sufficient condition for a matrix to be a primitive matrix is for the matrix to be a nonnegative, irreducible matrix with a positive element on the main diagonal.

The primitive property may be interpreted as follows. Since a direct path between j to i exists if and only if $P_{i,j} > 0$, and P^2 describes the two-step action of the graph, then a two-step path exists between j to i if and only if $P^2_{i,j} > 0$. In these terms, the primitive property demands that there is a time k when there is a path from everywhere to everywhere else in the graph. Conversely, if there is an i,j pair such that $P^k_{i,j} = 0$ for all times k, then as a graph structure, the two vertices must occupy different components of the graph. "Component" then describes much the same information as primitiveness for the corresponding graph.

Definition 4.9. A **component of a graph** G is a subgraph G' of G such that between every pair of vertices, there is a path between every vertex in the subgraph. A graph is called **connected** if it consists of exactly one component.

As an example, we see in Fig. 4.10 that the component including i, $i+1$, and j is apparently not connected to the component including vertices k, $k+1$. As such, the adjacency matrix A which generates this graph cannot be primitive.

Let $I = \{1, \ldots, N\}$ be an index set. For our specific application to the transition matrix P generated by the Ulam–Galerkin method, we define the set of vertices $V = \{v_i\}_{i \in I}$ to label the original boxes $\{B_i\}_{i \in I}$ used to generate the matrix P, and define the edges to be

[40] An epsilon-chain pseudo-orbitis a sequence of points, $\{x_i\}$, such that $\|T(x_i) - x_{i+1}\| \leq \epsilon$ for each i, and this specializes to a true orbit if $\epsilon = 0$. There are theorems from the shadowing literature [45, 46], especially for hyperbolic systems, that describe that near an epsilon chain there may be true orbits. In the context used here, however, those true orbits may not pass through the boxes corresponding to a particular refinement.

[41] An n-step path from vertex i to vertex j is the existence of n edges in order "end-to-end" (stated roughly to mean the obvious that each edge ends at a vertex where the next begins) beginning at i and ending at j.

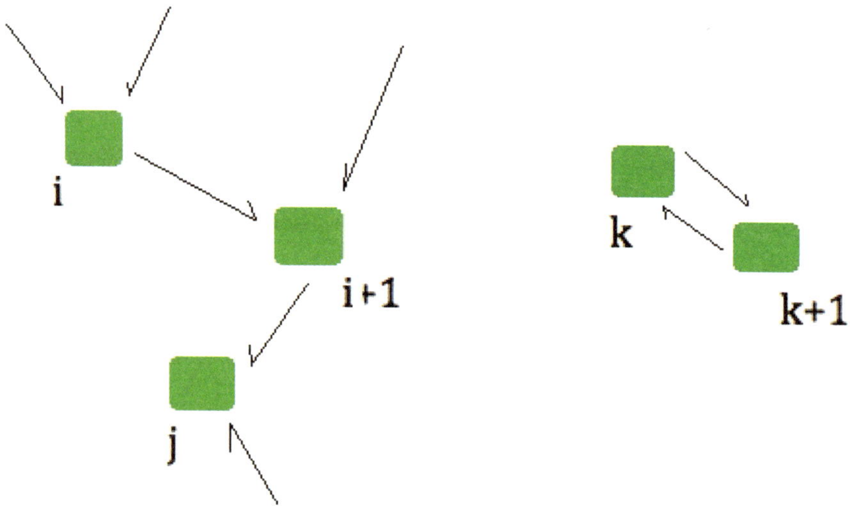

Figure 4.10. *A segment of a larger directed graph. We see that a one-step walk from vertex i to $i+1$ is possible but a walk to j from i requires two steps. No walk from the component containing vertices $i, i+1$, and j to the k, $k+1$ component is possible, at least along the edges shown.*

the set of ordered pairs of integers $E = \{(i, j) : i, j \in I\}$ which label the vertices as their starting and ending points.

Another concept which is useful in the discussion of partitioning a graph into dynamically relevant components is to ask if the graph is reducible.

Definition 4.10. A graph is said to be **reducible** if there exists a subset $I_o \subset I$ such that there are no edges (i, j) for $i \in I_o$ and $j \in I \setminus I_o$; otherwise, it is said to be **irreducible** [22, 308].

This condition implies that the graph is irreducible if and only if there exists only one connected component of a graph, which is G_P itself. In terms of the transition matrix P, G_P is reducible if and only if there exists a subset $I_o \subset I$ such that $P_{ij} = 0$ whenever $i \in I_o$ and $j \in I \setminus I_o$. Furthermore, P is said to be a reducible matrix if and only if there exists some permutation matrix S such that the result of the similarity transformation

$$R = S^{-1}PS \tag{4.26}$$

is block upper triangular:

$$R = \begin{pmatrix} R_{1,1} & R_{1,2} \\ 0 & R_{2,2} \end{pmatrix}. \tag{4.27}$$

This means that G_P has a decomposition into a partition,

$$V = V_1 \bigcup V_2, \tag{4.28}$$

such that V_1 connects with V_1 and V_2, but V_2 connects only with itself. When $R_{1,2} = 0$, P is said to be **completely reducible** [308],

$$R = \begin{pmatrix} R_{1,1} & 0 \\ 0 & R_{2,2} \end{pmatrix}. \tag{4.29}$$

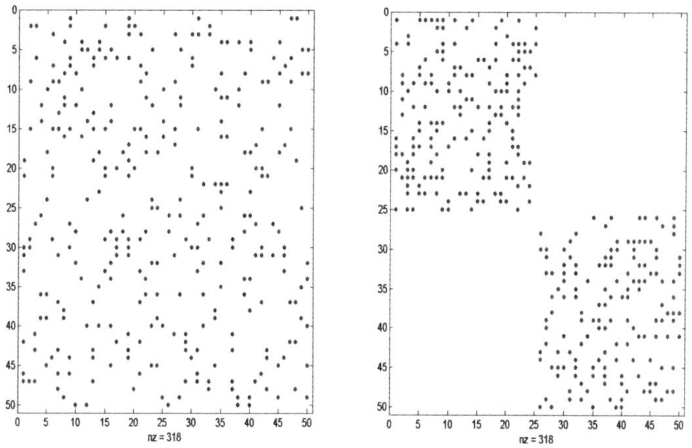

(a) The transition matrix before sorting, in arbitrary configuration.

(b) The transition matrix after sorting to reveal a block-diagonal form which informs that there are two components that do not communicate with each other. When arising from a dynamical system, this is typical of a bistable system, meaning two basins as two invariant sets and no transport between them.

Figure 4.11. *The matrix in this figure is reducible. However, the members of the original transition matrix before sorting are placed haphazardly. After using a diagonalizing similarity transformation* (4.26) *to the form* (4.29) *(when it exists), then the block-diagonal form is revealed. The corresponding graph description of the same is seen in Fig.* 4.12.

An instructive observation when relating these concepts back to dynamical systems is that in the case that G_P is generated from a *bistable dynamical system* as from Example 7.1, the transition matrix A will be completely reducible. Therefore, $R_{1,1}$ and $R_{2,2}$ correspond to the two basins of attractions of the system. The off-diagonal elements $R_{1,2}$ and $R_{2,1}$ give information regarding transport between these partition elements if there are any nonzero off-diagonal elements. Also, in a general multistable dynamical systems the transition matrix $\mathbf{A}$ has a similarity transformation into a block (upper) triangular form—emphasizing many components which may not communicate with each other.

A key point relevant to the theme of this writing is that most (randomly or arbitrarily realized) indexing of the configuration resulting from an Ulam–Galerkin matrix makes it difficult to observe even simple structures, like community structure or reducibility, of the graph and the corresponding transition matrix. That is, the indexing that may come from a sensible and suitable covering of a phase space by rectangles, or perhaps of triangles, will not generally be expected to directly reveal a useful structure in the associated Ulam–Galerkin matrix. The goal is then to reveal reducibility or at least community structure when it is there. Figs. 4.11 and 4.12 illustrate the kind of graph that results from a bistable system coming from a bistable system, when usefully sorting to reveal the structure. Notice that there are two disjoint components that do not communicate with each other. Corre-

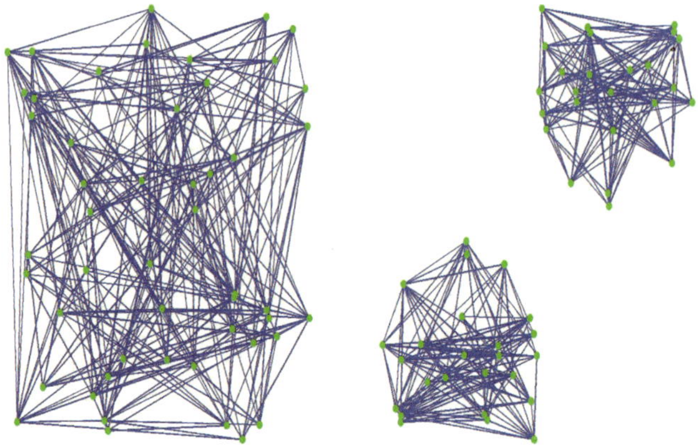

(a) The original graph. Embedding in the plane the vertices in a arbitrary configuration conceals simple structure.

(b) Sorted graph. The same graph subject to appropriate permutation reveals reducible structure community between components.

Figure 4.12. *Graph representation of the adjacency matrices shown in Fig.* 4.11.

spondingly this informs that the bistable system has two basins that do not communicate with each other. A more difficult problem of sorting a transition matrix comes from the scenario when there may be some off-diagonal elements even in a suitable sorting, as would arise when there may be some "leaking" or small transport between two basins which may no longer be invariant sets, but rather simply almost invariant.

In Figs. 5.3(a) and 5.3(b) we show an example of a transition matrix for a 30-vertex random community-structured graph with three communities whose members are placed haphazardly, but after a proper permutation this matrix is transformed into an "almost" block-diagonal form. This is the sort of Ulam–Galerkin matrix which would be expected from a bistable dynamical system. Figs. 5.4(a) and 5.4(b) illustrate the associated graph of the matrix in the above example. Before sorting the vertices into three separate communities, the graph looks like a random graph that has no community structure. However, after sorting, the community structure of the graph becomes obvious. This is central to the problem of this topic, to find the sorting that reveals the useful partition illustrative of transport.

There are several techniques for finding an appropriate permutation if the matrix is reducible. In the language of graph theory, we would like to discover all connected components of a graph. The betweenness methods in [241] and the local method in [9] are examples of numerous methods as reviewed in [73, 241] that can be successfully used to uncover various kinds of community structures.

Remark 4.3. Lack of reducibility of a graph relates to the measure theoretic concept of ergodicity (Definition 3.7), and the topological dynamical concept of transitivity (see Definition 6.1), in that when the graph representation is sufficient, they all describe the reduction of the system either into noncommunicating components, or not in the case of the complement.

Remark 4.4. An essential question is whether the represented dynamical system is well coarse grained by the graph and the action of the matrices. The question can be settled in terms of discussion of those dynamical systems which are exactly represented by a coarse grain, which are those called Markov dynamical systems, and then the density of these. When a coarse-grained representation is used, the associated transition matrix may reveal some off-diagonal elements when sorted, suggesting some transitivity in the dynamical system, but this could just arise as the error of the coarse-grain estimation.

4.3.2 Convergence Rates, the Power Method, and Stochastic Matrices

Analysis of the result of iterating a matrix is very instructive regarding the long time behavior of the transfer operators, and the matrix discussion is a simplifying discussion. It is a well-known feature in matrix theory that certain matrices are composed such that large powers tend to point arbitrary vectors toward the dominant eigenvector. This feature is the heart of the power method in numerical analysis [298] which has become crucial in computing eigenvalues and eigenvectors of very large matrices—too large to consider a computation directly by the characteristic polynomial $\det(A-\lambda I)=0$. This discussion is meant to give some geometric notion of the action of the Ulam–Galerkin matrices in the vector spaces upon which they act.

We will begin this discussion with a simple illustrative example.

Example 4.6 (power method). Let

$$A=\begin{bmatrix} 1/3 & 1/3 & 1/3 & 0 \\ 0 & 0 & 1 & 0 \\ 0 & 0 & 1/2 & 1/2 \\ 1/4 & 1/4 & 1/4 & 1/4 \end{bmatrix}, \tag{4.30}$$

which is again a matrix from the Ulam–Galerkin method example (4.21), derived from Fig. 4.4(a). Here we discuss it simply in terms of its matrix properties. Checking

$$A\begin{bmatrix} 1 \\ 1 \\ 1 \\ 1 \end{bmatrix}=\begin{bmatrix} 1 \\ 1 \\ 1 \\ 1 \end{bmatrix} \tag{4.31}$$

confirms that $\lambda=1$ is a **right eigenvalue** with a corresponding **right eigenvector**, $v=[1,1,1,1]'$, (the prime symbol denotes transpose), in the usual definition of an eigenvalue/eigenvector pair,

$$Av=\lambda v. \tag{4.32}$$

Likewise, a substitution,

$$\begin{bmatrix} 1,1,4,\frac{8}{3} \end{bmatrix}A=\begin{bmatrix} 1,1,4,\frac{8}{3} \end{bmatrix}, \tag{4.33}$$

which is an example of the **left eigenvalue/eigenvector** equation,

$$uA=\tau u, \tag{4.34}$$

wherein $\tau=1$ is a left eigenvalue with corresponding left eigenvector, $u=[1,1,4,\frac{8}{3}]$.

These particular eigenvalue/eigenvector pairs have special significance. It is easy to check that A row sums to 1,

$$\sum_j P_{i,j} \quad \forall i, \tag{4.35}$$

which defines A to be a stochastic matrix with probabilistic implications to be defined later in this section. In particular, the left side of Eq. (4.31) may be directly interpreted as a row-by-row summation, and matching the right side of the equation means that the result must be 1. Thus, we are able to summarize.

Remark 4.5. ($\lambda = 1, v = [1,1,1,1]'$) will be a *right* eigenvector/eigenvector pair if and only if A is row-wise stochastic.

That ($\tau = 1, u = [1,1,4,\frac{8}{3}]$) is a *left* eigenvalue/eigenvector pair may be derived from the characteristic polynomial,

$$\det(A' - \tau I) = 0. \tag{4.36}$$

That we may directly solve this 4th degree polynomial, with up to four roots due to the fundamental theorem of algebra, is possible primarily because the matrix is small. In principle these roots may be found directly by a direct computation akin to the quadratic formula [203], or at least by numerical root solvers. However, numerically approaching the spectrum of large matrices is feasible only through the power method cited in many texts in numerical analysis [298]. Furthermore, specifically considering the power method in this simple case allows us an easy presentation of the behavior of the evolution of the corresponding Markov chain, as we shall see.

Consider an arbitrary vector $w \in E^n$, where E^n denotes the vector space which serves as the domain of A, which in this case we generally choose to be $\mathbb{R}^n$. By arbitrary we generally mean almost any. A full measure of those "good" vectors w shall suffice, and those which are not good shall become clear momentarily. Let A act on w from the left,

$$wA = (c_1 u_1 + c_2 u_2 + \cdots + c_n u_n)A, \tag{4.37}$$

where we have written as if we have a canonical situation in which there is a spanning set of n-eigenvectors $\{u_i\}_{i=1}^n$, but all we require is that

- one eigenvalue is unique and largest, and
- the subspace corresponding to the rest of the spectrum may be more degenerate.

Of course in general there can be both algebraic and geometric multiplicities, but for now, for sake of presentation we describe the simplest situation of unique eigenvectors, and further assume that the corresponding eigenvalues are such that one eigenvector is unique and largest:

$$\tau_1 > \tau_i \quad \forall i > 1. \tag{4.38}$$

By linearity of A, and further by resorting to the definition of a left eigenvector for each of the u_i, Eq. (4.37) becomes

$$\begin{aligned} wA &= c_1 A u_1 A + c_2 u_2 A + \cdots + c_n u_n A \\ &= c_1 \tau_1 u_1 + c_2 \tau_2 u_2 + \cdots + c_n \tau_n u_n. \end{aligned} \tag{4.39}$$

Then proceeding similarly, applying A m times, we get,

$$wA^m = c_1\tau_1^m u_1 + c_2\tau_2^m u_2 + \cdots + c_n\tau_n^m u_n. \tag{4.40}$$

Therefore, we see roughly that[42]

$$\tau_1 > \tau_i \implies \tau_1^m \gg \tau_i^m \tag{4.41}$$

for large m, from which follows

$$wA^m \approx c_1\tau_1^m u_1. \tag{4.42}$$

This says that repeated application of the matrix rotates arbitrary vectors toward the dominant (u_1) direction.

The general power method from numerical analysis does not proceed in this way alone because, while it may be true that $\tau_1^m \gg \tau_i^m$, for large m both (all) of these numbers become large, and the computation becomes impractical on a computer. In general it is better to renormalize at each step [298] as follows. Let s_0 be chosen arbitrarily as was stated in Eq. (4.37), but then

$$s_{k+1} = \frac{s_k A}{\|s_k A\|}, \tag{4.43}$$

stated in terms of left multiplication of left eigenvectors and where $\|w_k A\|$ is a renormalization factor at each step in terms of a vector norm. Similar arguments to those stated in Eqs. (4.37)–(4.42) can be adjusted to show that

$$s_k \to w_1 \text{ as } k \to \infty \tag{4.44}$$

in the vector norm, $\|s_k - w_1\| \to 0$. The general statement is that a subsequence of s_k converges to w_1 because a general scenario is that the eigenvalue largest in magnitude may be complex. Rotations upon application of A may occur, complicating discussion of convergence. This will not be an issue for our specific problem of interest, which involves Frobenius–Perron matrices and in particular stochastic matrices, since such matrices have a positive real eigenvalue, as we will expound upon below.

The three most common vector norms we will be interested in here are

1. $\|v\|_1 = \sum_{i=1}^n |v_i|$, the sum of the absolute values of the entries of a vector v,
2. $\|v\|_2 = \sqrt{\sum_{i=1}^n, v_i^2} = \sqrt{v'v}$, the Euclidean norm, and
3. $\|v\|_\infty = \max_{i=1,2,\ldots,n} |v_i|$, the infinity norm, also known as the max norm,

in terms of an $n \times 1$ column vector. We shall discuss the power method in terms of the Euclidean norm $\|\cdot\|_2$. Further, by similar arguments, the (left) spectral radius[43] follows from the Raleigh quotient

$$\frac{s_k A s_k'}{s_k s_k'} \to \tau_1 \text{ as } k \to \infty. \tag{4.45}$$

---

[42]The notation "$\gg$" denotes "exceedingly larger than," which can be defined formally by the statement $\tau_1^m \gg \tau_i^m \equiv \lim_{m\to\infty} \frac{\tau_i^m}{\tau_1^m} = 0$.

[43]The spectral radius is the largest eigenvalue in complex modulus.

Also from the discussion of the power method, we get an idea regarding the rate of convergence. Again we start with

$$\begin{aligned} wA &= c_1 A u_1 A + c_2 u_2 A + \cdots + c_n u_n A \\ &= c_1 \tau_1 \left(u_1 + \frac{c_2}{c_1}\frac{\tau_2}{\tau_1} u_2 + \cdots + \frac{c_n}{c_1}\frac{\tau_n}{\tau_1} u_n \right), \end{aligned} \tag{4.46}$$

from which

$$wA^n = c_1 \tau_1^n \left(u_1 + \frac{c_2}{c_1}\left[\frac{\tau_2}{\tau_1}\right]^n u_2 + o\left(\left[\frac{\tau_2}{\tau_1}\right]^n \right)\right). \tag{4.47}$$

Thus follows geometric convergence as,

$$r = \left|\frac{\lambda_2}{\lambda_1}\right|. \tag{4.48}$$

See a geometric presentation of the power method in Fig. 4.13. Really, the only part of the terms after the first two are that they form $\text{proj}_\perp[w, \text{span}(u_1, u_2)]$, where $\text{proj}_\perp[w, \text{span}(u_1, u_2)]$ denotes the orthogonal projection of w onto the subspace spanned by u_1 and u_2, meaning $w - c_1 u_1 - c_2 u_2$. The details of the multiplicities and degeneracies of that main subspace are irrelevant to our discussion since we are interested in Frobenius–Perron matrices. See a caricature representation of the power method in Fig. 4.13 and its spectrum in Fig. 4.14.

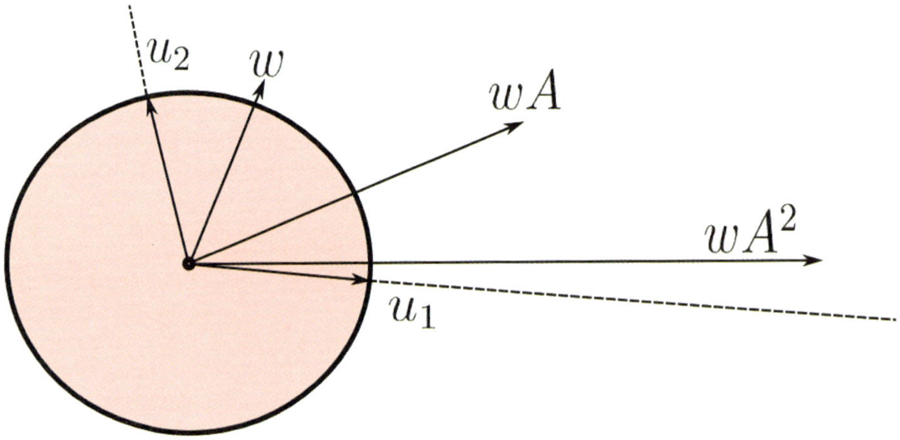

Figure 4.13. *Geometric representation of the power method representation of Eqs.* (4.46) *and* (4.47).

Note that usually this discussion of the power method is carried out regarding right eigenvectors in the numerical analysis literature, which we adapted here for our interest in row-stochastic matrices. In particular, in the case of a stochastic matrix when $\tau_1 = 1$, then the renormalization step will not be zero, and the convergence rate will be geometric with rate $r = |\lambda_2|$. While the expectation of the projection of Frobenius–Perron operators may be

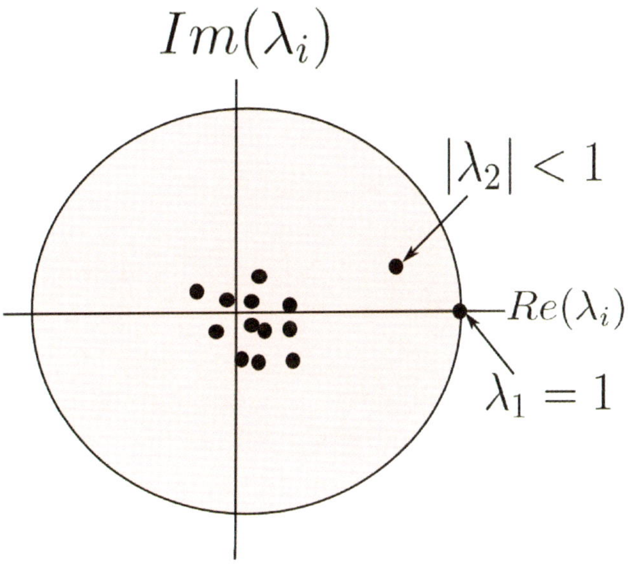

Figure 4.14. *Spectrum of the stochastic matrices in the complex plane. We expect the dominant eigenvalue to be* $\lambda_1 = 1$, *and the second eigenvalue,* $|\lambda_2| < \lambda_1$, *to describe the geometric convergence of the power method according to Eqs.* (4.47) *and* (4.48), $r = |\frac{\lambda_2}{\lambda_1}|$.

a row-stochastic matrix, with projection errors (due to truncation of the would-be infinite set of basis functions) and computational errors due to finite precision in the computers, typically the computation behaves as if $\tau \approx< 1$ as if there were a small mass leak. There will be more on this leak issue in Section 4.3.4 (see Eq. (4.58)).

4.3.3 Frobenius–Perron Theory of Nonnegative Matrices

There are several important and succinct statements that can be proven regarding stochastic matrices, since they are nonnegative. There are many useful statements which are part of the Frobenius–Perron theory, but we will highlight only a few here—those that are most relevant to the stochastic matrices which result from Ulam–Galerkin's projection of Frobenius–Perron operators. These matrices are nonnegative. As such, they have special properties that complement the description in the previous subsection, regarding the (non)possibility of multiplicity and complex value of the largest eigenvalue.

Where the previous section on the power method was meant to give the geometry of the action of an Ulam–Galerkin matrix, this section should be considered complementary since it is meant to give some algebraic information regarding the same action. We will start by stating a particular form of the Frobenius–Perron theorem, following the necessary definitions to interpret it, and then examples. We omit the proofs of this theorem since it is standard from matrix theory; refer to [13]. To interpret this theorem, we state the following definitions with brief examples.

Definition 4.11. An $n \times n$ square matrix $A_{n\times n}$ is called a **positive matrix** if $A_{n\times n} > 0$, meaning that each and every term in the matrix is nonnegative, $[A_{n\times n}]_{i,j} > 0$ for all i, j.

We present the definition in terms of square matrices since we are interested only in spectral properties. However, the theory of positive matrices is more broadly interesting.

Definition 4.12. An $n \times n$ square matrix $A_{n\times n}$ is called a **nonnegative matrix** if $A_{n\times n} \geq 0$, meaning that each and every term in the matrix is positive, $[A_{n\times n}]_{i,j} \geq 0$ for all i, j.

Example 4.7 (positive and nonnegative matrices). Inspecting

$$A = \begin{pmatrix} 1 & 2 & 3 \\ 4 & 5 & 6 \\ 7 & 8 & 9 \end{pmatrix}, B = \begin{pmatrix} 1 & 2 & 0 \\ 4 & 5 & 6 \\ 7 & 8 & 9 \end{pmatrix}, C = \begin{pmatrix} 1 & -1 & 3 \\ 4 & 5 & 6 \\ 7 & 8 & 9 \end{pmatrix}, D = \begin{pmatrix} 1 & 2 & 3 & 3.1 \\ 4 & 5 & 6 & 6.2 \\ 7 & 8 & 9 & 9.2 \end{pmatrix}, \tag{4.49}$$

A is a positive matrix, but B is nonnegative because of the single nonzero entry. C is neither because of the single negative entry, and D is not considered in the discussion of positivity since it is nonsquare and therefore irrelevant when discussing spectral properties.[44]

Remark 4.6. A nonnegative matrix has a unique positive and real dominant eigenvalue due to the Frobenius–Perron theory. Therefore, the power method proceeds as described in Eqs. (4.37)–(4.42) without the possible complications regarding multiplicities either geometric or algebraic when discussing convergence and convergence rate.

Remark 4.7. A stochastic matrix is nonnegative. We have already noted that $\tau_1 = 1$ is an eigenvalue since it corresponds to the definition that the matrix must row sum to 1. Further, any larger eigenvalue would correspond to a larger row sum, so it must be largest. Therefore, by the Frobenius–Perron theory, this is the unique largest eigenvalue of a stochastic matrix.

Theorem 4.1 (see [13]). If $A_{n\times n}$ is nonnegative and irreducible, then the following hold:

1. The spectrum of eigenvalues of $A_{n\times n}$ includes one real positive eigenvalue which is uniquely the largest in absolute value—meaning that it is the only eigenvalue on the outermost spectral circle. See Fig. 4.14.

2. The eigenvector corresponding to that eigenvalue has entries which are all positive real numbers.

3. Further, the largest eigenvalue λ is algebraically unique, meaning that it is a simple root of the characteristic polynomial, $\det(A - \lambda) = 0$.

4.3.4 Stochastic Matrices

It is easy to see that a stochastic matrix is a Frobenius–Perron matrix.

Definition 4.13. A (row) **stochastic matrix** is a square matrix $A_{n\times n}$ such that

1. each row sums to 1, $\sum_{j=1}^{n} A_{i,j} = 1$ for all $i = 1, 2, \ldots, n$, and

2. $0 \leq A_{i,j} \leq 1$ for all $i, j = 1, 2, \ldots, n$.

---

[44] D has no eigenvalues, but there is a related discussion of the singular spectrum through the SVD decomposition.

A convenient way to state the row-sums-to-1 property is through a matrix natural norm.

Definition 4.14. Given a square matrix $A_{n\times n}$, its **matrix natural norm** is defined as

$$\|A_{n\times n}\| = \sup_v \frac{\|Av\|}{\|v\|} = \sup_{w:\|w\|=1} \|Aw\|, \tag{4.50}$$

which is also often called an induced norm since the matrix norm is induced (inherited) by the vector norm $\|\cdot\| : E \to \mathbb{R}^+$, where E is the vector space domain of $A_{n\times n}$. Often $E = \mathbb{R}^n$.

In terms of the popular matrix norms listed in Section 4.3.2, the matrix natural norms are especially conveniently computed:

1. $\|A\|_1$ is the matrix 1-norm induced by the vector 1-norm, $\|A\|_1 = \max_{i=1,\dots,n} \sum_{j=1}^n |A_{i,j}|$, which is the maximum row sum.

2. $\|A\|_\infty$ is the matrix infinity-norm induced by the vector sup-norm, $\|A\|_\infty = \max_{j=1,\dots,n} \sum_{i=1}^n |A_{i,j}|$, which is the maximum column sum.

3. However, $\|A\|_2$ is conveniently but not as conveniently computed as $\|A\|_2 = \sqrt{\rho(A'A)}$, where $\rho(A'A)$ is the spectral radius of $A'A$.

In this notation, a stochastic matrix must satisfy the properties

$$\|A\|_1 = 1,\ 0 \le A_{i,j} \le 1\ \forall i, j. \tag{4.51}$$

The purpose of defining a stochastic matrix in such a manner is that it may be interpreted as reflecting the transition probabilities of a finite state Markov chain, which is a special case of a discrete time stochastic process.

Definition 4.15. A **discrete time stochastic process** is a sequence of random variables, $X_1, X_2, X_3, \dots$.

Definition 4.16. A **Markov chain** is a discrete time stochastic process of random variables, $X_1, X_2, X_3, \dots$ such that the conditional probability of each next state is independent of the prior history,

$$P(X_{m+1} = x | X_1 = x_1, X_2 = x_2, \dots, X_m = x_m) = P(X_{m+1} = x | X_m = x_m), \tag{4.52}$$

where $P(\cdot)$ denotes the probability of enunciated state of the random variable.

In each of these, we have referred to the concept of random variable; see Definition 3.3.

We may consider the random variable as a measurement device that returns a real number, or the random experiment in our language, for a given subset of Ω. Recall that for a measure space $(\Omega, \mathcal{F}, \mu)$, a measure μ is called a probability measure if $\mu : \mathcal{F} \to [0,1]$ and $\mu(\Omega) = 1$; hence a measure space $(\Omega, \mathcal{F}, \mu)$ will also be accordingly referred to as a probability space. With a probabilistic viewpoint in mind, the random variable tells us that the probability to observe a measurement outcome in some set $A \in \mathcal{B}(\mathbb{R})$ based on a probability measure μ is precisely $\mu(X^{-1}(A))$, which makes sense only if X is measurable.

When the state space is a finite set, $\Omega = \{x_1, x_2, \ldots, x_n\}$ or simply write $\Omega = \{1, 2, \ldots, n\}$, we have a **finite state Markov chain**, from which the set of all transition probabilities in Eq. (4.52) form a finite-sized table of possibilities, which may be recorded in a matrix:

$$P_{i,j} = P(X_{m+1} = x_j | X_m = x_i) \, \forall i, j \in 1, 2, \ldots, n. \tag{4.53}$$

To draw the connection to our larger theme, in some sense any Ulam–Galerkin method is simply an accounting of these probabilities for transitions between states identified by energy in each mode, as represented by a given chosen basis function. When a full (countable) basis set is used, then the resulting "matrix" would be infinite, but the truncation involved in Galerkin's method corresponds to ignoring the "less important states," to be discussed further shortly. Generally, this ignoring of less important states leads to a leak of measure in the states.

Further, evaluation of probability of events is a measure of the relative volume of a state in the set of outcomes. That is, given a probability space $(\Omega, \mathcal{F}, \mu)$, then formally the assignment of probability to an event $\omega \in \Omega$ corresponds to a measure which may be described by integration,

$$P(\cdot) : \Omega \to [0,1], \tag{4.54}$$

$$P(B) = \int_\Omega \chi_B(x) d\mu(x) = \int_B d\mu(x), \tag{4.55}$$

in terms of the indicator function $\chi_B(x)$ of a Borel set B. Here we follow a conventional abuse of notation so that $P(B)$ is technically equivalent to $P(\{\omega \in \Omega : X(\omega) \in B\})$, where in the above case the random variable X is χ_B.

That a finite state Markov chain results in a stochastic matrix is a direct consequence of independence,

$$P(X_{m+1} = x_j) = \sum_{i=1}^{n} P(X_{m+1} = x_j | X_m = x_i) P(x_i) = 1 \, \forall j \in 1, 2, \ldots, n, \tag{4.56}$$

simply because the union of all the states forms the full set of possibilities which is the full probability:

$$\Omega = \cup_{j=1}^{n} B_j. \tag{4.57}$$

Here is where we see the consequence of truncation in the Galerkin method in regard to leak of measure in terms of the probabilities. If Eq. (4.56) is a finite, and the true state space is infinite, but a small fraction of the states account for the majority of the probability, then we may write

$$
\begin{aligned}
P(x_j) &= \sum_{i=1}^{n} P(X_{m+1} = x_j | X_m = x_i) \\
&< \sim 1 = \sum_{i=1}^{\infty} P(X_{m+1} = x_j | X_m = x_i) \\
&= \sum_{i=1}^{n} P(X_{m+1} = x_j | X_m = x_i) \\
&\quad + \sum_{i=n}^{\infty} P(X_{m+1} = x_j | X_m = x_i),
\end{aligned} \tag{4.58}
$$

when

$$\sum_{i=n}^{\infty} P(X_{m+1} = x_j | X_m = x_i) \ll 1, \tag{4.59}$$

which will happen when

$$\cup_{j=1}^{n} B_j \text{ is a proper subset of } \Omega, \tag{4.60}$$

but measures most of the set. Formally, the way to extend the notions of finite state Markov chains even to uncountable state chains is through Harris chains [158]. We already noted in Section 4.2 how the Markov property of a map results in the Galerkin method yielding a finite truncation Ulam–Galerkin matrix, which we now describe as representing the transitions of a finite state Markov chain.

We now revert to a discussion of uniqueness of dominant states, in terms of matrix issues such as reducible matrices, primitive matrices, and the relationship to the dynamical systems analogue of the dominant state(s) to (unique) ergodicity.

4.4 Exact Representations Are Dense, and the Ulam–Galerkin Method

We close the chapter with presentations of some results from [19] of how Markov maps are dense. In this manner, analysis of approximation of the matrix approximations of the Frobenius–Perron operator can be developed. The error in an approximation of the infinite-dimensional operator can be described in two ways. The more common way is by discussion of an approximation due to finite truncation of the series in the Galerkin method, Eq. (4.4). A lesser known but also useful description of the same question is as follows. Each finite series approximation can be understood as an exact presentation of another map, a map which is Markov on exactly the finite partition used. As such, the approximation can be analyzed by a description of how far the approximating Markov map is from the non-Markov map under description. Finally, the expectation is that there is always a nearby Markov map for such a description. The concept we will discuss in this section is that Markov maps are expected to be dense in the space of maps, and therefore an error analysis of a finite truncation can be considered by the nearby map where the representation is exact. We can prove this statement at least for a certain class of systems. The results here are drawn from our work [19], where the fuller details of the brief presentation here can be found.

Consider a family of chaotic skew tent maps. The skew tent map is a two-parameter, piecewise linear, unimodal map of the interval. We show that these maps are Markov for a dense set of parameters and we find the exact probability density function (PDF) for any of these maps.

Remark 4.8. The central concept to this approach is the following: It is well known [50] that when a sequence of transformations has a uniform limit F and the corresponding sequence of invariant PDFs has a weak limit, then that invariant PDF must be F invariant.

As a case in point [19], we show for a family of skew tent maps that not only does a suitable sequence of convergent sequences exist, but also that these can be constructed entirely within the family of skew tent maps. Furthermore, such a sequence can be found among the set of Markov transformations, for which PDFs are easily and exactly calculated. While the theorems in the following sections assume the family of one-dimensional transformations, we will discuss extensions to higher dimensions at the end of the chapter.

Before we begin the main theme of this section, note that the discussion requires sharpening our analysis so that the underlying topology is clearer as to what we mean by dense as well as in what sense functions and invariant densities may converge.

4.4.1 Background Analysis and Topology

In discussing convergence, there is an implication of an underlying topology with respect to which the convergence may be defined. Analysis [271], topology [235], and functional analysis [191, 8] are the background fields here, and of course these are major areas of mathematics for which there are many excellent textbooks; we have just listed some of our favorites for each. For a quick start relevant to this discussion we can recommend [23]. This becomes an appropriate place to sketch in brief details the different kinds of convergence relevant here, for the sake of completeness of our discussion. We avoid an overly structured presentation of this would-be deep branch into extensive and interesting mathematics, which undoubtedly would require a large volume. For a complete introduction we recommend one of the cited texts. When discussing convergence of sequences of functions, a **uniform limit** is a "stronger" form of convergence than simply **convergence**, which is often used as the short form for the synonymous **pointwise convergence.** Therefore, we define both, and in terms of sequences of functions. Given a sequence of functions $\{f_n(x)\}_{n=0}^{\infty}$ for each $f_n : D \to R$ defined on a domain $x \in D$ and range R, f_n converges (pointwise) to f if each and every sequence of values $f_n(x) \in R$ (you get values by plugging x into each f_n) converges to a value which we shall label $f(x)$ one for each x. Stated formally, $y_n = f_n(x)$ is a sequence of numbers that converge $y_n \to y \equiv f(x)$ if for every $\epsilon > 0$ there exists $N(x) > 0$ such that $\|y - y_n\| < \epsilon$ if $n > N$, which is in terms of the metric topology of $\|\cdot\|$ on the range space $y_n \in R$. Note that as stated there is allowed an $\epsilon > 0$ for each x, and thus pointwise, $\epsilon(x)$ implies N is a function of x, $N(x) > 0$.

In terms of uniform convergence, the change relative to pointwise convergence is when we identify that there exists an $N > 0$ to discuss pointwise convergence, but that N may be found independently of x. N is not a function of x but still allows the convergence throughout the domain and range. A primary theorem regarding the notion of uniform convergence is that a sequence of continuous functions that converge uniformly must have a limit function which is also continuous. Pointwise limits of continuous functions need not be continuous functions, as exemplified by the common example

$$f_n(x) = x^n \text{ on } [0,1], \tag{4.61}$$

from which

$$f_n(x) \to f(x) \text{ if } (0, x < 1, \text{ else } 1) \text{ as } n \to \infty, \tag{4.62}$$

but only pointwise, and correspondingly $f(x)$ is not continuous, specifically at the point $x = 1$. However, the limit function $f(x)$ is continuous on a restricted domain $[0, L]$ for any $0 < L < 1$, and correspondingly the convergence $f_n(x) \to f(x)$ is uniform there.

A **weak limit** of a sequence of functions $f_n(x)$ is a separate notion of convergence of functions which is distinct from either pointwise convergence or uniform convergence. Written in terms of test functions $h(x)$, in a Hilbert space $\mathcal{H}(D)$ (defined as a complete inner product space),

$$f_n(x) \to^w f(x) \tag{4.63}$$

(note the superscript w), if

$$\langle f_n(x) - f(x), h(x) \rangle := \int_D |f_n(x) - f(x)| h(x) dx \to 0 \; \forall h(x) \in \mathcal{H}(D). \tag{4.64}$$

Since the convergence is in terms of integration, it is possible for a set of zero measure points not to converge despite weak convergence, but (strong) convergence implies convergence.

4.4.2 Example Tent Maps for Dense Markov Transformations

As a case in point, we present here a family of skew tent maps for which it can be explicitly proven that Markov maps are dense. Let $F_{a,b}$ denote the two-parameter, piecewise linear map on the interval $[0,1]$ satisfying

$$F_{a,b}(x) = \begin{cases} b + \frac{1-b}{a} x & \text{if } 0 \le x < a, \\ \frac{1-x}{1-a} & \text{if } a \le x \le 1, \end{cases} \tag{4.65}$$

with $0 < a < 1$ and $0 \le b \le 1$. It has been shown [14] that in the following region of the parameter space,

$$D = \{(a,b) : b < 1/(2-a) \text{ and } ((b < 1-a) \text{ or } (b \ge 1-a \text{ and } b > a))\}, \tag{4.66}$$

$F_{a,b}$ has chaotic dynamics, and in the parameter subset

$$D_0 = \{(a,b) : b < 1/(2-a) \text{ and } (b < 1-a)\}, \tag{4.67}$$

chaotic dynamics occur on the entire interval $0 \le x \le 1$. See Bassein [14] for a complete classification of the dynamics in parameter space $(a,b) \in (0,1) \times [0,1]$.

Specifically, $F_{a,b}$ is Markov for a dense set of (a,b) in D_0. It follows that the PDF for any one of the Markov transformations in this set can be found exactly to be a piecewise constant function in [50]. Therefore, these exact results approximate the PDF for any other transformation in D_0.

Sufficient conditions for a piecewise linear map on the interval to be Markov can be developed partly through symbolic dynamics. (See Chapter 6 for a discussion of symbolic dynamics.) Then proof that Markov maps are dense in the parameter set D_0 follows. PDFs computed by stochastic matrix methods quickly follow.

The form of $F_{a,b}$ from [14] allows a concise discussion of the full range of topological dynamics in the parameter space $(a,b) \in (0,1) \times [0,1]$. Piecewise linear transformations on the interval have been widely studied, and under different names: broken linear transformations [136] and weak unimodal maps [227]. These maps also allow explicit study of the chaotic behavior including the stability of the associated Frobenius–Perron operator [198], the topological entropy [228], and kneading sequences [61, 19].

4.4.3 Markov Skew Map Transformations

In the special but important case that a transformation of the interval are Markov (definition of Markov transformations are given in Section 4.4), the symbol dynamic is simply presented as a finite directed graph. The following result describes a set in D_0 for which $F_{a,b}$ is Markov.

Theorem 4.2 (see [19]). For a given $(a,b) \in D_0$, if $x_0 = 1$ is a member of a periodic orbit, then $F_{a,b}$ is Markov.

Proof. Set $F = F_{a,b}$. Assume $x_0 = 1$ is a member of a period n orbit $(n > 1)$. Next, form a partition of $[0,1]$ using the n members of the periodic orbit. The two endpoints of the interval are included since for all $(a,b) \in D_0$, $F(1) = 0$. Order these n points so $0 = c_0 < c_1 < \cdots < c_{n-1} = 1$, regardless of the iteration order. For $i = 1,\dots,n-1$, let $I_i = (c_{i-1}, c_i)$ and denote the restriction of F to I_i by F_i.

For a given $I_i = (c_{i-1}, c_i)$, the endpoints c_{i-1} and c_i will map exactly to two members of the partition endpoints by definition of the periodic orbit. Let these points be c_j and c_k, with $c_j < c_k$ and $j,k \in \{0,\dots,n-1\}$. The only turning point of the map is $x = a$, and for all $(a,b) \in D_0$, $F(a) = 1$. Therefore, $x = a$ must always be part of the period orbit and a member of the partition endpoints, implying each F_i is linear and, hence, a homeomorphism. Also $F_i(I_i) = (c_j, c_k)$, a connected union of intervals of the partition. By definition, F is Markov. □

At this point, symbolic notation becomes useful. In particular, some elements of kneading theory from the study of symbolic dynamics is useful. Details on kneading theory can be found in original form in [223] and nicely presented in [95]. The point $x = a$ is the critical point at the "center" of the interval, denoted by the letter C. All $a < x \leq 1$ is right of a, represented by R, and all $0 \leq x < a$ is left of a, represented by L. Represent each step of the iteration map by one of these three symbols. All parameter sets for which $F(x)$ is Markov must have a period-n orbit containing the point $x = a$ and be of the form $\{\overline{a,\dots,F^{n-1}(a)},\dots\}$. For example, the period-3 orbit has the form $\{\overline{a,1,0},\dots\}$ and occurs for any parameter set (a,b) on the curve $b = a$. It repeats the pattern CRL, which we call the kneading sequence $K(F_{a,b}) = (CRL)^\infty$.

If a periodic orbit contains the point $x = a$ and $a \neq b$, the point $x = b$ will either be greater or less than a. For a period-4 orbit with $a > b$, the symbolic sequence must repeat CRLL, or if $a < b$, CRLR. Therefore, a period 4 is found two ways.

Repeating this method for period 5 and higher, we see that there are 2^{n-3} possible combinations of the C-L-R sequences for a period-n orbit which includes the critical point.

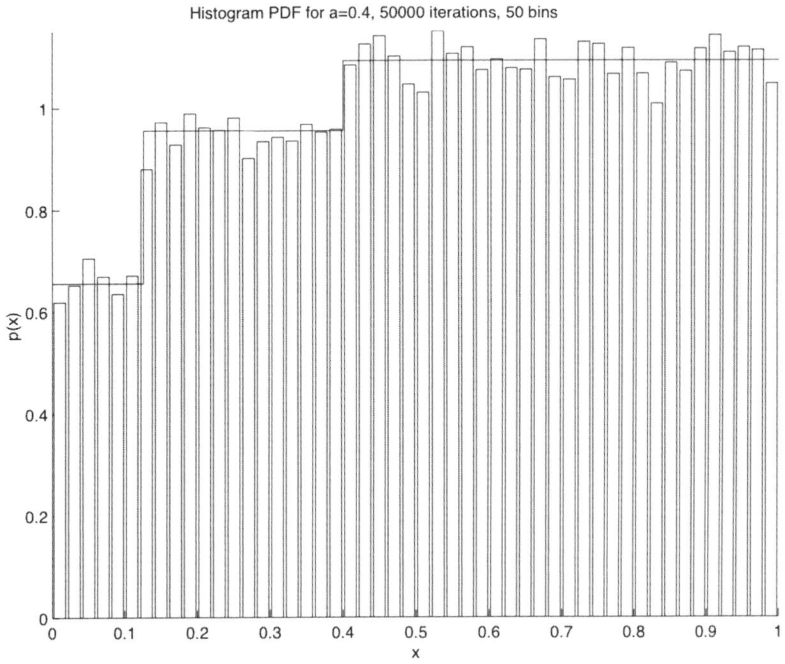

Figure 4.15. *PDF (histogram bars) compared to the exact solution (solid line). The calculation used* 50,000 *iterations and* 50 *intervals* [19].

The exponent $n-3$ reflects the necessary three-step prefix CRL. A full binary tree with 2^{n-3} leaves on each tier is possible, each implying a condition on the parameters and forming a countable set of curves in parameter space [228]. We can now restate Theorem 4.2 in terms of kneading sequences.

Corollary 4.1. If $K(F_{a,b})$ is periodic, then $F_{a,b}$ is Markov.

From this follows that the functions $F_{a,b}(x)$ (see (4.65)) which are Markov occur on a dense set of curves in the (a,b) parameter space D_0 (see (4.67)). Taking Σ_2 as the space of symbol sequences containing the full family of kneading sequences for two symbols and the kneading sequence $\sigma = \sigma_0\sigma_1\sigma_2\ldots$, with the metric $d(\sigma,\hat{\sigma}) = \sum_{i=0}^{\infty} \frac{|\sigma_i - \hat{\sigma}_i|}{2^i}$ and the norm $||\sigma|| = \sum_{i=0}^{\infty} \frac{\sigma_i}{2^i} = \sigma_0 + \frac{\sigma_1}{2} + \frac{\sigma_2}{2^2} + \cdots$, we have the following lemma.

Lemma 4.1. Periodic σ are dense in Σ_2.

Proof. For all $\epsilon > 0$ and given without loss of generality $\phi = \phi_0\phi_1\phi_2\ldots$, which is not periodic, there exists $N > 0$ large enough so that $\sigma_i = \phi_i$, $i \le N$, and $\sigma = (\phi_0\phi_1\ldots\phi_N)^{\infty}$ for $||\sigma - \phi||_{\Sigma_2} < \epsilon$. □

The result follows from combining Lemma 4.1 and Corollary 4.1 [19].

4.4.4 Probability Density Functions of Markov Maps Are Piecewise Constant

A useful property of Markov chaotic functions is that the PDFs can be determined quite easily. In fact, expanding piecewise linear Markov transformations have piecewise constant invariant PDF [19].

Theorem 4.3 (see [50, Theorem 9.4.4]). Let $\tau : I \to I$ be a piecewise linear Markov transformation such that for some $k \geq 1$,

$$|(\tau^k)'| > 1,$$

wherever the derivative exists. Then τ admits an invariant [probability] density function which is piecewise constant on the partition $\mathcal{P}$ on which τ is Markov.

4.4.5 Invariant Measures of Non-Markov Transformations as Weak Limits from Markov Maps

First we must prove that Markov maps are dense in D_0. Then, recalling results concerning weak limits of invariant measures of a sequence of transformations, we note that the Markov techniques can be applied, in a limiting sense, to describe statistical properties for all $F_{a,b}$ with $a, b \in D_0$. The map $f_{\lambda,\mu}$ studied in [228] is conjugate to the piecewise linear, interval map $F_{a,b}$ we are studying.

$$f_{\lambda,\mu}(x) = \begin{cases} 1+\lambda x & \text{if } x \leq 0, \\ 1-\mu x & \text{if } x \geq 0, \end{cases}$$

with $\lambda \leq 1$, $\mu > 1$, $0 < a < 1$, and $0 \leq b \leq 1$.

Using the conjugacy

$$\vartheta(x) = (x-a)/(1-a), \tag{4.68}$$

show that

$$\vartheta^{-1} \circ f_{\lambda,\mu} \circ \vartheta(x) = F_{a,b}(x). \tag{4.69}$$

Since $\vartheta(x)$ is a linear function of x, it is a homeomorphism and ϑ^{-1} is uniquely defined as $\vartheta^{-1}(x) = (1-a)x + a$:

$$\vartheta^{-1}(f(\vartheta(x))) = \begin{cases} (1-a\lambda)+\lambda x & \text{if } x \leq a, \\ (1+a\mu)-\mu x & \text{if } x \geq a. \end{cases} \tag{4.70}$$

Let $\lambda = (1-b)/a$ and $\mu = 1/(1-a)$. Then

$$1 - a\lambda = 1 - a\left(\frac{1-b}{a}\right) = b, \tag{4.71}$$

$$1 + a\mu = 1 - \mu\left(\frac{1}{1-a}\right) = \frac{1}{1-a}. \tag{4.72}$$

Therefore, $\vartheta^{-1} \circ f_{\lambda,\mu} \circ \vartheta(x) = F_{a,b}(x)$ by the homeomorphism ϑ and the two maps $F_{a,b}$ and $f_{\lambda,\mu}$ are topologically conjugate.

Call $\mathcal{M}$ the class of sequences $\mathbf{M}$ which occur as kneading sequences of $F_{a,\frac{1-2a}{1-a}}$ for $0 < a \leq 1/2$, also known as the primary sequences. Misiurewicz and Visinescu proved the following theorems.

Theorem 4.4 (see [228, Theorem A]). If (a,b), $(a',b') \in D$ and $(a,b) > (a',b')$, then $K(a,b) > K(a',b')$.

Theorem 4.5 (see [228, Theorem B]). If $(a,b) \in D$, then $K(a,b) \in \mathcal{M}$.

Theorem 4.6 (see [227], intermediate value theorem for kneading sequences). If a one-parameter family G_t of continuous unimodal maps depends continuously on t and $h(G_t) > 0$ for all t, then if $K(G_{t_0}) < K < K(G_{t_1})$ and $K \in \mathcal{M}$, there exists t between t_0 and t_1 with $K(G_t) = K$.

Using these previous results, we can prove the following.

Theorem 4.7. For all $(a,b) \in D_0$, one of the following is true:

1. $F_{a,b}$ is Markov.
2. There exists $(a^*,b^*)D_0$ such that F_{a^*,b^*} uniformly approximates $F_{a,b}$.

Proof. If $F_{a,b}$ is Markov, we are done. Otherwise, choose F_{a_0,b_0} such that $K(F_{a_0,b_0})$ is nonperiodic. Given a small $\epsilon > 0$, let $a_0 = a^*$ and $b_1 = b_0 + \epsilon$. By Theorem 4.4, $K(F_{a^*,b_0}) < K(F_{a^*,b_1})$; by Theorem 4.5, $K(F_{a^*,b_0})$, $K(F_{a^*,b_1}) \in \mathcal{M}$. Without loss of generality, we choose indices of b_0 and b_1 to create this ordering. Recall from Lemma 4.1 that periodic sequences are dense in Σ_2. Therefore, we may choose a sequence $M \in \mathcal{M}$ such that $K(F_{a^*,b_0}) < M < K(F_{a^*,b_1})$. Since $F_{a,b}$ does vary continuously with parameters a and b, then Theorem 4.6 implies an intermediate value b^* such that $b_0 < b^* < b_1$, and this intermediate map has the decimal kneading $K(F_{a^*,b^*}) = M$. Therefore, in any given neighborhood of a non-Markov map in D_0, there exists a Markov map M. □

Hence, we can construct a sequence, in D_0, of Markov maps that converges to any $F_{a,b}$ with $(a,b) \in D_0$. Considering Theorem 4.7, and the following result [50], we conclude that any transformation $F_{a,b}$ with $(a,b) \in D_0$ is either a member of the Markov set which we constructed, and the invariant density function can be calculated directly as described earlier in this paper, or if $F_{a,b}$ is not in that set, then a sequence of uniformly convergent Markov transformations, $F_{a_i,b_i} \to F_{a,b}$ and $(a_i,b_i) \to (a,b)$, each have easily calculated invariant densities which converge to the invariant density of $F_{a,b}$.

Define Q as the set $\{c_0, c_1, \ldots, c_{p-1}\}$ and $\mathcal{P}$ the partition of I into closed intervals with endpoints belonging to $Q: I_1 = [c_0, c_1], \ldots, I_{p-1} = [c_{p-2}, c_{p-1}]$ [19].

Theorem 4.8 (see [50, Theorem 10.3.2]). Let $f : I \to I$ be a piecewise expanding transformation, and let $\{f_n\}_{n\geq 1}$ be a family of Markov transformations associated with f. Note $Q^{(0)} = Q$, and

$$Q^{(k)} = \bigcup_{j=0}^{k} f^{-j}(Q^{(0)}), \quad k = 1, 2, \ldots.$$

Moreover, we assume that $f_n \to f$ uniformly on the set

$$I \setminus \bigcup_{k\geq 0} Q^{(k)}$$

and $f_n' \to f'$ in L^1 as $n \to \infty$. Any f_n has an invariant density h_n and $\{h_n\}_{n\geq 1}$ is a precompact set in L^1. We claim that any limit point of $\{h_n\}_{n\geq 1}$ is an invariant density of f.

4.4.6 Density of Markov Maps Beyond One Dimension

In the previous sections we have presented a rigorous treatment proving density, and that near the one-dimensional skew tent maps we can find a Markov map, and therefore a nearby map has an exact finite representation of the Frobenius–Perron operator as a stochastic matrix. Further, by the theorem summarized in Theorem 4.7, we know that the dominant eigenvectors of these resulting stochastic matrices can be used as a sequence of estimates to the invariant density, which converge weakly. While we cannot back the generalizations with theorem, intuition suggests that the idea is more general. For specific generalizations of piecewise affine maps of the plane, the proofs will generalize in a straightforward if tedious manner. Maps of the interval that are not piecewise linear can be continuously approximated by piecewise linear maps, from which a generalization can again be made. For several variable maps a similar strategy can be imagined, but we leave it as conjecture that the idea generalizes.

Chapter 5

Graph Partition Methods and Their Relationship to Transport in Dynamical Systems

5.1 Graphs and Partitions

A major theme throughout this book is that appropriately partitioning the phase space is a critical aspect of analyzing transport. The Ulam–Galerkin method discussed in Chapter 4 gives a way to represent the transfer operator of a dynamical system as a directed graph (either exactly in the case of a Markov representation in a sense of Section 4.3 or otherwise approximately in the case of a fine set of basis functions). Therefore, the topic of this chapter will be how we can analyze questions of transport in terms of transport on a graph. The major aspect of transport analysis, partitioning into invariant sets, if they exist, or almost-invariant sets allows us to borrow well-developed theories for partitioning graphs, as we will review here. We will include a discussion of strengths of each approach for our interest in the dynamical systems which they represent.

5.2 Weakly Transitive

A common scenario in advective turbulent dynamical systems is for the long time behavior to be transitive (see Definition 6.1), or correspondingly ergodic in terms of measurable dynamics, but the short time behavior may reveal only slow transport between almost invariant sets. Such behavior has been called **weakly transitive** [122, 125]. The following example is meant to emphasize almost-invariance and related graphs.

Example 5.1 (map with two invariant components). In Fig. 5.1, we see a one-dimensional dynamical system which has two invariant components that are the intervals labeled A and B. Using this partition, we get a representation by the directed graph also shown at the bottom of the figure. This graph is completely reducible, as in Eq. (4.27). Compare to the example in Example 3.9 studied for different tools.

By contrast, the dynamical system shown in Fig. 5.2 has two almost-invariant components, labeled A and B. In fact, the escape set from A to B after one iteration is the interval labeled y and colored yellow in a way which is representative of a cobweb diagram for the entire interval. That is,

$$e_A := \{x \in A : T(x) \in B\}, \tag{5.1}$$

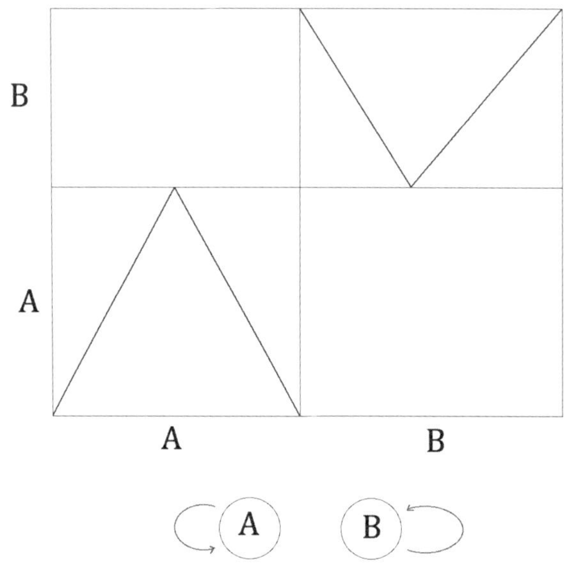

Figure 5.1. *A one-dimensional dynamical system with two invariant components. As such this dynamical system lacks transitivity. Each of the intervals, labeled A and B, is invariant under this mapping. (Bottom) A directed graph representation relative to the partition shown. The resulting graph is completely reducible. Compare to the form of the generating matrix* (4.27). *Contrast this example to that in Fig.* 5.2.

where T is the dynamical system. In the above example, we have $e_A = y$ and further colored is the preimage $T^{-1}(e_A)$, of e_A, which is the two thinner yellow subintervals shown, again in a manner suggestive of a cobweb diagram. As we usually denote here, T^{-1} for these one-dimensional maps denotes the preimages since the inverse does not exist; it has the two branches shown and is thus multivalued. Clearly, for any point $x \in T^{-n}(e_A)$, we have $T^{n+1}(x) \in B$. Note that it is not necessarily true that $T^{-n}(e_A) \subset A$ in general; that is, the orbit might escape from A to B and then return to A again before the nth iteration. Nevertheless, we are interested in defining a subset of A that escapes to B in terms of the preimage T^{-n}. To achieve this, we may define the (first) escape time $\xi(x)$ to be the smallest positive integer such that $T^{\xi}(x) \in B$. Then we could consider the set of points remaining in A up to the nth iteration of T, called A^n:

$$A^n = \{x \in A : \xi(x) > n\}. \tag{5.2}$$

So, the definition of A^n does not care about the fate of x after it escapes from A. It is easy to check that

$$A^n = T^{-n}(A) \cap T^{-n+1}(A) \cap \cdots \cap A. \tag{5.3}$$

In this notation, we can also define the set of points remaining in A indefinitely by

$$A \supset A^1 \supset \cdots \supset A^n, \qquad A^\infty := \bigcap_{n \geq 0} A^n. \tag{5.4}$$

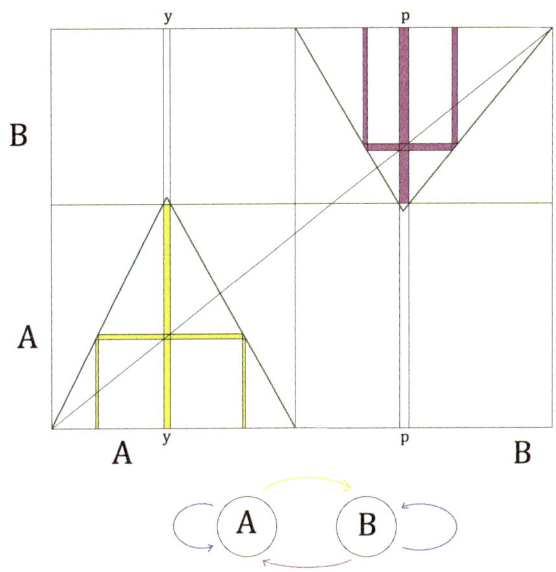

Figure 5.2. *A one-dimensional dynamical system with two almost-invariant components. Each of the intervals, labeled* A *and* B*, is almost-invariant under the mapping shown. The yellow region shown labeled* y *and its preimage(s) are the set of points which leave* A *in one and two iterates, respectively. Likewise,* p *is the subset of points which leave* B*. (Bottom) The directed graph representation includes the weak transport between* A *and* B *as the yellow and purple edges, respectively. Compare to the form of the generating matrix in* (5.9)*. Contrast this example to that in Fig.* 5.1.

Generally, A^∞ is a Cantor set of measure zero descriptive of the invariant set of A, but depending on the details of the map, there can be missing branches complicating its character [29]. The decay rate of a probability measure of A^n is of importance in understanding stability of the open systems with holes; see [90, 62] for a review. Such a quantity is referred to as the **escape rate** of a reference measure m from A and is given by

$$E_m(A) := -\lim_{n\to\infty} \frac{1}{n} \log(m(A^n)). \tag{5.5}$$

Clearly, the escape rate lies between 0 and ∞. For discussion here, the actions of the map on y and p are depicted in Fig. 5.2 (bottom) by the yellow and purple edges, respectively. The fact that y is (depicted to be) relatively small compared to A, stated in terms of Lebesgue measure $m(\cdot)$,

$$\frac{m(y)}{m(A)} \ll 1, \tag{5.6}$$

and likewise

$$\frac{m(p)}{m(B)} \ll 1, \tag{5.7}$$

denotes the weak transport descriptive of the almost-invariance of each of A and B.

As can be checked by Definition 4.1, $\{A, B\}$ is a Markov partition for Fig. 5.1, and as such the directed graph and corresponding matrix representations are exact in the sense of Section 4.3. Consequently the corresponding symbolic dynamics will be conjugate. However, by the same definition, $\{A, B\}$ is not a Markov partition for Fig. 5.2. To make a good approximating representation by a directed graph, a fine grid covering may be used. As previously discussed, this is what we call the Ulam–Galerkin method.

A dynamical system may be such that no partition exists that would reduce any graph representation to a completely reducible form. As shown in Figs. 5.3 and 5.4, and in Section 4.3.1, a graph may be such that it can be brought "nearly" to a reducible form, all except for "a few stray edges." A graph which is "almost reducible" is one which appears as shown in Fig. 5.4(b). The corresponding transition matrix is seen in Fig. 5.3(b). Comparing Figs. 5.3(a) and 5.4(a) to Definition 4.10 of completely reducible yields a form Eq. (4.27) by an appropriate sorting. This is another way of saying that there is a diagonalizing permutation similarity transformation, Eq. (4.26). The major problem here is that a graph in arbitrary configuration that may result from an arbitrary inspection[45] of the dynamical system cannot be expected to present the resulting graph approximation, and its corresponding transition matrices, in the convenient almost-block-diagonal form as illustrated in Figs. 5.3(b) and 5.4(b). Bringing the representation to such a convenient form is essentially equivalent to the problem of analyzing and understanding the transport mechanisms of the corresponding dynamical system.

We realize, therefore, the need for a notion of almost reducible.

Definition 5.1. A graph shall be called **almost reducible** if there exists a subset $I_o \subset I$ such that there are "relatively few edges" (i, j) for $i \in I_o$ and $j \in I \setminus I_o$.

The above condition of relatively few edges can be made more precisely in the following sense. In terms of the matrix representation of a graph by its transition matrix A, we can sharpen this notion by stating that this definition is equivalent to the statement that an almost-reducible matrix A must have a permutation matrix P such that

$$R = P^{-1} A P, \tag{5.8}$$

into a matrix R which is of the form

$$R = \begin{pmatrix} R_{1,1} & E_{1,2} \\ E_{2,1} & R_{2,2} \end{pmatrix}, \tag{5.9}$$

where $E_{1,2}$ and $E_{2,1}$ are relatively "small matrices," where small is to be measured in terms of the Frobenius norm. Recall that the Frobenius norm is a matrix norm, which for a matrix B is defined as [140]

$$\|B\|_F = \sqrt{Tr(B \cdot B^T)} = \sqrt{\sum_{i=1}^{n}\sum_{j=1}^{n} |b_{i,j}|^2}. \tag{5.10}$$

[45]We use the phrase "inspection" of a dynamical system as the choice and ordering of basis functions used in sampling the action of the dynamical system through the Ulam–Galerkin method; see (4.2)–(4.4). Notice that even a simple permutation of the basis functions (4.2) could easily change a configuration such as that in Fig. 5.4(b) in the **dark star** appearance in Fig. 5.4(a). In fact, a random embedding of a graph of any large number of vertices, say larger than $N = 50$, from basis functions in arbitrary configuration is expected to look all mixed up, like the dark star in Fig. 5.4(a). In this type of Ulam–Galerkin graph analysis, we will often work with N on the scale of many thousands or more.

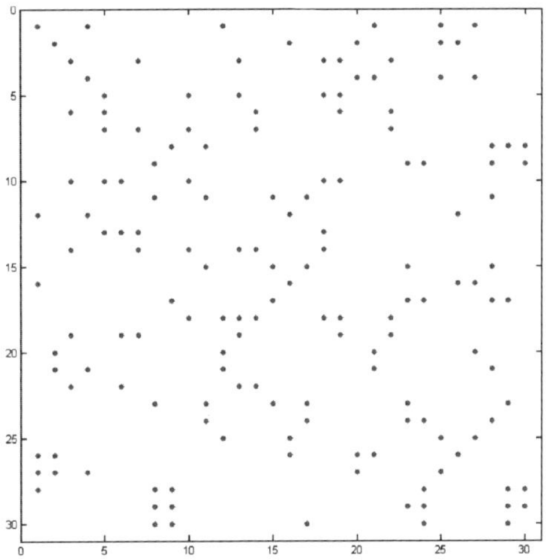

(a) A transition matrix before sorting.

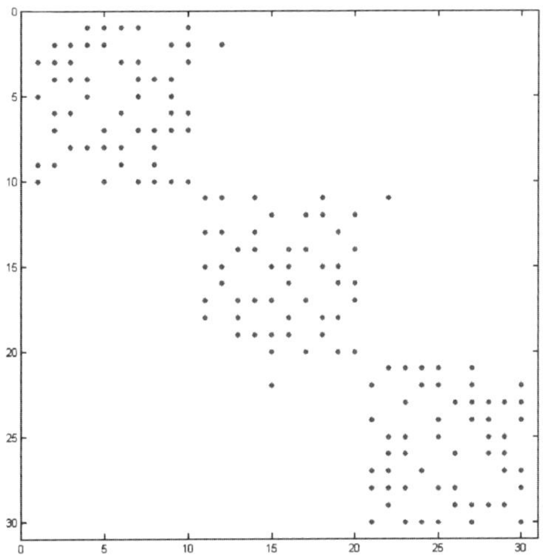

(b) The same transition matrix after sorting.

Figure 5.3. *The matrix in this figure is irreducible. Members of the original transition matrix before sorting are generally expected to be placed haphazardly and the structure is not obvious. After using the modularity method discussed in Section* 5.10.2, *the "almost" block-diagonal form is revealed. In this case, the sorted matrix indicates the existence of three communities and the few off-block-diagonal elements correspond to intrabasin diffusion.*

Therefore, to make precise the notion of relatively small we mean that $\|E_{1,2}\|_F$ and $\|E_{2,1}\|_F$ are small relative to $\|R_{1,2}\|_F$ and $\|R_{1,2}\|_F$.

We summarize this introductory section and lead into the algorithmic work of the chapter with the statement that the problem of partition is central to the study of transport mechanism, flux, and information flow in dynamical systems. Partitioning is squarely a

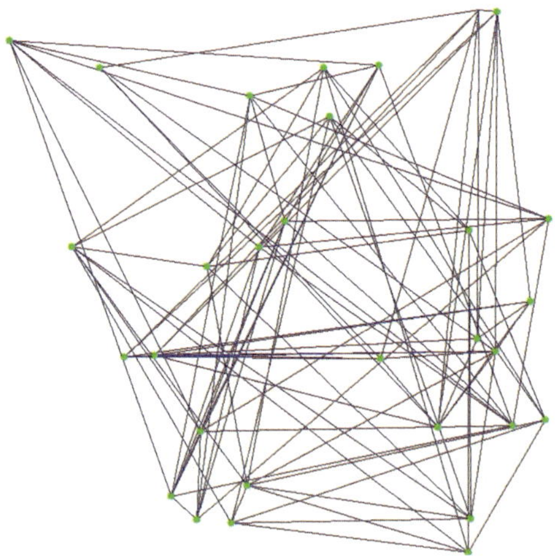

(a) The original graph. Embedding vertices into the plane in a random configuration conceals simple structure.

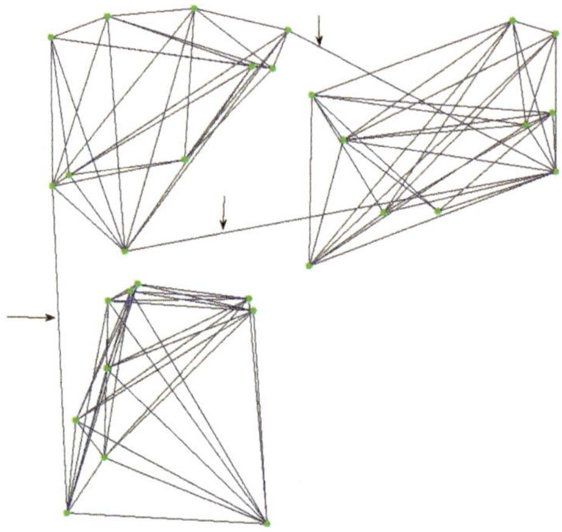

(b) Sorted graph. The same graph subject to appropriate permutation reveals "almost" block-diagonal structure community between components.

Figure 5.4. *An undirected graph before and after sorting. The arrows point at the intercommunity edges.*

problem of associating sets together, and as such, the matrix representation amounts to defining appropriate permutations.

5.3 Partition by Signs of the Second Eigenvector

There is a pleasingly simple but powerful method to partition an Ulam–Galerkin graph from a chaotic dynamical system to associate partition elements which are strongly transitive from other regions to which they may be only weakly transitive. Following the work of [91, 94] for Markov chains, and similarly [125] for stochastic matrices from Frobenius–Perron operators, we can inspect and threshold the eigenfunctions and corresponding discrete approximation of eigenvectors associated with secondary (not the largest) eigenvalues. The discussion here also has motivations from the early work with these methods in [93, 91].

We present here a simplified discussion of the thresholding concept in terms of a completely reducible stochastic system. Assume a stochastic matrix

$$P = \begin{pmatrix} P_{1,1} & 0 \\ 0 & P_{2,2} \end{pmatrix}. \tag{5.11}$$

As such, it can be confirmed that the submatrices $P_{1,1}$ and $P_{2,2}$ must each be stochastic, albeit smaller of size $N_1 \times N_1$ and $N_2 \times N_2$, $N = N_1 + N_2$, where P is $N \times N$. Then each of $P_{1,1}$ and $P_{2,2}$ have dominant eigenvalues $\lambda_1 = 1$ and $\lambda_2 = 1$, respectively, and for the sake of presentation simplicity, we assume that each matrix is algebraically and geometrically simple; it is sufficient to assume that they are each irreducible. Thus, we write

$$v_1 P_1 = v_1 \quad \text{and} \quad v_2 P_2 = v_2. \tag{5.12}$$

Therefore, we have the following lemma.

Lemma 5.1. A stochastic matrix P of the reducible form (5.11) with irreducible components $P_{1,1}$ and $P_{2,2}$ has a multiplicity-2 dominant eigenvalue, $\lambda = 1$, and corresponding eigenvector spanning a two-dimensional linear space, with eigenvector equations

$$\begin{pmatrix} v_1 \\ 0 \end{pmatrix} \cdot P = \begin{pmatrix} v_1 \\ 0 \end{pmatrix} \quad \text{and} \quad \begin{pmatrix} 0 \\ v_2 \end{pmatrix} \cdot P = \begin{pmatrix} 0 \\ v_2 \end{pmatrix}. \tag{5.13}$$

The proof of this statement is immediate, keeping in mind the appropriate size of the 0 vectors are $N_2 \times 1$ and $N_1 \times 1$, respectively.

In the span of $\{\binom{v_1}{0}, \binom{0}{v_2}\}$ is included the vector $w = \begin{pmatrix} |v_1| \\ -|v_2| \end{pmatrix}$. Note that Eq. (5.13) allows us to write

$$\left(\begin{pmatrix} |v_1| \\ 0 \end{pmatrix} - \begin{pmatrix} 0 \\ |v_2| \end{pmatrix} \right) \cdot S = \begin{pmatrix} |v_1| \\ -|v_2| \end{pmatrix} \cdot S = \begin{pmatrix} |v_1| \\ -|v_2| \end{pmatrix}. \tag{5.14}$$

Regarding the question then of a given stochastic matrix $\hat{P}$ which may not be in block-diagonal form, given a permutation,[46] and similarity transformation,[47]

$$\hat{P} = S^T \cdot P \cdot S, \tag{5.15}$$

that brings it there, we can make the following statement regarding signs and a bipartition.

[46] A permutation S is a square matrix that is simply a resorting of the identity matrix. An elementary permutation $S_{[i,j]}$ starts with the identity I, and we exchange the ith and jth rows. Therefore, applied as a similarity transformation, $S^T_{[i,j]} \cdot P \cdot S_{[i,j]}$ results in trading the positions of the ith and jth rows *and* columns of S. More complicated permutations can be formed as products (compositions) of such operations. $S = \Pi_{i,j} P_{[k_i,k_j]}$.

[47] Similarity transformations are especially relevant here since spectral properties are preserved.

Proposition 5.1. Suppose $\hat{P}$ is a stochastic matrix that is a permutation of a stochastic matrix P, $\hat{P} = S^T \cdot P \cdot S$, and P has components of the form Eq. (5.11). Then a dominant eigenvector v of $\hat{P}$ corresponding to $\lambda = 1$, which has opposite signs in its components, can be used to formulate a permutation $\hat{S}$ which diagonalizes $\hat{P}$. Let

$$s = sort(v),^{48} \tag{5.16}$$

and let

$$\hat{S} = perm(s) \tag{5.17}$$

be a corresponding permutation matrix[49]; then $\hat{S}^T \hat{P} \hat{S}$ has a block-diagonal form.

This proposition describes the situation of a completely reducible system. The following discussion is motivated in part from [93, 91] in what was called a "Gedanken-experiment" for the basic idea that the matrix representations of almost-invariant sets can be understood as perturbations of invariant sets. What if we perturb this completely reducible paradigm slightly? In considering

$$P \mapsto P + E, \tag{5.18}$$

as in comparing Eq. (5.11) to Eq. (5.9), we consider continuity results together with simple linear algebraic considerations. The weakly transitive case Eq. (5.9) of a stochastic matrix R and its partitioning problem can be well understood in terms of Proposition 5.1 and notions of continuity.

Theorem 5.1 (continuity eigenvalues). The eigenvalues of a matrix P are continuous with respect to variations of its matrix entries $a_{i,j}$.

The proof of this statement is standard, as can be found in [140], and is easily understood in terms of the statement that eigenvalues are roots of the polynomial $\det(P - \lambda I) = 0$, and these are continuous with respect to the coefficients. Alternatively and more typically, it can be understood in terms of the Gersgorin disc theory.

As for our problem (5.9), continuity results in the following statement.

Corollary 5.1. If a block-diagonalizable stochastic matrix is stochastically perturbed,[50]

$$P = \begin{pmatrix} P_{1,1} & 0 \\ 0 & P_{2,2} \end{pmatrix} \mapsto \overline{P} = \begin{pmatrix} P_{1,1} + \epsilon E_{1,1} & \epsilon E_{1,2} \\ \epsilon E_{2,1} & P_{2,2} + \epsilon E_{2,2} \end{pmatrix} \tag{5.19}$$

is continuous with respect to ϵ, or alternatively stated, in terms of each $\|\epsilon E_{i,j}\|_F$. Furthermore, since $\overline{P}$ is also assumed stochastic, ϵ: $0 < \epsilon \ll 1 \implies 0 \leq \lambda_1 - \lambda_2 \ll 1$.

---

[48]Let $sort(v)$ be the vector of reordered elements of v from largest to smallest, and let $sort(v)$ be the vector of integers whose ith element records the original position of the ith element of $sort(v)$. For example, if $v = \langle 1.6, 3.2, -2.2 \rangle$, then $sort(v) = \langle 3.2, 1.6, -2.2 \rangle$ and $sort(v) = \langle 2, 1, 3 \rangle$.

[49]If $s = \langle 2, 1, 3 \rangle$, then $S = perm(s) = [0,1,0;1,0,0;0,0,1]$ is a row restructuring of the 3×3 identity matrix, and the similarity transformation exchanges rows and columns of other matrices as specified.

[50]To remain stochastic, either $\overline{P}$ must be column rescaled or the matrices $\epsilon E_{1,2}$ and $\epsilon E_{2,1}$ must include the appropriate negative entries so that any variations to the total matrix S will be columnwise zero-sum changes.

By assumption that the perturbed matrix $\overline{P}$ remains stochastic, the statement is that the dominant eigenvalue remains $\lambda_1 = 1$. However, while the second eigenvalue of P may also be 1, generally the second eigenvalue of $\overline{P}$ may deviate from 1. Nonetheless, by continuity it will remain close to 1.

Example 5.2. Consider the matrix

$$P = \begin{pmatrix} 1 & 0 \\ 0 & 1 \end{pmatrix}. \tag{5.20}$$

Then $\lambda_1 = \lambda_2 = 1$. The eigenspace of P is any spanning set of $\mathbb{R}^2$. It is convenient to use the orthonormal basis, $\{\binom{1}{0}, \binom{0}{1}\}$, but it is just as suitable to use $\{\binom{1}{0}, \binom{1}{1}\}$ as the eigenvectors of 1. A stochastically perturbed matrix may be chosen, for example,

$$\overline{P} = \begin{pmatrix} 1+\epsilon & -\epsilon \\ -\epsilon & 1+\epsilon \end{pmatrix}, \tag{5.21}$$

where $-1 < \epsilon < 0$ to remain stochastic. As usual, $\overline{P}$ diagonalizes by a similarity transformation in terms of the matrix whose columns are the (new) eigenvectors,

$$\begin{pmatrix} 1+\epsilon & -\epsilon \\ -\epsilon & 1+\epsilon \end{pmatrix} = \begin{pmatrix} \frac{1}{\sqrt{2}} & \frac{1}{\sqrt{2}} \\ \frac{1}{\sqrt{2}} & -\frac{1}{\sqrt{2}} \end{pmatrix} \cdot \begin{pmatrix} 1+2\epsilon & 0 \\ 0 & 1 \end{pmatrix} \cdot \begin{pmatrix} \frac{1}{\sqrt{2}} & \frac{1}{\sqrt{2}} \\ \frac{1}{\sqrt{2}} & -\frac{1}{\sqrt{2}} \end{pmatrix}. \tag{5.22}$$

Notice that the diagonalizing similarity transformation is special,

$$P = VDV^{-1} = VDV^t, \tag{5.23}$$

since V, whose columns are the eigenvectors, happens to be a unitary matrix.[51]

There are certain points which may be interpreted from this example,

Remark 5.1. The eigenvalues are continuous with respect to variations in the matrix entries, as is promised by the general theory. We see specifically in this example that $\|P - \overline{P}\|_F = \epsilon$. Correspondingly,

$$\lambda_1(P) = \lambda_1(\overline{P}) = 1, \tag{5.24}$$

but

$$|\lambda_2(P) - \lambda_2(\overline{P})| = 2\epsilon. \tag{5.25}$$

However, the general theory only promises that eigenvectors must vary continuously if the eigenvalues are bounded away from a nontrivial multiplicity. In this case, since P has a $\lambda_1 = \lambda_2 = 1$ multiplicity-2, eigenvectors are not guaranteed to vary continuously. We see that while

$$\overline{v_1} = \begin{pmatrix} 1 \\ 1 \end{pmatrix}, \quad \overline{v_2} = \begin{pmatrix} 1 \\ -1 \end{pmatrix} \tag{5.26}$$

are eigenvectors of $\overline{P}$ (and also of P), these eigenvectors are not close to those eigenvectors of P,

$$v_1 = \begin{pmatrix} 1 \\ 0 \end{pmatrix}, \quad v_2 = \begin{pmatrix} 0 \\ 1 \end{pmatrix}, \tag{5.27}$$

which are (also) eigenvectors of P. This is the "wobble" problem [140] of lack of continuity associated with geometrically nonsimple eigenspaces due to geometric multiplicity.

[51] A matrix is called unitary if $V^t V = V V^t = I$.

Remark 5.2. The second eigenvalue of $\overline{P}$ is $\lambda_2 = 1+2\epsilon$, and its corresponding eigenvector (5.27), $v_2 = \begin{pmatrix} 1 \\ -1 \end{pmatrix}$, is notable in that the opposite signs are indicative of the partitions in more complicated almost-reducible problems. The important part of v_2 can be interpreted as,

$$Signs(v_2) = \begin{pmatrix} + \\ - \end{pmatrix}. \tag{5.28}$$

The generalized statement of the signs of the second eigenvector can be used to reveal almost-invariance, and associate the appropriate partition even if the partition is not obvious, as the vector is as good as scrambled with obvious pattern. The following is a variation on the statements in [91, 94].

Proposition 5.2. The signs vector of the second eigenvector,

$$Signs(\overline{v_2}) = \begin{pmatrix} + \\ + \\ + \\ .. \\ + \\ - \\ - \\ - \\ \dots \\ - \end{pmatrix}, \tag{5.29}$$

with m plus signs, and n minus signs, is associated with

$$\overline{P} = \begin{pmatrix} P_{1,1} + \epsilon E_{1,1} & \epsilon E_{1,2} \\ \epsilon E_{2,1} & P_{2,2} + \epsilon E_{2,2} \end{pmatrix} \tag{5.30}$$

from Eq. (5.19) in Corollary 5.1, when $P_{1,1}$ is $m \times m$ and $P_{2,2}$ is $n \times n$ and $\epsilon \ll 1$.

Now this proposition can be used to uncover a permutation which can be used to reveal a form $\overline{P}$ in Eq. (5.30). In other words, suppose P is a matrix presented in arbitrary configuration. That means that while we do not have the nice almost-block-diagonal form of $\overline{P}$, there exists a permutation matrix S which rearranges the rows and columns:

$$\overline{P} = SPS^t. \tag{5.31}$$

The difficulty of the problem is that when presented with a P but without an S, we need a method for finding whether a rearrangement to an almost-block-diagonal form $\overline{P}$ exists; the essence of the partition problem then is to construct a permutation S[52] if it exists.[53] The following constructs a useful partition, as a consequence of Proposition 5.2.

Corollary 5.2. Let

$$[Y, J] = Sort(Signs(v_2)),^{54} \tag{5.32}$$

[52]Such a partition is not unique in that elements in a component may be permuted.

[53]Note that such a P is not unique.

[54]We interpret the *Sort* function similarly to the *sort* function stated in Eq. (5.16). While *sort* results in a vector of entries listed from largest to smallest, we define *Sort* to have an output whose values in Y are those from *sort* listed from largest to smallest, and J is the corresponding rearranged indices.

where v_2 is the second eigenvector of a stochastic matrix P in arbitrary configuration. Then if there exists a permutation S such that P has an almost-block-diagonal form, (5.30), where $\epsilon \ll 1$, then a suitable permutation matrix may be formed by rearranging the rows of the identity matrix I according to the sort-indices J from Eq. (5.32),

$$S = I_{J,:}, \tag{5.33}$$

and this permutation can unfold the partition,

$$\overline{P} = SPS^t. \tag{5.34}$$

In summary, the statement is simple. While the first eigenvector may encode the steady state of the stochastic process, the second eigenvector reveals a partition associated with almost-invariance. If a block-diagonal form of the stochastic matrix in general configuration exists, the sort on the second eigenvector associates the partition which reveals the almost-invariant sets.

Example 5.3. Let

$$P = \begin{pmatrix} 0.1430 & 0.4293 & 0.7359 & 0 & 0 & 0 & 0 \\ 0.3007 & 0.4084 & 0.1111 & 0 & 0 & 0 & 0 \\ 0.5563 & 0.1623 & 0.1530 & 0 & 0 & 0 & 0 \\ 0 & 0 & 0 & 0.0886 & 0.3202 & 0.2036 & 0.3045 \\ 0 & 0 & 0 & 0.4557 & 0.0916 & 0.3345 & 0.3270 \\ 0 & 0 & 0 & 0.3333 & 0.0327 & 0.2510 & 0.3684 \\ 0 & 0 & 0 & 0.1224 & 0.5555 & 0.2109 & 0.0001 \end{pmatrix} \tag{5.35}$$

in the form of Eq. (5.19), and likewise let

$$\overline{P} = \begin{pmatrix} 0.1428 & 0.4290 & 0.7359 & 0 & 0 & 0 & 0.0001 \\ 0.3007 & 0.4084 & 0.1110 & 0.0001 & 0.0001 & 0.0001 & 0 \\ 0.5563 & 0.1623 & 0.1530 & 0 & 0 & 0 & 0 \\ 0 & 0 & 0 & 0.0886 & 0.3201 & 0.2035 & 0.3044 \\ 0.0001 & 0 & 0 & 0.4556 & 0.0916 & 0.3345 & 0.3270 \\ 0 & 0.0002 & 0 & 0.3333 & 0.0327 & 0.2510 & 0.3684 \\ 0.0001 & 0.0001 & 0.0001 & 0.1224 & 0.5555 & 0.2109 & 0.0001 \end{pmatrix} \tag{5.36}$$

also in the form of Eq. (5.19), the stochastically perturbed block almost-diagonal variation and scaled to remain stochastic. Then corresponding to P is the spectrum,

$$\begin{aligned} &eigs(P) \\ &\quad = \{1.0000, 1.0000, -0.5219, -0.2786 - 0.1294i, -0.2786 + 0.1294i, 0.2263, -0.0115\}, \end{aligned} \tag{5.37}$$

in which we see the multiplicity-2, $\lambda_1 = 1$, which agrees with the prediction of Eq. (5.24). The spectrum of the perturbed matrix $\overline{P}$ is close to that of P, but the repeated eigenvalues have separated, as promised by Corollary 5.1, with one of them remaining $\lambda_1 = 1$:

$$\begin{aligned} &eigs(\overline{P}) \\ &\quad = \{1.0000, 0.9997, -0.5220, -0.2786 - 0.1293i, -0.2786 + 0.1293i, 0.2264, -0.0115\}. \end{aligned} \tag{5.38}$$

The second eigenvector, $\overline{v_2}$ of $\overline{P}$, has signs which reveal with the block partitioning of $\overline{P}$,

$$\overline{v}_1 = \begin{pmatrix} -0.3596 \\ -0.2358 \\ -0.2814 \\ -0.4008 \\ -0.4963 \\ -0.4015 \\ -0.4096 \end{pmatrix}, \quad \overline{v}_2 = \begin{pmatrix} 0.5315 \\ 0.3482 \\ 0.4159 \\ -0.3040 \\ -0.3765 \\ -0.3045 \\ -0.3106 \end{pmatrix}. \tag{5.39}$$

Again we recall that the scale and sign of a given vector are not important since a scaled factor av_i of an eigenvector v_i is also an eigenvector. Many software packages normalize the vectors before they are presented, but they may be given as all negative values, for example, as we see v_1 here. To interpret these as probabilities, we would need to make sure to use it with positive entries and normalized.

If we hide the block almost-diagonal structure by scrambling $\overline{P}$,

$$S\overline{P}S' = \begin{pmatrix} 0.0886 & 0 & 0.3201 & 0.2035 & 0.3044 & 0 & 0 \\ 0 & 0.1428 & 0 & 0 & 0.0001 & 0.4290 & 0.7359 \\ 0.4556 & 0.0001 & 0.0916 & 0.3345 & 0.3270 & 0 & 0 \\ 0.3333 & 0 & 0.0327 & 0.2510 & 0.3684 & 0.0002 & 0 \\ 0.1224 & 0.0001 & 0.5555 & 0.2109 & 0.0001 & 0.0001 & 0.0001 \\ 0.0001 & 0.3007 & 0.0001 & 0.0001 & 0 & 0.4084 & 0.1110 \\ 0 & 0.5563 & 0 & 0 & 0 & 0.1623 & 0.1530 \end{pmatrix}, \tag{5.40}$$

as would be expected in a general configuration, in this case due to a similarity transformation by the permutation matrix

$$S = \begin{pmatrix} 0 & 0 & 0 & 1 & 0 & 0 & 0 \\ 1 & 0 & 0 & 0 & 0 & 0 & 0 \\ 0 & 0 & 0 & 0 & 1 & 0 & 0 \\ 0 & 0 & 0 & 0 & 0 & 1 & 0 \\ 0 & 0 & 0 & 0 & 0 & 0 & 1 \\ 0 & 1 & 0 & 0 & 0 & 0 & 0 \\ 0 & 0 & 1 & 0 & 0 & 0 & 0 \end{pmatrix}, \tag{5.41}$$

then the second eigenvector can still be used to reveal the structure. S here is a permutation of the rows of the identity matrix. The second eigenvector of the scrambled matrix, $S\overline{P}S'$, is

$$\overline{v_2} = \begin{pmatrix} -0.3040 \\ 0.5315 \\ -0.3765 \\ -0.3045 \\ -0.3106 \\ 0.3482 \\ 0.4159 \end{pmatrix}, \quad \text{sign}(\overline{v}_2) = \begin{pmatrix} -1 \\ 1 \\ -1 \\ -1 \\ -1 \\ 1 \\ 1 \end{pmatrix}. \tag{5.42}$$

Using Eq. (5.32), $[Y, J] = Sort(Signs(v_2))$, yields

$$J = \begin{pmatrix} 1 \\ 3 \\ 4 \\ 5 \\ 2 \\ 6 \\ 7 \end{pmatrix}. \tag{5.43}$$

Developing a new permutation matrix $\tilde{S}$ from this permutation as a permutation,

$$\tilde{S} = \begin{pmatrix} 1 & 0 & 0 & 0 & 0 & 0 & 0 \\ 0 & 0 & 1 & 0 & 0 & 0 & 0 \\ 0 & 0 & 0 & 1 & 0 & 0 & 0 \\ 0 & 0 & 0 & 0 & 1 & 0 & 0 \\ 0 & 1 & 0 & 0 & 0 & 0 & 0 \\ 0 & 0 & 0 & 0 & 0 & 1 & 0 \\ 0 & 0 & 0 & 0 & 0 & 0 & 1 \end{pmatrix}, \tag{5.44}$$

unscrambles the matrix in Eq. (5.40) and yields

$$\tilde{S}(S\overline{P}S')\tilde{S}' = \begin{pmatrix} 0.0886 & 0.3201 & 0.2035 & 0.3044 & 0 & 0 & 0 \\ 0.4556 & 0.0916 & 0.3345 & 0.3270 & 0.0001 & 0 & 0 \\ 0.3333 & 0.0327 & 0.2510 & 0.3684 & 0 & 0.0002 & 0 \\ 0.1224 & 0.5555 & 0.2109 & 0.0001 & 0.0001 & 0.0001 & 0.0001 \\ 0 & 0 & 0 & 0.0001 & 0.1428 & 0.4290 & 0.7359 \\ 0.0001 & 0.0001 & 0.0001 & 0 & 0.3007 & 0.4084 & 0.1110 \\ 0 & 0 & 0 & 0 & 0.5563 & 0.1623 & 0.1530 \end{pmatrix}, \tag{5.45}$$

which is again in a block almost-diagonal form, albeit not identically to the original $\overline{P}$, (5.36), due to nonuniqueness of almost diagonalizing permutations.

Example 5.4 (graphical presentation of almost-block-diagonal decomposition). The matrix shown as a spy plot in Fig. 5.3(a) (top) in an arbitrary configuration, and an arbitrary embedding of the corresponding graph in the plane, Fig. 5.4(a), both belie whether or not there may be any useful structure. The second eigenvector signs method of this section can be used to reveal the structure. The structure becomes obvious in the permuted version of the same matrix, Fig. 5.3(b), since it exists. Likewise, an appropriate embedding in the plane reveals the same fact; see Fig. 5.4(b).

While we have specialized the discussion of partitioning methods by inspection of the second eigenvector to stochastic matrices, as did [122, 125], the work in [91, 94] is more general in its application to transition matrices and Markov models. See also [116] for discussion of an SVD algorithm to almost-block-diagonal matrices, and see Section 5.7 for discussion of an SVD connection.

5.4 Graph Laplacian and Almost-Invariance

It has been previously demonstrated that the sign structure of the second eigenvector of the transition matrix suggests a bipartition of a graph. In this section, we will relate the optimal partitioning by the eigenvectors of the Laplacian matrix with the (measure-theoretic) almost-invariance.

We will begin by assuming that the transition matrix $P \in \mathbb{R}^{n\times n}$ is irreducible and symmetric with $P_{i,j} = P_{j,i}$ for $i, j \in \{1,\ldots,n\}$. In the language of the graph network, this transition matrix presents a weighted, undirected, and symmetric graph. This implies that e, a constant vector with a value $1/n$ on each component, is the stationary distribution. This implies there is a uniform probability to be in any state of the Markov chain represented by P. By symmetry, P is also doubly stochastic,[55] so the Markov chain represented by P is reversible.[56] Throughout this section, it will be assumed that there is no "isolated" state in the Markov chain (i.e., $P_{i,i} < 1$ for all i) and the matrix is row-stochastic so that each component represents a conditional probability of the corresponding Markov chain.

A Laplacian matrix is defined as

$$L = I - P, \tag{5.46}$$

where I is the identity matrix with the dimension of P. Then, for any $y \in \mathbb{R}^n$, we can show that

$$\begin{aligned} y^T Ly = y^T(I-P)y &= \sum_i y_i^2 - \sum_{i,j} y_i y_j P_{i,j} \\ &= \sum_i \left(\sum_j P_{i,j}\right) y_i^2 - \sum_{i,j} y_i y_j P_{i,j} \\ &= \frac{1}{2}\sum_{i,j}(y_i - y_j)^2 P_{i,j}. \end{aligned} \tag{5.47}$$

In going from the first to the second line, we use the fact that P is row-stochastic, and the symmetry of P allows us to write the quadratic form in the last line. Therefore, we can write the so-called Rayleigh quotient as

$$\frac{y^T Ly}{y^T y} = \frac{1}{2}\frac{\sum_{i,j}(y_i - y_j)^2 P_{i,j}}{\sum_i y_i^2}. \tag{5.48}$$

How is this related to almost-invariance? Suppose that we want to divide the n states of a Markov chain governed by P into two nonempty disjoint subsets I_1 and I_2. To describe this 2-partition, we consider a binary vector $y \in \pm 1$ such that $y_i = 1$ if $i \in I_1$, while if $i \in I_2$, $y_i = -1$. Substituting this binary vector y into (5.48) yields

$$\frac{y^T Ly}{y^T y} = \frac{2}{n}\left(\sum_{I_1\to I_2} P_{i,j} + \sum_{I_2\to I_1} P_{i,j}\right) = \frac{4}{n}\sum_{I_1\to I_2} P_{i,j}. \tag{5.49}$$

[55] Doubly stochastic is defined to be simultaneously row- and column-stochastic.

[56] A Markov chain represented by a transition matrix P is called *reversible* if $p_i P_{i,j} = p_j P_{j,i}$ for all i, j and p_i is a probability to be in the ith state. In particular, we will use the stationary distribution of P to be our probability measure p, i.e., $pP = p$. The reversibility condition implies that the probability to move from state i to j is the same as from j to i.

The subscript $I_1 \to I_2$ indicates that the sum is taken over the index $i \in I_1$ and $j \in I_2$, and similarly for $I_2 \to I_1$. Also, the constant term can be excluded in the optimization. It is now clear that minimizing the above Rayleigh quotient for $y \in \pm 1$ is equivalent to finding the optimal 2-partition that gives the minimal probability to move from one set to the other (from I_1 to I_2 and vice versa).

Remark 5.3 (optimal invariance). A pair of sets I_1 and I_2 that minimizes the Rayleigh quotient is maximally almost-invariant, with respect to the uniform probability measure as assumed for the present discussion.

The task now is to find such an optimal vector $y \in \pm 1$. Unfortunately, this problem is a difficult combinatorial problem if we wish to tackle it by brute force. Therefore, as a heuristic means of finding a good solution, we relax the binary restriction on y and allow continuous real values. Furthermore, we add a constraint that

$$\sum_i y_i = 0 \quad \text{(i.e., } y^T e = 0\text{)}. \tag{5.50}$$

This is equivalent to requiring that the partition I_1 and I_2 have the same weight sum, i.e., the sum of positive part equals to the negative part. The only reason for adding this constraint is not that we want to balance the size of I_1 and I_2,[57] but rather the constraint allows an application of the well-known Courant–Fischer theorem, which will be presented below. Based on the above discussion, our relaxed problem can be expressed by

$$\begin{aligned} &\text{Minimize } \frac{y^T L y}{y^T y}, \qquad y \in \mathbb{R}^n, \\ &\text{Subject to } y^T e = 0. \end{aligned} \tag{5.51}$$

To this end, we may apply the well-known Courant–Fischer theorem [140] to the minimization problem (5.51).

Theorem 5.2 (Courant–Fischer theorem). Let A be a real symmetric matrix with eigenvalues $0 \le \lambda_1 \le \lambda_2 \le \cdots \le \lambda_n$. Let X^k be a $(k-1)$-dimensional subspace of $\mathbb{R}^n$ and $x \perp X^k$ (i.e., $x \perp y$ for all $y \in X^k$). Then

$$\lambda_i = \min_{X^{n-i+1}} \max_{x \perp X^{n-i+1}, x \neq 0} \frac{x^T A x}{x^T x} = \max_{X^i} \min_{x \perp X^i, x \neq 0} \frac{x^T A x}{x^T x}. \tag{5.52}$$

The condition that A is real symmetric ensures the existence of a full set of orthogonal eigenvectors v_i corresponding to $\lambda_i \ge 0$ for $i = 1, \ldots, n$. Note that the X^{n-i+1} and x that result in λ_i in the first expression are indeed the span of the last $n-i+1$ eigenvectors and the $(n-i)$th eigenvector, respectively, while in the second expression λ_i is attained

[57]Nevertheless, this extra constraint will usually be relaxed since we will not strictly use the sign structure of the eigenvector for partitioning, but use it as a "level-set" in the thresholding algorithm, as a kind of one-dimensional optimization.

when X^i is the span of the first i eigenvectors and x is the $(i+1)$th eigenvector. These facts are normally shown in the standard proof of the Courant–Fischer theorem, which can be found in the popular graduate textbook in linear algebra [140].

To apply the Courant–Fischer theorem for our partitioning problem, we have to verify that $\lambda_1 = 0$, the smallest eigenvalue, and that the eigenvector corresponding to $\lambda_1 = 0$ is a constant vector; this can be easily checked as

$$eL = e(I - P) = e - e = 0, \tag{5.53}$$

where we use the fact that P is doubly stochastic. Thus, based on the above comments, we can calculate λ_2 by

$$\lambda_2 = \min_{y^T e=0, y\neq 0} \frac{y^T L y}{y^T y}. \tag{5.54}$$

It follows immediately from the above equation that the minimizing vector y that solves our relaxed problem is the eigenvector v_2 corresponding to λ_2. It should be noted that v_2 must have both positive and negative parts in order to satisfy the orthogonality with e. Consequently, the positivity of $\lambda_2 > 0$ is guaranteed by (5.47) and the presumption that P is irreducible; hence there will be at least one pair of i and j with opposite signs to make the summation in (5.47) positive. We close the case of a symmetric P here by briefly mentioning that the eigenvector v_2 can be used to create our partition by interpreting the value of v_2 as "fuzzy inclusion"; if the ith component of v_2 is strongly positive, the ith state of the Markov chain will very likely be in I_1, and for those with strong negativity, they are likely to be in I_2. For those near zero, the inclusion is less certain and an optimization algorithm is required to be the partition; see, for example, [122, 128].

It is fitting at this point to introduce an extension of the preceding technique to a general nonsymmetric transition matrix, which corresponds to a weighted, directed graph or a nonreversible Markov chain. By observing the steps in the previous analysis, it is evident that the conditions that P is reversible and symmetric are the key properties; the former allows us to write the Rayleigh quotient in the form of a quadratic cost function, while the latter is used to invoke the Courant–Fischer theorem. Therefore, a natural way to extend the previous technique is to first "reversibilize" the transition matrix P so that we obtain a new matrix representing a reversible Markov chain (the symmetry issue will be dealt with later on).

Before proceeding further, we need to introduce a number of variables required in the following analysis. Since P is nonsymmetric, in general e is not necessarily a stationary density of P; in fact, e is the stationary density only if P is still doubly stochastic. Therefore, we will use the eigenvector p such that $pP = p$ to be our measure; i.e., the probability to move from state i to state j is $p_i P_{ij}$.

Consider the transition matrix $\tilde{P}$ of the time-reversed chain defined by

$$\tilde{P}_{i,j} = p_j P_{j,i} / p_i \tag{5.55}$$

or, in a matrix form,

$$\tilde{P} = \Pi^{-1} P^T \Pi, \tag{5.56}$$

where Π is a diagonal matrix with $\Pi_{i,i} = p_i$. It is easy to check that $p\tilde{P} = p$. The *additive reversibilization* of P is defined by

$$R = \frac{1}{2}(P + \tilde{P}). \tag{5.57}$$

The Markov chain represented by the additively reversibilized transition matrix R is reversible and has p as the stationary distribution as well. That is, under the dynamic governing by the Markov chain R, the probability to go from states i to j is the same as going from states j to i (i.e., $p_i R_{i,j} = p_j R_{j,i}$). This has also been called detail balance in the language of statistical mechanics.

How does the matrix R relate to the our minimization? Similar to what we have done previously, our aim is to minimize the cost function

$$\begin{aligned} &\text{Minimize } \frac{\sum_{i,j}(y_i - y_j)^2 p_i P_{i,j}}{\sum_i y_i^2 p_i}, \qquad y \in \mathbb{R}^n, \\ &\text{Subject to } \sum_i y_i p_i = 0. \end{aligned} \tag{5.58}$$

The main difference between this cost function and the preceding one for the symmetric case is that the measure is now p instead of the uniform distribution e. To relate the above minimization problem to R, we rewrite the above cost function as

$$\begin{aligned} \sum_{i,j}(y_i - y_j)^2 p_i P_{i,j} &= \sum_{i,j}(y_i - y_j)^2 \frac{(p_i P_{i,j} + p_j P_{j,i})}{2} \\ &= \sum_{i,j}(y_i - y_j)^2 p_i R_{i,j}. \end{aligned} \tag{5.59}$$

Therefore, we can minimize the quadratic form related to R instead.

Next, we want to rewrite the above summation in a form similar to the Laplacian matrix we have used before. To achieve this, we can utilize the reversibility of R and the stationary measure p to show that

$$\begin{aligned} \sum_{i,j}(y_i - y_j)^2 p_i R_{i,j} &= \sum_i y_i^2 p_i - \sum_{i,j} y_i y_j p_i R_{i,j} \\ &= \sum_i x_i^2 - \sum_{i,j} x_i x_j (p_i^{1/2} R_{i,j} p_j^{-1/2}), \end{aligned} \tag{5.60}$$

where we substitute

$$x = \Pi^{1/2} y. \tag{5.61}$$

For such x, it follows that

$$\frac{x^T(I - \Pi^{-1/2} R \Pi^{1/2})x}{x^T x} = \frac{\sum_{i,j}(y_i - y_j)^2 p_i P_{i,j}}{\sum_i y_i^2 p_i}. \tag{5.62}$$

We now define the normalized Laplacian by

$$\mathcal{L} := I - \Pi^{-1/2} R \Pi^{1/2} = I - \frac{\Pi^{1/2} P \Pi^{-1/2} + \Pi^{-1/2} P^T \Pi^{1/2}}{2}.^{58} \tag{5.63}$$

An extensive study of the graph spectrum and their relations to properties and structures of a graph is covered in [58]. It is clear that $\mathcal{L}$ is symmetric, so we can now follow the same heuristic as in the case of the symmetric P in order to find a partition by applying the Courant–Fischer theorem to $\mathcal{L}$. More precisely, according to (5.62) and the definition (5.63), the optimization problem (5.58) can be rewritten as

$$\begin{aligned} &\text{Minimize } \frac{x^T(\mathcal{L})x}{x^T x} \\ &\text{Subject to } \sum_i x_i p_i^{1/2} = 0. \end{aligned} \tag{5.64}$$

Notice that due to the equation $x = \Pi^{1/2} y$ the optimal vector x has to be constrained to be orthogonal to $p^{1/2}$ for y to be orthogonal to p. Also, we note that after obtaining the optimal vector x, we have to revert it to y. To apply the Courant–Fischer theorem as we previously had done, we have to check that $p^{1/2}$ is indeed the eigenvector corresponding to $\lambda_1 = 0$. This can be easily verified as the following:

$$\begin{aligned} p^{1/2}\mathcal{L} &= p^{1/2}\left(I - \frac{(\Pi^{1/2} P \Pi^{-1/2} + \Pi^{-1/2} P^T \Pi^{1/2})}{2}\right) \\ &= p^{1/2} - \frac{p P \Pi^{-1/2} + e P^T \Pi^{1/2}}{2} \\ &= p^{1/2} - \frac{p \Pi^{-1/2} + e \Pi^{1/2}}{2} \\ &= p^{1/2} - \frac{p^{1/2} + p^{1/2}}{2} = 0. \end{aligned} \tag{5.65}$$

To derive the third line from the second line, we use the fact that $pP = p$ and $Pe = e$. Therefore, the optimal value of the problem (5.58) is

$$\lambda_2 = \min_{x^T p^{1/2} = 0, x \neq 0} \frac{x^T \mathcal{L} x}{x^T x} = \min_{y^T p = 0, y \neq 0} \frac{\sum_{i,j} (y_i - y_j)^2 p_i P_{i,j}}{\sum_i y_i^2 p_i}, \tag{5.66}$$

attained by the eigenvector $x = v_2$ (i.e., $y = \Pi^{-1/2} v_2$) corresponding to λ_2 of $\mathcal{L}$.

[58] $\Pi^{1/2}$ denotes the "square root" of the matrix Π in the sense that the square root of a square matrix A can be taken to be a matrix B such that $A = B^2$. Generally, the existence and computation of such a B requires conditions and an algorithm, but in the case that A is diagonal, as in fact Π is diagonal, then the square root of the matrix is especially simple to compute by the square root of all of its entries: $A^{1/2} = \text{diag}(\sqrt{A_{1,1}}, \ldots, \sqrt{A_{n,n}})$. Likewise the inverse of a diagonal matrix is particularly simple to compute by inverting the numbers on the diagonal. Therefore, $\Pi^{-1/2} := (\Pi^{1/2})^{-1} = \text{diag}(1/\sqrt{A_{1,1}}, \ldots, 1/\sqrt{A_{n,n}})$, assuming none of the diagonal elements are zero as the condition for nonsingularity.

Example 5.5. Let

$$P = \begin{pmatrix} 0.1 & 0.5 & 0.2 & 0.1 & 0.1 \\ 0.6 & 0.3 & 0.1 & 0.0 & 0.0 \\ 0.2 & 0.2 & 0.2 & 0.2 & 0.2 \\ 0.1 & 0.2 & 0.0 & 0.3 & 0.4 \\ 0.1 & 0.2 & 0.0 & 0.4 & 0.3 \end{pmatrix}. \tag{5.67}$$

It is quite clear that we have two separate groups $\{1,2\}$ and $\{4,5\}$, where a group is taken to be a highly connected subset of vertices. However, the group to which we should put $\{3\}$ in order to obtain the minimal transport is ambiguous. The stationary distribution p can be obtained by computing the left eigenvector corresponding to the unit eigenvalue of P as

$$p = \{0.5558, 0.6497, 0.2202, 0.3321, 0.3321\}^T.$$

Based on the measure p, we can calculate the normalized Laplacian as per Eq. (5.63) to be

$$\mathcal{L} = \begin{pmatrix} 0.9000 & -0.5556 & -0.2218 & -0.1033 & -0.1033 \\ -0.5556 & 0.7000 & -0.1441 & -0.0715 & -0.0715 \\ -0.2218 & -0.1441 & 0.8000 & -0.0814 & -0.0814 \\ -0.1033 & -0.0715 & -0.0814 & 0.7000 & -0.4000 \\ -0.1033 & -0.0715 & -0.0814 & -0.4000 & 0.7000 \end{pmatrix}. \tag{5.68}$$

The eigenvalues of $\mathcal{L}$ are $\lambda_1 = 0, \lambda_2 = 0.4320, \lambda_3 = 0.8907, \lambda_4 = 1.1000, \lambda_5 = 1.3773$. The optimal vector can then be obtained by calculating $y = \Pi^{-1/2} v_2$ as

$$y = [0.4486, 0.5571, 0.2577, -1.0059, -1.0059]. \tag{5.69}$$

It is worthwhile to check that $\sum_i y_i p_i = 0$, as required in the constrain optimization (5.58). Therefore, the 2-partition based on the parity of optimal vector y in this example is $I_1 = \{1,2,3\}$ and $I_2 = \{4,5\}$, which gives

$$\sum_{I_1 \to I_2} p_i P_{i,j} + \sum_{I_2 \to I_1} p_i P_{i,j} = 0.1992 + 0.1992 = 0.3984. \tag{5.70}$$

It can checked that the above partition does result in a lower value in the cost function than an alternative 2-partition $I_1 = \{1,2\}$ and $I_2 = \{3,4,5\}$, which gives

$$\sum_{I_1 \to I_2} p_i P_{i,j} + \sum_{I_2 \to I_1} p_i P_{i,j} = 0.2873 + 0.2873 = 0.5746. \tag{5.71}$$

So, this partition has a greater "leak."

Example 5.6. Consider the matrix P in (5.36). We will alternatively attempt to find the partition based on the eigenvector of the generalized Laplacian matrix. Since the setup of a stochastic matrix in this section is row-stochastic, we will use P^T instead of P.

$$P^T = \begin{pmatrix} 0.1428 & 0.3007 & 0.5563 & 0 & 0.0001 & 0 & 0.0001 \\ 0.4290 & 0.4084 & 0.1623 & 0 & 0 & 0.0002 & 0.0001 \\ 0.7359 & 0.1110 & 0.1530 & 0 & 0 & 0 & 0.0001 \\ 0 & 0.0001 & 0 & 0.0886 & 0.4556 & 0.3333 & 0.1224 \\ 0 & 0.0001 & 0 & 0.3201 & 0.0916 & 0.0327 & 0.5555 \\ 0 & 0.0001 & 0 & 0.2035 & 0.3345 & 0.2510 & 0.2109 \\ 0.0001 & 0 & 0 & 0.3044 & 0.3270 & 0.3684 & 0.0001 \end{pmatrix}. \tag{5.72}$$

The stationary distribution p of the above matrix is calculated to be

$$p = [0.3596, 0.2358, 0.2814, 0.4008, 0.4963, 0.4015, 0.4096]^T.$$

Based on the vector p, the normalized Laplacian is

$$\mathcal{L} = \begin{pmatrix} 0.8572 & -0.3594 & -0.6399 & 0 & 0 & 0 & -0.0001 \\ -0.3594 & 0.5916 & -0.1349 & -0.0001 & -0.0001 & -0.0001 & 0 \\ -0.6399 & -0.1349 & 0.8470 & 0 & 0 & 0 & 0 \\ 0 & -0.0001 & 0 & 0.9114 & -0.3828 & -0.2683 & -0.2144 \\ 0 & -0.0001 & 0 & -0.3828 & 0.9084 & -0.1686 & -0.4543 \\ 0 & -0.0001 & 0 & -0.2683 & -0.1686 & 0.7490 & -0.2904 \\ -0.0001 & 0 & 0 & -0.2144 & -0.4543 & -0.2904 & 0.9999 \end{pmatrix}.$$

The eigenvalues of $\mathcal{L}$ are $\lambda_i = \{0, 0.0003, 0.7730, 0.9453, 1.1543, 1.4690, 1.5226\}$. That λ_2 is nearly zero suggests a potential of almost-invariant partitioning. The optimal vector can then be calculated from the equation $y = \Pi^{-1/2} v_2$ as

$$y = [-0.8682, -0.8677, -0.8683, 0.4457, 0.4456, 0.4456, 0.4456]. \tag{5.73}$$

The sign structure of y clearly reveals the optimal 2-partition $I_1 = \{1,2,3\}$ and $I_2 = \{4,5,6,7\}$.

5.5 Finite Time Coherent Sets

Intensive study has been dedicated to understand how to define and/or detect coherent sets (or structures). Many of these definitions are based on physical quantities (e.g., using potential vorticity to define a polar vortex as a coherent set), which may not always be optimal. Alternatively, the approach reviewed here is to identify optimal coherent structures based only on the underlying flow [128]. For a general time-dependent or nonautonomous dynamical systems, it can be imagined, in analogy to the almost-invariant set, that a set moving collectively in a minimally dispersive manner is the subject of our interest. For instance, Fig. 5.5 shows a time sequence of a set that self-evidently moved coherently with the underlying flow. A set with this kind of behavior will be called here as a coherent set; a formalism will be given in what follows.

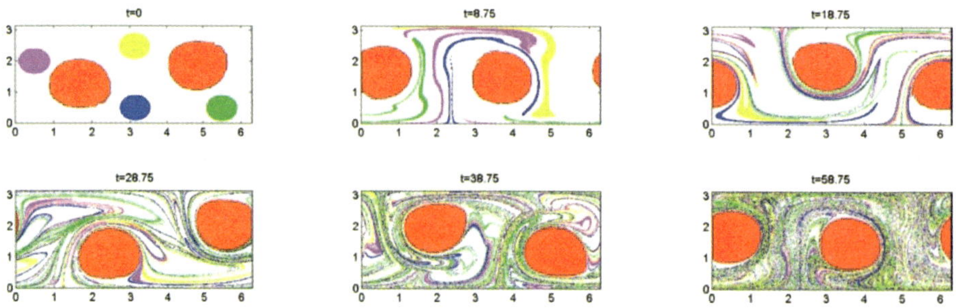

Figure 5.5. *An example of a coherent set in a time-dependent dynamical system. The equation of the underlying flow is not important here, but the distinct behavior of the coherent sets (shown in red) compared to other sets is clear; they move coherently along with the flow and rarely disperse.*

The concept of coherent sets may be deemed as a generalization of the almost-invariant set. Given a time-dependent flow map,

$$\Phi(z,t;\tau): M \times \mathbb{R} \times \mathbb{R} \to \mathbb{R}, \tag{5.74}$$

which described the terminal point time $t+\tau$ of an initial point z at time t, we desire to discover "coherent pairs" of subsets A_t and $A_{t+\tau} \subset M$ such that

$$\Phi(A_t,t;\tau) \approx A_{t+\tau}; \tag{5.75}$$

i.e., A_t is mapped mostly to $A_{t+\tau}$ by the flow Φ. This approximate statement is made precise in terms of measure.

Definition 5.2. We will call A_t, $A_{t+\tau}$ a (ρ_0,t,τ)-coherent pair if

1. $$\rho_\mu(A_t,A_{t+\tau}) := \frac{\mu(A_t \cap \Phi(A_{t+\tau},t+\tau;-\tau))}{\mu(A_t)} \geq \rho_0. \tag{5.76}$$

2. $\mu(A_t) = \mu(A_{t+\tau})$.

3. A_t and $A_{t+\tau}$ are "robust" to small perturbation.

In the above definition, the measure μ is called a reference probability measure at time t, which usually describes the mass distribution of the quantity we wish to study of the transport over the interval $[t,t+\tau]$. Note that the reference measure does not have to be invariant under the flow Φ. It is clear that the first two conditions in the above definition prescribe the coherency. However, a $(1,t,\tau)$-coherent pair can be trivially produced by choosing an arbitrary A_t and $A_{t+\tau} = \Phi(A_t,t;\tau)$. To eliminate this pathology, the third condition is used to select only those pairs that remain coherent under small diffusive perturbation of the flow.

In chaotic systems, it is usually the case that an arbitrarily chosen coherent pair will experience stretching and folding, and after a sufficiently long time τ they will become geometrically thin and irregular. Consequently, a small amount of diffusion is able to eject many particles from $A_{t+\tau}$ and hence reduce the coherence ratio $\rho_\mu(A_t,A_{t+\tau})$. Therefore, the requirement of robustness under perturbation favors a large-scale coherent set with a geometrically regular structure, which is of interest in most applications. Since the transfer operator involves a spatial discretization, hence discretization-induced diffusion is inevitable. This explains why the transfer operator approach tends to capture a large-scale regular structure. Our basic tool for identifying the coherent pair will again be the transfer (Frobenius–Perron) operator. For a given t and τ,

$$\mathcal{P}_{t,\tau} : L^1(M) \circlearrowleft, \tag{5.77}$$

defined by

$$\mathcal{P}_{t,\tau} f(z) := f(\Phi(z,t+\tau;-\tau)) \cdot |\det D\Phi(z,t+\tau;-\tau)|. \tag{5.78}$$

If $f(z)$ is a density of passive tracers at time t, then $\mathcal{P}_{t,\tau} f(z)$ is the tracer density at time $t+\tau$ induced by the flow Φ.

The above calculations involves constructing a Frobenius–Perron operator for the action of Φ on the *entire domain* M. In the time-dependent setting, we wish to study transport from $X \subset M$ to a small neighborhood Y of $\Phi(X,t;\tau) \subset M$. A global analysis would mean that $X = Y = M$ and a transfer operator would be constructed for all of M. However, often one is interested in the situation where the domain of interest X is "open," in contrast to a closed system where $X = Y$. By open we mean that trajectories may leave X in a finite time. Moreover, the subset X may be very small in comparison to M. In such instances, there are great computational savings if the analysis can be carried out using a nonglobal Frobenius–Perron operator defined on X rather than M. Therefore, we want to construct a numerical approximation of a Frobenius–Perron operator to allow precisely this.

We now describe a numerical approximation of the action of $\mathcal{P}_{t,\tau}$ from a space of functions supported on X to a space of functions supported on Y. We subdivide the subsets X and Y into collections of sets $\{B_1,\dots,B_m\}$ and $\{C_1,\dots,C_n\}$, respectively. We construct a finite-dimensional numerical approximation of the transfer operator $\mathcal{P}_{t,\tau}$, using a modification of Ulam's method [307]:

$$P^{(\tau)}(t)_{i,j} = \frac{\ell(B_i \cap \Phi(C_j, t+\tau; -\tau))}{\ell(B_i)}, \tag{5.79}$$

where ℓ is a normalized volume measure on M.

Note especially that P is a nonsquare matrix. Clearly, $P^{(\tau)}(t)$ is row-stochastic by its construction. The value $P^{(\tau)}(t)_{i,j}$ may be interpreted as the probability that a randomly chosen point in B_i has its image in C_j. We numerically estimate $P^{(\tau)}(t)_{i,j}$ by

$$P^{(\tau)}(t)_{i,j} \approx \#\{r : z_{i,r} \in B_i, \Phi(z_{i,r},t;\tau) \in C_j\}/Q, \tag{5.80}$$

where $z_{i,r}$, $r = 1,\dots,Q$, are uniformly distributed test points in $B_i(t)$ and $\Phi(z_{i,r},t;\tau)$ is obtained via a numerical integration. Fig. 5.6 shows a schematic diagram of the construction of $P^{(\tau)}(t)_{i,j}$. Probabilistically, $P^{(\tau)}(t)_{i,j} \approx \mu(B_i \cap \Phi(C_j, t+\tau; -\tau))/\mu(B_i)$. In fact, if μ is absolutely continuous with a positive density that is Lipschitz on the interior of each

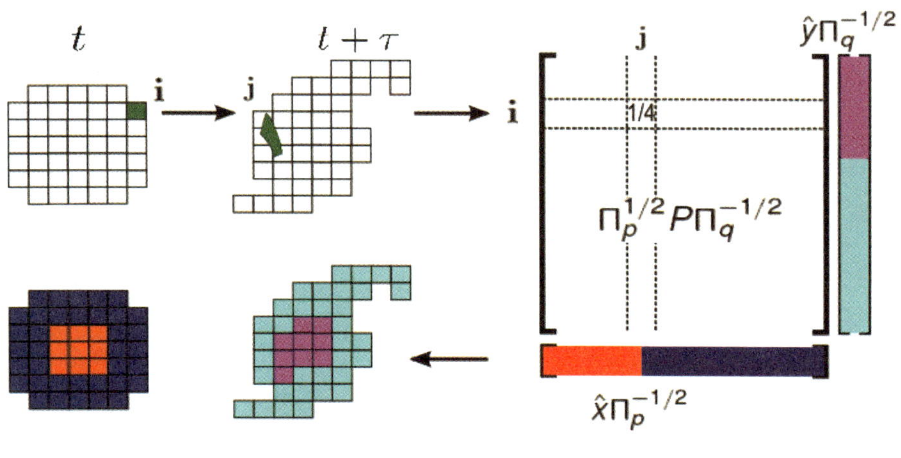

Figure 5.6. *A schematic demonstrating the construction of the transfer operator approximation. Notice that $X \neq Y$.*

B_i, then the error of this estimate goes to zero with decreasing diameter of B_i and C_j; see Lemma 3.6 of [120]. Also, the numerical discretization has the useful side benefit of producing a discretization-induced effective diffusion with magnitude the order of the image of box diameters (see Lemma 2.2 of [117]).

Recall that the coherent pair is a generalization of almost-invariant sets in autonomous systems to nonautonomous systems and has properties as given in Definition 5.2. For the remainder of this section we set $P(i,j) = P^{(\tau)}(t)_{i,j}$, fixing t and τ.

Suppose that we have constructed the $m \times n$ row-stochastic transition matrix $P^{(\tau)}(t)$ via the Ulam method. Let Π_1 be a diagonal matrix such that

$$\Pi_1(i,i) = \pi_1(i), \quad i = 1,\dots,m, \tag{5.81}$$

for an arbitrary vector π_1. In practice, we conventionally set

$$\pi_1(i) = \mu(B_i), \quad i = 1,\dots,m, \tag{5.82}$$

and assume that

$$\pi_i > 0 \; \forall i = 1,\dots,n. \tag{5.83}$$

If $\pi_i = 0$ for some i, then we remove those from the partition on X; i.e., we choose a reference probability measure that gives a mass $\pi_i > 0$ to the set B_i, $i = 1,\dots,m$ (if some sets have zero reference measure, we remove them from our collection). This reference probability measure describes the mass we wish to use to define our discrete version of coherence in (5.76). Now define

$$\pi_2 = \pi_1 P \tag{5.84}$$

to be the image probability vector on Y, and

$$\Pi_2(i,i) = \pi_2(i). \tag{5.85}$$

We also assume $q > 0$ (again, if not, we remove sets C_j with $q_j = 0$). The probability vector π_2 defines a probability measure ν on Y via

$$\nu(Y') = \sum_{j=1}^{n} \pi_2(j)\ell(Y' \cap C_j) \tag{5.86}$$

for measurable $Y' \subset Y$. We may think of the probability measure ν as the discretized image of μ.

Analogous to what we did in preceding sections, we define a **Laplacian matrix** $\mathcal{L}$ by

$$\mathcal{L} = \begin{pmatrix} \Pi_1 & -\Pi_1 P \\ -P^T \Pi_1 & \Pi_2 \end{pmatrix}. \tag{5.87}$$

Let I_1, I_2 partition $\{1,\dots,m\}$ and J_1, J_2 partition $\{1,\dots,n\}$, and set $X_k = \cup_{i \in I_k} B_i$ and $Y_k = \cup_{j \in J_k} C_j$, $k = 1,2$. To describe 2-partitions of X and Y we consider vectors $x \in \{\pm 1\}^m$, $y \in \{\pm 1\}^n$ and define $X_1 = \bigcup_{i:x_i=1} B_i, X_2 = \bigcup_{i:x_i=-1} B_i, Y_1 = \bigcup_{i:y_i=1} C_i, Y_2 = \bigcup_{i:y_i=-1} C_i$. Thus, the partitions I_1, I_2 and J_1, J_2 are described by the parity of x and y, respectively.

Based on the above notation, we can show the following estimate:

$$
\begin{aligned}
\alpha := x^T \Pi_1 P y = & \sum_{i\in I_1, j\in J_1} \pi_1(i)P(i,j) + \sum_{i\in I_2, j\in J_2} \pi_1(i)P(i,j) \\
& - \sum_{i\in I_1, j\in J_2} \pi_1(i)P(i,j) + \sum_{i\in I_2, j\in J_1} \pi_1(i)P(i,j) \\
& \approx (\mu(X_1 \cap \Phi(Y_1, t+\tau; -\tau)) + \mu(X_2 \cap \Phi(Y_2, t+\tau; -\tau))) \\
& - (\mu(X_1 \cap \Phi(Y_2, t+\tau; -\tau)) + \mu(X_2 \cap \Phi(Y_1, t+\tau; -\tau))) \\
& = \mu(X_1)\rho_\mu(X_1, Y_1) + \mu(X_2)\rho_\mu(X_2, Y_2) - \mu(X_1)\rho_\mu(X_1, Y_2) \\
& - \mu(X_2)\rho_\mu(X_2, Y_1).
\end{aligned}
\tag{5.88}
$$

Thus, maximizing α is a natural way to achieve our aim of finding partitions so that $\rho_\mu(X_k, Y_k) \approx 1$, $k = 1, 2$. However, for consistency with our previous formulation, we would like to develop a cost function in terms of the generalized Laplacian $\mathcal{L}$. To do this, let $p = [x, y]^T$. We can show that

$$
\begin{aligned}
p^T \mathcal{L} p &= x^T \Pi_1 x + y^T \Pi_2 y - (x^T \Pi_1 P y + y^T P^T \Pi_1 x) \\
&= \pi_1(X) + \pi_2(Y) - 2\alpha \\
&= 2(1-\alpha),
\end{aligned}
\tag{5.89}
$$

where the last line follows from a normalization of the reference measure. Therefore, to find the most coherent pair, we want to approximate a pair of binary vectors x and y that minimizes $p^T \mathcal{L} p$.

5.6 Spectral Partitioning for the Coherent Pair

Some standard graph-theoretical results for approximating balanced cut via spectral partitioning can be usefully applied to derive a cost function via the Rayleigh quotient. The following results are modified from [127] and will only be roughly sketched herein. The final result of this section will provide a heuristic mean to find the optimal pair of vectors that maximizes α, hence the coherency.

First, we define a block-diagonal matrix W to be

$$
W = \begin{pmatrix} \Pi_1 & 0 \\ 0 & \Pi_2 \end{pmatrix}.
\tag{5.90}
$$

Lemma 5.2. Let $\mathcal{L}$ and W be defined as above. Let $\eta_1 = \sum_{i\in I_1} \pi_1(i) + \sum_{i\in J_1} \pi_2(i)$ and $\eta_2 = \sum_{i\in I_2} \pi_1(i) + \sum_{i\in J_2} \pi_2(i)$. Then the generalized partition vector q with elements

$$
q(i) = \begin{cases} +\sqrt{\eta_2/\eta_1}, & i \in [1, \ldots, m] \quad \text{(called } I\text{)}, \\ -\sqrt{\eta_1/\eta_2}, & i \in [m+1, \ldots, m+n] \quad \text{(called } J\text{)}, \end{cases}
\tag{5.91}
$$

satisfies

$$
q^T W e = 0,
\tag{5.92}
$$

where e is a vector with $e_i = 1$ for all i, and

$$
q^T W q = \eta_1 + \eta_2.
\tag{5.93}
$$

Proof. The orthogonality between q and e with respect to the W-weight inner product can be readily seen from the expansion

$$q^T We = \sqrt{\frac{\eta_2}{\eta_1}} \sum_{i \in I} W(i,i) - \sqrt{\frac{\eta_1}{\eta_2}} \sum_{i \in J} W(i,i).$$

It is straightforward to show that

$$q^T Wq = \sum_i W(i,i) q^2(i) = \eta_1 + \eta_2. \quad \square$$

Lemma 5.3 (see [127]). Suppose that $\sum_{i=1,\dots,m} \pi_1(i) = \sum_{i=1,\dots,n} \pi_2(i) = 1$. The generalized Rayleigh quotient is given by

$$\frac{q^T \mathcal{L} q}{q^T Wq} = \frac{1}{\eta_1 \eta_2} - \alpha, \tag{5.94}$$

where α is defined in (5.88).

Proof. First, show that the generalized partition vector q can be rewritten as

$$q = \frac{\eta_1 + \eta_2}{2\sqrt{\eta_1 \eta_2}} p + \frac{\eta_2 - \eta_1}{2\sqrt{\eta_1 \eta_2}} e.$$

Then, using the fact, which can be readily derived, that

$$\mathcal{L} e = e^T \mathcal{L} = 0, \tag{5.95}$$

we can show that

$$q^T \mathcal{L} q = \frac{(\eta_1 + \eta_2)^2}{4\eta_1 \eta_2} p^T \mathcal{L} p, \tag{5.96}$$

where $p = [x, y]$ is the usual sign partition vector as used in (5.89). Finally, after substituting the values of $p^T \mathcal{L} p$ and $q^T Wq$, from Lemma 5.2, the desired result is achieved. $\square$

Note that the normalization in the above lemma gives the relation $\eta_1 + \eta_2 = 2$. Thus, if it happens that the measures of X_1, X_2, Y_1, and Y_2 are all equal, then (5.94) becomes

$$\frac{q^T \mathcal{L} q}{q^T Wq} = 1 - \alpha. \tag{5.97}$$

Therefore, in such case, the pair of X_1 and Y_1 minimizing the quotient in (5.97) will be our optimal pair, which also maximizes α. In general, the mass equivalency of the partition is, however, not directly achieved by minimizing the quotient in (5.94), and for this reason minimizing (5.94) will be called the relaxation problem. The relaxation to the optimal generalized partition vector can then be achieved by the following theorem.

Theorem 5.3. The problem

$$\min_{q \neq 0} \frac{q^T \mathcal{L} q}{q^T Wq}, \qquad \text{subject to } q^T We = 0, \tag{5.98}$$

is solved when q is the eigenvector corresponding to the second-smallest eigenvalue of the generalized eigenvalue problem,

$$\mathcal{L} z = \lambda W z. \tag{5.99}$$

Proof. This follows from a standard result from linear algebra; see [165]. $\square$

5.7 The SVD Connection

In the previous section, it was shown that a real-valued relaxation to the discrete optimization problem allows solution by the second eigenvector of the problem $\mathcal{L}z = \lambda W z$. However, given that P is $m \times n$ and $\mathcal{L}$ is $m+n \times m+n$, it would be preferable to solve the same problem based on P instead. Particularly, it will be shown in the following that the coherent partition can be obtained from the second-largest singular vector of $\Pi_1^{1/2} P \Pi_2^{-1/2}$. This result can also be derived from another heuristic approach based on the similarity transform of P presented in [277, 128].

To see this, we rewrite the generalized eigenvalue problem $\mathcal{L}q = \lambda_2 W q$ as

$$\begin{pmatrix} \Pi_1 & -\Pi_1 P \\ -P^T \Pi_1 & \Pi_2 \end{pmatrix} \begin{pmatrix} x \\ y \end{pmatrix} = \lambda_2 \begin{pmatrix} \Pi_1 & 0 \\ 0 & \Pi_2 \end{pmatrix} \begin{pmatrix} x \\ y \end{pmatrix}, \tag{5.100}$$

where

$$q = \begin{pmatrix} x \\ y \end{pmatrix}, \tag{5.101}$$

$(x, y \in \mathbb{R}^{N\times 1})$ denotes the eigenvector associated with the second-smallest eigenvalue λ_2.

Assuming that Π is nonsingular, we may rewrite the above equation as

$$\begin{aligned} \Pi_1^{1/2} x - \Pi_1^{-1/2} (\Pi_1 P) y &= \lambda_2 \Pi_1^{1/2} x \\ -\Pi_2^{-1/2} (P^T \Pi_1) x + \Pi_2^{1/2} y &= \lambda_2 \Pi_2^{1/2} y. \end{aligned} \tag{5.102}$$

Letting

$$u = \Pi_1^{1/2} x \quad \text{and} \quad v = \Pi_2^{1/2} y, \tag{5.103}$$

and by algebraic manipulation, we obtain

$$\begin{aligned} \Pi_1^{-1/2} (\Pi_1 P) \Pi_2^{-1/2} v &= (1-\lambda_2) u, \\ \Pi_2^{-1/2} (P^T \Pi_1) \Pi_1^{-1/2} u &= (1-\lambda_2) v. \end{aligned} \tag{5.104}$$

Letting

$$P_0 = \Pi_1^{1/2} P \Pi_2^{-1/2}, \tag{5.105}$$

we may write the above equation as

$$\begin{aligned} P_0 v &= (1-\lambda_2) u, \\ P_0^T u &= (1-\lambda_2) v. \end{aligned} \tag{5.106}$$

The above expression is precisely the SVD of the matrix P_0. Thus, instead of computing the eigenvalue of $\mathcal{L}$ in Theorem 5.3, we can compute the singular vectors of P_0 associated with the second-largest singular value and then retrieve

$$q = \begin{pmatrix} x \\ y \end{pmatrix} \text{ from } u \text{ and } v \text{ by } q = \begin{pmatrix} \Pi_1^{-1/2} u \\ \Pi_2^{-1/2} v \end{pmatrix}. \tag{5.107}$$

Furthermore, the largest singular value is $\sigma_1(P_0) = 1$ with the corresponding left singular vector $e\Pi_1^{1/2}$ and right singular vector $\Pi_2^{1/2} e$.

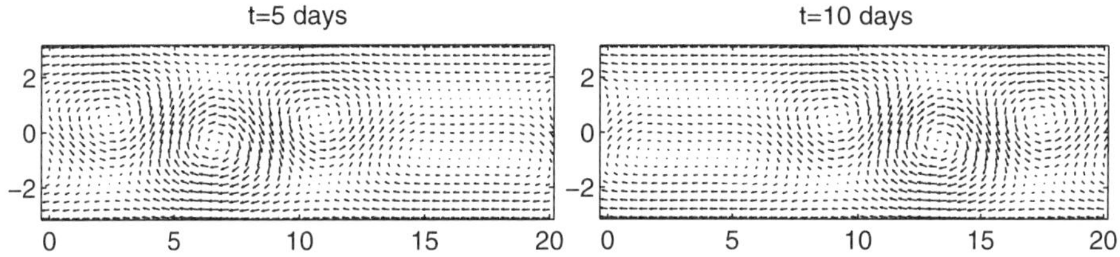

Figure 5.7. *Snapshots of velocity fields at $t = 5$ and $t = 10$. The key features of the flow is the a time-varying jet core oscillating in a band around $y = 0$ and three Rossby waves in each of the regions above and below the jet core.* [58]

5.8 Example 1: Idealized Stratospheric Flow

Following [128], we consider the Hamiltonian system,

$$\frac{dx}{dt} = -\frac{\partial \psi}{\partial y}, \qquad \frac{dy}{dt} = \frac{\partial \psi}{\partial x}, \tag{5.108}$$

where

$$\begin{aligned}\psi(x,y,t) &= c_3 y - U_0 L \tanh(y/L) + A_3 U_0 L \operatorname{sech}^2(y/L)\cos(k_1 x)\\ &+ A_2 U_0 L \operatorname{sech}^2(y/L)\cos(k_2 x - \sigma_2 t) + A_1 U_0 L \operatorname{sech}^2(y/L)\cos(k_1 x - \sigma_1 t).\end{aligned} \tag{5.109}$$

Snapshots of the velocity field are shown in Figure 5.7. This quasi-periodic system represents an idealized stratospheric flow in either the northern or southern hemisphere. Rypina et al. [274] show that there is a time-varying jet core oscillating in a band around $y = 0$ and three Rossby waves in each of the regions above and below the jet core. The parameters studied in [274] are chosen so that the jet core forms a complete transport barrier between the two Rossby wave regimes above and below it. We modify some of the parameters to remove the jet core band and allow transport between the two Rossby wave regimes [124]. We expect that the two Rossby wave regimes will form time-dependent coherent sets because transport between the two regimes is considerably less than the transport within regimes. We set the parameters as follows: $c_2/U_0 = 0.205$, $c_3/U_0 = 0.700$, $A_3 = 0.2, A_2 = 0.4$, and $A_1 = 0.075$, with the remaining parameters as stated in Rypina et al. [274].

Our initial time is $t = 20$ days and our final time is $t + \tau = 30$ days. At our initial time we set $X = S^1 \times [-2.5, 2.5]$ Mm, where S^1 is a circle parameterized from 0 to 6.371π Mm, and subdivide X into a grid of $m = 28200$ identical boxes $X = \{B_1, \ldots, B_m\}$. This choice of m is sufficiently large to represent the dynamics of a good resolution. We compute an approximation of $\Phi(X, 20; 30)$ by uniformly distributing $Q = 400$ sample points in each grid box and numerically calculating $\Phi(z_{i,r}, 20; 30)$ using the standard Runge–Kutta method. The choice of Q is made so that over the flow duration, the image of boxes is well represented by the Q sample points per box. These $Q \times m$ image points are then covered by a grid of $n = 34332$ boxes $\{C_1, \ldots, C_n\}$ of the same size as the B_i, $i = 1, \ldots, m$. We set $Y = \bigcup_{j=1}^{n} C_j$, covering the approximate image of X. The transition matrix $P = P_{20}^{(30)}$ is computed using (5.80).

As the flow is area preserving, a natural reference measure μ is the Lebesgue measure, which we normalize so that $\mu(X) = 1$. Thus, $\mu(B_i) = p_i = 1/m$, $i = 1, \ldots, m$, and thus $(\Pi_p)_{ii} = 1/m, i = 1,\dots,m$. The vector q is constructed as $q = pP$. We compute the second-largest singular value of $\Pi_p^{1/2} P \Pi_q^{-1/2}$ and the corresponding left and right singular vectors, and thus determine x and y from Eq. (5.107). The top two singular values were computed to be $\sigma_1 = 1.0$ and $\sigma_2 \approx 0.996$. We expect x to determine coherent sets at time $t = 20$ days and y to determine coherent sets at time $t + \tau = 30$ days. Figs. 5.8(a) and 5.8(b) illustrate the vectors x and y, which provide clear separations into red (positive) and green (mostly negative) regions.

We apply the thresholding algorithm from [128] to the vectors x and y to obtain the pairs (X_1, Y_1) and (X_2, Y_2)[59] shown in Figs. 5.9(a) and 5.9(b). To demonstrate that $Y_1 \approx \Phi(X_1, 20; 10)$, we plot the latter set in Fig. 5.9(c). When compared with Fig. 5.9(b) we see that there is very little leakage from Y_1—just a few thin filaments. Similarly, Figs. 5.9(d) and 5.9(b) compare Y_2 and $\Phi(X_2, 20; 10)$, again showing a small amount of leakage. This leakage is quantified by computing $\tilde{\rho}(X_1, Y_1) \approx \tilde{\rho}(X_2, Y_2) \approx 0.98$.

We compare our results with the attracting and repelling material lines computed via the finite time Lyapunov exponent (FTLE) field [153] with the flow time $\tau = 10$. The ridges of the FTLE fields are commonly used to identify barriers to transport, as discussed in detail in Chapter 8. Figs. 5.8(c) and 5.8(d) present an overlay of forward- and backward-time FTLEs at $t = 20$ and $t = 30$, respectively. In this example, there are several FTLE ridges in the vicinity of the dominant transport barrier across the middle of the domain, and several ridges far away from this barrier. The FTLE ridges do not crisply and unambiguously identify the dominant transport barrier shown in Figs. 5.8(a) and 5.8(b).

5.9 Example 2: Stratospheric Polar Vortex as Coherent Sets

In our second example, we use velocity fields obtained from the ECMWF Interim data set (http://data.ecmwf.int/data/index.html). We focus on the stratosphere over the Southern Hemisphere south of 30 degrees latitude. In this region, there are strong persistent transport barriers to midlatitude mixing during the austral winter; these barriers give rise to the Antarctic polar vortex. We will apply the methodology reviewed here to the ECMWF vector fields in two dimensions to resolve the polar vortex as a coherent set.

5.9.1 Two Dimensions

Our input data consist of two-dimensional velocity fields on a 121×240 element grid in the longitude and latitude directions, respectively. The ECMWF data provide updated velocity fields every 6 hours. The flow is initialized at September 1, 2008, on a 475K isentropic surface, and we follow the flow until September 14. To a good approximation isentropic surfaces are close to invariant over a period of about two weeks [179].

We set $X = S^1 \times [-90°, -30°]$, where S^1 is a circle parameterized from 0° to 360°. The domain X is initially subdivided into the grid boxes $B_i, i = 1,\dots,m$, where $m = 13471$ in this example. Based on the hydrostatic balance and the ideal gas law, we set the reference measure $p_i = \mathrm{Pr}_i^{5/7} \mathrm{a}_i$ for all $i = 1,\dots,m$, where Pr_i is the pressure at the center point of B_i and a_i is the area of box B_i.

[59]When determining X_1 and Y_1, Algorithm 1 in [128] produced values $b^* \approx 0.0077$ and $\eta(b^*) \approx 0.0005$.

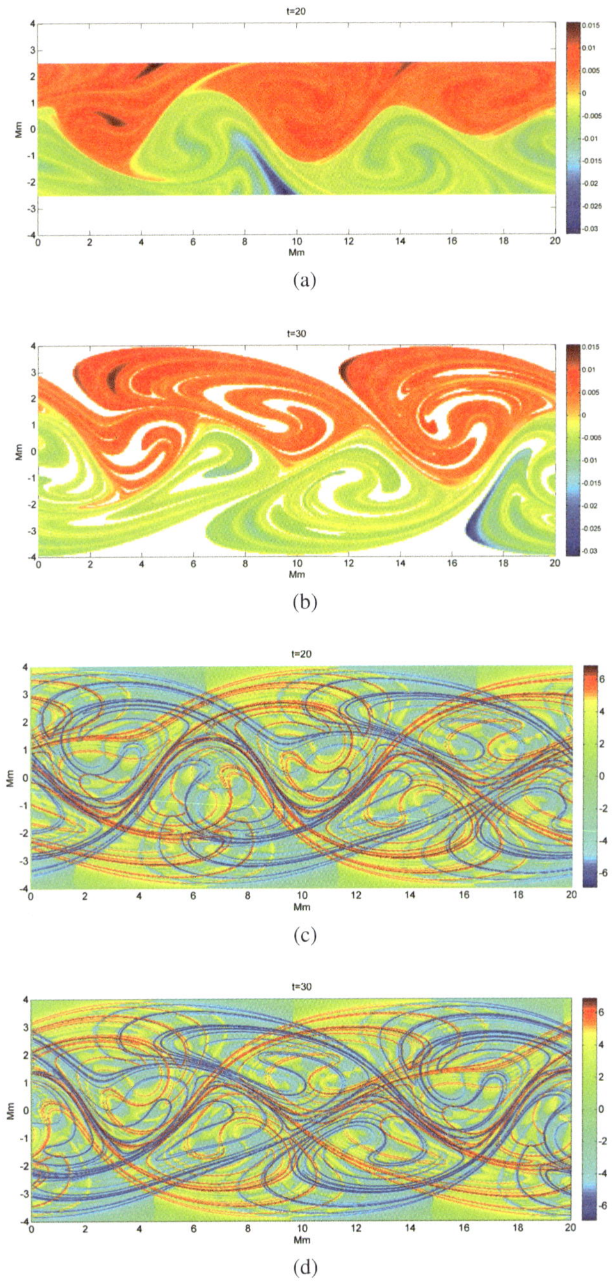

Figure 5.8. (a) *The optimal vector* x; (b) *the optimal vector* y; (c) *backward-time (blue) and forward-time (red) FTLEs at* $t = 20$ *days computed with the flow time* $\tau = 10$ *days;* (d) *backward-time (blue) and forward-time (red) FTLEs at* $t = 30$ *days computed with the flow time* $\tau = 10$ *days (from* [128]*).*

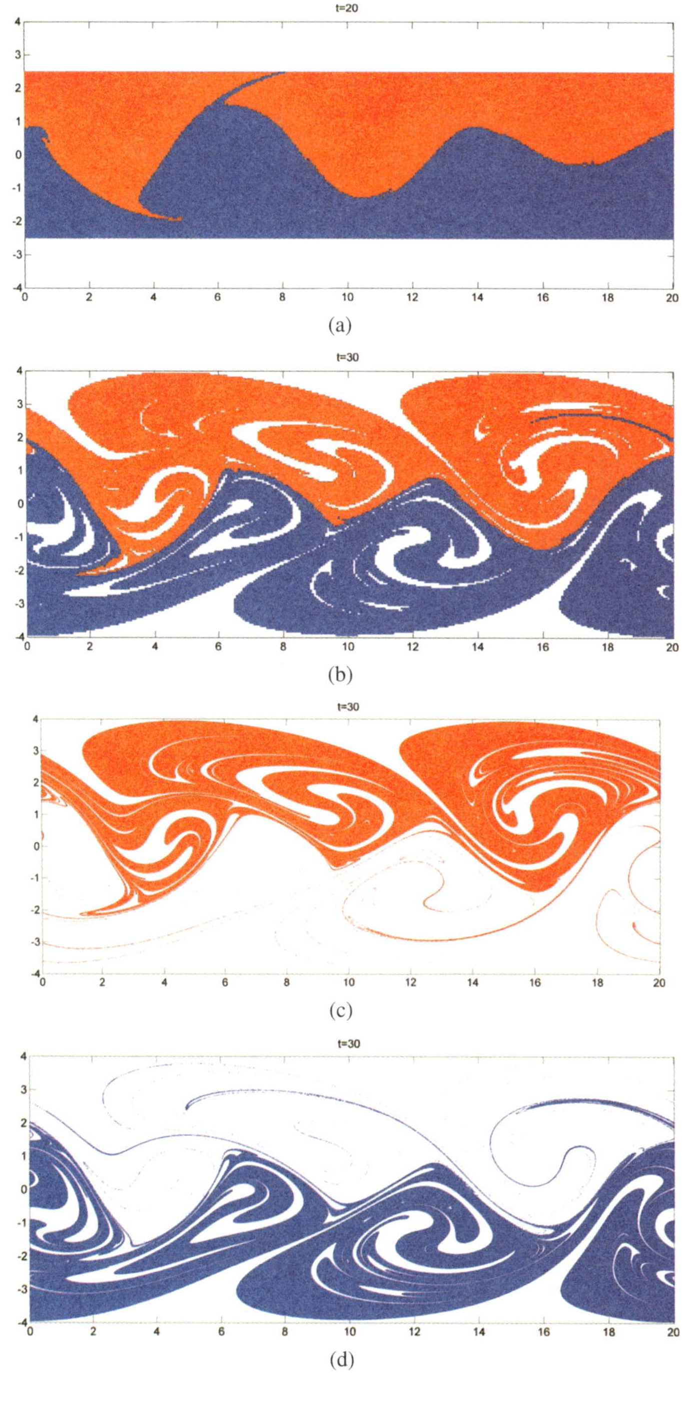

Figure 5.9. (a) *The sets* X_1 *(red) and* X_2 *(blue);* (b) *the sets* Y_1 *(red) and* Y_2 *(blue);* (c) *the set* $\Phi(X_1, 20; 10)$*;* (d) *the set* $\Phi(X_2, 20; 10)$ *(from* [128]*).*

Using $Q = 100$ sample points $z_{i,r}$, $r = 1,\ldots,$ with Q uniformly distributed in each grid box B_i, $i = 1,\ldots,m$, we calculate an approximate image $\Phi(X,t;\tau)$[60] and cover this approximate image with $m = 14395$ boxes $\{C_1,\ldots,C_n\}$ to produce the image domain Y. We construct $P = P_\tau^{(t)}$ as described earlier using the same $Q \times m$ sample points.

We compute x and y as described in Eq. (5.107); graphs of these vectors are shown in Fig. 5.10 (upper left and upper right). Fig. 5.10 (lower left and lower right) shows the result of Algorithm 1 in [128], extracting coherent sets A_t and $A_{t+\tau}$ from the vectors x and y. We calculate the coherent ratio $\rho_\mu(A_t, A_{t+\tau}) \approx 0.991$, which means that 99.1% of the mass in A_t (September 1, 2008) flows into $A_{t+\tau}$ (September 14, 2008), demonstrating a very high level of coherence.

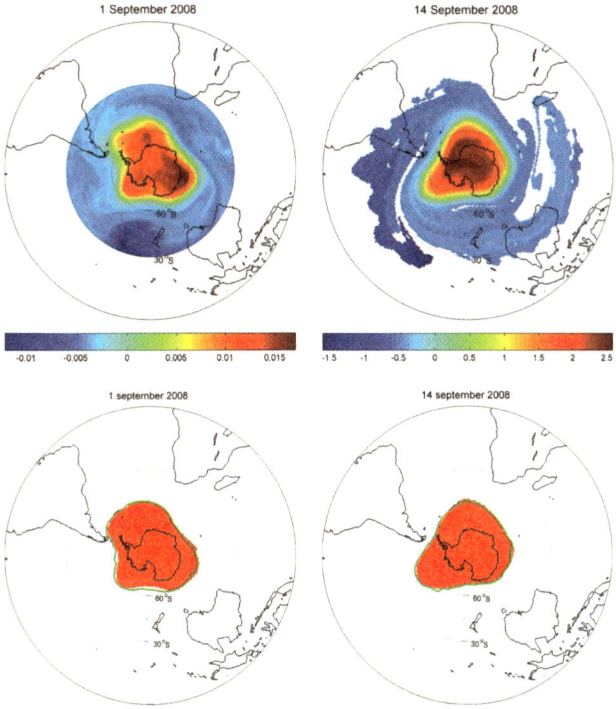

Figure 5.10. *upper left: Graph of x (September* 1, 2008*). upper right: Graph of y (September* 14, 2008*). lower left: The red set represents the coherent set A_t (September* 1, 2008*) obtained from Algorithm* 1 *in* [128]*. The green curve illustrates the vortex edge as estimated using PV. lower right: As for lower left plot at September* 14, 2008.

To benchmark the methodology as was done in [128], we will compare our result with a method commonly used in the atmospheric sciences to delimit the "edge" of the

[60] We use the standard Runge–Kutta method with step size of 3/4 hours. Linear interpolation is used to evaluate the velocity vector of a tracer lying between the data grid points in the longitude-latitude coordinates. In the temporal direction the data are independently affinely interpolated.

vortex. It has been recognized that during the winter a strong gradient of potential vorticity (PV) in the polar stratosphere is developed due to (1) strong mixing in the midlatitudes (resulting from the breaking of Rossby waves emerging from the troposphere and breaking in the stratospheric "surf zone" [216]) and (2) weak mixing in the vortex region. While PV depends only on the instantaneous vector field, PV is materially conserved for adiabatic, inviscid flow (both of which are good approximations in stratospheric flow over timescales of a week or two). Thus, PV may be viewed as a quantity derived from the Lagrangian specification of the flow and is therefore a meaningful comparator for these nonautonomous experiments that also use Lagrangian information. We used the method of Sobel et al. [297] to calculate a PV-based estimate of the vortex edge. The result is shown by the green curve in Fig. 5.10 (lower left and lower right). Notating the area enclosed by the green curve at September 1, 2008, by A_t^{PV} and at September 14, 2008, by $A_{t+\tau}^{PV}$, we compute $\rho_\mu(A_t^{PV}, A_{t+\tau}^{PV}) \approx 0.984$; 98.4% of the mass in A_t^{PV} flows into $A_{t+\tau}^{PV}$ over the 13-day period.

We see that a transfer operator methodology is clearly consistent with the accepted potential vorticity approach and in fact identifies a region that experiences slightly *greater* transport barriers across its boundary, indicated by the slightly larger coherence ratio: 99.1% versus 98.4%.

5.10 Community Methods

We close this chapter by relating two seemingly disparate fields, community finding and phase space partitioning. A topic of current interest in the area of network sciences has been the idea of communities and their detection; see the review article [261]. The applications of community finding range from social science to biological science to chemical interactions, to name a few. The betweenness methods in [241] and the local method in [9] are examples of numerous methods as reviewed in [73, 241] that can be successfully used to uncover various kinds of community structures.

A community could be loosely described as a collection of vertices within a graph that are densely connected among themselves while being loosely connected to the rest of the graph [311, 114, 266, 10]. This description, however, is somewhat vague and open to interpretation. This leads to the possibility that different techniques for detecting these communities may lead to slightly different yet equally interesting results.

Community finding methods are essentially structural analysis algorithms for the network structure (a graph is often described as a network in social sciences). Many techniques have been proposed to detect community structure inside a network. Clustering according to community structure in network science has many parallels to the problem of interest here of decomposing a dynamical system into almost-invariant sets when we realize that an Ulam–Galerkin matrix describes a graph structure.

The breadth of methods might be described as either agglomerative or as divisive. Rather than provide a survey of community finding [261], we will only touch on those methods which are relevant to our theme in this book, which regards partitioning the transfer operators so as to associate regions of phase space descriptive of transport in phase space. While most community methods were not designed explicitly to produce such a partition, some effectively do so. Only a true optimization cut method as discussed in previous sections addresses this directly, but these other methods can give comparable and sensible partitions.

5.10.1 Betweenness

Centrality measures serve as a sensible starting point to partitioning a network. Betweenness centrality has become a most popular partitioning tool [241]. Betweenness centrality is defined as the number of shortest (geodesic) paths passing through the edge.

Definition 5.3 (betweenness centrality). Assume a (directed) graph $G = (E,V)$, with vertices $V = \{v_1, v_2, \ldots, v_n\}$ and edges $E = \{e_1, e_2, \ldots, e_m\}$, and each edge can be written $e_k = (v_i, v_j), k = 1, \ldots, m, i, j = 1, \ldots, n$. A shortest path $p_{i,j}$ between any pair of vertices v_i and v_j exists and is computable. Let $\mathcal{P} = \{p_{i,j}\}_{i,j=1}^n$ be the set of all shortest paths between each pair of vertices. Therefore, the betweenness centrality of an edge e_k is the number of those shortest paths from $\mathcal{P}$ that pass through e_k.

The shortest path between a pair of vertices v_i and v_j can be computed by the Breadth First Search algorithm [22]. Betweenness centrality can be used to partition a network by the following process [241]. Given the graph $G = (E,V)$, remove from the edge set E the (an) edge with maximal betweenness centrality. Say it is e_k. Thus,

$$G = (E,V) \rightarrow G' = (E',V), \text{ where } E' = E - e_k. \tag{5.110}$$

This yields a new graph G'. This process is repeated until an initially irreducible graph is rendered reducible. The idea behind this algorithm closely mirrors the notion of congestion and bottlenecks.

Example 5.7. Consider the synthetic graphs illustrated in Fig. 5.11 [10]. Both of these synthetic examples in the figure illustrate the notion that betweenness serves as a measure of congestion in an unweighted graph. Thus, in this sense as a measurement of traffic flow between the components so identified, we have a network analogous concept to transport in a dynamical system.

Example 5.8. Consider the graph illustrated in Fig. 5.12 showing the social connections among a group of children in a karate club, called Zachary's Karate Club [241]. This network has become a paradigm for testing graph partition algorithms, and partitions well by the edge betweenness methods. See also the Amazon book example in Fig. 5.13, which naturally finds a politically left-leaning and right-leaning partition.

In terms of the betweenness centrality connection to the graph representation of dynamical systems, since two almost-invariant sets will communicate through a small number of connecting edges, the betweenness of those edges will be very high. Thus, cutting edges with the highest betweenness will render the graph partitioned into invariant sets if they were initially almost-invariant. The issue of both computing and presenting in an understandable format the huge data sets of a large network is a question of data mining.

The edge betweenness method [240], based on computing shortest paths, is a good robust candidate algorithm for detecting communities, but its run time is of the order $O(m^2n)$ for a graph with m edges and n vertices, or $O(n^3)$ for a sparse graph ($m \sim n$). Such expensive demands on computational resources make it infeasible for a large graph. Furthermore, a d-dimensional dynamical system with an h^d-element box (for box of order length h on a side) or triangle covering of the invariant set makes, therefore, an $O(h^{3d})$ algorithm, which is quite expensive for a fine grid, $h \ll 1$, in high dimension d. In the next section we will describe a method based on a modularity measure of a graph partition developed by Newman [238]. This method is far less expensive than the betweenness method and gives useful results in many applications.

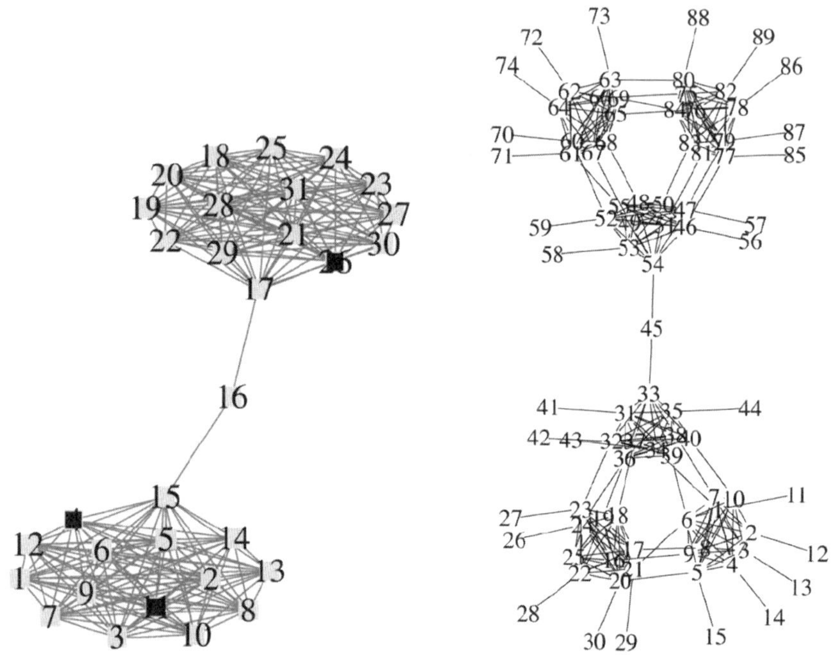

Figure 5.11. *Graphs with idealized betweenness centrality* [10]. *(Left) Ideal graph* 1: *two complete subgraphs of size* 15 *bridged by a common vertex. Initially this graph's highest betweenness centrality is carried by the edges* (v_{15}, v_{16}), *and likewise tied with* (v_{16}, v_{17}) *is this bridge which serves as a point of congestion/bottleneck in the graph. Removal of these renders a graph with two large components that can no longer communicate. (Right) Ideal graph* 3: *three complete subgraphs joined together by multiple edges with singletons attached bridged by a single vertex to another group of similar, but not identical, structure. Initially, maximal betweenness is carried by edges* (v_{33}, v_{45}) *tied with* (v_{45}, v_{54}). *Once these edges are removed the resulting graph in this example has further hierarchical structure.*

5.10.2 Modularity Method

First, we discuss the concept of the modularity as a measure of quality of a proposed community structure found by any graph partition algorithm. The modularity is a cost function associated with a partitioning of a given graph, $G(E, V)$,

$$Q : \mathcal{C}_G \to \mathbb{R}, \tag{5.111}$$

where $\mathcal{C}_G$ is the collection of all subpartitions C of a given graph G. Given a graph G and a (test) partition, $C \in \mathcal{P}_G$, $C = \{C_k\}$, each C is a set of subsets C_k, and C_k is a collection of vertices of G such that $\bigcup C_k = G$. We will refer to C_k as community-k. Newman [238] proposed that a good division of a graph network into communities could be justified only if the number of within-communities edges is significantly larger than the "expected number" of such edges when edges are randomly placed into a network. This requires that we must formally define what we mean by the expected number of edges for each vertex.

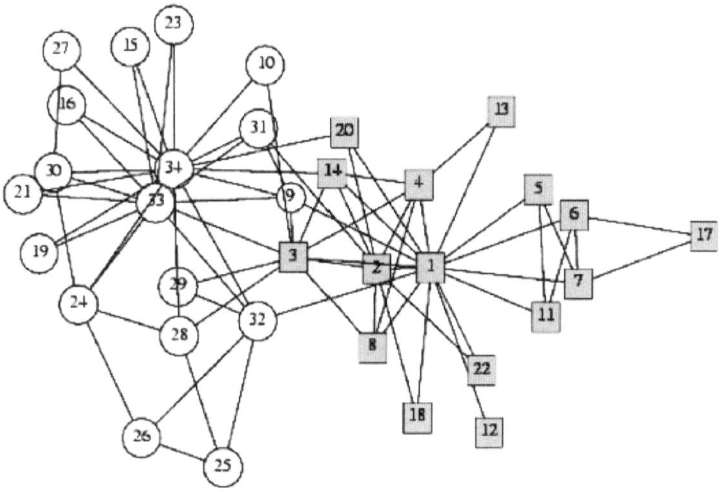

Figure 5.12. *Actual breakdown of the karate club from Newman.* [241]

For a given community partition C, let us denote by c_i the community that the vertex i belongs to and let P_{ij} be an element of the transition matrix $P \in \Re^{m \times m}$ of the network

$$Q = \frac{1}{m} \sum_{ij} \left(P_{ij} - \frac{k_i^{out} k_j^{in}}{m} \right) \delta(c_i, c_j), \tag{5.112}$$

where the delta-function $\delta(c_i,c_j) = 1$ if $c_i = c_j$ and 0 otherwise, $m = \sum_{ij} P_{ij}$, $k_i^{out} = \sum_j P_{ij}$, and $k_j^{in} = \sum_i P_{ij}$. Intuitively, this is the difference between the actual number of within-communities weighted edges and the expected number of such weighted edges. In the above definition, to see where the second term (the expected number of edges) comes from is to consider a weighted random graph model in which we give the vertex i the in- and out-degrees k_i^{in} and k_i^{out}, respectively, so the probability of finding the weighted $i \to j$ edge is $(k_i^{out} k_j^{in})/m$. When a strong community partition C exists, Q will be large. Put another way, Q measures how randomly the actual edges are distributed. If randomly distributed, Q will be small; otherwise it will be large.

For a particular case of undirected, nonweighted graph, we consider $P_{ij} = A_{ij}$, where the A_{ij} is an entry of the adjacency matrix given by

$$A_{ij} = \begin{cases} 1 & \text{if } (i,j) \in E, \\ 0 & \text{otherwise.} \end{cases} \tag{5.113}$$

The modularity Q can be expressed by

$$Q = \frac{1}{2m} \sum_{ij} \left(A_{ij} - \frac{k_i k_j}{2m} \right) \delta(c_i, c_j). \tag{5.114}$$

In this simplified case, the edges are counted twice, hence the use of $2m$ in the denominator, and there is no distinction between the in- and out-degrees, i.e., the probability of

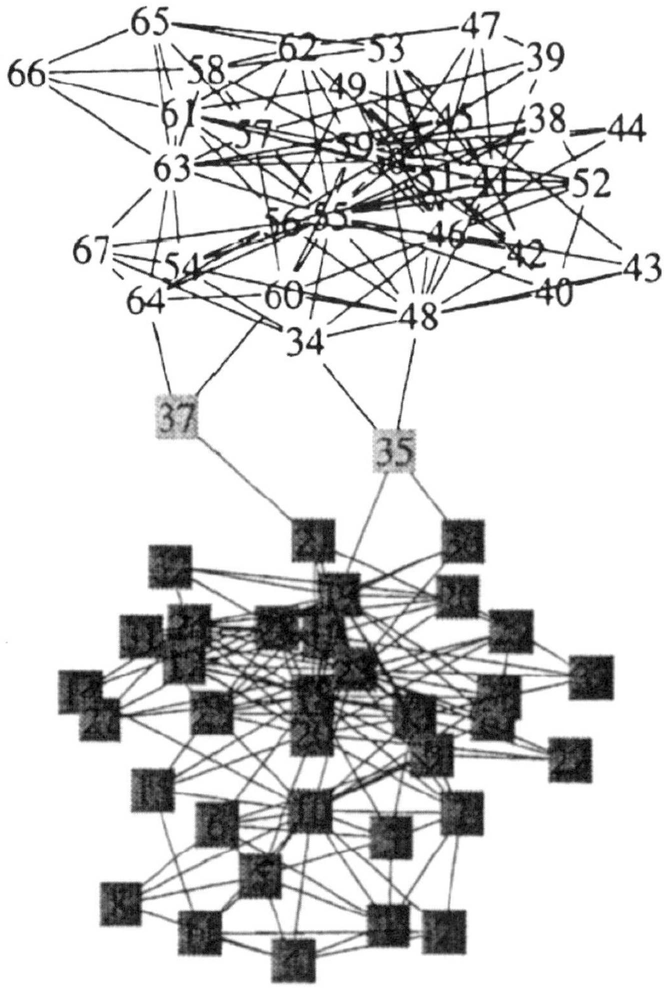

Figure 5.13. *The network of copurchased books on American politics as shown in* [10]. *Here a link is drawn between two vertices if those books were purchased together from a major online retailer. V. Krebs, Political Patterns on the WWW—Divided We Stand, http://www.orgnet.com/divided2.html.*

finding the edge connecting i and j is $\frac{k_i k_j}{2m}$. Since $\delta(c_i,c_j) = \sum_k \delta(c_i,k)\delta(c_j,k)$, the expression (5.114) can be rewritten as

$$\begin{aligned} Q &= \frac{1}{2m}\sum_{ij}\left(A_{ij} - R_{ij}\right)\sum_k \delta(c_i,k)\delta(c_j,k) \\ &= \sum_k \left(\frac{1}{2m}\sum_{ij} A_{ij}\delta(c_i,k)\delta(c_j,k) - \frac{1}{2m}\sum_i k_i\delta(c_i,k)\frac{1}{2m}\sum_j k_j\delta(c_j,k)\right). \end{aligned} \tag{5.115}$$

Now we define e_k to be the fraction of edges that connect vertices within the community P_k,

$$e_k = \frac{1}{2m}\sum_{ij} A_{ij}\delta(c_i,k)\delta(c_j,k), \tag{5.116}$$

and a_k to be the fraction of ends of edges that are attached to vertices in community P_k,

$$a_k = \frac{1}{2m}\sum_{i} k_i\delta(c_i,k). \tag{5.117}$$

Therefore, an alternative expression of Eq. (5.114) is

$$Q = \sum_k \left(e_k - a_k^2\right). \tag{5.118}$$

Example 5.9. Consider a graph network consisting of K identical communities without any intercommunity edges. Let m be the total number of edges in the graph network. If we choose a test partition to be a collection of these K communities, it is simple to see that $e_k = a_k = \frac{1}{K}$. So, we have $Q = 1 - \frac{1}{K}$.

Example 5.10. Consider again the graph network in the preceding example but allow an edge to connect between communities i and $i+1$. Thus we now have $2m+K-1$ edges in the network. Again, we choose the test partition to be the K communities. It follows that

$$\begin{aligned} Q &\le K\left(\frac{1}{K} - \left(\frac{\frac{2m}{K}+1}{2m+K-1}\right)\right)^2 \\ &= 1 - \frac{1}{K}\left(\frac{\frac{2m}{K}+1}{\frac{2m}{K}+1-\frac{1}{K}}\right)^2 \\ &< 1 - \frac{1}{K}. \end{aligned} \tag{5.119}$$

Recall that a good partition of a graph network into communities could be found by optimizing the modularity function Q. The true optimization is, however, very costly to implement in practice for very large networks. Clauset, Newman, and Moore [60] proposed an approximate algorithm based on greedy optimization. Suppose that we have a graph with n vertices and m edges. This algorithm starts with each vertex being the only member of one of n communities. At each step, the change in $Q(\mathrm{P})$ is computed after joining a pair of communities together and then choosing the pair that gives the greatest increase or smallest decrease in $Q(\mathrm{P})$. The algorithm stops after $n-1$ such joins in which a single community is left. This algorithm therefore runs in time $O((m+n)n)$, or $O(n^2)$ on a sparse graph ($n \sim m$). Note that using a more sophisticated data structure as introduced in [60] can reduce a run time to $O(n\log^2 n)$. Furthermore, the algorithm builds a dendrogram which represents the nested hierarchy of all possible community partitions of the network. Then we select the partition corresponding to local peaks of the objective function Q for our satisfactory community division. Fig. 5.14 shows an example of the dendrogram of the graph network shown in Fig. 5.4(a). We can see that the peak in the modularity corresponds to the correct identification of the community structure, and the modularity is dropped from the peak when two of the three communities are joined together at the final step. We note

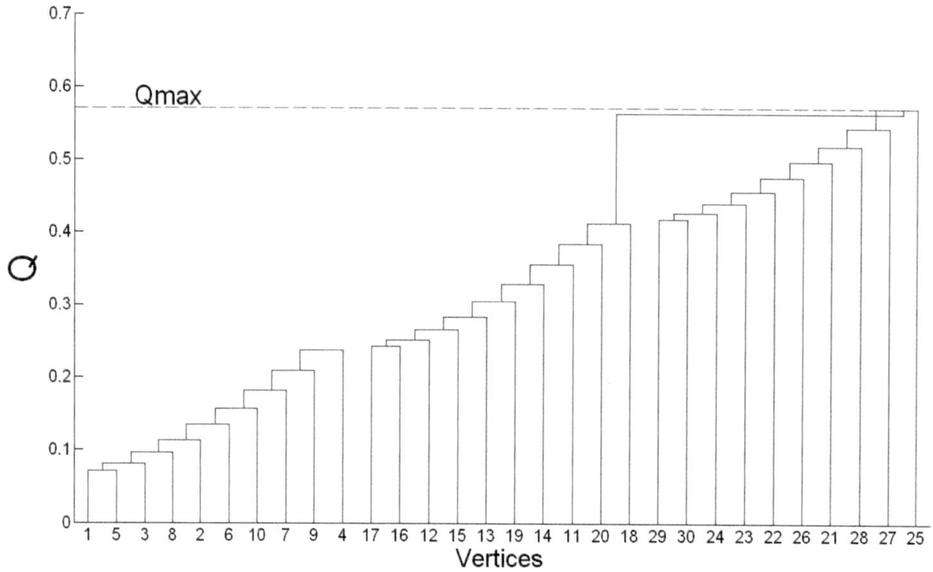

Figure 5.14. *Plot of dendrogram for a graph network in Fig.* 5.4(a). *As we can see, the peak in the modularity (dashed line) occurs where the network is divided into three communities.*

that the quality of the partition can be determined by the value of Q. A typical value of Q for a good division in many real-world networks is around 0.2–0.3. If the partition obtained from an optimization scheme shows a considerably smaller value of Q, say $Q < 0.1$, we assume that there are no significant partitions of the given network into communities.

5.10.3 Matrix Form of Modularity Method

Another approach to optimizing the Q functional reveals a spectral method. An idea of optimization schemes based on spectral methods has been traditionally used to determine a division of a network so that the minimum cut is achieved. In this sense, both in spirit and in some details, the community finding methods from the network literature has a great deal of coincidence between the spectral methods already highlighted here from the almost-invariant set work in Eq. (5.57) and the coherent set methods from Section 5.6, which at the root also have a notion of balanced cut and of reversible Markov chains.

To minimum cut methods in the graph theory literature, the eigenvalue spectrum of the Laplacian matrix is utilized to divide the network. A review of this method as applied to the minimum cut problem may be found in [239]. Min-cut problems have become a classic problem in graph theory [187, 138, 183, 299], and specifically there is a large literature by spectral methods [177, 178, 150]; see especially the text on spectral graph theory [59]. In particular, the second-smallest eigenvector of the graph Laplacian and corresponding eigenvector is called the Fiedler vector [113], and this gives rise to a classic bisection method of the graph into only two communities based on the sign of the corresponding vector entry [262]. These graph methods are known as algebraic connectivity.

For our purposes, we first consider the partition of a network into two communities. Let a vector **s** of n elements, where n is the number of vertices, be defined as

$$s_i = \begin{cases} 1 & \text{if } i \text{ is in community 1,} \\ -1 & \text{if } i \text{ is in community 2.} \end{cases} \tag{5.120}$$

Then we can rewrite Eq. (5.112) as

$$Q = \frac{1}{2m}\mathbf{s}^T B s, \tag{5.121}$$

where B is called the *modularity* matrix defined by

$$B_{ij} = P_{ij} - \frac{k_i^{out} k_j^{in}}{m}. \tag{5.122}$$

Therefore, we would like to determine the vector s that maximizes Q under the constraint that $s^T s = n$. In general, the modularity matrix B for a directed network is not symmetric, and so the spectral method cannot be directly applied. Instead, we consider the modularity Q of the following form:

$$Q = \frac{1}{4m} s^T (B + B^T) s. \tag{5.123}$$

Note that the matrix $B + B^T$ is not the same as the modularity matrix of the same network but ignores the directions. Also note the similarity in spirit to the reversible Markov chain discussed in Eq. (5.57). The idea of the spectral scheme is to write s as a linear combination of the eigenvectors v_i of $B + B^T$. Then we can express Q in terms of the eigenvectors v_i and their corresponding eigenvalues λ_i as

$$Q = \sum_i \lambda_i (v_i^T s)^2. \tag{5.124}$$

Then we can choose s to be as close to parallel with the leading eigenvector, say v_1, of the largest eigenvalue to maximize Q, i.e., maximizing the quantity

$$|v_1^T s| = \left| \sum_i s_i v_1^{(i)} \right| \le \left| \sum_i v_1^{(i)} \right|. \tag{5.125}$$

Under the constraint $s_i = \pm 1$, this can be achieved by choosing the sign of s_i to be the same as that of $v_1^{(i)}$.

We have only determined the optimal partition of the network into two communities. If the network comprises more than two communities, we have to further divide these two communities. Thus we repeat the bisection process to each community of the network using the spectral partitioning as described above and continue dividing the result further until the modularity of the network cannot be increased. In the course of this procedure, we have to compute the change in modularity denoted by ΔQ to decide whether or not the community k can be subdivided further. We define a vector s as before except that only vertices within the community g are taken into account, and note that ΔQ has to be computed based on the entire network. Then we have

$$\begin{aligned}\Delta Q &= \frac{1}{2m}\left[\sum_{i,j\in c_g}(B_{ij}+B_{ji})\frac{s_i s_j+1}{2} - \sum_{i,j\in c_g}(B_{ij}+B_{ji})\right] \\ &= \frac{1}{4m}\sum_{i,j\in c_g}\left[(B_{ij}+B_{ji}) - \delta_{ij}\sum_{k\in c_g}(B_{ik}+B_{ki})\right]s_i s_j \\ &= \frac{1}{4m}s^T(B^{(g)}+B^{(g)^T})s,\end{aligned} \tag{5.126}$$

where we set

$$B_{ij}^{(k)} = B_{ij} - \delta_{ij}\sum_{k\in g}B_{ik}. \tag{5.127}$$

To obtain the first line in the above derivation we subtract the modularity Q after subdivision from Q before subdivision, and notice that the term concerning vertices outside the community g can be canceled out. Then we use the fact that $s_i^2 = 1$ to derive the second line. Now, the spectral bisection scheme can be applied to determine the partition of the community g that maximizes ΔQ of the entire network. If ΔQ happens to be negative or smaller than some threshold value of ΔQ, it means that the community k should not be subdivided further.

5.11 Open Systems

In the previous sections of this chapter, the implication has been that the dynamical system is closed relative to the window of interest. It is natural that the window of interest may be smaller than the full phase space and it may not be an invariant set. While the discussion of coherent sets in Section 5.5 also includes notions and application to open systems, the emphasis is designed for "tracking" moving objects by moving the "windows" along with them. However, in discussions of transport properties confined within a fixed window, we take the perspective of an open system, which is the main subject of this section.

5.11.1 Open Systems and Holes

When the phase space of a dynamical system is covered, say by rectangles or a tessellation of triangles, as the support for characteristic functions as the basis set, then representing the Frobenius–Perron operator by the Ulam–Galerkin method requires that we decide how much of the phase space to inspect; we must choose a window of interest W. We can take a window W to be an open set relative to which we will consider the orbits of a dynamical system as they may enter or leave W, but generally it is simpler to discuss W's that are topological rectangles. While ideally, we would cover the entire phase space, that is not always possible for various reasons, either engineering, technical, or mathematical.

See Fig. 4.1, where some of the boxes (triangles) will partly or completely map outside the region of focus W. In Fig. 4.3, it is apparent that to study the Gulf of Mexico the data availability issues may be quite separate from the dynamical issue of invariance. The window focusing on the Gulf of Mexico naturally results in an open system according to the action of this dynamical system. The general issue is if W maps into itself, $T(W) \subseteq W$, then the system is closed.

For example, each of the maps shown in Figs. 5.1 and 5.2 are closed relative to $W = [0,2] = \overline{A} \cup \overline{B}$. But only the map shown in Fig. 5.1 is closed relative to the window choice, $W = [0,1] = \overline{A}$, whereas the map shown in Fig. 5.2 is open. This leads us to the following definition.

Definition 5.4. If $T(W) - T(W) \cap W = \emptyset$, then the dynamical system is **dynamically closed relative to** a window (set) W of interest. Otherwise the dynamical system is called **dynamically open relative to** a window (set) of interest W.

Dynamically closed relative to W is synonymous then with the statement that W is an "invariant set" but the word "closed" emphasizes the possibility of open or leaking windows W to a dynamical systems. Note that the use of the terms dynamically open and closed should not to be confused with open and closed as used in point set topology.

Example 5.11. Compare the smaller window W in Fig. 5.15 with those shown in Figs. 5.1 and 5.2. Here we choose a smaller window of inspection, $W = [0,1]$, as compared to previously, $W = [0,2]$. Otherwise the maps shown in Figs. 5.15 and 5.2 are identical in $[0,1]$; it is irrelevant whether they are different maps or not if we are inspecting a subregion and considering escape and invariance in $W = [0,1]$. In Fig. 5.15 we see that points that have orbits that are in the yellow strips, the preimages, escape W.

Those points that do not stay in W may be described as if they disappear into a hole. Such systems have been studied in many papers, including [256, 15, 206, 129, 28, 29] to name a few. It is not surprising that since the preimage of the middle yellow strip shown is a Cantor removal process; then those points that do not eventually escape form a Cantor set, and the dynamics restricted to this set is called an unstable chaotic saddle. More will be said on this topic regarding escape and unstable chaotic saddles in Chapter 7, in particular in Section 7.4.3. See in particular Fig. 7.29, which shows an unstable chaotic saddle in a Henon map example.

In the remainder of this section we will study the spectral properties of the corresponding transition matrix in systems such as those with holes. An extensive discussion of spectral properties of open systems can be found in [129, 126].

5.11.2 The Transition Matrix of an Open System

We will now mirror the analysis of Eqs. (5.11)–(5.19) in Section 5.3, particularly regarding a perturbation of a form $P \mapsto P + E$ as in (5.18), which we previously specialized for a closed system of the form (5.19) in Corollary 5.1. Now we will adjust the analysis for open systems. Specifically, we will perturb a state labeled 1 to describe the leaking window of interest W and 2 to describe the hole into which states evolve, as described by the leaky tent map dynamical system shown in Fig. 5.15.

In Section 5.3 we discussed that signs of the second eigenvector give some useful partition information when the first eigenvector is uniform. The analysis assumed that the matrix is almost block-diagonal of the form Eq. (5.11), and it is a closed system of the form (5.19).

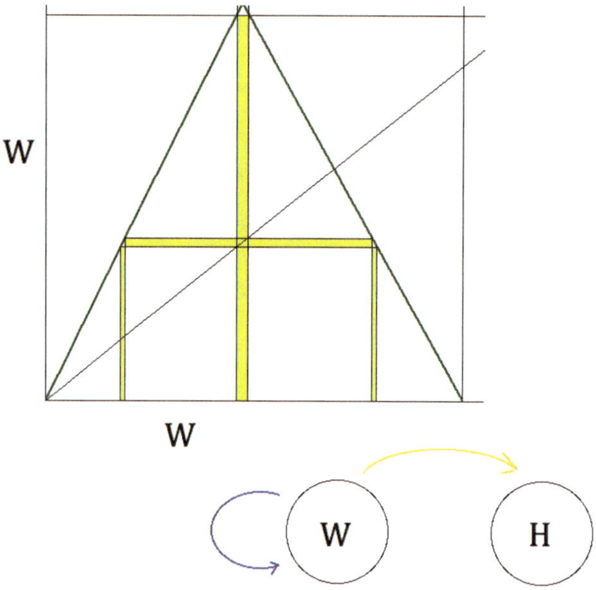

Figure 5.15. *In contrast to the map shown in Fig. 5.2, where the window of interest was chosen to be $W = [0,2]$, here the window of interest is chosen to be the smaller $W = [0,1]$; otherwise the maps are the same in this smaller window. Relative to W, the dynamical system behaves as if there were a hole. Dynamically, this is easy to see due to the fact that any orbit that falls in the center yellow-colored interval, or any preimage, will leave W. As far as W is concerned, these points have simply fallen into a hole outside the system, even if a larger windowed description in Fig. 5.2 shows further behavior of these points not discussed; relative to this W, they fell into the hole. Correspondingly a coarse graphical representation of the transfer operator is the directed graph shown below the map. Some orbits are invariant to W, but some leave W transitioning to what is called a hole, labeled H.*

Here akin to Eq. (5.19), assume that an open system with a hole can be written as a block-upper-diagonal system near a block-diagonal system, of the form

$$P = \begin{pmatrix} P_{1,1} & 0 \\ 0 & P_{2,2} \end{pmatrix} \mapsto \overline{P} = \begin{pmatrix} P_{1,1} + \epsilon E_{1,1} & \epsilon E_{1,2} \\ 0 & P_{2,2} + \epsilon E_{2,2} \end{pmatrix}. \tag{5.128}$$

Recalling Lemma 5.1, if the unperturbed block-diagonal form has block matrices which are algebraically and geometrically simple, then the dominant eigenvalue $\lambda = 1$ has a multiplicity-2 and partitioned eigenvectors as described by Eq. (5.13),

$$u = (v_1^T | 0^T)^T \quad \text{and} \quad w = (0^T | v_2^T)^T. \tag{5.129}$$

As such we recall (see Eq. (5.14)) that in the span of these two eigenvectors is the partitioned signed vector,

$$v = (v_1^T | - v_2^T)^T \tag{5.130}$$

such that a permutation on sorting the signs of a random permutation of P recovers the partition, analogous to Proposition 5.1.

We now describe how, according to Proposition 5.2 and Corollary 5.2, a sort on the signs of the second eigenvector of the perturbed matrix, (5.32), reveals that a useful partition generalizes to open systems.

Proposition 5.3. If in Eq. (5.128) P is stochastic, $E_{1,2} \geq 0$, and the perturbations $E_{1,1}$, $E_{1,2}$, and $E_{2,2}$ are such that $\overline{P}$ is also stochastic, then $\overline{P}$ has a dominant eigenvector,

$$v = (0^T_{n_1 \times 1} | \overline{v_2}^T)^T, \tag{5.131}$$

where $0_{n_1 \times 1}$ is a vector of zeros of the same size as v_1, and $\overline{v_2}$ is a vector of the same size and is near v_2.

This simply states that the dominant behavior describes the measure leaking out of the partition elements labeled 1 into those labeled 2.

Corollary 5.3. If $P_{2,2} = [1]_{1\times 1}$ is the 1×1 identity matrix, then $\overline{v_2} = v_2 = [1]_{1\times 1}$.

These statements can be compared to the leaky tent map dynamical system shown in Fig. 5.15, choosing a single element for H as per Corollary 5.3, $P_{2,2} = [1]_{1\times 1}$. The dominant eigenvector labeling those states labeled 1 covering W has 0's in the dominant eigenvector and the state labeled 2 describing H has a 1 in the eigenvector. This does not mean that *all* orbits fall into the hole, but only almost all orbits. This agrees with our knowledge that there is an invariant Cantor set in W which we have already described as the unstable chaotic saddle.

Note that to maintain that $P_{1,1} \to P_{1,1} + \epsilon E_{1,1}$ may describe a region W which leaks into a hole, but as $P_{1,1}$ is stochastic, then $\epsilon E_{1,1}$ should have negative entries so that $P_{1,1} \to P_{1,1} + \epsilon E_{1,1}$ will have a dominant eigenvalue less than 1 but the full system covers W and the hole H should still be stochastic. Therefore, to meet the hypothesis of Proposition 5.3, some entries of $E_{1,1}$, $E_{1,2}$, and $E_{2,2}$ will likely be negative.

Other interesting analyses regarding systems with holes come in the theory of Markov processes [296] and practice in statistical physics for diffusive systems and random walkers [308], most notably in the **first passage time** and the mean first passage time problems. These address the question, "How long does it take a random walker to reach a given target?" The mean first passage time $M_{i,j}$ from the state i to state j is the expected number of steps to arrive at destination j when the origin is i, and the derivations involve deliberately adding an extra hole vertex as a trap. Some related analysis for our dynamical systems interest here can be found in Section 7.4.2. All of the discussion in this section regarding holes and their Markov models should be compared to the traditional Markov process theory pertaining to **absorbing sets** [176], **hitting times,** and **absorption probabilities**.

5.11.3 Connecting Escape Rate and the Hole

Based on Eq. (5.128) it is certain that every transient states in $P_{1,1}$ will be eventually absorbed to the absorbing states, and it is important to estimate the rate at which the mass escapes from the set of transient states to the set of absorbing states. Without loss of generality, we can "lump" all the absorbing states and assume that we have the absorbing Markov chain with transition matrix

$$P = \begin{pmatrix} Q & R \\ 0' & 1 \end{pmatrix}, \tag{5.132}$$

where Q is $(N-1)\times(N-1)$ and $R, 0$ are both $(N-1)\times 1$. Then

$$P^n = \begin{pmatrix} Q^n & (I+Q+\cdots+Q^{n-1})R \\ 0' & 1 \end{pmatrix}. \tag{5.133}$$

We denote the index set of transient states, corresponding to the block Q in Eq. (5.132), by A. We will also assume that Q is irreducible. The probability that the transient states remain in the set of transient states after n steps, called $M^{(n)}$, is then given by

$$M^{(n)} := \sum_{i\in A, j\in A} Q^n(i,j)\pi(i) \tag{5.134}$$

for a given finite reference measure π, and $Q^n(i,j)$ is the (i,j) component of the matrix Q^n. Alternatively, we can write $M^{(n)} = \pi' Q^n \mathbf{1}$, where $\mathbf{1}$ is an $N-1\times 1$ vector with all elements equals to unity. We assume that Q is irreducible and substochastic. By the Frobenius–Perron theorem, we may arrange the eigenvalues of Q in the order $1 > \lambda_1 > |\lambda_2| > |\lambda_3| \geq |\lambda_4| \geq \cdots \geq |\lambda_{N-1}|$, and their corresponding normalized (right) eigenvectors are denoted by v_j for $j = 1,\ldots,N-1$ (i.e., $|v_j|_1 = 1$). We assume the strict inequality between $|\lambda_2|$ and $|\lambda_3|$ to simplify the analysis below. Then it can be readily shown, by writing $\mathbf{1}$ as a linear combination of eigenvectors, that

$$Q^n \mathbf{1} \propto \lambda_1^n v_1 + O\left(\left|\frac{\lambda_2}{\lambda_1}\right|^n\right). \tag{5.135}$$

In the limit of $n\to\infty$, the second term in the above equation vanishes, and we can show that

$$\begin{aligned} M^{(n)} &= \pi' Q^n \mathbf{1} \\ &\propto \lambda_1^n \sum_{j\in A} \pi(j) v_1(j) \leq \lambda_1^n \max[\pi(j)] \sum_{j\in A} v_1(j) \\ &\leq \lambda_1^n |v_1|_1 = \lambda_1^n. \end{aligned} \tag{5.136}$$

In other words, $M^{(n)} \propto \lambda_1^n$ in the limit and this is independent of the reference probability measure π. It is clear that $M^{(n)}$ is decreasing as n goes to infinity. The question is how fast it is decreasing as n is increasing. This motivates a notion of the escape rate w.r.t. a reference probability measure π, which is defined by

$$E(A) := -\lim_{n\to\infty} \frac{1}{n} \log(M^{(n)}). \tag{5.137}$$

Therefore, the escape rate is $E(A) = -\log\lambda$. The larger the escape rate, the faster the loss of the probability mass from the set of transient states A to the absorbing state.

Note that the Ulam matrix is usually not in the form of Eq. (5.132). However, when we consider a set of some states as a hole, we could eliminate the rows and columns associated with those states labeled as holes. This will effectively give rise to the Q matrix as in Eq. (5.132). More formally, the modified Ulam matrix for an open system T with a hole would be

$$\widehat{P} := \frac{m(B_i \cap T^{-1}(B_j) \cap H)}{m(B_i)}, \tag{5.138}$$

where H is the collection of B_i constituting the hole. The matrix $\widehat{P}$, as compared to Eq. (4.6), will be substochastic and is exactly the same as Q except for the zero rows and columns.

As mentioned earlier, this idea can be made analogous to the escape rate of the mass into the holes. In the case of general maps that do not admit a Markov partition, similar relations between the escape rate and the leading eigenvalue can also be obtained, but the discussion requires some technicalities beyond the scope of this book, so we refer to [90, 129, 256] for more details and reviews of this topic. Also, if the rate of escape is very slow, it could be of more relevance to consider the so-called absolutely continuous conditional invariant measure (ACCIM) (see [256, 90]), which is analogous to the quasi-stationary distribution in absorbing Markov chain; see [214].

5.12 Relative Measure and Finite Time Relative Coherence

In [128, Section 5.5], Definition 5.76 of finite time coherent pairs leads to a notion of optimal bipartition as computed by the spectral methods discussed in Section 5.6. See, for example, the optimal partition computed and displayed in Fig. 5.9 for the Rossby wave system, in Section 5.8. However, it may be relevant to consider finer-scaled coherent structures and even a hierarchy of coherent structures.

With the motivation of fine-scaled coherent structures, we will recast as in [37] a straightforward adjustment of definitions of the original measure-based notion of finite time coherent structures, using a relative measure instead of the global measure. Consider a measure space $(\Omega, \mathcal{A}, \mu)$ on (the window on) Ω, the phase space for which the finite time coherent sets are defined in Definition 5.2; notice that the coherence function $\rho_\mu(A_t, A_{t+\tau})$ in Eq. (5.76) explicitly includes the measure μ.

A standard definition of relative measure will be used in the following.

Definition 5.5. Given a measure space, $(\Omega, \mathcal{A}, \mu)$, and a μ-measurable subset $\omega \subset \Omega$, then the normalized inherited **relative measure** is defined on a measure space $(\omega, \mathcal{B}, \mu_\omega)$ by

$$\mu_\omega(B) = \frac{\mu(B \cap \omega)}{\mu(\omega)} \text{ for any } \mu\text{-measurable set } B \cap \omega \subset \mathcal{A}, \tag{5.139}$$

where $\mathcal{B}$ is the μ_ω-measurable sets.

If Ω is an invariant set, then the windowing is considered a closed system, but otherwise we are considering an open system. In any case, it is convenient for what follows to assume that $\mu : \Omega \to \Re+$ is a probability measure and as such $\mu(\Omega) = 1$.

With relative measure immediately follows a notion of **finite time relative coherence**, by slightly adjusting Definition 5.76 in [128] to use μ_ω in the restricted window ω instead of μ on the larger window Ω.

Definition 5.6. We will call A_t, $A_{t+\tau}$ a $(\rho_0, t, \tau) - \mu_\omega$-**relative coherent pair** if the following hold:

1.

$$\rho_{\mu_\omega}(A_t, A_{t+\tau}) := \frac{\mu_\omega(A_t \cap \Phi(A_{t+\tau}, t+\tau; -\tau))}{\mu_\omega(A_t)} \geq \rho_0. \tag{5.140}$$

2. $\mu_\omega(A_t) = \mu_\omega(A_{t+\tau})$.

3. A_t and $A_{t+\tau}$ are "robust" to small perturbations.

We take the term "robust" to mean simply that the observed coherent set varies continuously with respect to perturbations.

Relative measure suggests a straightforward refinement of the spectral-based balanced cut method from Section 5.6 to optimize coherence [128]. Suppose that we have already optimized coherence to produce a bipartition, $X \cup Y = \Omega$. Then we may repeat the process in each subset. Let $\omega = X$ and then use the same spectral method from Section 5.6 to optimize coherence $\rho_{\mu_\omega} := \rho_{\mu_X}$. This produces what we shall label $X_1 \cup Y_1 = X$. Likewise, repeat: let $\omega = Y$ to produce a $\rho_{\mu_\omega} := \rho_{\mu_Y}$-coherent partition, $X_2 \cup Y_2 = Y$. Thus the bipartition $X \cup Y$ produces a partition in four parts, $X_1 \cup Y_1 \cup X_2 \cup Y_2$. Then repeat, choosing successively $\rho_{\mu_\omega} := \rho_{\mu_{X_1}}$, $\rho_{\mu_{Y_1}}$, $\rho_{\mu_{X_2}}$, and then $\rho_{\mu_{Y_2}}$, to produce an eight-part partition, etc, in a deepening tree structured partitioning of the phase space into relatively coherent sets. See Fig. 5.16 for a depiction of the repetition aspect of this algorithm. A nat-

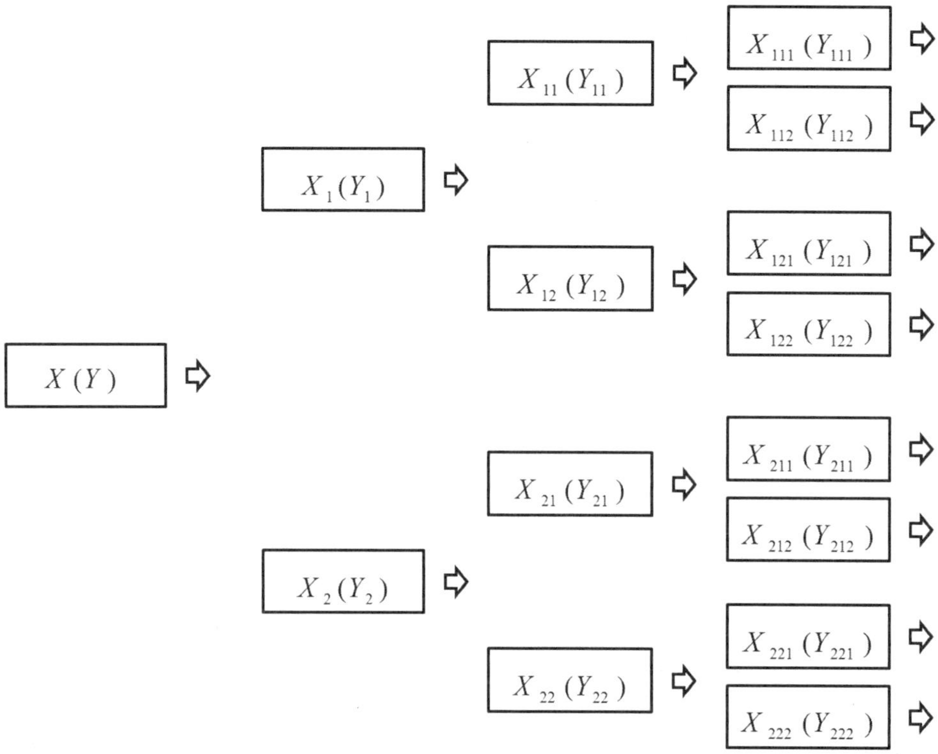

Figure 5.16. *Algorithm tree toward finite time relatively coherent sets. Using Definition* 5.76 *of finite time coherent sets from* [128] *and optimizing leads to a partition* $X \cup Y = \Omega$. *Further refinements using relative measure on successively smaller elements of the partition leads to finite time relatively coherent sets according to Definition* 5.6 *from* [37].

ural stopping criterion for the algorithm is to terminate any given branch of the tree when the spectrally computed optimal relative measure on the given branch produces a partition where the objective function is ρ_{μ_ω}, which even when optimized is nonetheless not large. This indicates that the test subpartition is not relatively coherent, in the sense of Eq. (5.6).

Example 5.12. Consider again the Rossby wave system already discussed in Section 5.8, with the finite time coherent partition shown as a bipartition X, Y colored red and blue in Fig. 5.8. In Fig. 5.16, we illustrate from [37] several subdivisions toward finite time relatively coherent sets as described in Fig. 5.16.

Again, we write in terms of the Hamiltonian,

$$dx/dt = -\partial\Phi/\partial y,$$
$$dy/dt = \partial\Phi/\partial x, \tag{5.141}$$

where

$$\begin{aligned}\Phi(x,y,t) = {} & c_3 y - U_0 L\tanh(y/L) \\ & + A_3 U_0 L\text{sech}^2(y/L)\cos(k_1 x) \\ & + A_2 U_0 L\text{sech}^2(y/L)\cos(k_2 x - \sigma_2 t) \\ & + A_1 U_0 L\text{sech}^2(y/L)\cos(k_1 x - \sigma_1 t).\end{aligned} \tag{5.142}$$

This is a quasi-periodic system that represents an idealized zonal stratospheric flow [128]. There are two known Rossby wave regimes in this system. Let $U_0 = 63.66, c_2 = 0.205U_0, c3 = 0.7U_0, A3 = 0.2, A2 = 0.4, A1 = 0.075$ and the other parameters.

Building a 32640×39694 matrix, we choose 20,000,000 points in the domain $X = [0,\ 6.371\pi \cdot 10^6] \times [-2.5 \cdot 10^6,\ 2.5 \cdot 10^6]$ of the flow and use 32640 triangles as the partition $\{B_i\}_{i=1}^{32640}$ for the initial status points and 39694 triangles as the partition $\{C_j\}_{j=1}^{39694}$ for the final status of the points. Note that this system is "open" relative to the domain X chosen, though it is an area-preserving flow. The two coherent pairs are colored blue and red which are defined as (X_1, Y_1) and (X_2, Y_2) in the first level of Fig. 5.17. Again, we now build the relative measures and tree of relatively coherent pairs. By applying the method as we have done with the previous two examples, we develop four and eight different coherent structures for the second level and the third level, respectively.

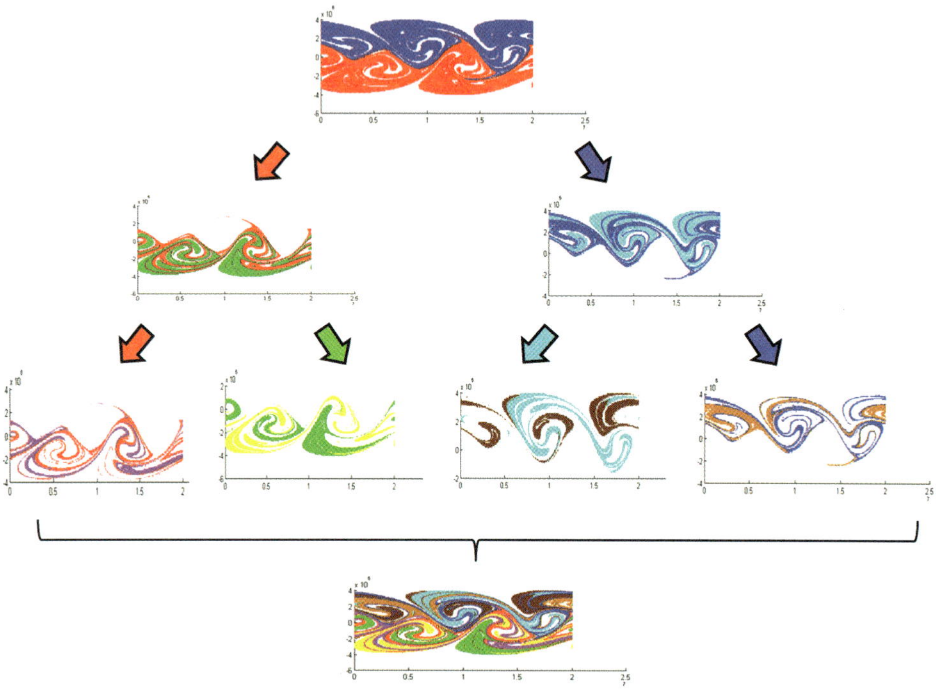

Figure 5.17. *Compare to Fig.* 5.8. *Tree of finite time relatively coherent sets according to the algorithm described in Fig.* 5.16.

Chapter 6

The Topological Dynamics Perspective of Symbol Dynamics

Why include a chapter on symbol dynamics in a book about measurable dynamics? After all, symbol dynamics can rightly be described as part of topological dynamics. That is the study of a dynamical system in a topological space (the set of open sets), meaning the action of the map not taking into consideration any measure structure. So how does this relate to measurable dynamics? The answer is in the shared tools of analysis and hopefully broader perspective and common methods. Specifically, we have already described that the Frobenius–Perron transfer operator is well approximated by stochastic matrices when it is compact (recall Sections 4.2 and 4.3), and there is a corresponding description by weighted directed graphs. Likewise, symbol dynamics, as we shall describe here, is well understood as approximated by an adjacency matrix (a matrix of 0's and 1's rather than weights summing to 1 as in the stochastic case), and the corresponding description by an unweighted directed graph. In fact, for a given dynamical system, the same directed graph can be used if we ignore the weights. Furthermore, the discussion regarding exactness of the representation by finite-sized directed graphs, which occurs when the dynamical system is Markov and the Markov partition is used, is identical between both the measurable and topological dynamical systems analysis.

6.1 Symbolization

Some of the most basic questions in dynamical systems theory do not require us to even consider measure structure. These are questions of loss of precision regarding state and information rather than description of how much and how big. Symbolic dynamics is a framework by which a number of theoretical and also practical global descriptions of a chaotic dynamical system may be analyzed, and often quite rigorously. Symbolic dynamics as defined in this chapter and the related topic of lobe dynamics defined in the next chapter can help lead to a better understanding of transport mechanisms. Then considering the corresponding measure structure can lead to escape rates, loss of correlation, and a partition-specific description of the Frobenius–Perron operator.

We take a naive perspective here for the sake of narration and include topological issues of lobe dynamics relevant to the measurable dynamics discussion of the rest of this book. For more in-depth discussion of these topological issues, we refer the reader to the excellent texts [268, 315].

Formally, the topic of symbolic dynamics is descriptive of topological dynamics, whereas a great deal of the theme of this book regards measurable dynamics. Without including measure structure, only the evolution of open sets and points therein are considered; topological dynamics does not worry about the size/weight of sets.

Symbolic dynamics may seem at first blush to be an abstract and even complicated topic without relevance to applied dynamical systems and certainly not useful to the experimental scientist. To the contrary, symbolic dynamics is in many ways a simplifying description of chaotic dynamical systems that allows us to lay bare many of the fundamental issues of otherwise complicated behavior. These include

- the complementary roles of information evolution (see further in Chapter 9), measurement precision, entropy, and state;
- the role of partitions;
- a better understanding of discrete time and discrete space descriptions of a dynamical system—leading naturally to the graph theoretic tools already discussed;
- mechanisms of chaos.

Even applied topics such as the widely popular topic of controlling chaos [245, 115, 288][61] become approachable within the language of symbolic dynamics, simply by forcing desirable itineraries [254, 65, 255, 26].

The next section of this chapter will serve as a quick-start tutorial describing the link between symbolic dynamics as a description of a chaotic attractor, and how a dynamical system in some sense constantly transmits information, and may even be manipulated to be an information-bearing signal. Some of this description is drawn roughly from our review [26]. A more detailed treatment will be given in the following sections.

6.1.1 Those "Wild" Symbols

The "wild" is a standard chaos toy that can be purchased from many science education outlets. One such realization of the toy is shown in the photograph in Fig. 6.1 (left). The device has a two-degree of freedom pendulum suspended from above which is free to oscillate along both angular axes, θ and ϕ. While we could certainly write equations of motion by the Lagrangian method [139] and details are beside the point of this discussion, we will simply state that the differential equation is available. Perhaps more realistically, it is possible that the data shown are collected from the actual physical experiment. The point is, we can refer to output of said variables as seen as time series in Fig. 6.1 (right).

A natural guess occurs to us as a partitioning from this system. There are 6 magnets placed symmetrically under the colored circular segments shown, which are placed to repel the pendulum arm. Over the top of each magnet, a colored segment is labeled one of the following:

$$\{\text{Go to Lunch, Maybe, Yes, Quit, Try Again, No}\}, \tag{6.1}$$

over the segments colored, in order,

$$\{\text{orange, red, green, yellow, purple, white}\}. \tag{6.2}$$

[61] Control of chaos can be described briefly as a different perspective on sensitive dependence to initial conditions and parameter variations—small feedback control inputs can be *designed* to yield dramatic output variations.

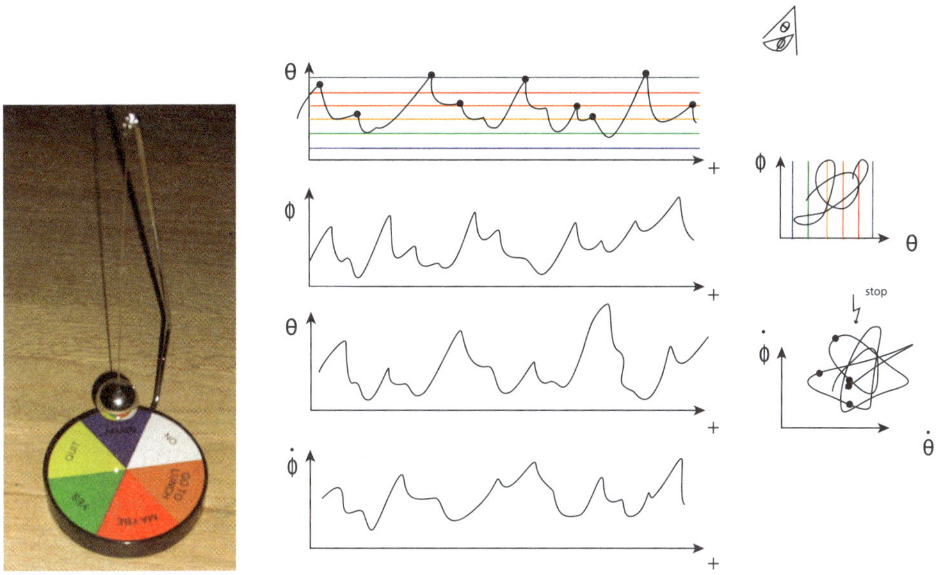

Figure 6.1. *(Left) Photograph of a chaotic "wild," which here is a realization of a chaotic desktop toy. Six magnets are in the base, one under each colored segment corresponding to symbols in Eqs.* (6.1)–(6.3). *Each magnet in the base has an opposing polarity and so repels to a magnet that is in the silver pendulum tip. (Middle) Time series in each of the free angles,* $\theta(t)$ *and* $\phi(t)$, *as well as angular velocities,* $\dot{\theta}(t)$ *and* $\dot{\phi}(t)$. *The top panel shows the colored segments in the wild—left corresponds to a partition in the time series easily seen in* $\theta(t)$ *as shown. (Right) The phase space of the chaotic trajectories in* $(\theta, \phi, \dot{\theta}, \dot{\phi})$ *suppresses the time as parameterized curves, here shown top and bottom as projections in the* (θ, ϕ), *and* $(\dot{\theta}, \dot{\phi})$, *respectively.*

Either way, these serve as perfectly good symbol sets, and no better than the simple indexed set,

$$\{1, 2, 3, 4, 5, 6\}. \tag{6.3}$$

Given these symbolic labels, one can ask the following simple question: Over which colored segment can the pendulum be found when it reaches its maximum angular displacement and reverses direction? Thus the labels become immediately connected to the dynamics, from which results a symbolic dynamics. In one run during this writing, we found, $1-5-1-5-1-5-4-3-5-\ldots$[62] which is equivalent to the labels Go to

[62] Recall that in standard dynamical systems language, an orbit is an infinite sequence describing the long-term behavior of the system. In symbolic dynamics that means the orbit must be an infinite itinerary, and the ellipsis, "..." denotes "keep going forever." However, the lack of such precision is encoded in the distance function of the symbolic dynamics equation (6.11), which rewards for more precision near the beginning, and loss of precision as time goes on. Furthermore, the representation as short time segments is in many senses the theme of this book that short time representations of a dynamical system are also useful. In this case, the long time behavior of this dissipative system is that the pendulum stops in the straight down position.

Lunch – Try Again – Go to Lunch – Try Again – Go to Lunch – Try Again – Quit – Yes – Try Again – .... This in a primitive form is an example of symbolic dynamics.

The amazing story behind symbolic dynamics is that with an appropriately chosen partition, the symbols alone are sufficient to completely classify all the possible motions of the dynamical system.[63] That is, there may be a conjugacy between the symbolic dynamical system and the dynamical system in natural physical variables. The Smale homoclinic theorem and horseshoe theory are the historical beginning of this correspondence [294, 295]. Defining these concepts and determining their validity is discussed in the rest of this chapter. As it turns out, the partition shown here is not likely to be one of the special (generating) partitions which allow such an equivalence. Nonetheless, interesting symbolic streams result, and such is the issue of the "misplaced" partition which we shall also discuss [41] in Section 6.4.6. Now we shall segue to the more detailed topic with a bit more analytic introductory example.

6.1.2 Lorenz Differential Equations and Successive Maxima Map

As another introductory example to symbolic dynamics, consider the now famous and favorite[64] Lorenz system [207],

$$\begin{aligned} \dot{x} &= 10(y-x), \\ \dot{y} &= x(28-z)-y, \\ \dot{z} &= xy-(8/3)z, \end{aligned} \tag{6.4}$$

as a benchmark example of chaotic oscillations. Those interested in optical transmission carriers may recall the Lorenz-like infrared NH3 laser data [312, 313, 168]. A time series of the $z(t)$ variable, of a "typical" trajectory, is seen in Fig. 6.2.

Edward Lorenz showed that his equations have the property that the successive local maxima of the z time series can be described by a one-dimensional, one-hump map,

$$z_{n+1} = f(z_n), \tag{6.5}$$

where we let z_n be the nth local maximum of the state variable $z(t)$. The chaotic attractor in the phase space $(x(t), y(t), z(t))$ shown in Fig. 6.3 corresponds to a one-dimensional chaotic attractor in the phase space of the discrete map $f(z)$, and hence the symbol dynamics are particularly simple to analyze; see Fig. 6.4. The generating partition for defining good symbolic dynamics is now simple as this is a one-dimensional map; it is the critical point z_c of the $f(z)$ function. A trajectory point with

$$z < z_c \; (z > z_c) \text{ bears the symbol } \mathbf{0} \; (\mathbf{1}). \tag{6.6}$$

The partition of this one-dimensional map, of successive $z(t)$ maxima, corresponds to a traditional Poincaré surface mapping, as the two leaves of the surface section can be seen in Fig. 6.3. Each bit roughly corresponds to a rotation of the $(x(t), y(t), z(t))$ flow around

[63] If an "inappropriate" partition is used, many orbits may give the same symbolic stream [41], and thus the symbolic dynamics representation may not be faithful.

[64] The Lorenz system was originally discussed as a toy model for modeling convection rolls in the atmosphere—a phenomenological weather model. However, following a now famous discovery regarding loss of precision and sensitive dependence, it has become a pedagogical favorite for introducing chaos.

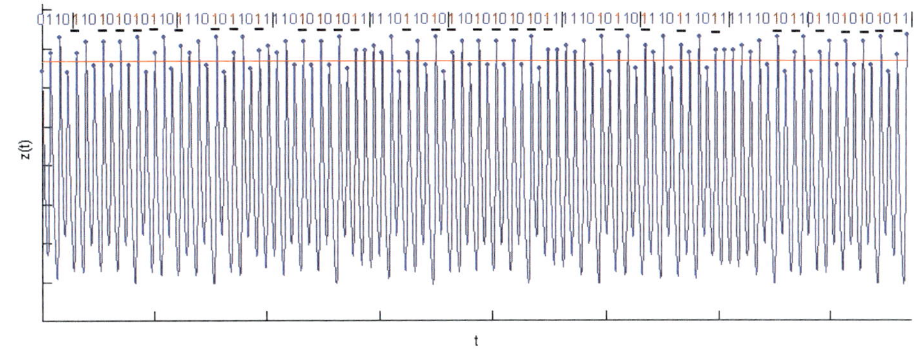

Figure 6.2. *A $z(t)$ time series from the Lorenz equations,* (6.4). *This particular $z(t)$ time series is from the $(x(t), y(t), z(t))$ trajectory shown in Fig.* 6.3. *The underlined bits denote non-information-bearing buffer bits, which are necessary either due to nonmaximal topological entropy of the underlying attractor or further code restrictions which were added for noise resistance, as discussed in* [28, 26]. *In Section* 9.4 *we will discuss further this example in the context of information theory in dynamical systems, and that these underlined bits correspond to a submaximal entropy.* [26]

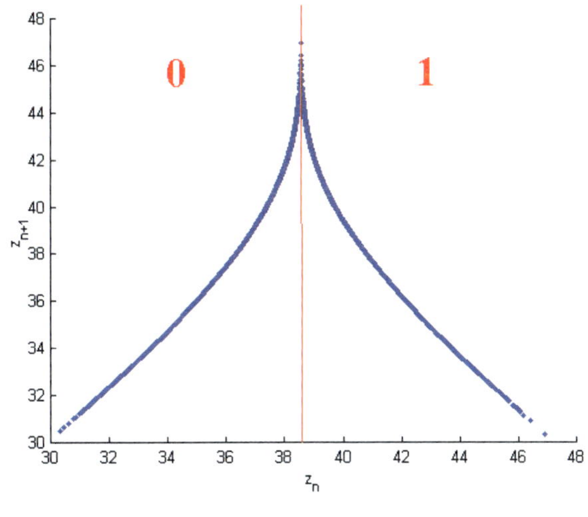

Figure 6.3. *Successive maxima map $z_{n+1} = f(z_n)$ from the measure $z(t)$ variable, of the Lorenz flow $(x(t), y(t), z(t))$ from Eqs.* (6.4). [26]

the left or the right wings of the Lorenz butterfly-shaped attractor. However, the Lorenz attractor does not allow arbitrary permutations of rotations around one and then the other wing; this translates to the statement that corresponding symbolic dynamics has a somewhat restricted grammar which must be learned. The grammar of the corresponding symbolic dynamics is a statement that completely characterizes the allowed trajectories of the map, and hence the flow, or equivalently classifies all periodic orbits. In fact, understand-

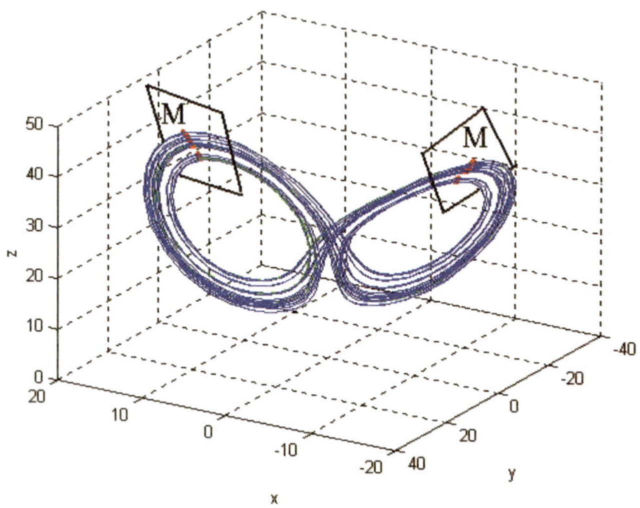

Figure 6.4. *Lorenz's butterfly attractor. This particular "typical" trajectory of Eqs.* (6.4) *shown in Fig.* 6.2 *when interpreted relative to the generating partition* (6.6) *of the one-dimensional successive maxima map* (6.5) *shown in Fig.* 6.4 *is marked by the red points, which relate to the Poincaré section manifold M.* [26]

ing how a chaotic oscillator can be forced to carry a message in its symbolic dynamics is not only instructive and interesting, but has also became a subject of study for control and transmission purposes [160, 28, 26, 67].

6.1.3 One-Dimensional Maps with a Single Critical Point

In this subsection we begin a slightly more detailed presentation of symbolic dynamics corresponding to a dynamical system, starting with the simplest case, a one-humped interval map, such as the situation of Lorenz's successive maxima map. Successively more complicated cases, multihumped and then multivariate, will be handled in the following subsections.

$$f:[a,b]\to[a,b]. \tag{6.7}$$

Such a map has symbolic dynamics [223, 89] relative to a partition at the critical point x_c. Choosing a two-symbol partition, labeled $\mathcal{A}$={0,1}, naming iterates of an initial condition x_0 according to

$$\sigma_i(x_0)=\begin{pmatrix} 0 \text{ if } f^i(x_0)<x_c \\ 1 \text{ if } f^i(x_0)>x_c \end{pmatrix}. \tag{6.8}$$

The function h labels each initial condition x_0 and corresponding orbit $\{x_0,x_1,x_2,\ldots\}$ by an infinite symbol sequence:

$$h(x_0)\equiv\sigma(x_0)=\sigma_0(x_0).\sigma_1(x_0)\sigma_2(x_0)\ldots. \tag{6.9}$$

Defining the "fullshift,"

$$\Sigma_2=\{\sigma=\sigma_0.\sigma_1\sigma_2\ldots,\text{ where }\sigma_i=0\text{ or }1\}, \tag{6.10}$$

to be the set of all possible infinite symbolic strings of 0s and 1s, then any given infinite symbolic sequence is a singleton (a point) in the fullshift space, $\sigma(x_0) \in \Sigma_2$.[65]

The usual topology of open sets in the shift space Σ_2 follows the metric

$$d_{\Sigma_2}(\sigma,\overline{\sigma}) = \sum_{i=0}^{\infty} \frac{|\sigma_i - \overline{\sigma}_i|}{2^i}, \tag{6.11}$$

which defines two symbol sequences to be close if they agree in the first several bits.

Eq. (6.8) is a good change of coordinates, or more precisely a homeomorphism,[66]

$$h : [a,b] - \cup_{i=0}^{\infty} f^{-i}(x_c) \rightarrow \Sigma_2',^{67} \tag{6.12}$$

under conditions on f,[68] such as piecewise $|f'| > 1$.[69] The Bernoulli shift map moves the decimal point in Eq. (6.32) to the right, and eliminates the leading symbol,

$$(s(\sigma))_i = \sigma_{i+1}. \tag{6.13}$$

All of those itineraries from the map f Eq. (6.7) in the interval correspond by Eq. (6.8) to the Bernoulli shift map restricted to a subshift,

$$s : \Sigma_2' \rightarrow \Sigma_2'. \tag{6.14}$$

Furthermore, the change of coordinates h respects the action of the map, meaning it commutes, and furthermore it is a conjugacy.[70]

In summary, the above simply says that corresponding to the orbit of each initial condition of the map (6.7), there is an infinite itinerary of 0s and 1s, describing each iterate's position relative the partition in a natural way, which acts like a change of coordinates such that the dynamical description is equivalent in either space, whether it be in the interval or in the symbol space.

---

[65]We are using the notation $\sigma_i(x_0)$ to denote the symbol of the ith of the initial condition according to Eq. (6.8) but $\sigma(x_0)$ is a function which assigns a full infinite symbolic sequence corresponding to the full orbit from the initial condition x_0 by $\sigma : [a,b] \rightarrow \Sigma_2$.

[66]A *homeomorphism* between two topological spaces A and B is a one-one and onto continuous function $h : A \rightarrow B$, which is formerly the equivalence relationship between two topological spaces.

[67]A *subshift* Σ_2' is a closed and Bernoulli shift map invariant subset of the fullshift, $\Sigma_2' \subset \Sigma_2$. A subshift describes the subset of the fullshift which is those infinite symbol sequences which actually do occur in the dynamical system as some periodic orbits may not exist, and correspondingly so must the corresponding symbolic sequences be absent from the representing subshift Σ_2'.

[68]Conditions are needed on the set to guarantee uniqueness of the symbolic representation. Some kind of fixed-point theorem is needed to guarantee that contraction to a unique point in $\mathbb{R}$ occurs when describing a symbol sequence of increasing length. In one-dimensional dynamics, the contraction mapping theorem is often used. In diffeomorphisms of the plane, often homology theory is used.

[69]Note that preimages of the critical point are removed from $[a,b]$ for the homeomorphism. This leaves a Cantor subset of the interval $[a,b]$. This is necessary since a shift space is also closed and perfect, whereas the real line is a continuum. This is an often overlooked technicality, which is actually similar to the well-known problem when constructing the real line in the decimal system (the ten-shift Σ_{10}) which requires identifying repeating decimal expansions of repeating 9's such as, for example, $1/5 = 0.1\overline{99} = 0.2$. The corresponding operation to the shift maps [12] is to identify the repeating binary expressions $\sigma_0.\sigma_1 \ldots \sigma_n 0\overline{N = 11} \equiv \sigma_0.\sigma_1 \ldots \sigma_n 1\overline{N = 11}$, thus "closing the holes" of the shift space Cantor set corresponding to the critical point of the map and its preimages.

[70]A *conjugacy* between two maps is a homeomorphism h between each phase space as topological spaces A and B, which commutes the maps on those two spaces, $\alpha : A \rightarrow A$, $\beta : B \rightarrow B$, then $h \circ \alpha = \beta \circ h$. Conjugacy can be considered as the major notion and the gold standard of equivalence used in dynamical systems theory when comparing two dynamical systems.

6.1.4 One-Dimensional Maps with Several Critical Points

In general, an interval map (6.7) may have n critical points,

$$x_{c,j}, \quad j = 1, 2, \ldots, n, \tag{6.15}$$

and hence there may be points $x \in [a,b]$ with up to $n+1$ preimages. See Fig. 6.5 for such an example of a one-dimensional map with many critical points. Therefore, the symbol dynamics of such is naturally generalized [204] by expanding the symbol set,

$$\mathcal{I} = \{0, 1, \ldots, n\}, \tag{6.16}$$

to define the shift space Σ_{n+1}. The subshift,

$$\Sigma'_{n+1} \subset \Sigma_{n+1}, \tag{6.17}$$

of itineraries corresponding to orbits of the map (6.7) follows the obvious generalization of Eq. (6.8),

$$\sigma_i(x_0) = j \text{ if } x_{c,j} < f^i(x_0) < x_{c,j+1}, \quad j = 0, 1, \ldots, n+1, \tag{6.18}$$

and taking $x_{c,0} = a$ and $x_{c,n+1} = b$. The characterization of the grammar of the resulting subshift Σ'_{n+1} corresponding to a map with n turning points is well developed following the kneading theory of Milnor and Thurston [223]. See also [89].

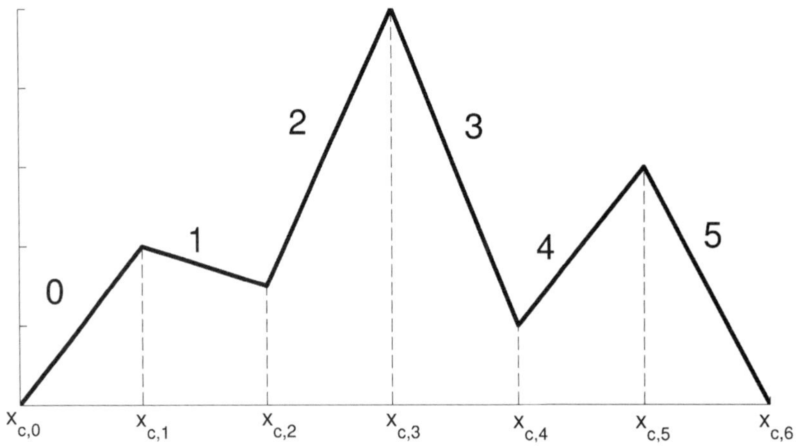

Figure 6.5. *A one-dimensional map with many critical points requires many symbols for its symbolic dynamical representation, according to Eq.* (6.18).

6.1.5 The Topological Smale Horseshoe

In the 1960s an American topologist named Stephen Smale developed [294] a surprisingly simplified model of the complex dynamics apparent in oscillators such as the Van der Pol oscillator. Smale's ingenious approach was one of simplification, stripping away the details and using the tools of topology, then often called rubber sheet geometry. He did so during an impressive summer at the beaches in Rio, about which he writes in his own words [295], leading to the results we review here, as well as separately a huge breakthrough leading eventually to his Field's Medal on the topic of the Poincaré conjecture. Despite the amazing success of the summer, the very location of his work efforts apparently led to a public inquiry over doubts that there was any actual work being done. The continuing importance even half a century later of the results of that "vacation" leave little doubt as to the strength of the summer's work.

The discussion of symbolic dynamics for one-dimensional transformations in the previous subsections appears first in this writing for the simpler issues involved with the one-dimensional phase space, and that noninvertible systems require singly infinite shifts. Historically, the more general case came first: bi-infinite shifts were used for diffeomorphisms, which arise naturally from describing differential equations. The mantra of topology/rubber sheet geometry is to ignore scale and size and reveal structures which are independent of such coordinate-dependent issues. The approach is well presented pictorially, as we can see here. In subsequent sections, in particular Section 7.1.3, we will discuss how it relates to applied dynamical systems.

The basic horseshoe mapping is a diffeomorphism of the plane,[71]

$$H : \mathbb{R}^2 \to \mathbb{R}^2. \tag{6.19}$$

However, we will be most interested in the invariant set of the (unit) box (square), which we will denote $B \subset \mathbb{R}^2$; see Fig. 6.6. The basic action of the horseshoe map shall be a stretching phase, which we denote

$$S : \mathbb{R}^2 \to \mathbb{R}^2, \tag{6.20}$$

and a folding phase,

$$F : \mathbb{R}^2 \to \mathbb{R}^2. \tag{6.21}$$

The detailed specifics of each of these two phases are not important; we wish only to show a rough sketch as drawn in Fig. 6.6 and subsequently Fig. 6.7. That is, stated loosely, $H(B)$ is meant to map across itself twice, and in such a manner that there is everywhere hyperbolicity (stretching) in the invariant set, and the no-dangling-ends property referred to in Fig. 4.5 and considered in the definition of Markov partition; see Section 4.2.3. The details are not important, but it is easy to see, for example, that S may be a linear mapping $S(x,y) = (ax, by)$, $a > 1$, $0 < b < 1$, and certain quadratic functions may be used for a specific realization of F. When we describe the horseshoe mapping as having elements of stretch+fold, we are referring to the definition

$$H = F \circ S. \tag{6.22}$$

In fact, J. Moser discussed such a decomposition of Henon-type maps [234], the composition of a rotation and a quadratic shear. We will review a Henon-type mapping [96] at the end of this section for specificity.

[71] A diffeomorphism, $F : M \to M$, is a homeomorphism which is also differentiable. That is, in the topic of smooth manifolds, a diffeomorphism is an invertible function that maps one differentiable manifold to another. Furthermore, both the function and its inverse are smooth. It serves as the equivalence relationship in smooth topological manifolds.

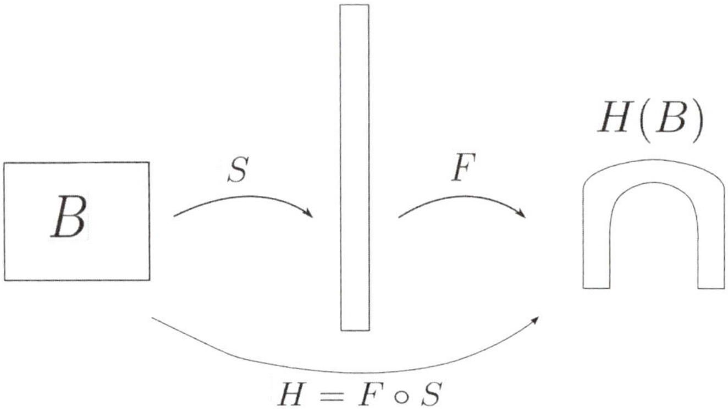

Figure 6.6. *The basic topological horseshoe mapping consists of the composition of two basic mapping: S, which is basically a stretching operation, and F, which is basically a folding operation.*

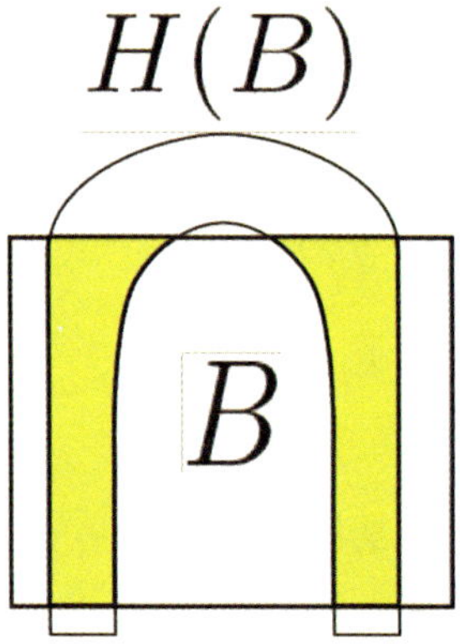

Figure 6.7. *The horseshoe mapping is designed to map the two "legs" of $H(B)$ over the original box B, since we are interested in the invariant set, Λ.*

The form of the horseshoe map decomposition, (6.23) simplifies the inverse mapping,

$$H^{-1} = S^{-1} \circ F^{-1}, \tag{6.23}$$

which is shown simply by

$$S^{-1} \circ F^{-1} \circ F \circ S = S^{-1} \circ I \circ S = S^{-1} \circ S = I, \tag{6.24}$$

where I is the identity function. Hence, starting with Eq. (6.23) and left operating (multiplying) by $S^{-1} \circ F^{-1}$,

$$S^{-1} \circ F^{-1} \circ H = S^{-1} \circ F^{-1} \circ F \circ S = I, \tag{6.25}$$

and therefore $S^{-1} \circ F^{-1}$ has the property of being the (left) inverse of H. Likewise, it can be shown to be the right inverse. Such is shown in Fig. 6.8, in which the inverse of the

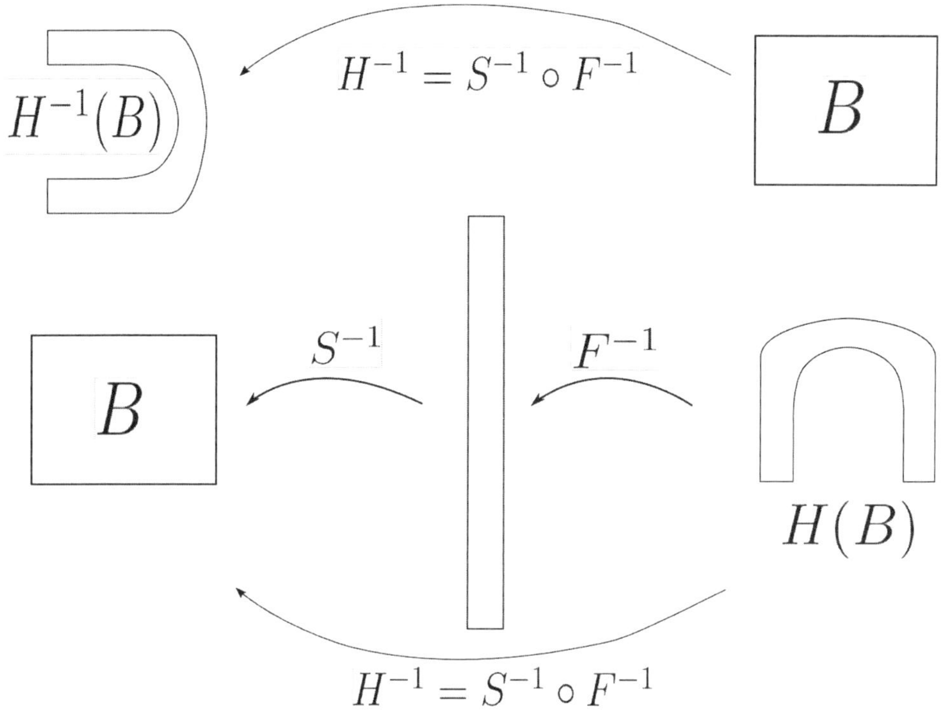

Figure 6.8. *The inverse of the horseshoe map simply reverses the fold, then the stretch, according to Eq.* (6.25). *(Bottom) Thus the folded horseshoe is first straightened (F^{-1}), and then unstretched/shortened (S^{-1}). (Top) Meanwhile, considering what these two operations must do to the original square, a horizontally folded u-bended version of the box must result.*

horseshoe reverses the fold and then the stretch of the forward mapping. That H^{-1} is the inverse of the horseshoe mapping and is itself a horseshoe mapping is argued geometrically by the pictures in Fig. 6.8.

We shall be concerned with the invariant set Λ of the box B. As such, define

$$\Lambda_i = \{z : z \in B, H^j(z) \in B \; \forall \, 0 \le j \le i\}, \tag{6.26}$$

where the notation $z = (x, y) \in \mathbb{R}^2$ denotes an arbitrary point in the plane. Thus Λ_i denotes all those points in B which remain in B for each and every iterate of H at least through the ith iterate. Likewise, define

$$\Lambda_{-i} = \{z : z \in B, H^{-j}(z) \in B \; \forall \, 0 \le j \le i\}. \tag{6.27}$$

The invariant set is written simply as

$$\Lambda = \cap_{i=-\infty}^{\infty} \Lambda_i. \tag{6.28}$$

In Figs. 6.9–6.14, we see various stages of the eventual invariant set, Λ_1, $\Lambda_{-1} \cap \Lambda_1$, Λ_2, Λ_{-2}, and $\Lambda_{-2} \cap \Lambda_2$, respectively, labeled according to itinerary relative to the partition

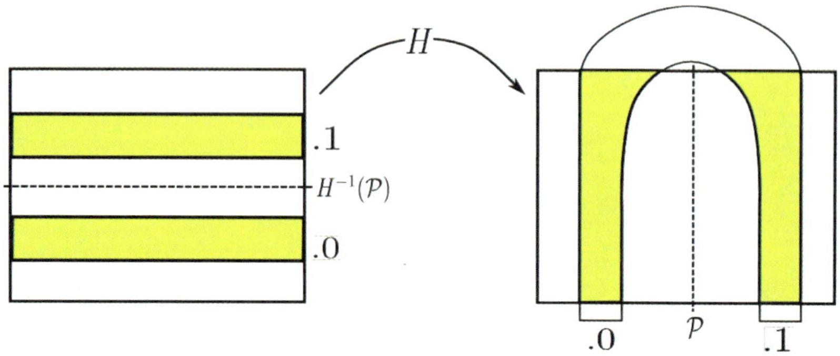

Figure 6.9. *(Right) The symbol dynamics generating partition $\mathcal{P}$ is shown. (Left) Those points which will next land in* 0. *are labeled* .0, *and likewise* .1 *to* 1.; *thus on the right we see Λ_1, the one-step invariant set colored yellow, and its preiterate on the right, also in yellow.*

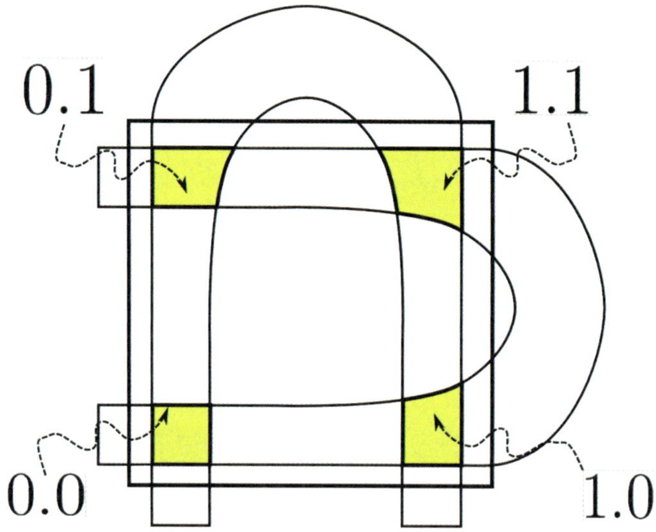

Figure 6.10. *The iterate and preiterate of B, which are the vertically oriented and horizontally oriented folded sets $H(B)$ and H^{-1}, are shown, and the one-step forward-backward set $\Lambda_{-1} \cap \Lambda_1$ is labeled according to itinerary and colored yellow.*

shown. This leads naturally to symbolic dynamics, which in some sense is just a method for bookkeeping itineraries of each initial condition relative to the partition.

Analogous to the development in Section 6.1.3, the horseshoe map admits a symbolic dynamics. A key difference here as compared to the one-dimensional case is that the one-dimensional mappings got their "fold" step from a two-to-one-ness almost everywhere,[72] which is just another way to state the noninvertibility. This noninvertibility is reflected in

[72]Recall that is the logistic map $x_{n+1} = \lambda x_n(1 - x_n)$ when $\lambda = 4$.

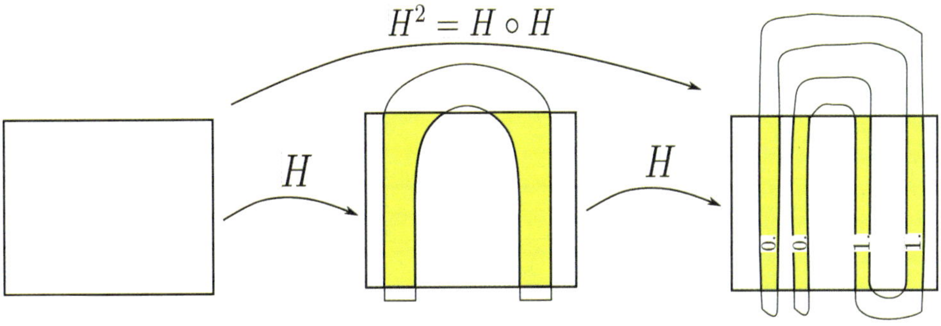

Figure 6.11. *Vertical strips shown describe Λ_{-1} and Λ_{-2}, respectively.*

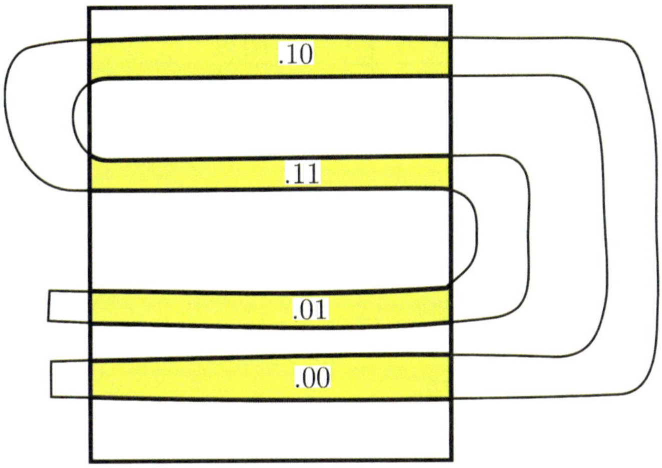

Figure 6.12. *Horizontal stripes represent Λ_2, and labels show the four possible symbolic states of itineraries through the partition $\mathcal{P}$.*

the one-sided shift of the symbolic dynamics. However, the horseshoe map is invertible. The invertibility correspondingly must be reflected in the symbolic dynamics, which is why a two-sided shift must be used. See Section 6.2.1 for an analogy of the logistic map to actual card shuffling.

Define a symbolic partition $\mathcal{P}$ to be any curve in between the two legs of the fold, as shown in Fig. 6.9. Then a symbolic dynamics may be defined in terms of itineraries:

$$\sigma_i(z_0) = \begin{pmatrix} 0 \text{ if } H^i(x_0) < \mathcal{P} \\ 1 \text{ if } H^i(x_0) > \mathcal{P} \end{pmatrix} \text{ for any } -\infty < i < \infty. \tag{6.29}$$

Now function h labels each initial z_0 according to its bi-infinite orbit,

$$\{\dots, z_{-2}, z_{-1}, z_0, z_1, z_2, \dots\}, \tag{6.30}$$

where

$$z_i = H^i(z_0) \tag{6.31}$$

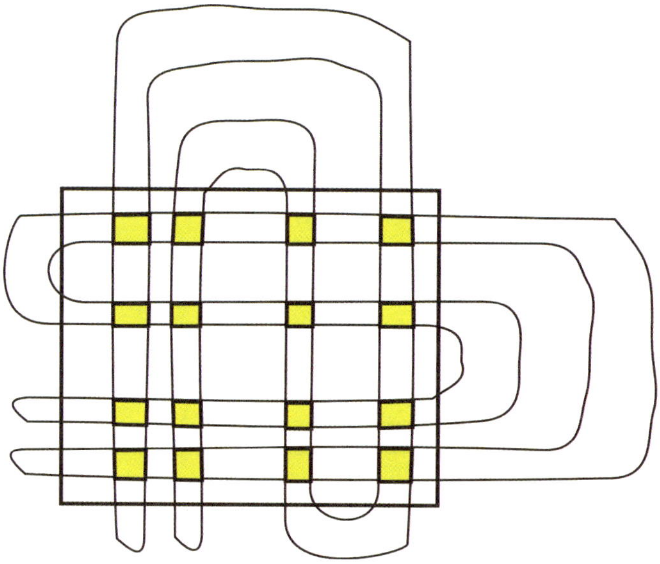

Figure 6.13. *Intersection of vertical and horizontal stripes, $\cap_{i=-2}^{2}\Lambda_i$, yield the small yellow squares which would be labeled with the* 2*-step future symbols of the vertical stripes Λ_{-2}, and* 2*-bit and prehistories of the horizontal strips. For example, one of the rectangles is labeled* 01.01.

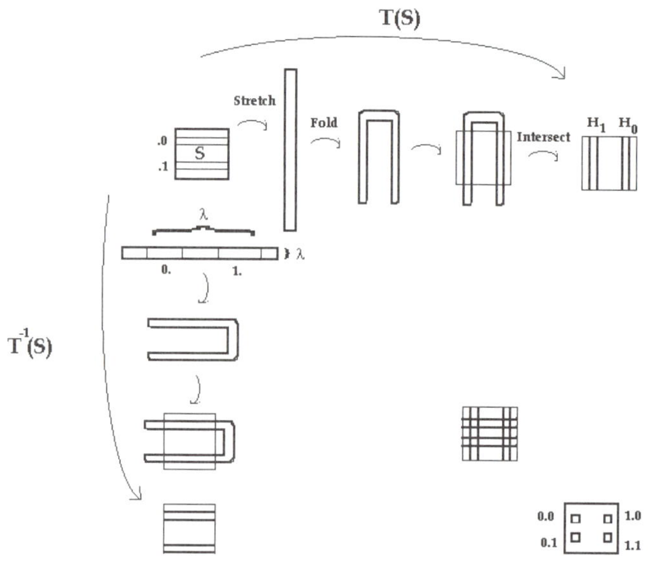

Figure 6.14. *The Smale horseshoe map in summary. Collecting the geometry already shown in Figs.* 6.6–6.12 *gives the following summary of the stretch+fold process.*

for any initial $z_0 \in \mathbb{R}^2$ and $-\infty < i < \infty$. Analogous to Eq. (6.32) we have

$$h(z_0) \equiv \sigma(z_0) = \ldots \sigma_{-2}(z_0)\sigma_{-1}(z_0)\sigma_0(z_0).\sigma_1(z_0)\sigma_2(z_0)\ldots. \tag{6.32}$$

For this reason, the symbol space must be bi-infinite in the sense

$$\Sigma_2 = \{\sigma = \ldots\sigma_{-2}\sigma_{-1}\sigma_0.\sigma_1\sigma_2\ldots, \text{ where } \sigma_0 = 0 \text{ or } 1\}, \tag{6.33}$$

since there is an infinite number of symbols before the decimal point, representing the prehistory of a trajectory, and another infinity following representing the future. The set Σ_2 is the set of all possible such bi-infinite symbol sequences. In what follows we explore several properties of Σ_2 (uncountable, chaotic, infinitely many periodic orbits, positive Lyapunov exponent, etc.).

Open sets in this shift space Σ_2 again follow the metric topology, but this time by a bi-infinite symbol comparison,

$$d_{\Sigma_2}(\sigma,\overline{\sigma}) = \sum_{i=-\infty}^{\infty} \frac{|\sigma_i - \overline{\sigma}_i|}{2^i}, \tag{6.34}$$

rewarding agreement for the first several bits near the decimal point. This defines so-called cylinder sets which are the symbolic version of open sets in Σ_2.

Under a Markov partition property for hyperbolic diffeomorphisms, stated in Section 4.2.3, again a good "change of coordinates" exists [46], but this time by

$$\begin{aligned} h : \Lambda &\to \Sigma_2, \\ z &\mapsto h(z) = \sigma. \end{aligned} \tag{6.35}$$

The resulting conjugacy describes an equivalence between the dynamical systems,

$$s \circ h(z) = h \circ H(z), \tag{6.36}$$

meaning each point $z \in \lambda \subset B \subset \mathbb{R}^2$ can allow the change of coordinates and mapped s in Σ_2. Alternatively this gives the same thing as mapping in $\mathbb{R}$ by the horseshoe H and then changing coordinates to symbols by h. This is equivalently stated by the commuting diagram,

$$\begin{array}{ccccc} H & : & \Lambda & \to & \Lambda \\ & h & \downarrow & & \downarrow \\ s & : & \Sigma & \to & \Sigma \end{array}. \tag{6.37}$$

We recall that s is the Bernoulli shift, which simply moves the decimal,

$$s(\sigma_i) = \sigma_{i+1}, \tag{6.38}$$

but as a bi-infinite shift, no symbols are dropped. Instead, symbols are simply forgotten into a fading past in terms of the metric topology inherited from Eq. (6.34). This means the focus on the $m+n+1$ bits before and after the decimal point is shifted to the right. Remembering that those symbols near the point define the neighborhood (the N-cylinder) in which the symbol sequence is found, this implies a loss of precision, or alternatively a loss of initial information regarding the initial state; the rate at which this happens shall be explored in Sections 9.4 and 9.5, concerning entropy and Lyapunov exponents. Also see further discussion in Example 6.1.

Example 6.1 (shift of focus and loss of precision—the symbolic metric topology and the Bernoulli shift map). Consider a specific point σ from the set of all possible bi-infinite symbol sequences Σ_2, which we write so as to emphasize only m bits before the current bit ahead of the decimal point, descriptive of the prehistory, and n bits after the decimal point, descriptive of the future of the orbit of σ relative to the Bernoulli shift. We choose the specific point to be

$$\begin{aligned} \sigma &= \ldots ??\sigma_{-6}\ \overbrace{\sigma_{-5}\sigma_{-4}\sigma_{-3}\sigma_{-2}\sigma_{-1}\sigma_0.\sigma_1\sigma_2\sigma_3\sigma_4\sigma_5}^{N=m+1+n=5+1+5=11}\ \sigma_6??\ldots \\ &= \ldots ??1\ \overbrace{010011.10010}^{N=11}\ ??\ldots . \end{aligned} \tag{6.39}$$

The overbrace is shown to indicate precision of $N = m+n+1$ points, which is in fact a neighborhood in symbol space. That is, any other point that agrees to that many bits,

$$\sigma' = \ldots\sigma_{-7}\sigma'_{-6}\ \overbrace{010011.10010}^{N=11}\ \sigma_6\sigma_7\ldots, \tag{6.40}$$

has a distance from σ within the symbol norm (6.34),

$$d_{\Sigma_2}(\sigma,\sigma') \le \sum_{-\infty}^{6}\frac{1}{2^i} + \sum_{6}^{\infty}\frac{1}{2^i} = \frac{2}{2^6}\frac{1}{1-\frac{1}{2}} = \frac{1}{16}, \tag{6.41}$$

assuming the worst-case scenario in the inequality in which all the other bits are opposite outside the bracketed window. The question marks shown in Eqs. (6.39) and (6.40) outside of the braces emphasize to a given N-bit precision that those bits are essentially unknown. Similarly as in the physical world, claimed precision beyond the capabilities of any experiment should be considered fiction.

The Bernoulli shift map of the decimal point σ gives a new point in Σ_2 which we will call $\hat{\sigma}$,

$$\hat{\sigma} = s(\sigma) = \ldots ??10\ \overbrace{100111.0010?}^{N=11}\ ??\ldots, \tag{6.42}$$

and likewise the second iterate is a point which we shall call $\hat{\hat{\sigma}}$,

$$\hat{\hat{\sigma}} = s^2(\sigma) = \ldots ??101\ \overbrace{001110.010??}^{N=11}\ ??\ldots . \tag{6.43}$$

Our use of the notation "??" in these annotations of cylinder sets is meant to emphasize the unspecified/unknown information corresponding to focusing on just the few bits corresponding to a neighborhood in symbol space. Really, however, when measuring to a precision of $m = 5$ bits of prehistory and $n = 5$ bits of future fate, it is unfair to maintain the bits outside of the braces, as we have shown in Eqs. (6.42)–(6.43). Instead, one should write

$$\hat{\sigma} = s(\sigma) = \ldots ??\ \overbrace{100111.0010?}^{N=11}\ ??\ldots,\ \hat{\hat{\sigma}} = s^2(\sigma) = \ldots ??\ \overbrace{001110.010??}^{N=11}\ ??\ldots . \tag{6.44}$$

The resulting distance between the iterates of σ and σ' increases as the unspecified digits become prominent, and the specified digits become "forgotten":

$$d_{\Sigma_2}(s(\sigma), s(\sigma')) \leq \sum_{-\infty}^{6} \frac{1}{2^i} + \sum_{5}^{\infty} \frac{1}{2^i} = \frac{2}{2^6} \frac{1}{1-\frac{1}{2}} = \frac{3}{64},$$

$$d_{\Sigma_2}(s(\sigma), s(\sigma')) \leq \sum_{-\infty}^{6} \frac{1}{2^i} + \sum_{4}^{\infty} \frac{1}{2^i} = \frac{2}{2^6} \frac{1}{1-\frac{1}{2}} = \frac{5}{64}.$$

The loss of precision occurs at an exponential rate definitive of the concepts of entropy, and also Lyapunov exponents, which will be explored in what follows.

This example illustrates that in the symbol space metric, "close" means matching the first many bits, and this precision is lost with iteration of the Bernoulli shift. The continuity of the homeomorphism[73] $h : \mathbb{R}^2 \to \Sigma_2$ is reflected by the ever-decreasing width and nesting of the vertical stripes, and likewise the horizontal stripes in Figs. 6.9–6.14, whose intersections form small squares. Where continuity means small sets correspond to small sets under the mapping, said succinctly as a topologist, the preimage under h of any open set in Σ_2 must be an open set in $\mathbb{R}^2$.

The vertical stripes shown in Figs. 6.7, 6.9 and 6.11 are Λ_{-1} and Λ_{-2}, respectively. These represent the future fate of the orbit of those z-values therein through the symbolic partition. Inspection of these pictures and Eqs. (6.26)–(6.27) suggests that

$$\Lambda_i \subset \Lambda_j, \text{ when } 0 < j < i, \text{ or } i < j < 0. \tag{6.45}$$

Likewise, the horizontal stripes shown in Fig. 6.12 represent prehistories. The intersection of horizontal and vertical stripes (6.28) makes for the rectangles shown. The nesting of each rectangle of

$$\Lambda_{-i} \cap \Lambda_i \subset \Lambda_{-j} \cap \Lambda_j, \text{ when } 0 < j < i, \tag{6.46}$$

reflects the corresponding symbolic nesting that j-bit neighborhoods are finer and more precise than i-bit neighborhoods when $0 < j < i$ definitive of the continuity of h.

The limit set Λ in Eq. (6.28) is a Cantor set, whose properties are shared by C_∞ in Fig. 6.15, but let us mention here that as such Λ

- is closed,
- is perfect,
- is totally disconnected, and
- has cardinality of the continuum.

Also, this Cantor set has zero measure. In this sense there is an oxymoron; it can be said that the set "counts big," but "measures small."

The power of this symbolic dynamics description is that we can actually prove many of the basic tenants of chaos theory. That is, this symbolic dynamical system has infinitely

[73] As used here comparing dynamical systems, the homeomorphism should be thought of as a change of coordinates to compare to dynamical systems.

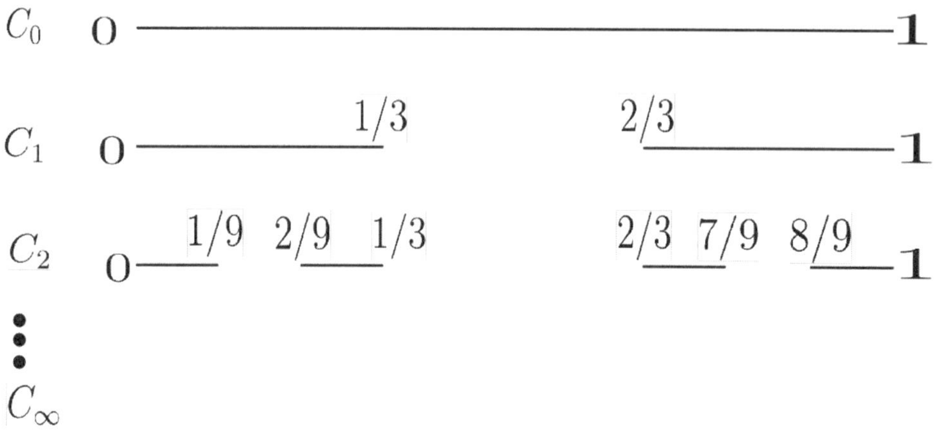

Figure 6.15. *The middle thirds Cantor set $C_\infty = \bigcap_{n\to\infty} C_n$ is a useful model set since invariant sets in chaotic dynamical systems, and in particular of horseshoes, are generalized Cantor sets, sharing many of the topological properties as well as many of the metric properties, such as self-similarity and fractal nature.*

many periodic orbits, sensitive dependence to initial conditions, and dense orbits. These shall be included in Section 6.2. The trick then to make a proof regarding the original dynamics is a proof in which the representation is correct, and that is often the difficult part, even while at least numerical evidence may be excellent and easily obtainable.

For now, we shall present the central theorem due to Smale. With this theorem, the seemingly abstract topic of symbolic dynamics becomes highly relevant to even certain differential equations that may be found in practical engineering applications.

6.2 Chaos

Chaos theory is the face of dynamical systems theory which has appeared most popularly in the media and cultural outlets [137] and includes a broad list of applications.[74] This popularity has been a two-edged sword, in that while it attracts interest, attention, and eager students, the very term "chaos" has encouraged misunderstanding. A pervasive popular misconception is that the mathematical property of chaos is somehow a philosophical pondering and even that it means the same as it does in its English definition, i.e., "disorder." In mathematics, chaos denotes a kind of order behind what seems to be irregularity. Further, chaos is a property like any other mathematical property in that there must be a specific mathematical definition.

Specifically, there are two popular definitions of chaos, which we highlight and contrast here. Both are approachable and checkable when we have horseshoes or otherwise symbolic systems.

[74] Applications of chaos theory range from biology, epidemiology, astronomy, chemistry, medicine, and even finance and economics, to name a few. It is hardly an exaggeration to say that any science which includes a time-evolving component and oscillation has the propensity to display chaos as one of its natural behaviors.

The popular **Devaney definition of chaos**[75] [95] is equivalent to the following.

Definition 6.1 (Devaney chaos). Let X be a metric space. A continuous map $T : X \to X$ on a metric space X is chaotic on X if

1. T is transitive,[76]
2. the set of periodic points of T are dense in X,
3. T has sensitive dependence to initial conditions.[77]

These properties are easy to check by constructive proof for a symbolic system, and hence by conjugacy, for the models which are equivalent to a symbolic shift map, such as the horseshoe or, similarly, for certain one-dimensional mappings such as the logistic map with $r = 4$.

Property 1: Transitive. To show the transitive property, it is sufficient to construct a dense orbit. Certainly not all points are dense in Σ_2. For example, each of the (countably) infinitely many periodic points are not dense. E.g.,

$$\sigma = 0.000\ldots, \quad \text{and} \quad \sigma = 1.111\ldots \tag{6.47}$$

are the two fixed points. Likewise,

$$\sigma = 0.1010\ldots \tag{6.48}$$

represents the period-2 orbit, and any other periodic orbit can be written simply with a repeating symbol sequence in the obvious way. However, the point

$$\sigma = \overline{0.\,1}_1\,\overline{00\;01\;10\;11}_2\,\overline{000\;001\;\ldots\;111}_3\,\overline{0000\;0001\;\ldots\;1111}_4\,\ldots \tag{6.49}$$

is an example of a dense orbit. Each overbar groups symbols of one, then two, then three, etc., symbols of all permutations on those n symbols. There are 2^n n-bit strings in each grouping. Appended together in *any* order, including the specific one shown, allows for a dense orbit. Why? Any other $\sigma' \in \Sigma_2$ is approached to any arbitrary precision, since precision in the symbolic norm means agreeing to m bits for some (possibly large) m, and those specific m bits are found somewhere among the grouping of all possible m-bit permutations encoded in σ. In fact, the set of all such dense orbits can be shown to be an uncountable set, as they are formed by all possible orderings of these permutations, and thus a variation on Cantor's diagonal argument applies [190].

---

[75] The original Devaney definition includes the requirement that periodic orbits of T are dense in X, but several papers eventually showed that two of the three original requirements are sufficient to yield the third [12].

[76] It is sufficient to state that a map T is *transitive* if there exists a dense orbit. That is, a set A (the orbit of a point is a set—let $A = orbit(x_0)$) is dense in another B (in this case X if the closure of A includes B, $B \subset \mathrm{cl}(A)$). For example, the rational numbers are dense in the set of real numbers because any real number is approximated to arbitrary precision by an appropriately chosen fraction.

[77] A map T on a metric space is said to have *sensitive dependence on initial conditions* if there is an $r > 0$ such that, given a point $\boldsymbol{x}$ and arbitrary $\epsilon > 0$, there is a point $\boldsymbol{y}$ such that $d(\boldsymbol{x},\boldsymbol{y}) < \epsilon$ and a time k when $d(T^k(\boldsymbol{x}), T^k(\boldsymbol{y})) \geq r$ [268]. That is, there is always a nearby point whose orbit will end up far away.

If instead we are discussing a bi-infinite shift arising from a horseshoe, the construction is very similar, as repeating the same permutations in both bi-infinite directions is sufficient.

Property 2: Dense Periodic Orbits. Any periodic orbit in Σ_2 is a symbol sequence of m bits which repeats. For example, the period-4 orbits include the 4-bit sequence $\sigma = 0.001000100010001\ldots$. Therefore, given any other symbol sequence σ', periodic or not, it is sufficient to identify a periodic point which agrees with the first m bits, no matter what $m > 0$ may be chosen. All we need to do is select the first m bits of σ' and then construct σ as a repetition of those bits. For example, if

$$\sigma' = 0.0011001010100011\ldots \tag{6.50}$$

and $m = 4$ is selected, then the same,

$$\sigma = 0.001000100010001\ldots \tag{6.51}$$

will suffice, repeating the first 4 symbols of σ'.

Property 3: Sensitive Dependence to Initial Conditions. Consider a point σ and any arbitrary (close) precision, meaning a specific (large) number of symbols m. Then it is necessary to demonstrate that we can always find some point σ', such that σ and σ' agree to at least the first m bits. But no matter how far we wish, $r > 0$, there is a time when they iterate at least that far apart. For notation, given σ, let us define $\overline{\sigma_i}$ to be the opposite symbol of σ_i in the ith position of σ. That is, if $\sigma_i = 0$, then $\overline{\sigma_i} = 1$, and otherwise $\overline{\sigma_i} = 0$. Therefore, for the m-bit precision, if

$$\sigma = \sigma_0.\sigma_1\sigma_2\ldots\sigma_m\sigma_{m+1}\sigma_{m+2}\ldots, \tag{6.52}$$

then choose

$$\sigma = \sigma_0.\sigma_1\sigma_2\ldots\sigma_m\overline{\sigma_{m+1}\sigma_{m+2}}\ldots. \tag{6.53}$$

That is, make the first m bits agree, and reverse the rest of the bits. It is a direct check by the geometric series and Eq. (6.11) that such construction yields an $r > 0$, within diameter of the space Σ_2, at least by the $k = m$ iteration of the shift map.

These constructions are symbolic, but that is the simplifying point. Thus, with a conjugacy to the horseshoe, recall that each symbolic sequence addresses specific points or neighborhoods as suggested in Figs. 6.10 and 6.14.

Another well-accepted definition of chaos is found in the popular book by Alligood, Sauer, and Yorke [2] (**ASY**). It defines a chaotic orbit (although it is stated therein for maps of the interval, $T : \mathbb{R} \to \mathbb{R}$) rather than a chaotic map.

Definition 6.2 (ASY chaos). A *bounded* orbit, $orbit(x_0) = \{x_0, x_1, x_2, \ldots\} = \{x_0, T(x_0), T^2(x_0), \ldots\}$ of a map T is a **chaotic orbit** if

1. $orbit(x_0)$ is not asymptotically stable, and

2. the Lyapunov exponent of T from x_0 is strictly positive.

We will not confirm these here for the symbolic dynamical case since we will need to refer to Lyapunov exponents to be defined in Chapter 8, but it is easy to state roughly that Lyapunov exponents are descriptive of stretching rates or, in some sense, the rate of the sensitive dependence found in the Devaney definition. So it allows for stretching, but in a bounded region by assumption. Further assuming the orbit is not converging to some periodic orbit leaves only that it must be wandering in a way that may be thought of as analogous to the transitivity of the Devaney definition.

On the other hand, as descriptive contrast, where the Devaney definition states that the map is chaotic or not on some invariant set, the ASY definition is stated for single orbits. Whereas the Devaney definition is popular for the mathematical proof, it allows for certain systems.[78] By contrast, the ASY definition allows for a single long trajectory, perhaps even from an orbit measured from an experimental system to be checked—it is a bit closer to the popular physicists' notion of chaos based on positive Lyapunov exponents, and the quantities are more directly estimated with numerics.

6.2.1 Stretch+Fold Yields Chaos: Shuffling Cards

While mathematical chaos may be misunderstood by many in the popular public to mean random, the mathematician refers to a deterministic dynamical system which satisfies the definition of chaos (one of the favorite definitions at least, Definition 6.1 or 6.2). Many so-called random processes such as coin flipping or shuffling cards are also deterministic in the sense that a robot can be built to flip a coin in a repeatable manner [97] and a machine can be made to shuffle cards in a repeatable manner, and even purchased as standard equipment in the croupier's[79] arsenal. These are illustrated in Figs. 6.16 and 6.17. While these devices are deterministic in the sense that identical initial conditions yield identical results, theoretically it is practically impossible to repeat initial conditions *identically*, and they have such a high degree of sensitivity to initial conditions that such error quickly grows to swamp the signal. Thus the randomness is not in the dynamical system, so goes the argument. Rather, the randomness is in the small initial imprecision—the inability to specify initial conditions exactly.

Focusing on shuffling cards as an analogy, the dynamics of the logistic map (1.2), which we repeat,

$$x_{n+1} = \lambda x_n(1 - x_n), \tag{6.54}$$

when $\lambda = 4$, can be described literally as card shuffling-like dynamics. This is a useful analogy for understanding the interplay between chaos, determinism, and randomness; see Fig. 6.18. If we imagine 52 playing cards laid uniformly along the unit interval so that half are before the partition point $x_c = \frac{1}{2}$ and half after, then the action on the first half is to lay them along their range. $range([0, \frac{1}{2}]) = [0, 1]$. Thus the cards are shown spread out vertically along the entire unit interval. Likewise, $range([\frac{1}{2}, 1]) = [0, 1]$. It has the same range, and those cards are also laid out along the same unit interval. (But this is not a good shuffling, since the cards on the right are placed right-side up, whereas those on the left remain upside down. This is denoted by the orientations of the arrows. Any

[78]Usually for symbolic systems, namely, the horseshoe, the Devaney definition is quite strong, and especially popular since the Melnikov method [146, 316, 315] can be used to show that certain periodically forced integrable flows have embedded horseshoes.

[79]A croupier is a professional card dealer such as employed at Las Vegas blackjack tables.

Figure 6.16. *Deterministic random-like motion: chaos. A card-mixing machine which may be purchased from Jako-O. Reprinted with permission from Noris-Spiele.*

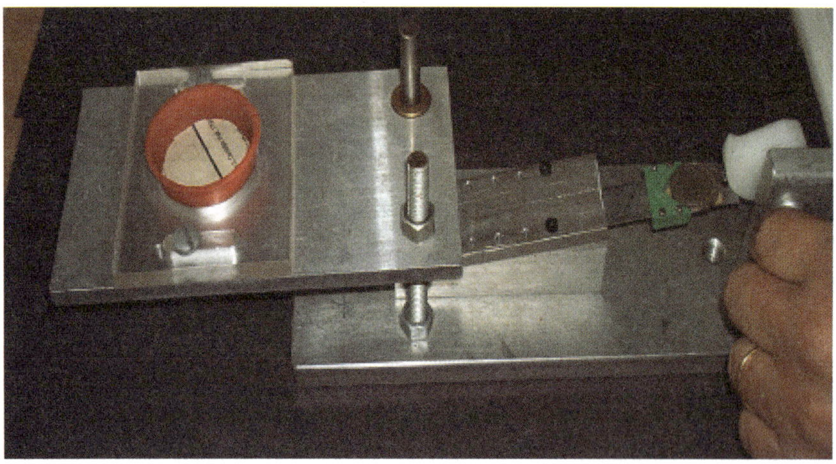

Figure 6.17. *Deterministic random-like motion: chaos. A coin-flipping robot. This figure appeared as Fig.* 1a *in* [97].

croupier who shuffles like this would not keep his job for long.) Other than that, the double covering reflects the two-one-ness of the map. Then the cut deck is pushed together due to the mapping, since upon the next iteration, the origin of left side or right side is "forgotten."

Continuing with the analogy between the logistic map and the card shuffling: As to the role of sensitive dependence to initial conditions, following the two blue cards shown in $[0, \frac{1}{2}]$ in Fig. 6.18, we see them spread vertically under the action of the map. This is related to the unstable (positive) Lyapunov exponent. Then the shuffle pushes the extra card

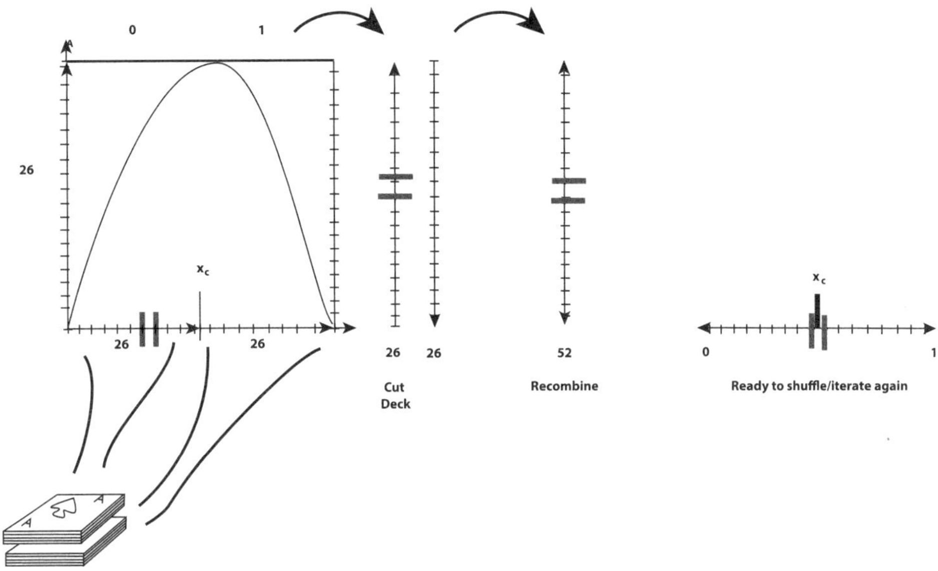

Figure 6.18. *The logistic map has dynamics analogous to shuffling cards. The stretch and folding process mechanism underlying chaos is illustrated here as if we are shuffling cards, which consists of cutting and then recombining a deck of cards.*

between them. Then these two cards may end up falling on opposite sides of the partition x_c, which means they will be cut to opposite sides of future shuffles. Their initial closeness will be lost and forgotten. The determinism is nonetheless descriptive of a stretching and folding process which naturally forgets initial imprecision. Thus the mechanism behind the chaos, which is described rigorously and mathematically by symbolic dynamics, is nothing other than an analogy to the stretch+fold process of a card shuffle.

6.2.2 Horseshoes in Henon

According to previous sections, when a horseshoe topology can be proven, just by considering a single iterate of a carefully chosen subset, then chaos has been proven in an embedded subset by preceding theory. Devaney and Nitecki have shown that exactly this strategy is straightforward for the Henon map. We will also discuss a similar strategy for a Poincaré map derived from a continuous flow—the Duffing oscillator.

The Henon map [163] is an extremely famous example, both for pedagogy and research, of simple mapping of the plane, $T : \mathbb{R}^2 \to \mathbb{R}^2$, which gives rise to complicated behavior and which apparently has a "strange attractor." This mapping written in the form

$$T(x,y) = (1 - ax^2 + y, bx) \tag{6.55}$$

gives rise to a chaotic attractor,[80] as shown in Fig. 6.19 using the parameters $a = 1.4, b =$

[80]The attractor set is apparently chaotic, but this turns out to be difficult to prove, and remains open for specific typical parameter values [326, 16].

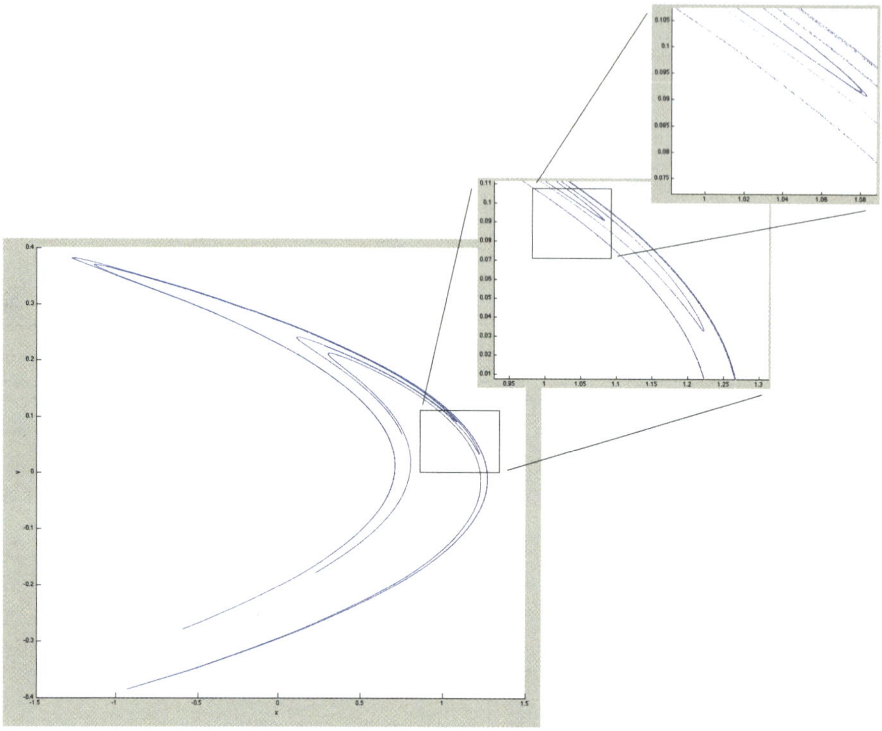

Figure 6.19. *The Henon attractor, from a long-term trajectory of the Henon map* (6.55), *with the usual parameter values $a = 1.4, b = 0.3$, gives rise to the form shown using* 100,000 *iterates. The stretch and folding nature of the resulting attractor is highlighted by a few blow-up insets.*

0.3. These are the most common parameter values and give rise to this familiar form seen in so many other presentations.

In a slightly different form, Devaney and Nitecki showed explicitly that the Henon equations can give rise to a Smale horseshoe [96]. They proved existence of an embedded horseshoe dynamics for a form of the Henon mapping which may be obtained by a coordinate transformation from the usual (6.20),

$$T(x,y) = (a - by - x^2, x), \tag{6.56}$$

which as a diffeomorphism of the plane has an inverse,

$$T^{-1}(x,y) = (y, (a - x - y^2)/b). \tag{6.57}$$

Specifically, their result may be stated as the following theorem.[81]

[81] We have abbreviated a more detailed form found in [268] to highlight the feature of the embedded horseshoe.

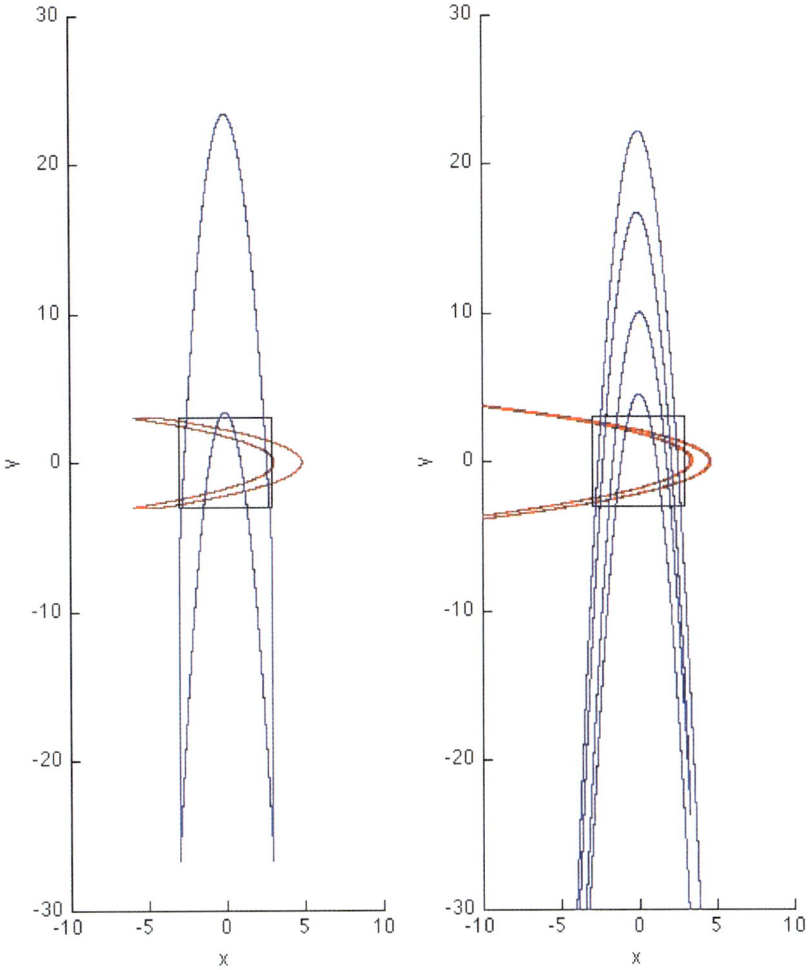

Figure 6.20. *A numerical iteration of the square S in* (6.58) *(black) by a special form of the Henon map T and its preiterate* T^{-1}, *Eqs.* (6.56)–(6.57). *(Left) The forward iterate (red) and preiterate (blue) reveal the topological horseshoe geometry.*

Theorem 6.1. If $b \neq 0$ and $a \geq (5+2\sqrt{5})(1+|b|)^2/4$, then there exists an embedded horseshoe which may be found by considering the invariant set of T in (6.56) in the square,

$$S = \{(x,y) : -s \leq x, y \leq s\}, \text{ where } s = \left(1+|b|+\sqrt{(1+|b|)^2+4a}\right)/2. \tag{6.58}$$

It can be checked that $a = 4, b = 0.3$ are parameters that admit a horseshoe according to the Devaney–Nitecki theorem. Simply stated, the proof consists of a direct check that the stated square maps across itself in the geometric way described by the topological horseshoe; see Section 6.1.5 and Fig. 6.6. Rather than repeat that analysis, we will simply refer to the picture of the numerical iteration, and preiterate of the square shown in Fig. 6.20.

This folding is sufficient to cause an invariant set Λ which is a Cantor set on which $T|_\Lambda$ is conjugate to the Bernoulli shift map on symbol space, $s|_{\Sigma_2}$, as promised by Smale's horseshoe theory. Furthermore on that set with the stated geometry, we have already repeated the proof that there is chaos. Of course, as usual this specific set is a Cantor set of measure zero. In a much larger set, the attractor, if it exists, is not addressed by the horseshoe or, in this case, the attractor includes that almost every point diverges to infinity. This leads to the following remark.

Remark 6.1. Where it turns out that the invariant set Λ for the Henon map (6.56) is an *unstable* Cantor set, chaos is proven, but the proof is only relevant almost nowhere in a measure theoretic sense for Fig. 6.20! It could be in some systems that either a measure zero set is the only invariant set, or, on the other hand, there may be a larger invariant set on which there is also chaos, but the chaos may not be that of a fullshift.[82] In the case of the Henon map (6.56), almost every initial condition does not have a fate related to the chaotic set; these points have orbits which behave quite simply. They diverge to infinity. The chaotic set of the Henon map (6.56) is called a **chaotic saddle** since the chaotic set Λ has a stable manifold $W^s(\Lambda)$ which does not contain any open discs in $\mathbb{R}^2$—it has dimensionality less than 2.

On the other hand, the Henon map (6.55) as shown in Fig. 6.19 has the attractor $\mathcal{A}$ shown. This follows from demonstrating a trapping region $\mathcal{T}$, which in this case may be demonstrated by a trapezoid shown (in blue) in Fig. 6.21, where we demonstrate that $T(\mathcal{T})$ (in red) properly maps into $\mathcal{T}$. This means that *every* initial condition from $\mathcal{T}$ maps into $\mathcal{T}$. For relevance, we also show the attractor set in blue, $\mathcal{A} \subset T(\mathcal{T}) \subset \mathcal{T}$.

6.2.3 Horseshoes in the Duffing Oscillator

The direct construction method of Section 6.2.2 of finding a topological horseshoe in the Henon map by finding a topological rectangle which maps across itself in the appropriate geometry described in Section 6.1.5 may be attempted for other maps, and even for Poincaré maps derived from a flow. Here we show the example of the Duffing map derived from a Duffing flow of the differential equations, Eqs. (1.36)–(1.39), with an attractor already shown in Fig. 1.12.

In Fig. 6.22, we show an oriented disc which maps across itself in a manner suggestive of the topological horseshoe. The orientation is relevant for the topological horseshoe, and we have taken the liberty of demonstrating the orientation with a happy face. This figure, while indicative of embedded complex behavior, also shows some shortfalls.

- The figure hides this fact since we have shown a disc instead of a rectangle, but the happy face does not demonstrate a fullshift horseshoe. Further investigation reveals that only an incomplete horseshoe is described by this set and its iterations, such as the incomplete horseshoes shown in Fig. 6.23. Such incomplete horseshoes still indicate some stretch and folding behavior, but a fullshift does not result, as explained by the picture and caption of Fig. 6.24. Rather, some missing words in the grammar are typical of a subshift, as discussed further in Section 6.4. The rigorous pruning theory can be found in [71, 144, 81] as it applies to certain special cases.

- No matter how complex the invariant set of the oriented disc shown, only two symbols can be accounted for by a simple horseshoe. The Duffing map, however, has

[82]The restricted grammar of a subshift will be discussed in Section 6.4.

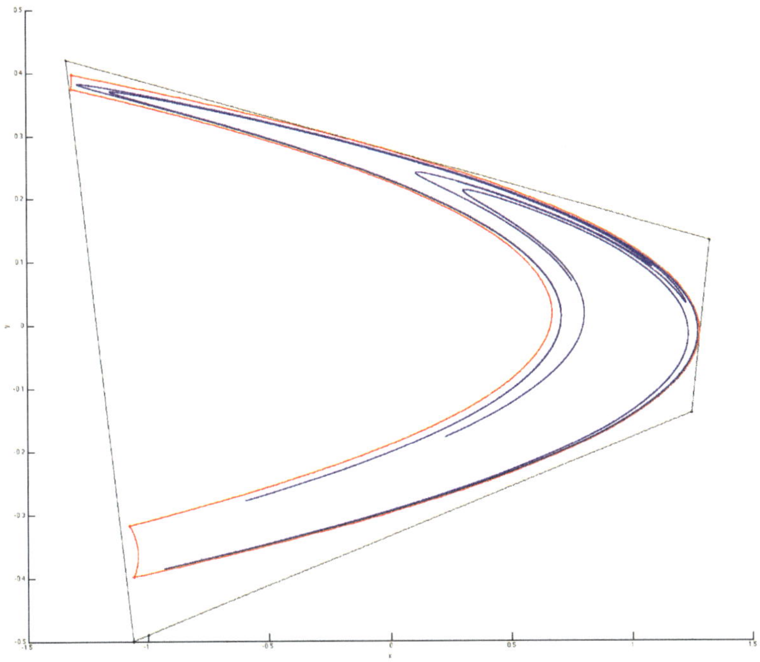

Figure 6.21. *That there is an invariant set of the Henon map* (6.55) *follows by demonstrating a set that maps entirely into itself. The trapezoid set shown in black,* $\mathcal{T}$, *maps to* $T(\mathcal{T})$ *which is the red set shown and which can be confirmed to be properly contained. Also shown is the attractor set* $\mathcal{A}$ *in blue already seen in Fig.* 6.19.

many folds and pleats which requires partial use of many symbols at least. See Fig. 6.23 (right). More discussion toward learning the symbolic dynamics (the grammar) of the attractor, rather than simply of a measure zero unstable chaotic saddle, is discussed in Section 6.4. While it is exceedingly difficult to prove that the representation of such a complex symbolic dynamics is faithful, it is at least computationally quite feasible and straightforward to use empirically.

6.3 Horseshoe Chaos by Melnikov Function Analysis

Melnikov's method is a powerful analytic tool which has a role both in verifying existence of chaos as well as sometimes characterizing transport activity in certain dynamical systems for which the appropriate setup can be achieved. Note that it as an alternative to the more computational-based methods discussed as a centerpiece in most of this book. Rather than getting too sidetracked into another large field, we will give a very short presentation of this rich method here, and then cite several references to excellent sources for Melnikov analysis.

Direct construction of a horseshoe as described above in Sections 6.2.2 and 6.2.3 for the Henon mapping, and then the Duffing oscillator Poincaré mapping in Figs. 6.20,

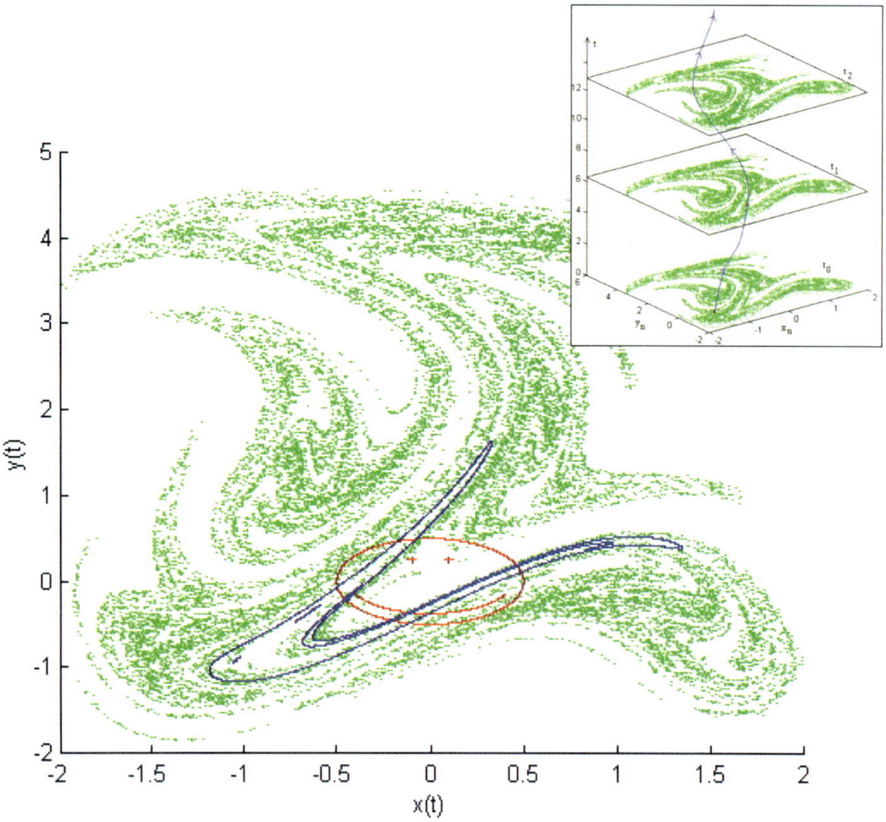

Figure 6.22. *A Duffing map derived by the stroboscopic method of Poincaré mapping can be used as an example to investigate an embedded horseshoe. (Inset) The Poincaré section shown as successive* 2π *and copies of the attractor are painted on each such surface between which each orbit flights (stretch+fold in the Duffing attractor). The oriented disc shown (happy face) maps across itself in a manner suggestive of a horseshoe. However, as discussed the covering is not complete enough to reveal a fullshift horseshoe. See Fig.* 6.24 *for further discussion of the incompleteness of this horseshoe.*

6.21, and 6.22 cannot be a general method due to the difficulty of producing the appropriate regions. However, the Melnikov analysis applies more broadly to certain differential equations; it also allows us to check for the existence of horseshoes, and further for parametric study to discuss homoclinic (and heteroclinic) bifurcations which produce such chaos. Recall that Smale proved (Theorem 7.1 in [294]) that a transverse homoclinic point of a hyperbolic periodic point $\mathbf{w}$ of a C^r diffeomorphism, $r \geq 2$, implies an embedded horseshoe. It is well known and straightforward to prove that a horseshoe is chaotic. The Melnikov function gives a measure of the distance between stable and unstable manifolds, $W^s(\mathbf{w})$ and $W^u(\mathbf{w})$, with respect to a parameterization of these curves when this distance is small. In this way, the Melnikov function can be used to decide the existence of a transverse intersection. We follow most closely [315].

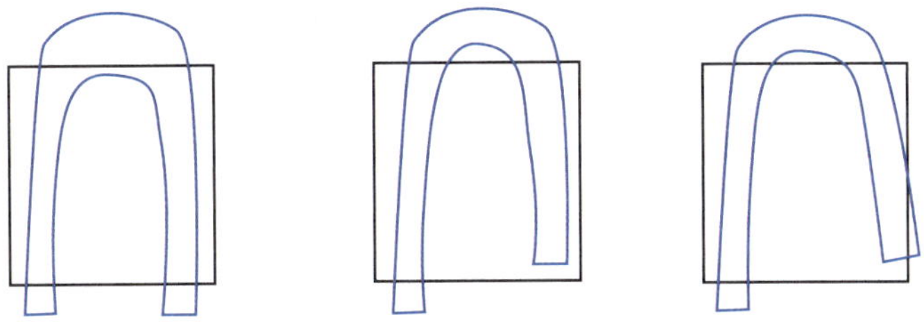

Figure 6.23. *Incomplete horseshoes. The first horseshoe is complete, but the next two are incomplete in that they do not fold completely across the region. See also Fig.* 6.24 *for a discussion of how such missing words could arise from tangency bifurcation, and Fig.* 6.31 *for an illustration of the consequence of missing words when a tangency bifurcation occurs. See* [81] *for a full discussion of pruning horseshoes. (Right) The third picture suggests that at least a* 2*-bit symbol space* Σ_2 *would be required, but incomplete folding suggests a submaximal entropy,* $h_T < \ln 2$.

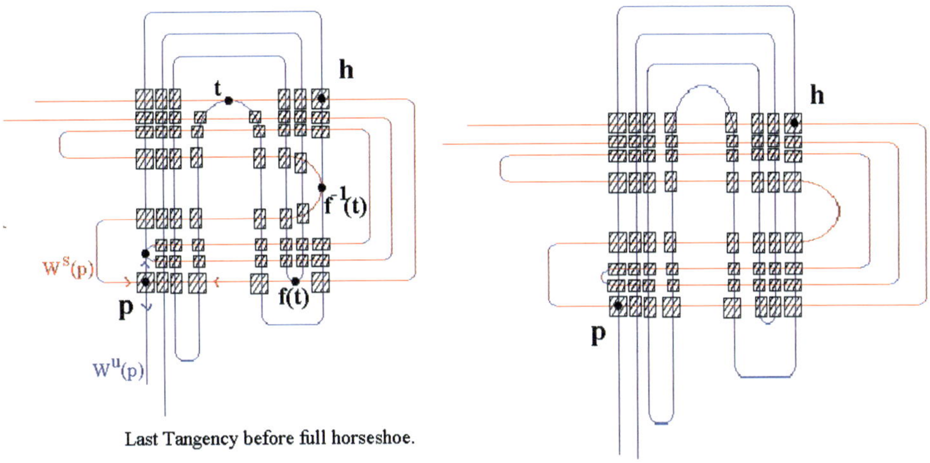

Figure 6.24. *(Right) A caricature of a homoclinic tangle due to a hyperbolic fixed point* $\boldsymbol{p}$ *with stable and unstable manifolds* $W^s(\boldsymbol{p})$ *and* $W^u(\boldsymbol{p})$*, causing a principle intersection point (p.i.p.) homoclinic point* $\boldsymbol{h}$*. A full folding horseshoe allows for the Markov partition property—no dangling ends—suggested in Section* 4.2.3 *and Fig.* 4.5*. A fullshift on all symbols of* Σ_2 *follow. (Left) Some of those words are lost when the Markov partition property is lost, shown here by an unfolding of a tangency bifurcation. The consequence of this is that* 2 *of the* 2^8 8*-bit words draw so close that they annihilate at* t*, and likewise their iterates annihilate at* $f(t)$ *and* $f^{-1}(t)$*. Thus the entropy decreases accordingly,* $h^T < \ln(2)$*, which may be calculated by the* ln *of the largest eigenvalue of a* 2^8 *de Bruijn graph with* 6 *missing transitions. As further tangency bifurcations unfold, generally more words will be lost.* [41]

For the sake of simplicity of presentation, we will restrict our discussion to the most straightforward version of the Melnikov method. We will assume here an autonomous Hamiltonian system of the plane $\tilde{H}(q, p)$, under the influence of a small time periodic perturbation,

$$g(q,p,t) = g(q,p,t+T) \text{ for some } T > 0. \tag{6.59}$$

The Melnikov analysis we use assumes a dynamical system of the form

$$\frac{dq}{dt} = \frac{\partial \tilde{H}}{\partial p} + \epsilon g_1(q,p,t), \tag{6.60}$$

$$\frac{dp}{dt} = -\frac{\partial \tilde{H}}{\partial q} + \epsilon g_2(q,p,t), \tag{6.61}$$

or

$$\dot{\mathbf{z}} = J \cdot \nabla \tilde{H}(\mathbf{z}) + \epsilon \mathbf{g}(\mathbf{z},t), \tag{6.62}$$

where

$$J = \begin{pmatrix} 0 & 1 \\ -1 & 0 \end{pmatrix}, \quad \nabla \tilde{H} = \left\langle \frac{\partial \tilde{H}}{\partial q}, \frac{\partial \tilde{H}}{\partial p} \right\rangle^t, \quad \mathbf{g} = \langle g_1, g_2 \rangle^t, \quad \mathbf{z} = \langle q, p \rangle^t \tag{6.63}$$

[316]. Furthermore, the unperturbed system,

$$\dot{\mathbf{z}} = J \cdot \nabla \tilde{H}(\mathbf{z}), \tag{6.64}$$

must have a hyperbolic fixed point $\mathbf{w}$ with a homoclinic connection orbit, which we call $\mathbf{z}^*(t)$, which surrounds a continuous family of nested periodic orbits. Note that in the extended system, the hyperbolic fixed point of the unperturbed vector field becomes a period point, and under a sufficiently small perturbation of period T, we may assume the existence of unique hyperbolic period orbit $\mathbf{w}^*$ of period T. Under these assumptions, the Melnikov function

$$M(t_0) = \int_{-\infty}^{\infty} \mathbf{g}(\mathbf{z}^*(t), t+t_0) \cdot \nabla \tilde{H}(\mathbf{z}^*(t)) dt \tag{6.65}$$

measures the distance between the stable and unstable manifolds of $\mathbf{w}^*$ the direction of the $\nabla \tilde{H}$ in the time T-stroboscopic Poincaré section phase plane, where t_0 parameterizes the unperturbed homoclinic orbit $\mathbf{z}^*(t)$. The Melnikov function $M(t_0)$ is proportional to the distance between the stable and unstable manifolds of $\mathbf{w}^*$ at $\mathbf{z}^*(-t_0)$. Under the above assumptions, the result is that existence of a zero (a t_0 such that $M(t_0) = 0$), which is simple ($\frac{\partial M}{\partial t}|_{t_0} \neq 0$), implies that the dynamical system (6.60) has a transverse homoclinic point and hence possesses an embedded horseshoe. The depiction in Fig. 6.25 suggests no intersection between the stable and unstable manifolds and therefore the resulting $M(t_0)$ would have no roots. On the other hand, a tangency between the stable and unstable manifolds would result in nonsimple roots. Finally, transverse intersections result in simple roots, as focused on by Melnikov's theorem.

For the sake of brevity, we mention only a specific example here, but without details. Perhaps the most standard example is for a Duffing oscillator of the form $\dot{x} = y, \dot{y} = x - x^3 + \epsilon(\gamma \cos \omega t - \delta y)$ [251, 202, 315, 146]. This is a very special (rare) example in that it results in a Melnikov integral which can actually be computed in closed form, and the state of existence of simple roots can be decided explicitly and with respect to variation of the system parameters. In this way, a complete global bifurcation analysis can be performed. However, Melnikov integrals are generally nontrivial to evaluate and one must resort to numerical evaluation, as in [35]. The difficulty is due to an infinity of oscillations

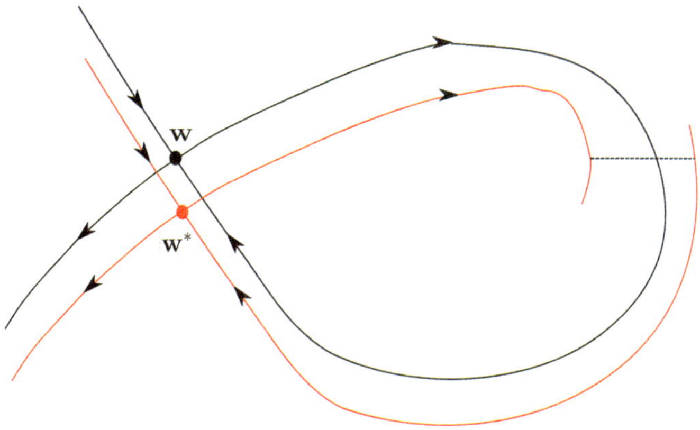

Figure 6.25. *The Melnikov function $M(t_0)$, Eq. (6.65), as applied to flows of the form (6.60), identifies the distance between stable and unstable manifolds of $\mathbf{w}^*$ of the perturbed system (in red) at a reference time t_0. This perturbative analysis starts with a homoclinic connection of $\mathbf{w}$ of the autonomous system $\epsilon = 0$, shown in black. This depiction suggests that this Melnikov method will never reach zero since the stable and unstable manifolds do not cross.*

of the integrand in the finite space, the length of the parameterized curve of the homoclinic connection of the autonomous system; this should be expected since the integrand is a description of a homoclinic tangle.

There are many generalizations of this basic Melnikov analysis which include allowing for higher-dimensional problems [320], stochastic problems [290], subharmonic analysis [189], as well as analysis of the area of lobes and turnstiles and thus discussion of transport. The power of the Melnikov method is that it is a more analytically based approach for global analysis, and it is capable of including parametric study, but it does require the dynamical system to be presented in some variation of the basic form, Eqs. (6.59)–(6.62). We assert that the transfer operator methods and the FTLE methods are both more empirically oriented and capable of handling dynamical systems known only through observations, and that this is very possible when the Melnikov method setup is possible. The empirical methods of this book and the Melnikov methods have complementary information to offer.

6.4 Learning Symbolic Grammar in Practice

6.4.1 Symbol Dynamics Theory

Symbol dynamics is a detailed and significant theory unto itself, with important connections to dynamical systems as discussed here, but also to coding theory and information theory. We will summarize here these concepts in a manner meant to allow computational approximation relevant to empirically investigated dynamical systems.

One might ask what the discrete mathematical notions of grammar and subshifts have to do with a flow such as the Lorenz equations, (6.4). In terms of the Lorenz butterfly, and corresponding one-dimensional map (6.7), if a 0 bit represents a flight around one of the two butterfly wings, and a 1 represents a flight around the other wing, before piercing the Poincaré surface, then a specific infinite string of these symbols indicates a corresponding trajectory flight around the attractor. On the other hand, suppose there is a missing word. For example, suppose no infinite word has the string 00 in it. This would indicate that no initial condition $(x(0), y(0), z(0))$ of the Lorenz equations exists which makes two successive laps around the left wing. In fact, on the Lorenz attractor corresponding to standard parameter values $p_0 = (10, 28, 8/3)$, there do exist initial conditions which permit both a 00 string and a 11 string, but not all finite length strings do occur. The missing minimal forbidden words tend to be somewhat longer than two bits, but not necessarily. The point is that we generally expect only a "subshift" of the fullshift of all possible words.

Definition 6.3. Given a Σ_n', a subset of an n-bit symbol space Σ_n, Σ_n' is a **subshift** if it is

1. topologically a closed set (contains all of its limit points), and
2. closed with respect to the action of the Bernoulli shift map; that is, if $\sigma \in \Sigma_n'$, then $s(\sigma) \in \Sigma_n'$.

In a physical experiment, corresponding to the one-dimensional map such as Eq. (6.7), it is possible to approximately deduce the grammar of the corresponding symbolic dynamics by systematic recording of the measured variables. Note that any real measurement of an experiment consists of a necessarily finite data set. Therefore, in practice, it can be argued that there is no such thing as a grammar of infinite type in the laboratory.[83] Thus, without loss of generality, we may consider only grammars of finite type for our purposes. Such a subshift is a special case of a sofic shift [188, 204]. In other words, there exists a finite digraph which completely describes the grammar. All allowed words of the subshift Σ_2' corresponding to itineraries of orbits of the map correspond to some walk through the graph.

Definition 6.4 (see [204, 188]). A **sofic shift** is a subshift Σ_n' which is generated by a digraph (directed graph) G.[84] If each edge from each vertex is labeled by one of the symbols from the symbol set $\{1, 2, \ldots, n\}$, then an infinite walk through the graph generates a symbol sequence $\sigma \in \Sigma_n$. Let S be the set of all possible symbol sequences so generated by the graph G. If $S = \Sigma_n'$, then the subshift is generated by G. A sofic shift is a subshift generated by some finite graph G.

Not all subshifts are sofic, but any subshift can be well approximated by a sofic shift, which is convenient for computational reasons. For example, the full 2-shift shift is generated by the graph in Fig. 6.27(a), but this is not the minimal graph generating Σ_2. The graph in Fig. 6.26 (left) also generates Σ_2. Likewise, the "no two zeros in a row" subshift Σ_2' is generated by all possible infinite walks through the digraph in Fig. 6.27(b), in which

[83] It could be further argued that there is no such thing as measuring chaos in the laboratory, since most popular definitions of chaos [95, 2] are asymptotic in that they require sensitive dependence and topological transitivity, both of which would require time going to infinity to confirm. In fact, it is valid to argue that a mathematical dynamical systems is defined in terms of orbits as infinite trajectories, and therefore no such thing exists in physical experiments. In this sense, the premises of many points in this book regard what can be learned from finite samples of orbits.

[84] A directed graph $G = (E, V)$ is a set of edges $E = \{e_1, e_2, \ldots, e_M\}$ and vertices consisting of specific edge pairs, $\{e_i, e_j\}$.

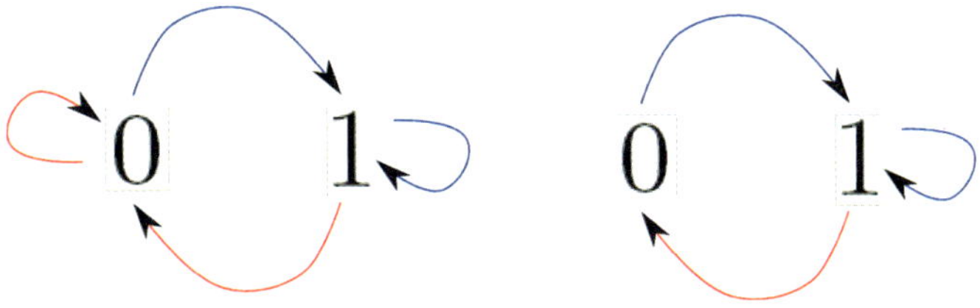

Figure 6.26. *The full 2-shift Σ_2 is generated by all possible infinite walks through the (left) digraph. (Right) All walks through the graph describe a subshift Σ_2' which is the set of all symbol sequences which can be described by a grammar in which the symbol* 0 *never occurs more than once in a row. Contrast the presentation of these shifts to lifted versions of the same or larger graph presentations in Fig.* 6.27, *but the larger graphs would also allow more nuanced and finer restrictions of the grammar.*

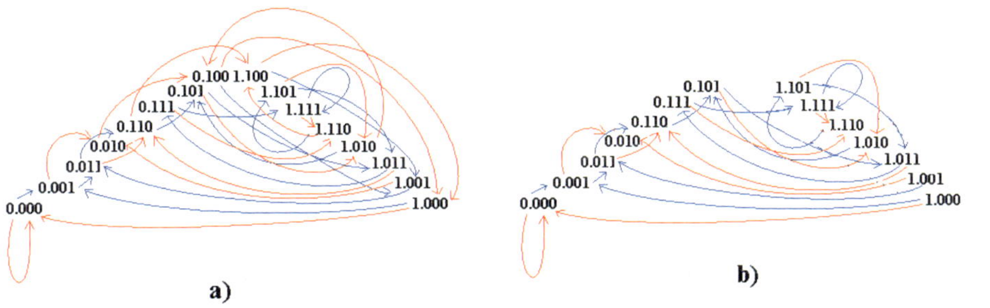

Figure 6.27. *The full 2-shift Σ_2 is generated by all possible infinite walks through the graph in* (a). (b) *The "no two zeros in a row" subshift grammar Σ_2' is generated by all possible infinite walks through the above digraph, in which the only two vertices corresponding to* 00 *words have been eliminated, together with their input and output edges. Contrast this figure to a minimal graph presentation of the same subshift in Fig.* 6.26. [26]

the only two vertices corresponding to 00 words have been eliminated, together with their input and output edges.

Entropy is a way of comparing different subshifts. While subshifts are generally uncountable, we may describe the entropy of a subshift as the growth rate of the number of words of length n. There are strong connections to information theory, which will be explored in Chapter 9. Meanwhile, for our discussion here, we will state simply that the topological entropy of a subshift Σ_k' on k-symbols Σ_k' can be defined by cardinality (counting when finite) [188, 204, 268],

$$h_T(\Sigma_k') = \limsup_{n\to\infty} \frac{\ln N_n}{n}, \tag{6.66}$$

where $N_n \leq 2^n$ is the number of $(\{\mathbf{0}\},\{\mathbf{1}\})$ binary sequences (words) of length n.

Definition 6.5. A sofic shift is defined as **right resolvent** if each "out edge" from a vertex of the graph presentation is uniquely labeled.

By theorem [188, 204, 268], when the subshift is a right resolvent shift, then a spectral computation can be used to compute entropy of shift spaces of many symbols.

$$h_T(\Sigma'_k) = \ln \rho(A), \tag{6.67}$$

where $\rho(A)$ is the spectral radius[85] of an adjacency matrix A,[86] and A generates the sofic shift Σ'_k.

Concepts of entropy are useful in considering evolution with respect to the bifurcations which occur when a parameter is changed. For example, consider the entropy of the symbolic dynamics from the logistic map, Eq. (6.54). When $\lambda = 4$, we have already discussed that a fullshift results. Since

$$A_1 = \begin{pmatrix} 1 & 1 \\ 1 & 1 \end{pmatrix} \tag{6.68}$$

is the adjacency matrix for the graph G shown in Fig. 6.26 (left) generates the fullshift Σ_2, and $\rho(A) = 2$, then $h_T(\lambda) = \ln 2$ when $\lambda = 4$. Notice that now we are writing the argument of $h_T(\cdot)$ to be the parameter of the logistic map. Comparing also to the graph and corresponding adjacency matrix in Fig. 6.29(a), it can be confirmed that $\rho(A_1) = \rho(A_4) = 2$ and in fact the generated shift spaces are the same fullshift Σ_2. The graphs G_{A_1} and G_{A_4} are both called de Bruijn graphs.

6.4.2 Symbolic Dynamics Representation on a Computer Representation of a Dynamical System

In practice, beyond the horseshoe map and beyond a one-dimensional one-hump map, some work must be done to learn the grammar of a symbolic dynamics corresponding to the dynamical system which generated the data. We will discuss here some of the bookkeeping involved to develop a useful symbolic model of the underlying dynamical system, but we do not claim that this representation is exactly correct. Several possible errors of doing this work on the computer can easily creep into the computations, including incorrect partition or at least inexact computer representation of a partition even if known exactly, finite word bookkeeping when longer or infinite representations of the grammar should be used, and non-Markov partitions used, which is related to the errors already stated. A great deal of excellent work has been carried out to produce computer-assisted proofs of a symbolic dynamics representation of a dynamical system, and underlying periodic orbit structure, in part using interval arithmetic, such as the work in [133, 131, 135, 224, 326] and including proof of chaos. Our discussion here will simply be heuristic in that we will not discuss computer proof but rather simply imply that the methods presented suggest refining and improving symbolic models of a dynamical system rather than exact representations.

6.4.3 Approximating a Symbolic Grammar in a One-Dimensional Map

Given a finite-measure data set $\{x_i\}_{i=0}^N$, simply by recording all *observed* words of length n corresponding to observed orbits, and recording this list among all possible 2^n such words,

[85] Largest eigenvalue.

[86] A is an adjacency matrix of a graph A if A is a matrix of entries of 0's and 1's where $A_{i,j} = 1$ if there is an edge from vertex j to i and otherwise $A_{i,j} = 0$.

the appropriate digraph can be constructed as in Fig. 6.27. One should choose n to be the length of the observed minimal forbidden word. Sometimes n is easy to deduce by inspection, as would be the case if it were 00 as in Fig. 6.27(b), but difficult to deduce for larger n. The minimal forbidden word length follows a subgroup construction [41, 40] related to the "follower-set" construction [188, 204].

Since the data set $\{x_i\}_{i=0}^{N}$ is finite, then if the true minimal forbidden word corresponding to the dynamical system is longer than the data sample size, $n > N$, only an approximation of the grammar is possible. Therefore, the corresponding observed subshift is expected to be a subset of whatever might be the true subshift of the model map (6.7) or experiment. This is generally not a serious problem for our purposes since some coarse-graining always results in an experiment, and this sort of error will be small as long as the word length is chosen to be reasonably large without any observed inconsistencies. As a technical note of practical importance, we have found link lists to be the most efficient method to record a directed graph together with its allowed transitions.

6.4.4 As a Parameter Varies, the Grammar Also Varies as Told by the Kneading Theory Story

Now it is both interesting and instructive to consider what happens to the shift spaces and the corresponding entropy as λ is decreased from 4. First we will discuss this in terms of changes to the graph representations of the grammar of the symbolic dynamics, and then we will relate that to the elegant kneading theory of Milnor and Thurston [223].

A well-known bifurcation sequence of the logistic map, Eq. (6.54), as we vary $0 \leq \lambda \leq 4$, gives rise to the famous Feigenbaum diagram shown in Fig. 6.28. The bifurcation analysis summarized in this figure has been written about extensively elsewhere, such as in [95, 268, 2]. This is the story of the period doubling bifurcation route to chaos, including the "period three implies chaos" story [257] and the detailed elegant theory of Šarkovs'kiĭ's theorem [278] which completely characterizes the order in which the periodic orbits appear as the parameter λ is increased. Therefore, we will not focus on these features here. Rather, we will focus on just the changes to symbolic dynamics as learned from a finite graph, as this keeps the focus on the general topic of this writing.

In [29] we considered changes to a finite representation of the grammar of a logistic-like map (one-hump map) as a parameter is varied, equivalent[87] to lowering the peak, such as reducing λ from 4 in the logistic map. Consider a 4-bit representation of the symbolic dynamics on two symbols regarding Σ_2 as illustrated by the de Bruijn graph shown in Fig. 6.29 (top) which is an unrestricted 2-shift, and as such the generated grammar is equivalent to the simpler 2-vertex graph presentation in Fig. 6.26 of the same grammar. Compare also to Fig. 6.27.

Inherently, reducing λ from 4 results successively in losing access to words in the shift space. The key to making such a statement computation is to know the *order* in which words are lost, and this is with respect to the so-called lexicographic order, otherwise called the Gray-code order. This order is depicted to 4-bit representation in the graphs in Figs. 6.26 and 6.29, as these words are laid out left to right from 0.000 to 1.000 in the order which is monotonic with the standard order of the unit interval. Formerly, it is one

[87]We were interested in communication with chaos, and the formulation in [29] resulted in increased noise resistance in the transmission of messages on a chaotic carrier by avoiding signals that wander near the symbol partition.

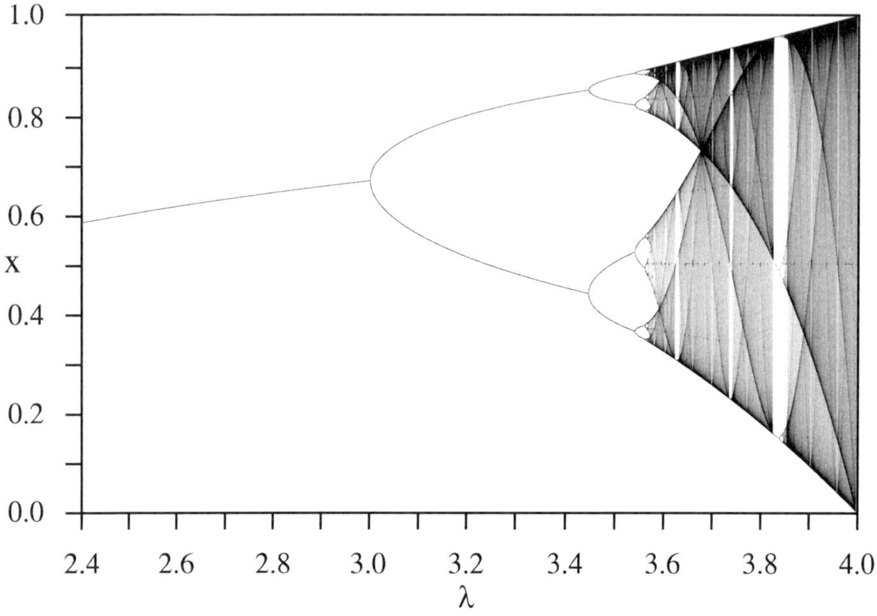

Figure 6.28. *The Feigenbaum diagram illustrating bifurcations of the logistic map* (6.54) *as we vary* $0 \leq \lambda \leq 4$.

of the major results of kneading theory that these two orders are monotonic with respect to each other [223]. That is, symbol sequences ordered $\sigma(x') \prec \sigma(x)$ implies $x' < x$, where $\sigma(x)\Sigma_2$ is the symbol sequence by Eqs. (6.29) and (6.32) corresponding to real values $x \in [0,1]$, $\prec$ is the Gray-code order in the symbol space, and $<$ is the standard order in $\mathbb{R}$. Furthermore, in the kneading theory there are conditions under which as the parameter λ varies in a "full family" of maps (such as the logistic map since it gives the fullshift), the corresponding grammar reduces continuously with respect to the order topology from the Gray-code order. We can see the result of removing words in order with respect to $\prec$ in Fig. 6.30 in terms of decreasing topological entropy. The computation is in terms of a finite representation using the graph of size $N = 2^n$ vertices, $n = 14$ bits, and the corresponding spectral computation by Eq. (6.67), in which words are removed one by one from the graph in the same order by known monotonicity by $\prec$ that they disappear by reducing λ from 4. The coarser presentation of the same idea is depicted in the left column of Fig. 6.29 for $2^4 = 16$-bit de Bruijn graph presentations of the grammars resulting in removing 4-bit words in order monotone by $\prec$ to reducing λ. The resulting transition matrices are shown in the right column of Fig. 6.29. Removing a word from the graph (a vertex) must result in removing all the ingoing edges and outgoing edges to that vertex (for the shift-closed property in the definition of a subshift), and therefore sometimes some words not directly removed (x'ed) effectively disappear. Correspondingly, all the rows and columns from a removed word must be zeroed. Thus the spectral radius and hence topological entropy must be monotone nonincreasing. This is what is seen in Fig. 6.30, where the entropy is indeed monotone nonincreasing with respect to words removed and hence also with respect

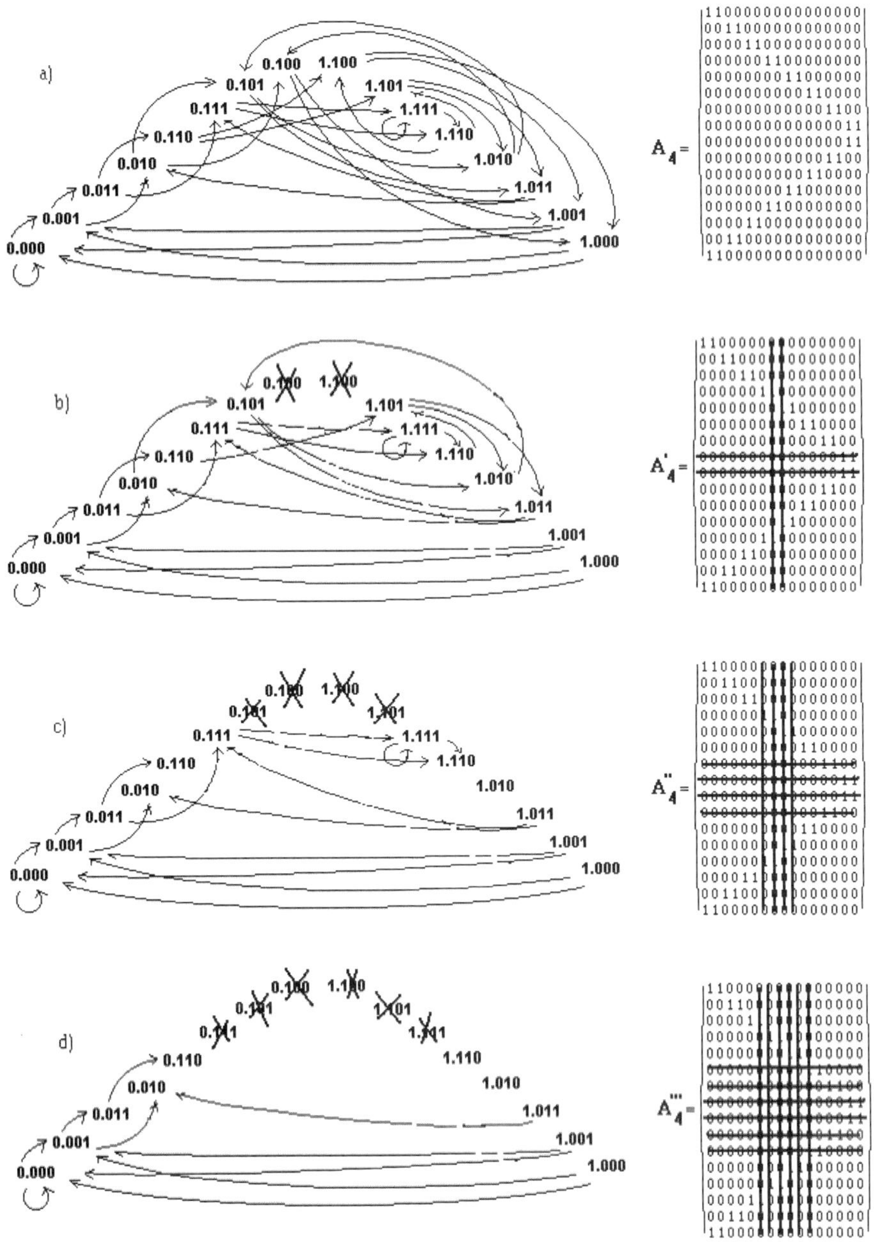

Figure 6.29. *Restricting the grammar either by reducing the height of the one-hump map (for example, by reducing r in the logistic map $x_{n+1} = \lambda x_n(1 - x_n)$) corresponds to removing words in this dyadic graph presentation, shown here ordered to the Gray-code order* [29] *as dictated by the kneading theory* [223]. *Compare also to Fig.* 6.27. [29]

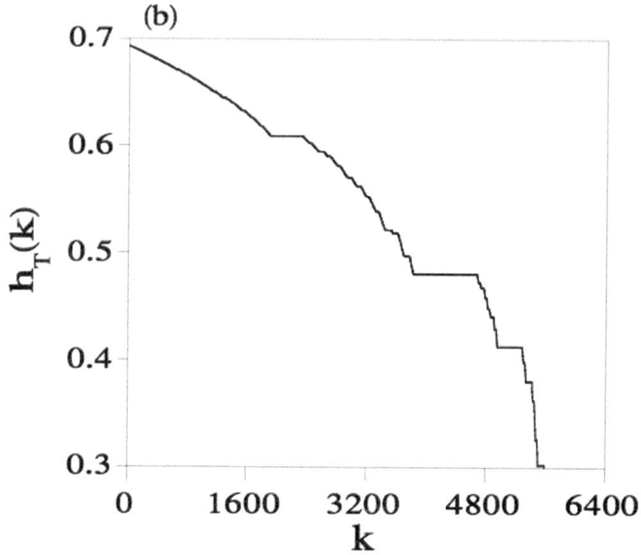

Figure 6.30. *As the grammar becomes more restrictive, the topological entropy decreases. The devil's staircase aspect of this function is described in* [29]. *Grammar restrictions describe removed words from the grammar as illustrated in Fig.* 6.29. *Here for a fine representation of the grammar with a graph of size* 2^n, *for a large n, and a corresponding transition matrix* $A_{2^n \times 2^n}$, *entropy is computed by the spectral formula Eq.* (6.67). $n = 14$. [26]

to λ. Another enticing feature of the figure is that entropy has flat spots—apparently the function reminds us of a devil's staircase function, which indeed is the result of the limit of the process to the continuum of a fine grammar representation as $n \to \infty$. As discussed in [29], the flat spots result because those words which are removed by losing paths to them—of having directly removed some other word—cannot be removed directly when it might be their turn in order to be removed; such is a common mechanism, as we see, resulting in the flat spots, and in fact the width of the flat spot represents the order n of the presentation of the grammar as a finite graph of a given scale. The smaller the n where this phenomenon may occur, the wider the flat spot.

6.4.5 Approximating a Symbolic Grammar and Including a Multivariate Setting

Empirically, the multivariate setting of learning a symbolic grammar has a great deal in common with the single variate map case of the previous subsection. At its simplest, the problem reduces to good bookkeeping. For a given sample orbit segment, $\{z_n\}_{n=0}^{N}$, record observed m-bit words relative to a given partition, and the transitions between them. This is just a matter of keeping the observed generating graph as a link list, since a link list is particularly efficient; therefore, the bookkeeping work here has almost no issues that are dependent on dimension, other than that dimensionality effects how long the orbit must be to usefully saturate toward a full set of observations (fill the space). However, the main difficulty is in determining an appropriate generating partition, and this becomes nontrivial in more than one dimension. The issue of partition is discussed in Section 6.4.6. In any case, given a partition there is a corresponding symbolic dynamics relative to that partition with a grammar that may be useful to learn, regardless of whether the appropriate partition is used from which the dynamical system may be conjugate to the symbolic dynamics.

Example 6.2 (symbol dynamics of the Henon map approximation and recorded as a link list). Consider the Henon map, Eq. (1.1). The attractor is shown in Fig. 6.31 together with a symbolic partition that is useful for our discussion. A well-considered conjecture [144, 71] regarding the symbol partition of the Henon map is that a generating partition must be a curve that passes through homoclinic tangencies. As we see in Fig. 6.31, the darkened "horizontal" w-shaped curve near $y = 0$ is just such a curve passing through primary tangencies. Call this curve C. Above C we label the region 0 and below we label 1. Iterates and preiterates of C define a refining partition on the phase space with labels partially shown in Fig. 6.31 up to two (pre)iterates. As an aside, it is easy to see that the partition is not Markov in two iterates and preiterates, at least as is most clear by inspecting the example of the 11.10 labeled branch, as it has the dangling ends property forbidden of

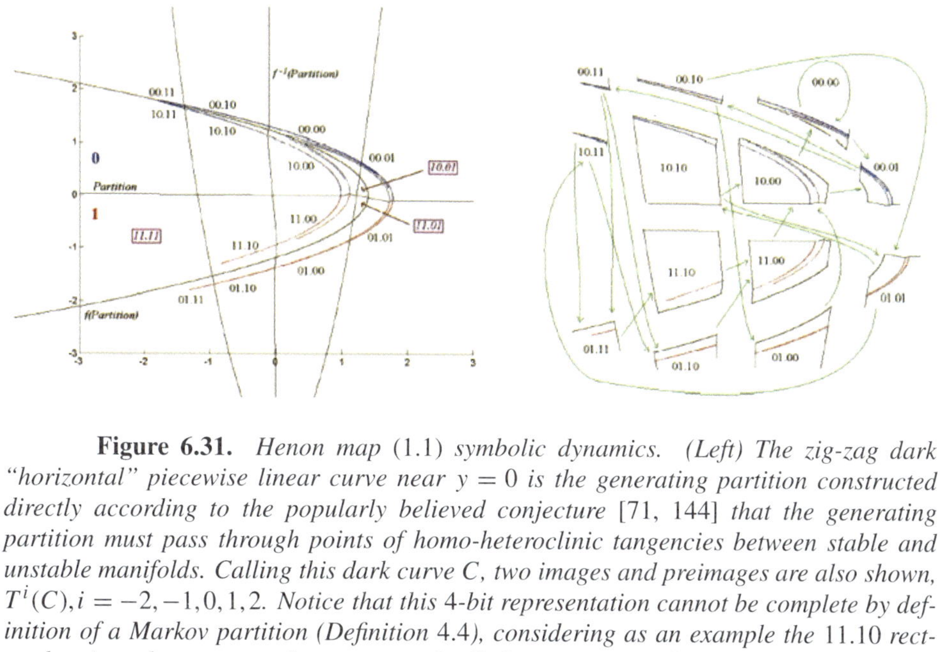

Figure 6.31. *Henon map* (1.1) *symbolic dynamics. (Left) The zig-zag dark "horizontal" piecewise linear curve near* $y = 0$ *is the generating partition constructed directly according to the popularly believed conjecture* [71, 144] *that the generating partition must pass through points of homo-heteroclinic tangencies between stable and unstable manifolds. Calling this dark curve C, two images and preimages are also shown,* $T^i(C), i = -2, -1, 0, 1, 2$. *Notice that this 4-bit representation cannot be complete by definition of a Markov partition (Definition* 4.4*), considering as an example the* 11.10 *rectangle, since the attractor does not stretch all the way across the* 11.10 *region. Further images and preimages* $T^i(C)$ *have the possibility of closing all the conditional dependencies (conditional probabilities when associated with probability measure) to produce a Markov partition, but no such completion is known for this Henon map with parameter values* $(a,b) = (1.4, 0.3)$. *Thus the partition may be generating but does not produce a Markov partition. Compare to Definition* 4.4 *of higher-dimensional Markov partitions and Fig.* 4.5. *(Right) The action of trajectories through the few symbols shown describes the directed graph here.*

a Markov partition. Compare to Definition 4.4 of higher-dimensional Markov partitions, and to Figs. 4.5 and 6.23.

Observed word	$\to$	Shift 0 observed	Shift 1 observed
00.11	$\to$	01.10	
10.11	$\to$	01.10	
10.10	$\to$	01.00	01.01
00.10	$\to$	01.00	01.01
00.00	$\to$	00.00	00.01
10.00	$\to$	00.00	00.01
00.01	$\to$	00.10	00.11
11.00	$\to$	10.00	10.01
01.01	$\to$	10.10	10.11
01.00	$\to$	10.00	10.01
11.10	$\to$	11.00	11.01
01.10	$\to$	11.00	11.01
01.11	$\to$	11.10	

(6.69)

Notice that there are only 13 4-bit words out of the $2^4 = 16$ feasible words. Likewise there are missing transitions. As we can see from Fig. 6.31, the branches of the attractor fail to extend to so-labeled regions. For example, immediately above the region labeled 01.11 is a region that would be labeled 11.11 if it were occupied. But since the attractor does not intersect that region, no points on the attractor have orbits whose symbolic words have four 1's in a row. This is reflected in the table of allowed words.

Similarly, consider the same map under a somewhat arbitrary partition. We should not a priori expect that a partition of arbitrary rectangles is a Markov or even a generating partition. Consider the coarse partition of rectangles that was used to produce the directed graph in Fig. 1.1. Nonetheless, a symbolic dynamics is induced, this one on 15 symbols in Σ_{15}. Reading directly from the directed graph in Fig. 1.1, the corresponding link list could be written

From	$\to$	To			
1	$\to$	13			
2	$\to$	10	13	15	
3	$\to$	10			
4	$\to$	7	8		
5	$\to$	3	4	5	
6	$\to$	7	8		
7	$\to$	5	6	7	
8	$\to$	1	2	3	
9	$\to$	6	7		
10	$\to$	2	3	4	6
11	$\to$	1	2	3	
12	$\to$	9	12	13	
13	$\to$	9	10		
14	$\to$	7			
15	$\to$	2	3	4	6

(6.70)

Likewise, a link list of the 2-bit words on these 15 symbols could be formed, to include 1.1, 1.2, 1.3, etc.

These observations beg the question as to which partition is correct. Why do we need a generating partition? Can an arbitrary partition be used, or at least a very fine arbitrary partition? These questions are addressed in Section 6.4.6. At least the homoclinic tangency conjecture is satisfied by the partition, as shown in Fig. 6.31, it is believed to be generating. As such, the resulting symbolic dynamics generated by "typical orbits" are believed to well represent the dynamical system as approximated by the 4-bit representation in Eq. (6.69). Of course this is a coarse representation here for illustration only, and in principle many symbols could be used as far as the length of the test orbit would support.[88]

As an aside, using link lists as an efficient structure in the case of symbolic dynamics, we need only record whether a transition occurs or not, which is boolean information. In contrast to a transition matrix, or to a stochastic matrix, link lists are useful memory structures and easy for programming the systematic recording of transitions while scanning through a sampled orbit, whether it be for symbolic dynamics or measurable dynamics applications.

This empirical discussion of symbolic dynamics is useful and straightforward insofar as we have a useful partition. The analysis is similar irrespective of dimension, other than the difficulty of filling the space with a good test orbit. A more analytic analysis to considerations of allowable symbolic dynamics is as already addressed in the case of one-dimensional maps by the kneading theory [223] discussed above. A semirigorous pruning theory can be found in [71, 144, 81] for higher-dimensional cases for a few special cases, including the Henon map. It indicates a partial order in a symbol plane representation of the symbol space to indicate which words occur within a grammar from the dynamical system, much in analogy to the order-based description of the symbolic grammar from the kneading theory.

6.4.6 Is a Generating Partition Really Necessary?

We studied in [41, 40] the detailed consequences of using an arbitrarily chosen partition, which is generally not generating. Our work was motivated by the many recent experimentalists' studies who, with measured time series data in hand but in the absence of a theory leading to a known generating partition, simply choose a threshold crossing value of the time series to serve as a partition of the phase space.

On the experimental side, there appears an increasing interest in chaotic symbolic dynamics [77, 26]. A common practice is to apply the threshold-crossing method, i.e., to define a rather arbitrary partition, so that distinct symbols can be defined from measured time series. There are two reasons for the popularity of the threshold-crossing method:

1. It is extremely difficult to locate the generating partition from chaotic data.

2. Threshold-crossing is a physically intuitive and natural idea.

Consider, for instance, a time series of temperature $T(t)$ recorded from a turbulent flow. By replacing the real-valued data with symbolic data relative to some threshold T_c, say a

[88] A long enough orbit should be used so that there is high probability relative to the invariant measure that each symbolic bin of the n-bit representation will have an iterate in it, so that the observed symbolic link list such as Eq. (6.69) and each of the actual allowed transitions will be observed when it should be observed for a long enough orbit.

$\{\mathbf{0}\}$ if $T(t) < T_c$ and a $\{\mathbf{1}\}$ if $T(t) > T_c$, the problem of data analysis can be simplified. A well-chosen partition is clearly important: for instance, T_c cannot be outside the range of $T(t)$ because, if it is, the symbolic sequence will be trivial and carry no information about the underlying dynamics. Similarly, an arbitrary covering of a dynamical system with small rectangles leads to a directed graph representation as a Markov model of the system and, correspondingly, a symbolic dynamics, as was already seen for the Henon map, for example, in Fig. 1.1 with symbolic transitions shown in Eq. (6.70). It is thus of paramount interest, from both the theoretical and experimental points of view, to understand how misplaced partitions affect the goodness of the symbolic dynamics, such as the amount of information that can be extracted from the data.

As a model problem, we chose to analyze the tent map

$$f : [0,1] \to [0,1],\ x \mapsto 1 - 2|x - 1/2|, \tag{6.71}$$

for which most of our proofs applied. Our numerical experiments indicated that the results were indicative of a much wider class of dynamical systems, including the Henon map, and experimental data from a chemical reaction.

The tent map is a one-humped map, and it is known that the symbolic dynamics indicated by the generating partition at $x_c = 1/2$ by Eq. (6.8) gives the full 2-shift Σ_2 on symbols $\{0,1\}$. The topological entropy of $\Sigma_2^{\mathbf{0,1}}$[89] is ln(2) since it is a fullshift. Now misplace the partition at

$$p = x_c + d, \text{ where } d \in [-1/2, 1/2] \tag{6.72}$$

is the misplacement parameter. In this case, the symbolic sequence corresponding to a point $x \in [0,1]$ becomes

$$\phi = \phi_0.\phi_1\phi_2\ldots, \text{ where } \phi_i(x) = \mathbf{a}(\mathbf{b}) \text{ if } f^i(x) < p(> p), \tag{6.73}$$

as shown in Fig. 6.32. The shift so obtained, $\Sigma_2^{\{\mathbf{a,b}\}}$, will no longer be a fullshift because not every binary symbolic sequence is possible. Thus, $\Sigma_2^{\{\mathbf{a,b}\}}$ will be a subshift on two symbols $\mathbf{a}$ and $\mathbf{b}$ when $d \neq 0$ ($p \neq x_c$). The topological entropy of the subshift $\Sigma_2^{\{\mathbf{a,b}\}}$, denoted by $h_T(d)$, will typically be less than $h_T(0) = \ln 2$. Numerically, $h_T(d)$ can be computed by using the formula [268]

$$h_T(d) = \limsup_{n\to\infty} \frac{\ln N_n}{n}, \tag{6.74}$$

where $N_n \leq 2^n$ is the number of $(\mathbf{a}, \mathbf{b})$ binary sequences (words) of length n. In our computation, we choose 1024 values of d uniformly in the interval $[-1/2, 1/2]$. For each value of d, we count N_n in the range $4 \leq n \leq 18$ from a trajectory of 2^{20} points generated by the tent map. The slopes of the plots of $\ln N_n$ versus n approximates h_T. Fig. 6.33 shows $h_T(d)$ versus d for the tent map, where we observe a complicated, devil's staircase–like, but clearly nonmonotone behavior. For $d = 0$, we have $h_T(0) \approx \ln 2$, as expected. For $d = -1/2$ $(1/2)$, from Fig. 6.32 we see that the grammar forbids the letter $\mathbf{a}$ ($\mathbf{b}$) and, hence, $\Sigma_2^{\{\mathbf{a,b}\}}(-1/2)$ $[\Sigma_2^{\{\mathbf{a,b}\}}(1/2)]$ has only one sequence: $\phi = \mathbf{b}.\bar{\mathbf{b}\mathbf{b}}$ ($\phi = \mathbf{a}.\mathbf{a}\bar{\mathbf{a}}$). Hence, $h_T(\pm 1/2) = 0$.

[89] We will denote the symbol sequence space resulting from the generating partition to be $\Sigma_2^{\mathbf{0,1}}$ and the symbol space resulting from misplaced partition to be $\Sigma_2^{\mathbf{a,b}}$, both being 2-bit symbol sequences, so both share the notation Σ_2 and the superscript just reminds us of which shift space is being discussed.

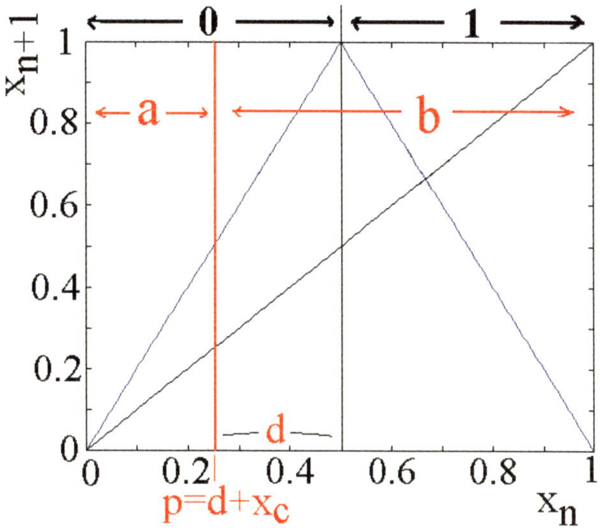

Figure 6.32. *Tent map and a misplaced partition at $x = p$.* [26]

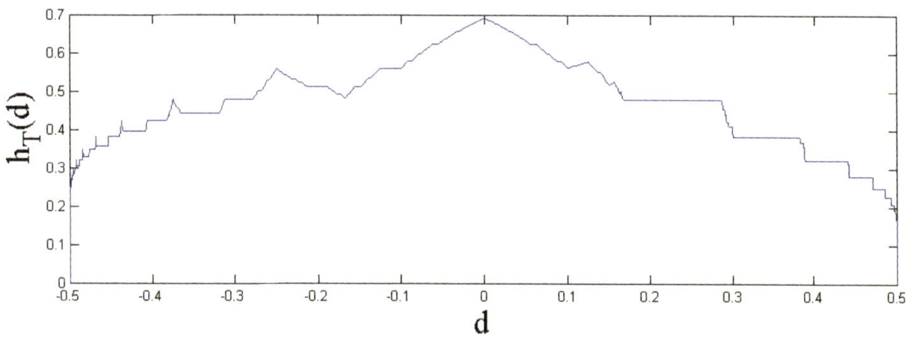

Figure 6.33. *For the tent map: numerically computed $h_T(d)$ function by following sequences of a chaotic orbit.* [26]

Many of our techniques were somewhat combinatorial, relying on a simple idea that a dense set of misplacement values (in particular if d is "dyadic," of the form $d = p/2^n$) allows us to study the related problem of counting distinctly colored paths through an appropriate graphic presentation of the shift, in which vertices have been relabeled according to where the misplacement occurs. See Fig. 6.34.

One of our principal results was a theorem that the entropy can be a nonmonotone and devil's staircase–like function of the misplacement parameter. As such, the consequence of a misplaced partition can be severe, including significantly reduced topological entropies and a high degree of nonuniqueness. Of importance to the experimentalist who wishes to characterize a dynamical system by observation of a bit stream generated from the measured time series, we showed that interpreting any results obtained from a threshold-crossing type of analysis should be exercised with extreme caution.

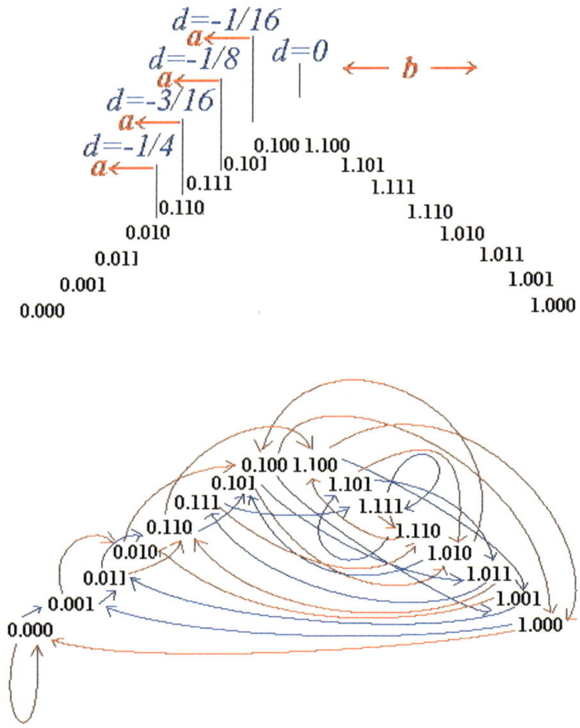

Figure 6.34. *Graphic presentation for the Bernoulli fullshift and some dyadic misplacements.* [26]

Specifically, we proved that the splitting properties of a generating partition are lost in a severe way. We defined a point x to be p-undistinguished if there exists a point $y \neq x$ such that the p-named $a-b$ word according to Eq. (6.73) does not distinguish the points, $\phi(x) = \phi(y)$. We defined a point x to be uncountably p-undistinguished if there exists uncountable many such y. We proved a theorem in [40] that states that if $p = q/2^n \neq 1/2$, then the set of uncountably p-undistinguished initial conditions is dense in $[0,1]$. In other words, the inability of symbolic dynamics from the "bad" nongenerating partition to distinguish the dynamics of points is severe. We described the situation as being similar to that of trying to interpret the dynamical system by watching a projection of the true dynamical system. In this scenario, some "shadow" or projection of the points corresponds to uncountably many suspension points. In our studies [41, 40], we also gave many further results both describing the mechanism behind the indistinguishability and further elucidating the problem.

6.5 Stochasticity, Symbolic Dynamics, and Finest Scale

In the table in (6.69) we used 4 bits to approximate the symbolic dynamics representing the dynamical system, and we suggested that more bits should be used since more is better. Is more always better? It is natural to ask what is the appropriate resolution scale to estimate a symbolic dynamics and, correspondingly, any discrete coarse-grained representation of a dynamical system, whether it be a topological perspective of symbol dynamics or a measurable dynamics perspective of Ulam's method.

In the presence of noise, there is a blurring effect of the randomized input which makes it unreasonable to discuss arbitrary precision of an evolution. Likewise, an infinite symbolic stream corresponding to an initial condition would imply infinite precision regarding knowledge of an infinite future itinerary. That is beyond measurement possibility in the case of a stochastic influence, and similarly the story is the same when limited by any finite-precision arithmetic representation on a computer, where round-off error acts as a noise-like influence. The question of appropriate scale in the presence of noise was the subject of the recent work by Lippolis and Cvitanovic [205], who ask (as entitled by their paper), "How well can one resolve the state space of a chaotic map?" We review this question briefly in this section.

In Lippolis and Cvitanovic [205], the stochastically perturbed dynamical system in the form of an additive Gaussian noise term was discussed,

$$x_{n+1} = f(x_n) + \xi_n, \tag{6.75}$$

for normal ξ_n with mean 0 and variance $2D$.[90]

This gives a Frobenius–Perron operator whose form we have written (3.46) as a special case of that found in [198] of the general multiplicative and additive stochastic form (3.44). In [205], a Gaussian stochastic kernel was chosen:

$$g(y - f(x)) = \frac{1}{\sqrt{4\pi D}} e^{-\frac{(y-f(x))^2}{4D}}. \tag{6.76}$$

The smallest noise-resolvable state space partition along the trajectory of a point x_a was determined by the effect of noise on the points preceding x_a. This is achieved by the backward transfer operator P_{F_ν}, which describes the density preceding the current state, $\rho_{n-1} = P^\dagger_{F_\nu} \rho_n$. The Gaussian form allows the form

$$\rho_{n-1}(x) = P^\dagger_{F_\nu}(\rho_n(y)) = \frac{1}{\sqrt{4\pi D}} \int_M \rho_n(y) e^{-\frac{(y-f(x))^2}{4D}} dy. \tag{6.77}$$

Following evolution of densities along orbits, and specifically along periodic orbits, allows the discussion of dispersion of our ability to measure the state *with confidence*. In [205] the following partition optimality condition was stated:

> the best possible of all partitions hypothesis, as an algorithm: assign to each periodic point x_a a neighborhood of finite width $[x_a - \sigma_a, x_a + \sigma_a]$. Consider periodic orbits of increasing period n_p, and stop the process of refining the state space partition as soon as the adjacent neighborhoods overlap.

Here, σ_a was taken to correspond to variance from the Gaussian computed as $\sigma_a = \frac{2D}{1-\Lambda_p^{-2}}(\frac{1}{(f'_a)^2} + \cdots + \frac{1}{\Lambda_p^2})$, where $\Lambda_p = f_a^{\prime n_p}$ is the Floquet multiplier of an unstable periodic

[90] In [205] this stochastic form was called a discrete Langevin equation, whereas following [198] we call this form a discrete time with constantly applied stochastic perturbation system. Likewise, its corresponding transfer operator was called a Fokker–Planck operator, whereas we call it a stochastic Frobenius–Perron operator.

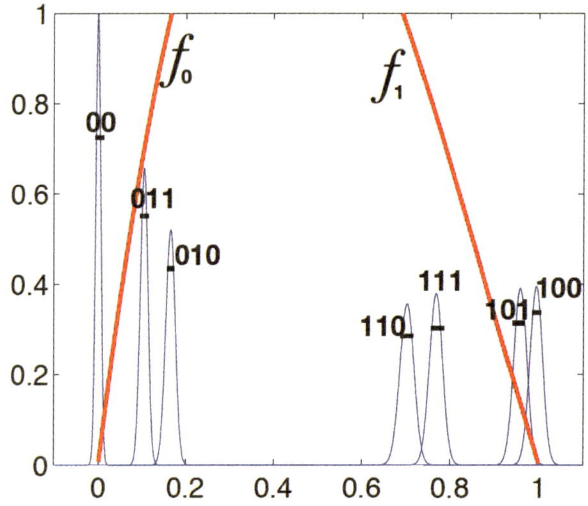

Figure 6.35. *Periodic orbits of the stochastically perturbed map* (6.78) *are symbolized up to a period such that neighborhoods about them blur to the point that they overlap other periodic orbits which also blur in both forward- and backward-time histories, according to the best possible of all partitions hypothesis. In red is shown f_0 and f_1 branches of the deterministic map which is stochastically perturbed with Gaussian noise, $D = 0.001$ of the form* (6.76). *Following the effect of noise on the points preceding points on periodic orbits by the backward operator* (6.77), *from intervals $[x_a - \sigma_a, x_a + \sigma_a]$, gives overlapping regions at the symbolization with the seven regions shown. This, according to the hypothesis, is the optimal stochastic partition.* [205]

orbit $x_a \in p$ of period n_p, $|\Lambda_p| > 1$. In Fig. 6.35 we reprise the main result from [205], which was for an example of a stochastically perturbed one-dimensional map,

$$f(x) = \Lambda_0 x(1-x)(1-bx), \tag{6.78}$$

with parameters chosen, $\Lambda_0 = 8, b = 0.6$, and branches of the deterministic map f_0 and f_1 shown in red. Some further description of a $D = 0.001$ is given in Fig. 6.35 leading to a seven-element partition. Further resolution does not lead to more knowledge but rather just ambiguity. This is a confidence-based description of the Markov chain model where we have used the word confidence in analogy to a statistics confidence interval.

A major note made in [205] is that the stochastic symbolic dynamics coming from this hypothesis is a finite state Markov chain, whereas the no-noise version may be infinite. We further emphasize here that this formalism is equally applicable to a measurable dynamics Markov chain description of the dynamical system and, correspondingly, an Ulam-like method would be applicable by a finite rank matrix computation.

We close this section by pointing the reader to a very nice alternative method of ascribing meaning to a stochastic version of symbolic dynamics, due to Kennel and Buhl [186]. That method is called symbolic false nearest neighbors, which in detail it is quite different from what is described here, but in spirit there is similarity. A statistic is introduced to accept or reject a hypothesis of false near neighbors which should be respected by a good symbolic partition according to the notion of a generating partition. In this sense, the work reviewed in this section, which demands no overlapping smeared symbolic neighborhoods, is comparable.

Chapter 7

Transport Mechanism, Lobe Dynamics, Flux Rates, and Escape

7.1 Transport Mechanism

A good understanding of transport mechanism will inform us regarding global analysis of topological dynamics, symbolic dynamics, and measurable dynamics questions such as escape rates. This understanding can eventually lead to control strategies to develop dynamical systems to yield results as engineered and desirable.

7.1.1 Preliminaries of Basic Transport Mechanism

Any discussion of transport must include a definition of "inside" and "outside" and thus a barrier in between, relative to which the transport can be referenced. To understand transport, we restrict the discussion to continuous dynamical systems of orientable manifolds.[91] The reason for this will become clear shortly. In addition, we will discuss orientation-preserving maps, which are so defined in terms of the tangent map $DT|_z$,

$$\det(DT|_z) > 0 \; \forall z. \tag{7.1}$$

We can also easily develop transport mechanisms for other types of maps, such as orientation reversing maps ($\det(DT|_z) < 0$ for all z), as long as the property is consistent.

Two-dimensional transport is particularly well understood and will be the subject of most of this discussion. Furthermore, the drawings used for our presentation will be much clearer in the two-dimensional setting. A two-dimensional discrete time map can, as usual, result from a flow on a three-dimensional manifold. Consider a Jordan curve C enclosing a region A.[92] We will investigate the relative orientation of forward and backward iterations of these sets (see Fig. 7.1). There are four basic cases of iterations that may result:

1. $T(A) \cap A$ is empty.
2. $T(A)$ is completely contained in A.

[91]Two-dimensional, orientable manifolds include the plane, the sphere, and the torus, but not the Klein bottle or the Möbius strip.

[92]The Jordan curve theorem is a basic theorem in topology that states that every Jordan curve (a non-self-intersecting loop) divides the plane into an inside and an outside, meaning that any path connecting a point of one region to a point of the other intersects that loop somewhere.

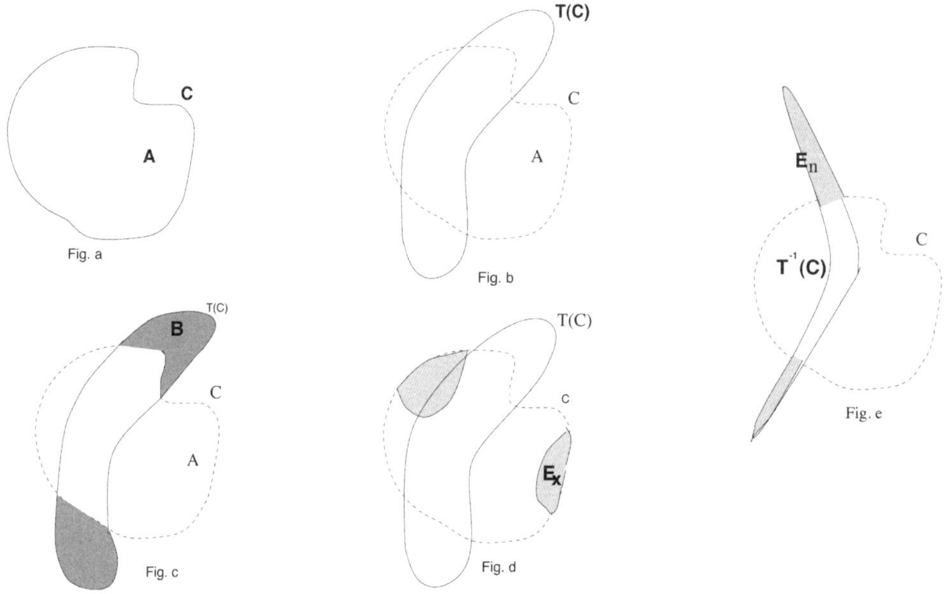

Figure 7.1. (a) *Jordan curve C enclosing a region A.* (b) *The first iterates of C and A intersect C and A, respectively.* (c) *The region* $B = T(A) - T(A) \cap A$ *contains all points which will enter A on one application of the inverse map.* (d) *The region* $E_x = T^{-1}(B)$ *contains all the points in A which will leave A upon one application of the map, and hence will be called the "exit region."* (e) *The "entrance region,"* $E_n = T^{-1}(A) - T^{-1}(A) \cap A$, *contains all points which will enter A upon one iteration of the map. Compare this figure, which discusses arbitrary enclosing curves C, to the special case in Fig. 7.6, which uses a carefully chosen enclosing curve of stable and unstable manifold segments.*

3. A is completely contained in $T(A)$.

4. $T(A) \cap A$ is nonempty and neither set is completely contained in the other.

Proceeding in a manner to emphasize an exploration of consequences of scenarios, we will emphasize the final case because it is typical of the "nice" barriers we will define in the next section and descriptive of the horseshoe dynamics from Section 6.1.5. Consider, for example, that C is chosen so that there exists a fixed point z^* on C. Then $T(C) \cap C \neq \emptyset$. Similarly, a fixed point $z^* \in A$ is sufficient for a nontrivial intersection, but not necessary.

To define the subset of A that leaves A on one iteration of the map, consider the first iterate of the curve, $T(C)$, enclosing $T(A)$ as an illustrative example. Notice in particular the region

$$B = T(A) - T(A) \cap A, \tag{7.2}$$

shaded in Fig. 7.1(c). The set B contains all those points that left A after one iteration. Alternately, it is the set that will enter A in one iteration of the inverse map T^{-1}. Thus, B defines the entrance set of T^{-1}. In this sense, we can say that the points in B cross the barrier C, thinking of C as a barrier.

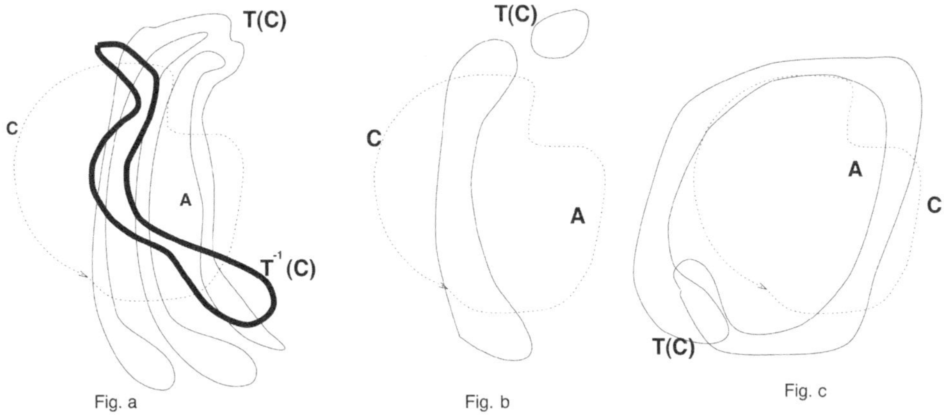

Figure 7.2. (a) *A possible, more complicated iterate (and back-iterate) of the region A. The implied entrance and exit regions in this example can intersect, which simply means that some subset of points entering A will immediately exit A on the next iterate.* (b) *This configuration of $T(C)$ is not possible because it violates assumptions of continuity. (A continuous function must map a connected set to a connected set.)* (c) *This configuration of $T(C)$ overlaps itself, and so violates single-valuedness of $T^{-1}(T(C))$.*

A typical inverse iterate scenario of B is shown in Fig. 7.1(d), labeled

$$E_x = T^{-1}(B). \tag{7.3}$$

E_x is the subset of A that will leave A on one iteration of T and hence is called the exit set. *The only* way for an orbit initially contained in A to leave A is for an iterate of the orbit to land in E_x. The map moves all the contents of E_x outside the closed curve on each iteration.

Repeating this discussion in reverse, we may similarly construct the entrance set E_n outside of C, which is defined as $T^{-1}(A) - T^{-1}(A) \cap A$. It is the set outside of C which is moved inside of C on each iteration. It is the only way in, across C.

In summary, the entrance and exit sets are defined as

$$\begin{aligned} E_n &= T^{-1}(A) - T^{-1}(A) \cap A, \\ E_x &= T^{-1}[T(A) - T(A) \cap A]. \end{aligned} \tag{7.4}$$

These definitions apply to all four of the intersection types listed. The fourth is shown in Fig. 7.1(d), but the other three are just as valid. In the first case, for example, if $T(A)$ is disjoint from A, then $E_x = A$. In the second case, where $T(A)$ is completely contained in A, we see that $T(A) - T(A) \cap A = T(A) - T(A) = \emptyset$, and therefore $E_x = \emptyset$.

There are certainly more complicated configurations possible for $T(A)$, relative to a general set A, than are implied by the previous figures. Some of these are indicated by Fig. 7.2. Nonetheless, we can uniquely define E_n as the set that enters A in one iteration, and E_x as the set that leaves A in one iteration. Eq. (7.4) defines the entrance and exit lobes with no limitations on the amount of folding possible. A configuration such as Fig. 7.2(a) presents no contradictions despite the overlap of $T(C)$ and $T^{-1}(C)$; it simply implies that once leaving A, the subset $T(E_x) \cap E_n \cap \bar{A}$ will immediately re-enter A on the next iteration. Eq. (7.4) implies no statement regarding two iterations. Configurations such as in Figs. 7.2(b) and 7.2(c), which may present problems, are not possible due to violations of assumptions of continuity and single-valuedness, respectively.

As a matter of discussion, we ask what it takes for a barrier to divide a manifold, even if the manifold may not be homeomorphic to the plane. The key property for discussing transport is the **topological partition,** already defined formally in Definition 4.2. The question of transport is to ask how and where, in each of the open sets defining the topological partition, points move from one element to the other, and this description is the same regardless of dimensionality. A most interesting special case is when the elements of the topological partition are connected open sets. Specifically in two dimensions, the Jordan curve case discussed above consists of a two-element topological partition, labeled "inside" and "outside" and the boundary curve is so named Jordan. **Barriers** are generally the boundary points of the topological partition. Of course the whole discussion can be carried forward in more than two dimensions, but we have limited the presentation for simplicity of discussion as the main idea becomes clear.

On certain manifolds, it is possible to describe transport across a barrier C which is not a Jordan curve. The role which the closed curve serves in the above discussion is that it divides the space in two—an inside and an outside. On the other hand, if a curve does not completely divide the space, transport can occur "across" the barrier by going around it (or by going the "other way" around the cylinder $S^1 \times \mathbb{R}$ to avoid an infinite line barrier in the case of a cylinder, for example). Without becoming involved in an out-of-scope discussion in defining the genus of a manifold,[93] and ways to partition such manifolds, consider the example of a cylinder that can be divided in two (a top part and bottom part) by a closed curve (a "belt" wrapped around the "waist"). See Fig. 7.3, and notice in particular in Fig. 7.3(c) that C_2 divides the top from the bottom.

The description of transport across any barrier is made by forward and backward iterating the barrier, then finding the regions bounded by C and $T^{-1}(C)$ (or C and $T(C)$) and asking which region crosses the barrier on the next iterate (back-iterate). We found lobe-like structures in Figs. 7.1 and 7.2 because we illustrated the situation where $T(A) \cap A \neq \emptyset$, and neither set is completely contained in the other. We will see this situation in the next section, where there will typically be a fixed point z^* on C.

7.1.2 Chaotic Transport Mechanism and Lobe Dynamics

In the previous section, we described that questions of transport are only relevant to a topological partition and its barrier; see Definition 4.2. In this section, we ask the related question, "What are the most natural barriers in chaotic transport (and the topological partition whose closure includes these barriers)?" The arbitrarily chosen barriers in the previous section typically evolve and deform upon iteration. The situation is typically exponentially exaggerated with continued iterations. We continue this discussion in terms of maps for the sake of specificity; a similar discussion could be stated for flows.

A natural barrier can be constructed of segments of stable and unstable manifolds on a homoclinic orbit. Given a period-n point z of a map T, recall (Section 2.3) the stable

[93] Genus is a notion from algebraic topology which in the case of two-dimensional manifolds may be roughly described as the number of hoops attached to the manifold.

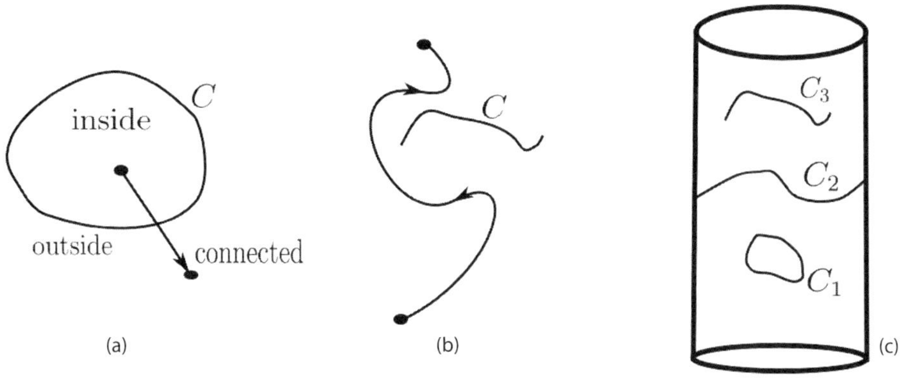

Figure 7.3. (a) *A Jordan curve C in the plane has an inside and an outside, and any path connection from inside to outside must intersect the curve according to the Jordan curve theorem.* (b) *C is not a Jordan curve, and likewise there is a path around it.* (c) *A cylinder $S^1 \times \mathbb{R}$ has the possibility of a Jordan curve C_1, or a belt-like curve C_2 that induces a partition between those points above and below. In the cylinder, C_3 does not permit a topological partition.*

subspace $W^s(z)$, and $W^u(z)$, the unstable subspace, defined as follows:

$$W^s(z) \equiv \{\boldsymbol{x} : T^{jn}(\boldsymbol{x}) \to z \text{ as } j \to \infty\},$$
$$W^u(z) \equiv \{\boldsymbol{x} : T^{jn}(\boldsymbol{x}) \to z \text{ as } j \to -\infty\}. \tag{7.5}$$

We repeat that a point z is defined to be hyperbolic when the tangent space at that point is decomposable as the direct sum

$$M = E^s(z) \oplus E^u(z), \tag{7.6}$$

where $E^s(z)$ (or $E^u(z)$) is the linear subspace of the tangent space at z, spanned by the eigenvectors corresponding to eigenvalues with modulus strictly different than one.

Summarizing Section 2.3, some standard hyperbolicity results from the mathematical theory, under suitable hypothesis requiring regularity of the vector field and hyperbolicity following either expanding or contracting eigenvalues.

- The stable manifold theorem [268] implies that these eigenvectors can be continued to the global stable (unstable) manifolds.
- The Hartman–Grobman theorem [251, 146] states that for a diffeomorphism T^n and a small enough neighborhood $N(z)$ of z, there is a homeomorphism between the dynamics of the linearized mapping DT^n on the corresponding tangent space $E^s(z) \oplus E^u(z)$ and $T^n|_{N(z)}$ on U, which is a neighborhood of z in the original phase space. See Fig. 7.4 (right).
- A saddle point, a special case of a hyperbolic point, is categorized by having all of the eigenvalues λ_i of the tangent map at z such that $\lambda_i \in \mathbb{C}, |\lambda_i| \neq 1$ for all i, using the complex modulus $|\cdot|$.

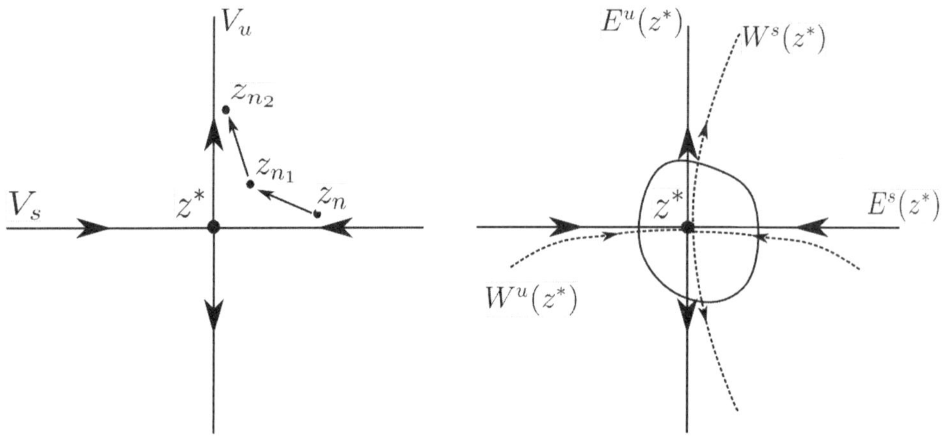

Figure 7.4. *(Left) A hyperbolic linear (linearized) mapping DT has a stable space E^s spanned by eigenvectors with eigenvalues of modulus less than one and an unstable space E^u spanned by eigenvectors with eigenvalues of modulus less than one. Drawn here is the scenario that all eigenvalues are real and positive, $0 < \lambda_s < 1 < \lambda_u$, since it is a saddle. Iterates of z_n successively jump pointing progressively along E^u. (Right) The Grobman–Hartman theorem provides that a hyperbolic fixed point z^* has the property that there is a neighborhood $N(z^*)$ such that in that neighborhood, the linearized map is conjugate to the nonlinear map, and furthermore the corresponding stable and unstable manifolds $W^s(z^*)$ and $W^u(z^*)$ become tangent to the manifolds of linearized map, E^s and E^u.*

A hyperbolic saddle fixed point of a two-dimensional map is shown in Fig. 7.4. It should be stressed that in the case of a discrete time map, the smooth curves shown are not flows of a single point. Each point "jumps" upon application of the map to another location on the curve; see Fig. 7.4 (left). Continuity of T implies that a nearby point jumps close by.[94]

Certain rules must be obeyed by such dynamics on such manifolds.

- By definition, these manifolds are invariant; a point on the stable (unstable) manifold remains on the manifold.
- Single-valuedness forbids that a stable (unstable) manifold intersect itself or the stable (unstable) manifold of another point.
- It is allowed, however, for the stable manifold to intersect the unstable manifold.

Recall the following from Definition 2.11.

Definition 7.1. A point p on an intersection of $W^s(z_i)$ and $W^u(z_j)$ is called a **homoclinic point** if $i = j$ or a **heteroclinic point** if $i \neq j$.

[94]It may seem paradoxical that a chaotic dynamical system can nonetheless be continuous. Sensitivity to initial conditions and exponential spreading of nearby points may seem to some to exclude continuity, since it suggests that nearby points *eventually* map far away, but the key is the emphasis on that word *eventually*. Continuity is a property of *single* applications of the map, and sensitivity to initial conditions describes the evolution of nearby points under *many* applications of the map.

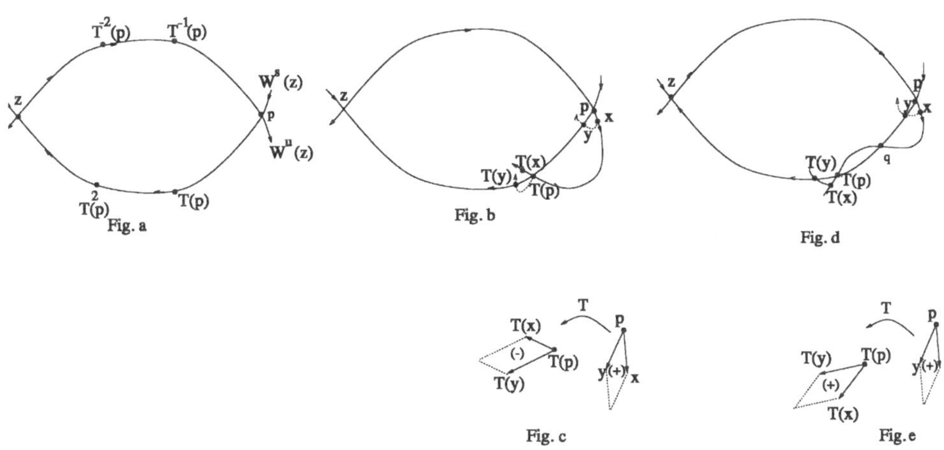

Figure 7.5. *Principle intersection points.* (a) *A transverse homoclinic connection at point* $\boldsymbol{p}$, *and a few of its iterates and preiterates.* (b) *A single lobe between* $\boldsymbol{p}$ *and* $T(\boldsymbol{p})$ *causes an illegal (by assumption) orientation change from* $\boldsymbol{p},\boldsymbol{x},z$ *to* $T(\boldsymbol{p})$, $T(\boldsymbol{y})$, $T(\boldsymbol{x})$. (c) *The "orientation of surface" (or "signed area") of the parallelogram described by the vectors* $\boldsymbol{p}-\boldsymbol{y}$ *and* $\boldsymbol{p}-\boldsymbol{x}$ *has opposite sign to that of parallelogram* $T(\boldsymbol{p})-T(\boldsymbol{y})$, $T(\boldsymbol{p})-T(\boldsymbol{x})$. (d) *Inserting one more transverse homoclinic point* $\boldsymbol{q}$ *yields a legally oriented image of* $\boldsymbol{p},\boldsymbol{x},\boldsymbol{y}$. (e) *Here we can see that the sign of the area of the nearby parallelogram is preserved by* T.

Fig. 7.5 illustrates homoclinic points p of stable and unstable manifolds of a fixed point z. In fact, it is a transverse homoclinic point. By definition, the orbit of $\boldsymbol{p}$ accumulates on z_i in forward time, and on z_j in backward time. Thus, iterates of homoclinic (heteroclinic) points are homoclinic (heteroclinic) points: The existence of one intersection implies infinitely many intersections.

Fig. 7.5(a) shows a homoclinic orbit with transverse intersection.[95] Also shown is part of the family of points corresponding to the orbit of $\boldsymbol{p}$. The stable and unstable manifolds must intersect at each point in this family, but it is easy to show that there exists another homoclinic point $\boldsymbol{q}$ between $\boldsymbol{p}$ and $T(\boldsymbol{p})$, following an assumption of orientation preservation. To see this, consider two arbitrary nearby points: $\boldsymbol{x}$ near $\boldsymbol{p}$, where $\boldsymbol{x}$ is on $W^u(z)$ "farther" along $W^u(z)$ before $T(\boldsymbol{p})$,[96] and $\boldsymbol{y}$, also near $\boldsymbol{p}$, but on $W^s(z)$ closer to z, but again before $T(\boldsymbol{p})$. The relative configurations of $\boldsymbol{x}$, $\boldsymbol{y}$, and $\boldsymbol{p}$ are drawn in Fig. 7.5(b). Reading clockwise, the order is $\boldsymbol{p}$, $\boldsymbol{x}$, and $\boldsymbol{y}$. $T(\boldsymbol{x})$ must still be farther along than $T(\boldsymbol{p})$, and likewise so must $T(\boldsymbol{y})$ occur after $T(\boldsymbol{p})$. Again, reading clockwise, we get $T(\boldsymbol{p})$, $T(\boldsymbol{y})$, and $T(\boldsymbol{x})$, which is in violation of orientation preservation. We can see this in Fig. 7.5(c),

[95] As one varies parameters of a parameter-dependent dynamical system, $z_{n+1} = T_\lambda(z_n)$, the manifolds $W^s(z_i)$ and $W^u(z_j)$ tend to move, and may intersect either transversally or tangentially. Tangent-type intersections are not structurally stable (stable to arbitrarily small perturbations to the mapping itself perhaps by varying a parameter, versus the usual notion of stability which is a discussion concerning perturbation to the initial condition), unlike the transverse-type which will be the subject of the discussion to follow.

[96] An ordering on $W^u(z)$ is possible since the invariant manifold is one-dimensional. A point is defined as farther away from z than another parameterized point in the sense of the arc length along the unstable manifold. An ordering on $W^s(z)$ can be similarly defined in terms of the arc length closeness to z.

where the area of the parallelogram, described by the vectors $\boldsymbol{p}-\boldsymbol{y}$ and $\boldsymbol{p}-\boldsymbol{x}$, is opposite the parallelogram $T(\boldsymbol{p})-T(\boldsymbol{y})$, $T(\boldsymbol{p})-T(\boldsymbol{x})$. However, we can see in Fig. 7.5(d) that inserting an additional transverse homoclinic intersection at $\boldsymbol{q}$ preserves the orientation, shown in Fig. 7.5(e). Hence, there must be at least one more homoclinic point $\boldsymbol{q}$.

It is convenient to choose $\boldsymbol{p}$ to be what Wiggins [315] defines as a principle intersection point). Any point on $W^s(z_i) \cap W^u(z_j)$ is a heteroclinic (or homoclinic) point.

Definition 7.2. Using the ordering implicit along these invariant manifolds, we can define a **principle intersection point (p.i.p.)** as a heteroclinic (homoclinic) point for which the stable manifold segment between z_i and $\boldsymbol{p}$ has no previous intersections with the unstable manifold segment between z_j and $\boldsymbol{p}$.

These segments of the stable and unstable manifolds have also been called "initial segments" [108].

In Fig. 7.6(c), the shaded regions are labeled E_x and E_n, describing their transport roles. These lobes have infinitely many (pre)images, whose endpoints are the (pre)images of $\boldsymbol{p}$ and $\boldsymbol{q}$. We may now define a Jordan curve C using the unstable manifold initial segment between z and $\boldsymbol{p}$, and the stable manifold initial segment between z and $\boldsymbol{p}$, for any p.i.p. $\boldsymbol{p}$.[97] See Fig. 7.6(a). There is a well-defined inside and outside for this barrier C. Eq. (7.4), defining transport across an arbitrary barrier, applies to this special choice of C.

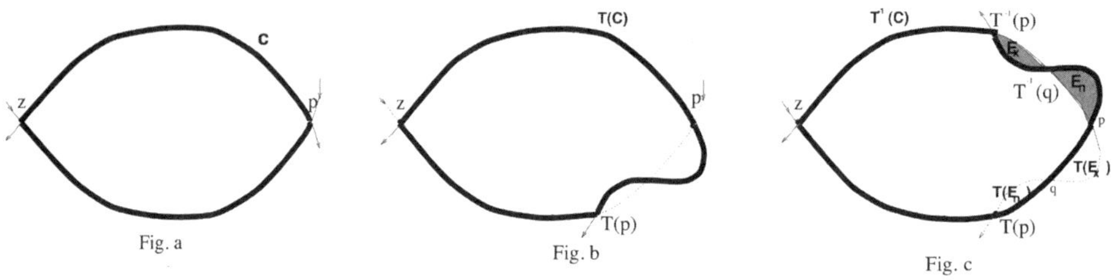

Figure 7.6. *Definition of a homoclinic barrier in the orientation-preserving case.* (a) *Defining the barrier by initial segments of the stable and unstable manifolds between the fixed point z and p.i.p. $\boldsymbol{p}$.* (b) *The iterate C lies largely on top of C, as much of the curve stretches over itself.* (c) *The exit and entrance lobes E_x and E_n, which together are called the turnstile. Compare to Fig.* 7.1, *where we use a carefully chosen enclosing curve in terms of stable and unstable manifold segments.*

The commonly chosen barrier C of principle segments of stable and unstable manifold segments illustrated in Fig. 7.6 is quite natural because orbits on the manifolds stay on the manifolds. Following the discussion in the previous section, in Fig. 7.6(b) we draw $T(C)$, and in Fig. 7.6(c), we draw $T^{-1}(C)$. In terms of the original barrier C, we see that the shaded region E_n in Fig. 7.6(c) iterates to the region $T(E_n)$ inside C, which we easily see by following the iterates of $\boldsymbol{p}$, $\boldsymbol{q}$, and the manifold segments in between.

The only alteration in the overall form of C is the "growth" of the lobes E_n and E_x upon application of T^{-1}. In this sense the choice of the barrier C is minimal. Mackey, Meiss, and Percival coined the term "turnstile" [209] to describe the two lobes E_n and

[97] In fact, as long as $\boldsymbol{p}$ is a p.i.p., any of its iterates are just as legitimate, and the resulting entrance and exit lobes can be used to define transport.

E_x in that they act like the rotating doors of the same name used in underground railroad transportation, transporting across area C.

Remark 7.1. P.i.p.'s are not unique. Iterates of p.i.p.'s are also p.i.p.'s. Both families of points, shown in Fig. 7.5(d), are examples of p.i.p.'s. Starting with p.i.p.'s, nonprinciple intersection points arise from the stretching and folding typical with transverse heteroclinic (homoclinic) intersections. See Fig. 7.7. The resulting tangle quickly generates infinitely many other families of heteroclinic (homoclinic) points which are not p.i.p.'s. More will be said about the tangling process in the next section since the generation of infinitely many further families results due to issues of measure evolution.

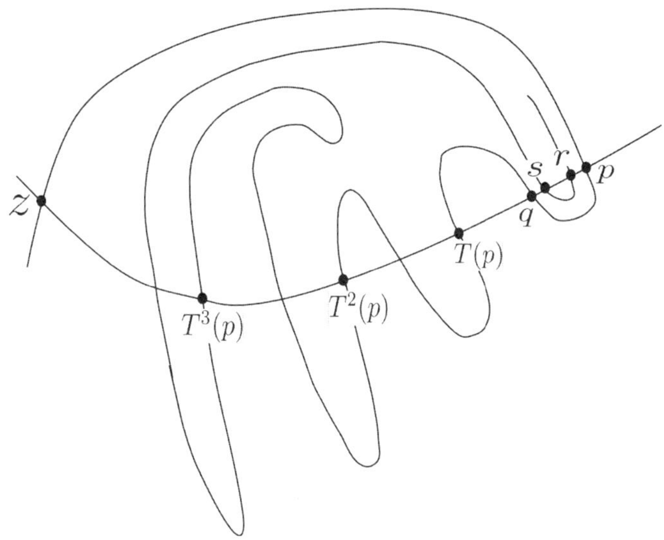

Figure 7.7. *P.i.p.'s are not unique. Note that both **p** and **q** are p.i.p.'s but so are all the iterates shown and reiterates as well. The stretching along unstable manifolds and compression along stable manifolds cause the stretching of the turnstiles, which under considerations of growth of measure eventually causes further non-p.i.p. intersections such as **r** and **s** (nontangential) as shown. Then all the points on their orbits are homoclinic.*

In Section 7.11, we present a homoclinic tangle generated by an area-preserving Henon map; the stable and unstable manifolds reveal the story so described. See further discussion of entrance and exit steps for this map in Section 7.2.

In closing this subsection, we can describe transport as in some sense just an illusion which is clarified only by the choice of barrier. Studying Fig. 7.6, there is another perspective on transport to be made. Forgetting our barrier C for a moment, let us focus on a point in the entrance set E_n "outside"[98] of the manifold segment of $W^s(z_i)$ between $\boldsymbol{p}$ and $\boldsymbol{q}$. The role of iterating the map is to cause that segment of $W^s(z)$ to push in (relative to C). Points outside that segment may be viewed as still outside. In this perspective, there is no transport at all; it is just an illusion of the outside punching in further and further upon

[98] Of course, a curve segment is not enough to define a barrier; a fully closed curve is required.

iteration. This description only makes use of the stable manifold. Of course, only in terms of the full barrier C can we truly describe transport across the barrier.

7.1.3 Lobes, the Homoclinic Tangle, and Embedded Horseshoes

The discussion in the previous sections about turnstiles concerned the one-step action of the map relative to a barrier. In this section, we ask what is the long-term fate of

- the barrier C in forward and backward time,
- the points inside and outside of C,
- the points in E_n and E_x.

These are the fundamental questions that lead to understanding of **chaotic transport across barriers,** and the notion of the **homoclinic tangle.** The answers will also lead us to find horseshoes embedded within the tangle, the prototypical example of chaos.

Taking a simple two-p.i.p. family generated by $\boldsymbol{q}$ and $\boldsymbol{p}$ above as a case study (e.g., in Fig. 7.6), we see that the arc length between $\boldsymbol{q}$ and $\boldsymbol{p}$ along $W^s(z)$, labeled $\ell(W^s_{[\boldsymbol{q},\boldsymbol{p}]})$, must eventually shrink upon repeated applications of the map, as the two points eventually accumulate at the fixed point. The arc length at time n is the line integral of the nth iterate of $W^s_{[\boldsymbol{q},\boldsymbol{p}]}$. Likewise, the arc $W^u_{[\boldsymbol{q},\boldsymbol{p}]}$ iterates with $\boldsymbol{q}$ and $\boldsymbol{p}$. Hence, the curve $W^u_{[\boldsymbol{q},\boldsymbol{p}]} \cup W^s_{[\boldsymbol{q},\boldsymbol{p}]}$ is a dynamically varying boundary of $T(E_n)$.

Throughout the recent sections, the symbolic dynamical descriptions of the topological horseshoe have been entirely topological, meaning no measure structure was needed or assumed. Only action of the mapping on the structure of the open sets was assumed, definitive of the point set topology. To quantify the fate of lobes, we will now resort to measure and distance.

Assume

$$T : X \to X \text{ occurs on a measure space } (\mu, X, \mathcal{A}) \tag{7.7}$$

on measurable sets $\mathcal{A}$, and there is a distance function:

$$d : X \to \mathbb{R}^+. \tag{7.8}$$

For narrative description, we will refer to a popular example, the area-preserving maps which arise commonly when studying Hamiltonian systems [209, 219]. If the measure μ is descriptive of area,

$$Area(A) = \mu(A) \tag{7.9}$$

for each measurable $A \in \mathcal{A}$, then the map is called **area preserving** if

$$\mu(T(A)) = \mu(A) \text{ for each } A \in \mathcal{A}. \tag{7.10}$$

Each A may deform under T, but its area always remains the same. If $|\lambda_u| > 1$ is the expanding exponent along unstable manifolds and, likewise, $0 < |\lambda_s| < 1$ is the contracting exponent, then in the case of area-preserving maps it quickly follows that

$$\lambda_u = \lambda_s^{-1}. \tag{7.11}$$

The lobes E_n and E_x, specifically $T^n(E_n)$, may stretch exponentially and transversally contract, resulting in a long and narrow lobe for large n. See Fig. 7.7. Likewise, $T^n(E_x)$ becomes narrow and long. To make these statements rigorous, we would resort to the lambda lemma from hyperbolicity theory [268], but we will continue with our narrative description for now. It is half of the story that stretching is one of the main components that can result in chaos, the other part of the story being the folding;[99] this story is revealed in full by the horseshoe example.

Remark 7.2. In the area-preserving case, it is easy to see that *the finite area of region A, bounded by C, cannot completely contain all future iterates of* E_n. There is a time m when

$$\sum_{i=1}^{m} \mu(T^i(E_n)) \geq \mu(A), \tag{7.12}$$

i.e., the first time that

$$m \geq \frac{\mu(A)}{\mu(E_n)}. \tag{7.13}$$

In terms of transport, some of the points in E_n which enter A must leave A by the mth iterate, implying that there exists an

$$l \leq m \text{ such that } T^l(E_n) \cap E_x \neq \emptyset. \tag{7.14}$$

This follows since the only way out is through an exit lobe, and we already concluded by Eqs. (7.12)–(7.13) that some points coming in by E_n must soon leave. Almost all of the points[100] must eventually leave. Once this intersection occurs, a new family of homoclinic points is implied. See points $\boldsymbol{r}$ and $\boldsymbol{s}$ in Fig. 7.7. Considering the history of the lobe E_x, which also becomes long and narrow (as $n \to \infty$), we see that a homoclinic point is implied each of the times m and $-n$ that

$$T^m(E_n) \cap T^{-n}(E_x) \neq \emptyset. \tag{7.15}$$

The segment $W^u_{[\boldsymbol{q},\boldsymbol{p}]}$ accumulates at z as $n \to -\infty$, and $W^s_{[\boldsymbol{q},\boldsymbol{p}]}$ accumulates at z as $n \to \infty$. Of course, a "new" family of intersections implies infinitely more intersections as the homoclinic point iterates in forward and backward time, and so forth for further families. This is the "homoclinic tangle."

As presented in Section 6.1.5 and Figs. 6.6–6.14, the horseshoe construction implies a set which remains trapped inside the region A for all time. Smale invented the topological horseshoe and showed its relevance to applied dynamical systems [294].

Theorem 7.1 (Smale [294]; see also [315]). A diffeomorphism T with a transverse homoclinic point $\boldsymbol{p}$ has a time $m > 0$ such that the composition map T^m has an invariant Cantor set $\Lambda \in A$. Further, there exists a conjugacy

$$h : \Lambda \to \Sigma_2 \tag{7.16}$$

such that

$$h \circ T^m|_\Lambda = \alpha \circ h. \tag{7.17}$$

The conjugacy is with the dynamics of the Bernoulli shift map α on the space of bi-infinite sequences of countably many symbols σ_i. In the simple horseshoe, we let

$$\sigma_i = 0 \text{ or } 1. \tag{7.18}$$

[99] A popularly described (necessary) recipe for chaos is "stretch+fold yields chaos."

[100] "Almost all" and "almost every" are stated in the usual measure theoretic terms, meaning all but a set of measure zero.

This theorem tells us that a horseshoe and all of its implied complex and chaotic dynamics is relevant to a standard setting due to a transverse homoclinic point, a geometric configuration of stable and unstable manifolds that occurs even in real-world and physical systems; see Fig. 7.8. The horseshoe may be constructed for Fig. 7.6 by drawing a thin curved strip S over $W^s_{[z,\boldsymbol{p}]}$ as shown in Fig. 7.8. As $\boldsymbol{p}$ iterates closer to z, it drags the strip with it. Meanwhile, the point s, defined as the intersection $W^u_{[z,\boldsymbol{p}]} \cap S$, marches away from z upon repeated application of T. Define m as the first time that $T^m(s)$ is ordered on the arc segment *after* $\boldsymbol{p}$. By time m, the short side of the strip has stretched and folded over to the strip $T^m(S)$ along $W^u_{[z,\boldsymbol{p}]}$ which intersects S by construction. Here again we see the stretch and fold, which can be thought of as the ingredients for chaos.

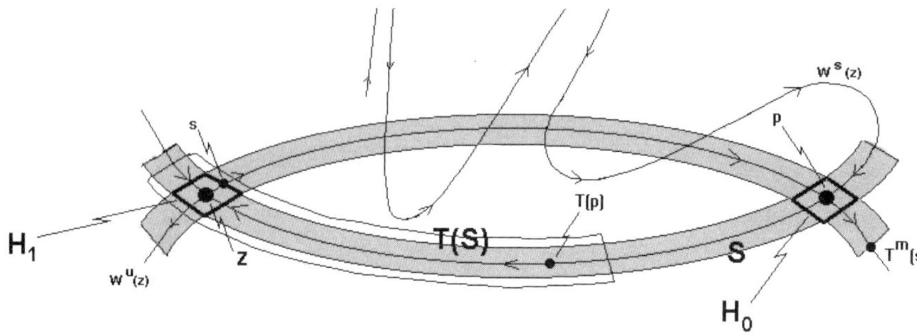

Figure 7.8. *Constructing a horseshoe on a homoclinic orbit, as per Theorem* 7.1. *The strip S contracts along the stable manifold, and expands along the unstable manifold to the shorter, wider strip $T(S)$. By the mth iterate, the point $T(s)$ has passed p; the long and short sides of the strip $T^m(S)$ are reversed from the long and short sides of the original strip S. The invariant sets H_0 and H_1 are the first steps in generating the invariant Cantor set Λ.*

Define

$$H_0 = T^m(S) \cap S \tag{7.19}$$

at $\boldsymbol{p}$ and

$$H_1 = T^m(S) \cap S \tag{7.20}$$

at z, and define

$$H = H_0 \cup H_1. \tag{7.21}$$

We see that the invariant set of T^m is contained in $T^m(H) \cap H$, which defines two vertical strips in H_0, and the two vertical strips in H_1. Similarly, the invariant set of T^{-m} is $T^{-m}(H) \cap H$, which forms two horizontal strips in H_0 and H_1. Define

$$\Lambda = \bigcap_{i=-\infty}^{\infty} T^{im}(H). \tag{7.22}$$

Λ is the invariant set of the horseshoe,[101] which we see is the Cartesian product of two

[101]The Smale horseshoe is so named because the horseshoe is constructed by stretching and folding a square into horseshoe shapes (again and again and ...). The process is perhaps more akin to building a Japanese samurai sword, whose building process includes thousands of stretches and folds.

Cantor sets, one in forward and one in backward time. For a thin enough strip S, the invariant set is hyperbolic [268, 208].

The address of a point in H can be labeled .0 if it is in H_0 or .1 if it is in H_1. On iteration, the point (say it is .0) lands in either H_0 or H_1, and hence is labeled .00 or .01, which defines which vertical strip in H_0 contains the point. Similarly, the address to the left of the decimal determines in which square the point lands, H_0 or H_1. From this point, the theory of the topological horseshoe presented in Section 6.1.5 requires further rigor to prove that the representation is correct [268].

Horseshoes have been explicitly constructed for a number of examples, separately from this Smale theorem that discusses transverse intersection of stable and unstable manifolds. As revealed in Section 6.2.2, it is possible to construct a horseshoe explicitly for the Henon map (Figs. 6.20 and 6.21), the standard map when $k > 2\pi$, (see [208]), and the Poincaré mapping from the Duffing equations (Fig. 6.22), to name a few. Horseshoes can also be constructed for heteroclinic cycles. As discussed further in Section 6.4, other grammars besides the simple left shift[102] on many symbols can also be useful in the more general scenario of many folding and partially folded systems.

In terms of transport, the horseshoe serves only as an incomplete model of the dynamical system, for two reasons:

1. Typically, one may be interested in the transport of more than a measure zero set of points. The horseshoe set Λ is a topological Cantor set,[103] and one of zero measure.

2. A serious deficiency for transport study is that the horseshoe models those points invariant to the horseshoe, i.e., those points which never transport out of the horseshoe set. Transport within Λ is completely described by the horseshoe model, but no more is said by the analysis of the action of the broader map on the rest of its phase space.

The measure zero property is easy to show for the middle thirds Cantor set. Inspecting Fig. 7.9, since

$$C_\infty = \cap_{n\to\infty} C_n, \tag{7.23}$$

then

$$l(C_0) = 1, \quad l(C_1) = \frac{2}{3} l(C_0), \quad l(C_2) = \frac{2}{3} l(C_1), \tag{7.24}$$

and so

$$l(C_\infty) = \lim_{n\to\infty} \frac{2}{3}^n l(C_0) = 0. \tag{7.25}$$

A similar reasoning applies to the invariant set of the horseshoe as depicted in Figs. 6.6–6.14, and it has similar measure properties.

Given a complicated transport problem, say from near some point $\boldsymbol{a}$ to near a point $\boldsymbol{b}$, where only a long, convoluted heteroclinic connection may exist, one may be successful in finding a complex grammatical rule on a long list of symbols, if $\boldsymbol{a}$ and $\boldsymbol{b}$ happen to be

[102] The Bernoulli left shift grammar on two symbols can be described by the directed graph $0 \leftrightarrow 1$, which is equivalent to the 2×2 identity transition matrix. Other grammars on n symbols have directed graphs describable by more general $n \times n$ identity matrices.

[103] A generalized Cantor set is defined to be compact, nonempty, perfect (has no isolated points), and totally disconnected, and it will be uncountable. See Fig. 7.9 főr a illustration of the famous middle thirds Cantor set, which will remind us of the invariant sets of a horseshoe.

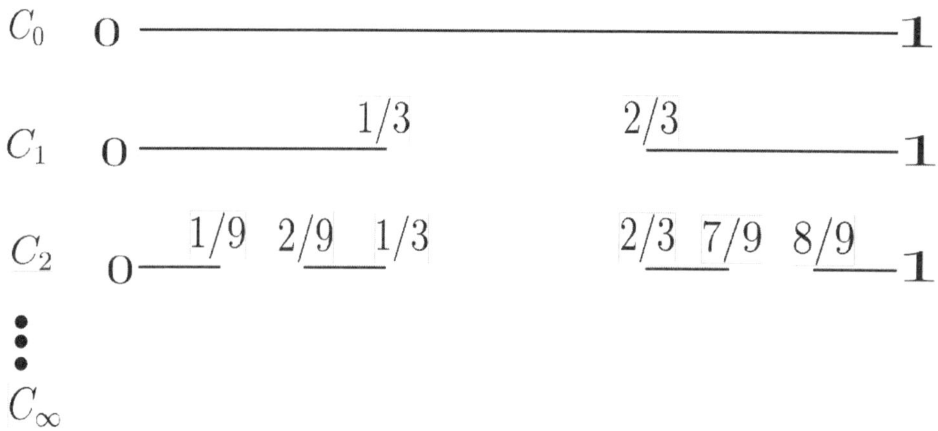

Figure 7.9. *The middle thirds Cantor set $C_\infty = \bigcap_{n\to\infty} C_n$ is useful model set since invariant sets in chaotic dynamical systems, and in particular of horseshoes, are generalized Cantor sets, sharing many of the topological properties as well as many of the metric properties such as self-similarity and fractal nature.*

in some invariant set of the dynamics. But, in general, only heteroclinic cycles are homeomorphic to the horseshoe, and hence have a reasonably easy-to-find symbol dynamics. As more of the phase space is to be described, then more complicated symbolic dynamics may be required for each fattening of the corresponding unstable chaotic saddle Cantor set.

7.2 Markov Model Dynamics for Lobe Dynamics: A Henon Map Example

Traditionally, following the Smale horseshoe example, symbol dynamics has been incorporated into dynamical systems theory in order to classify and investigate the behavior of dynamics on invariant sets. As we have pointed out, these invariant sets are often of measure zero. There is often a great deal of other behaviors that the dynamical system is capable of demonstrating. In particular, we will give an example here of how symbolic dynamics can be used to characterize transport behavior and lobe dynamics. Such modern approaches are presented for more sophisticated and complete examples, including homoclinic and clinic transport [219, 106, 217, 315, 317] and, in particular, symbolic characterization of lobe dynamics in [63] and [229].

7.2.1 An Example Area-Preserving Henon Map

Take the area-preserving Henon map [162],

$$\begin{aligned} z_{i+1} = T(z_i) &= \begin{pmatrix} x_i\cos\alpha - y_i\sin\alpha + x_i^2\sin\alpha \\ x_i\sin\alpha + y_i\cos\alpha - x_i^2\cos\alpha \end{pmatrix} \\ &= \begin{pmatrix} \cos\alpha & -\sin\alpha \\ \sin\alpha & \cos\alpha \end{pmatrix}\begin{pmatrix} x_i \\ y_i - x_i^2 \end{pmatrix}, \end{aligned} \tag{7.26}$$

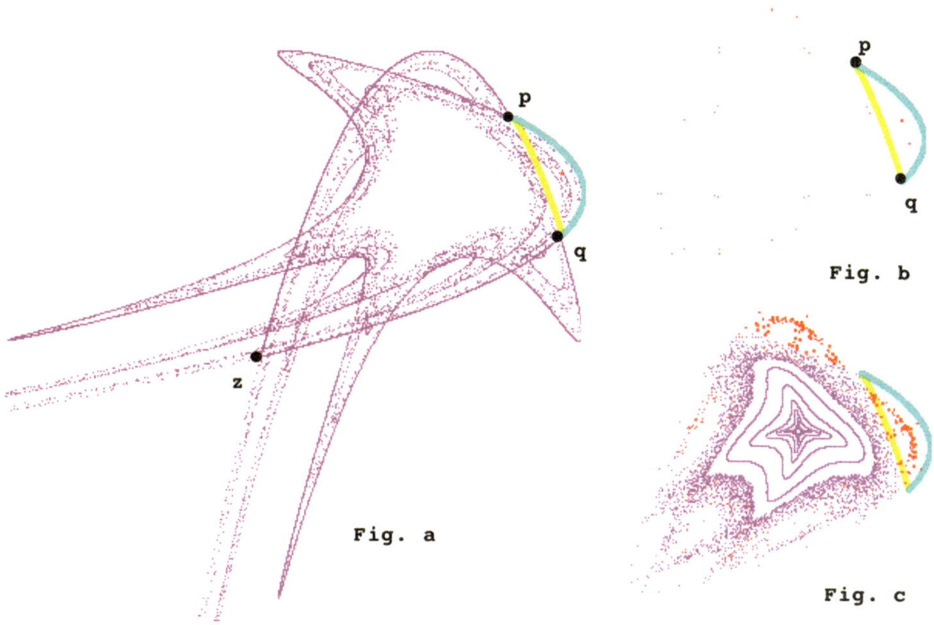

Figure 7.10. *Henon map turnstile and invariant sets. Compare to Figs.* 7.11 *and* 7.12. (a) *The homoclinic tangle of the Henon map. Shown are the fixed point z, the two p.i.p.'s $\boldsymbol{p}$ and $\boldsymbol{q}$, and the exit lobe E_x.* (b) *The exit lobe E_x of the Henon map.* (c) *Controlled dynamics. We reflect a point z_i through the symmetry $S: y = x$ into the entrance lobe whenever it enters the exit lobe. The area-preserving Henon map is written* (7.27).

where α is a parameter. As usual, we will consider this as a mapping of the plane, $T: \mathbb{R}^2 \to \mathbb{R}^2$, which is again a diffeomorphism. Furthermore, this is a special example of a quadratic map in that it is an area-preserving[104] Hamiltonian system.[105] We may interpret this mapping according to Moser [234] as the composition of a rotation and a shear. This map is perhaps the prototypical example for its conjugacy to the horseshoe, whose stretch and fold dynamics are visibly apparent and proved in [208].

This map is in a form which is particularly elegant to invert. In Fig. 7.10(c), we see the orbits of several initial conditions under the influence of the altered Henon dynamics

$$z_{i+1} = \begin{pmatrix} T(S(z_i)) & \text{if } z_i \in E_x \\ T(z_i) & \text{otherwise} \end{pmatrix}. \tag{7.27}$$

Here is an interesting consequence of the symmetric nature of the dynamics, $S: y = x$.[106] Let us recall the definition (see [208]) that a map T has a symmetry S if S is an orientation-reversing involution such that

$$S^2 = (TS)^2 = I. \tag{7.28}$$

[104]Each Lebesgue measurable set A has the property that $area(A) = area(T(A))$.

[105]Formerly there is a Hamiltonian from which the dynamics follows, but we shall take area-preserving Hamiltonian here to be more loosely descriptive of "not dissipative."

[106]If we fold the paper through the line $y = x$, we have still captured the whole orbit.

Figure 7.11. *A "typical" homoclinic tangle. Compare to Figs.* 7.10 and 7.12 *from the area-preserving Henon map* (7.27).

It follows that

$$T^{-1} = STS^{-1}. \tag{7.29}$$

Those points which eventually escape the barrier outlined in Figs. 7.10 and 7.12 under the Henon map (7.27). As shown in Fig. 7.10(c), these are now periodic under Eq. (7.27). All those points are forever bounded inside the barrier, including the quasi-periodic invariant circles,[107] the regions they bound, and other chaotic orbits, including those which are part of the horseshoe invariant set Λ (which we know is embedded since we see a transverse homoclinic point at $\boldsymbol{p}$ as well as at $\boldsymbol{q}$).

7.2.2 The Lobe Dynamics

The typical homoclinic tangle shown in Fig. 7.11 is from the area-preserving Henon map, Eq. (7.27). A detailed barrier illustration is shown in Fig. 7.12 for the lobe dynamics in

[107] There is an interesting and deep story regarding KAM theory, invariant tori, and resonance pertaining to this and many other examples of area-preserving maps, but this story leads us too far away from the subject of this book. See [219].

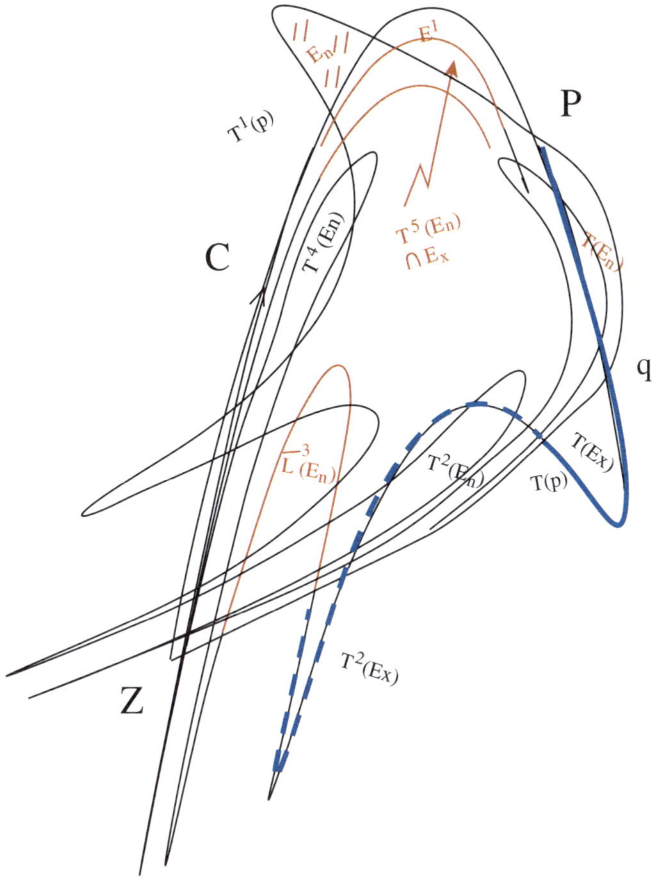

Figure 7.12. *Transport in a "typical" homoclinic tangle as seen in Fig.* 7.11 *related to our discussion of horseshoe invariant sets is now illustrated to discuss lobe dynamics and escape. The partition associated an inside and outside relative to the entrance and exit lobes corresponding to iterates and preiterates of the partitioned region. Shown here is a particularly convenient partition corresponding to segments of stable and unstable manifolds and the implied entrance and exit lobes labeled E_n and E_x. Compare to Figs.* 7.10 *and* 7.13. *The map used here as a case example is the area-preserving Henon map in Eq.* (7.27), *with a fixed point z and p.i.p. homoclinic point **p** as shown in Fig.* 7.10.

a manner analogous to Fig. 7.6, with the coloring and details regarding entrance and exit lobes, E_n and E_x. The particular details of this map, with parameters as chosen, reveal that $E_x \cap T^5(E_n)$ is the first such intersection. This information yields a Markov model of the escape dynamics which may be symbolized by the graph(s) shown in Fig. 7.15.

Once the inside and outside (Fig. 7.13) are defined, we have lobe dynamics as follows. As shown in Fig. 7.12, the unstable manifold segments of $W^u(z)$ from z to p.i.p. $\boldsymbol{p}$ and $W^s(z)$ from z to $\boldsymbol{p}$ define a boundary curve C. This is a particularly convenient choice of boundary curve to discuss the transport mechanism. The entrance set E_n could be called

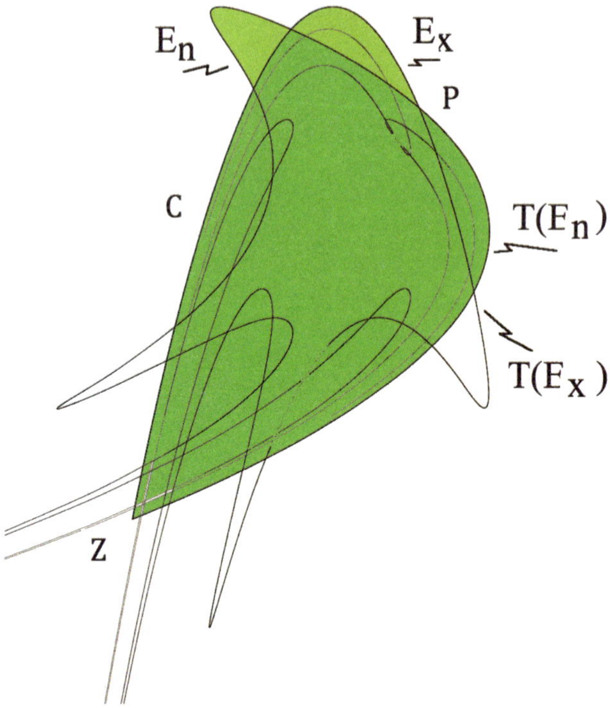

Figure 7.13. *Entrance and exit lobes in the area-preserving Henon map* (7.27). *The darker green set together with the light green lobe labeled E_x, which we call $\mathcal{C}$ the interior of the barrier C, described by the stable and unstable manifold segments $W^s(z)$ and $W^u(z)$ from z, meeting at p.i.p.'s* $\mathbf{p}$. *The entrance set E_n is outside this barrier, but iterates to $T(E_n) \subset \mathcal{C}$. Likewise E_x starts inside $\mathcal{C}$, but leaves in one iterate. Compare to Fig.* 7.6. *Iterates of these lobes allow us to follow the transport activity of this map relative to the green barrier. These are outlined by stable and unstable manifolds, and we see that since $E_x \cap T^5(E_n)$, then some points will escape the region after* 5 *iterates, but some will remain. See a Markov model of the same in Fig.* 7.15.

in standard lobe dynamics notation

$$L_{1,2} = \{z | z \in R_1, T(z) \in R_2\}, \tag{7.30}$$

where in Fig. 7.13 we will name R_1 the set of all points outside the curve C, and R_2 will be the set of those points inside C. Likewise, the exit lobe E_x can be denoted

$$L_{2,1} = \{z | z \in R_2, T(z) \in R_1\}. \tag{7.31}$$

In this notation for the lobe dynamics, the discussion of transport, flux, and escape is all in terms of the orbits of these lobes as sets,

$$\{T^i(L_{1,2})\}_{i=-\infty}^{\infty} \quad \text{and} \quad \{T^i(L_{2,1})\}_{i=-\infty}^{\infty}. \tag{7.32}$$

Especially interesting to the complex behaviors becomes the various higher-order intersections which may occur:

$$\text{any } i,j \text{ such that } T^i(L_{1,2}) \cap T^j(L_{2,1}) \neq \emptyset. \tag{7.33}$$

Remembering that orbits are indexed infinitely in forward and backward time, we must acknowledge that if there is one such overlap, there must be *infinitely many such intersections, at least in this family* of intersections. In this case, it is possible that there is a set of points whose fate is to enter the region R_2 for some sojourn time. Then such points exit after some time, and then re-enter, repeating indefinitely and forever. Such a scenario is another view of the homoclinic tangle, which can be said to have been befuddling to founders of the field.[108] By previous discussion regarding the possibility and likelihood of infinitely many such families of intersections, one should expect the lobe dynamics to allow for points which enter and exit the region in infinitely many different patterns, meaning infinitely many different i, j pairings denoted (7.33). Not only are such complicated behaviors expected, under mild sufficient assumptions such as area-preserving maps, we can prove the existence of this scenario as per Eqs. (7.12)–(7.14).[109]

The orbits of the $L_{1,2}$ entrance lobe and the $L_{2,1}$ exit lobe are illustrated as colored blue and red, respectively, in Fig. 7.14. The discussion of transport rates then becomes a matter of considering the relative measure of these sets, the rate of change of their measure with respect to iteration, and the measure of nontrivial overlaps. See in particular the special case of resonance overlap in Section 7.3.

Markov Model Dynamics and the Symbol Dynamics of Lobes

The labeling of lobes, such as the example (7.30)–(7.31), can be considered to imply a Markov model and likewise to generate a symbolic dynamics. Markov modeling of escape dynamics is a type of symbol dynamics where the symbolization is relative to inside and outside a barrier. These states could just as well serve as a symbol set. Including a symbol for escape to ∞ allows us the description shown in Fig. 7.15. This leads to the discussion of transport rates, flux, and resonance overlap whose lobe dynamics are discussed in the next section, leading to interesting escape rate sets.

7.3 On Lobe Dynamics of Resonance Overlap

The lobe dynamics defined by Eqs. (7.30)–(7.31) can be readily generalized to discuss more complicated geometries of homoclinic and heteroclinic tangles corresponding to partitions with many elements. We draw here from elements which can be found in Wiggins [315] and Meiss [209, 219], among other sources. Notably some of the more complicated scenarios

[108] Henri Poincaré, who was the founding father of the field of dynamical systems, understood hints of the homoclinic tangle in his work, but was not able to draw this complicated structure [259]. "If one seeks to visualize the pattern formed by these two curves and their finite number of intersections, each corresponding to a doubly symptotic solution, these intersections for a kind of lattice-work, a weave, a chain-link network of infinitely fine mesh; each of the two curves can never cross itself, but it must fold back on itself in a very complicated way so as to recross all the chain-links an infinite number of times. One will be struck by the complexity of this figure, which I am not even attempting to draw."

[109] A lobe has area; therefore, an infinite forward orbit of such a lobe cannot remain in a bounded region forever, at least for an area-preserving map.

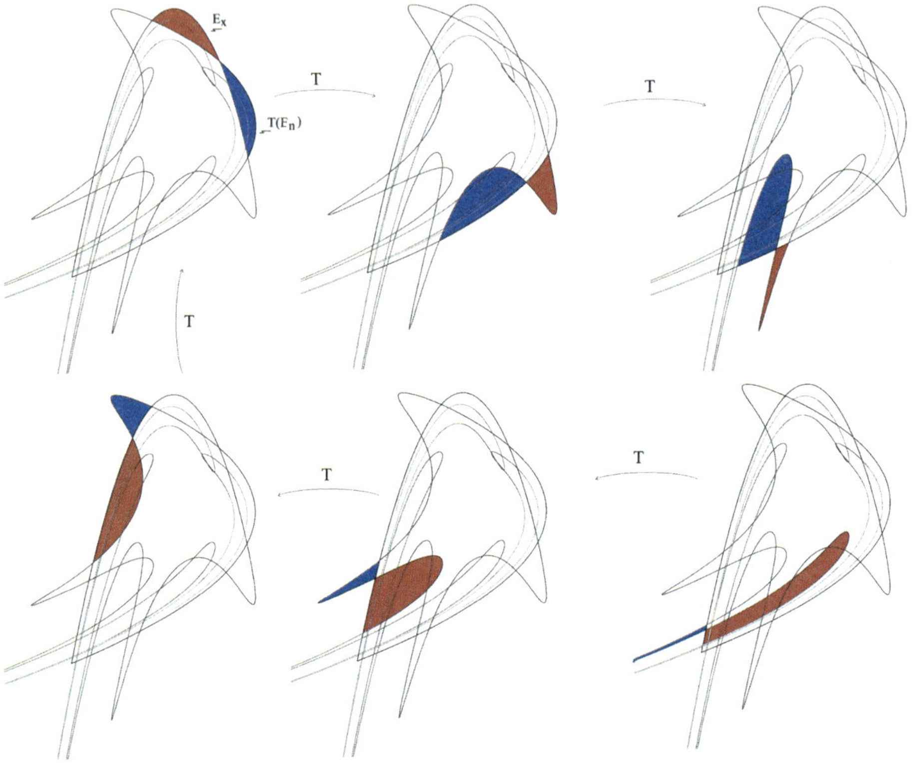

Figure 7.14. *(Upper left) Entrance (blue) and exit (red) lobes labeled E_n and $T(E_x)$ in Fig. 7.10 are colored. By lobe dynamics notation in Eqs.* (7.30)–(7.31), *$L_{1,2}$ is colored blue in the lower left image, and $L_{2.1}$ is colored red in the upper left figure. Iterations of these regions by the Henon map T are shown along the upper row, where the action of the map pushes those initial conditions inside the colored red set E_x. Compare to the dark green colored region from Fig.* 7.13. *Those points inside the blue entrance set E_n are shown to enter the dark green region shown in Fig.* 7.13 *by the action of T, as seen here along the top row. Along the bottom row is shown the action of the inverse of the map T^{-1} successively on the sets E_n and E_x. Compare to Figs.* 7.10, 7.12, *and* 7.13.

are discussed most recently in Mitchell [231, 229], the latter having the enticing title, "The Topology of Nested Homoclinic and Heteroclinic Tangles."

For the sake of presentation, in Fig. 7.16 we illustrate a particular scenario of resonance overlap due to a pair of heteroclinic orbits of fixed points z_1, z_1' and z_2, z_2', respectively. We repeat that the key to any transport study starts with the definition of a partition relative to which the transport may be discussed. Now we need pairs of open sets, one pair for each barrier to be crossed. The complete barrier will be the "meets," $\wedge$, of these four sets.[110] Consider the open regions Ω_1, Ω_1^c and also Ω_2, Ω_2^c by segments of stable and unstable manifolds as defined further in the caption of Fig. 7.16. In this drawing, the curves

[110] $\wedge$ is called the "meets" and $\vee$ is called the "join" in comparing sets.

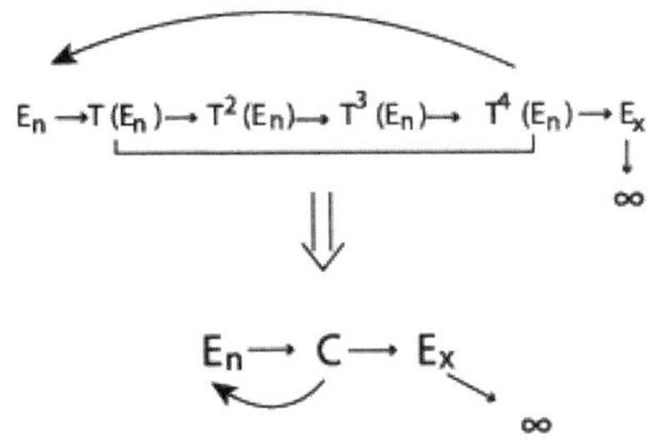

Figure 7.15. *The action of ensembles of initial conditions relative to the barrier C discussed in Fig. 7.12 can be summarized by the top graph, which is a Markov model describing the first intersection, $E_x \cap T^5(E_n)$. Summarizing further the quotient that if we are only interested that inside is inside, then a quotient vertex C gives the bottom graph. Implications for escape rate of measure is discussed in Eq.* (7.12).

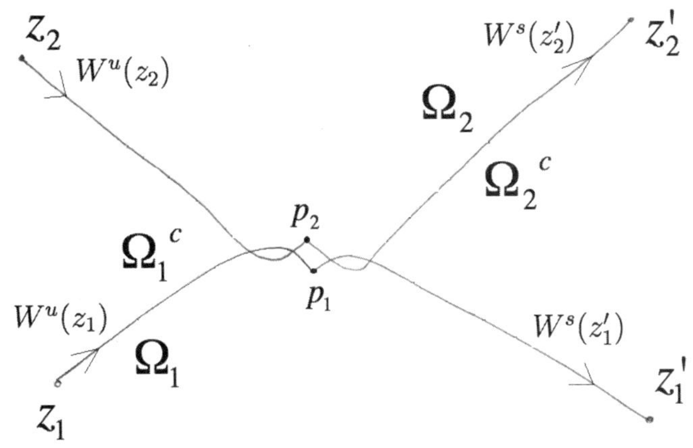

Figure 7.16. *A typical resonance overlap scenario leading to transport and escape. A heteroclinic orbit of fixed points z_1 and z_1' with stable manifold $W^s(z_1)$ and unstable manifold $W^u(z_1')$, respectively are shown. $W^u(z_1)$ and $W^s(z_1')$ are not shown, only principle segments from the fixed points to the p.i.p. p_1. Assume dynamics on a cylinder, so $z_1 = z_1'$ or $z_2 = z_2'$. Denote by Ω_1 and Ω_1' the two regions of the topological partition such that Ω_1 and Ω_1^c form an open topological partition of the region. Likewise Ω_2 and Ω_2^c are formed by the heteroclinic connection of segments of $W^u(z_2)$ and $W^s(z_2')$ to the p.i.p. p_2, defined by the boundary curve of the open partition elements Ω_2 and Ω_2'. See Figs. 7.17–7.19 for more discussion of this lobe dynamics.*

shown do not form a Jordan curve to partition the space. In order to permit the partition as outlined by these curves of stable and unstable manifold segments, we should allow that either the dynamics is on a cylinder, or otherwise the region not shown outside the figure also acts to dynamically delineate the space.

The dynamics of entrance and exit lobes is stated here in the same format as Eqs. (7.30)–(7.31),

$$L_{1,1^c} = \{z | z \in \Omega_1, T(z) \in \Omega_1^c\}, \tag{7.34}$$

and likewise

$$L_{1^c,1} = \{z | z \in \Omega_1^c, T(z) \in \Omega_1\}, \tag{7.35}$$

and the lobes $L_{2,2^c}$ and $L_{2,2^c}$ are similarly defined. These lobes, and the dynamics of their first few iterates and preiterates, are shown in Figs. 7.17–7.18. The most interesting feature in discussing transport between resonance layers is the overlap of these lobes. When it occurs that

$$T^i(L_{1,1^c}) \cap T^j(L_{2^c,2}) \neq \emptyset \text{ for some } i, j, \tag{7.36}$$

interesting orbits, and particularly interesting heteroclinic orbits, occur. The example of the fate of this resonance overlap is illustrated in Fig. 7.19.

The orbit of these sets and their intersections explains the *geometric description of transport*. Just as shown in Fig. 7.15, one can pursue a symbolic dynamics description of transport following a Markov model. Some detailed topological discussion of lobe dynamics and symbolic dynamics can be found in [231, 229]. An elegant special example of transport in a celestial mechanics system described in terms of symbolic dynamics can be found in [192]. Discussion of the *measure theoretic description* of these evolving lobes addressed questions of *transport rates and flux across barriers,* to be discussed in the next section.

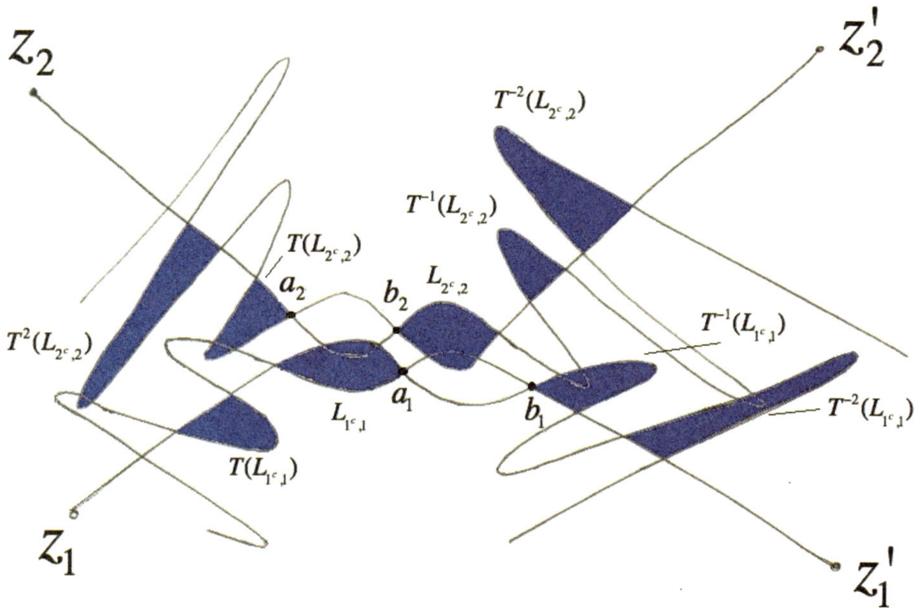

Figure 7.17. *Lobe dynamics continuing the general resonance overlap scenario of Fig. 7.16, as defined in Eqs. (7.34)–(7.36). Here the lobe $L_{1^c,1}$, consisting of those points which cross the heteroclinic orbit segment curve defined of p.i.p. segments, crossing from Ω_{1^c} to Ω_1, is shown in blue. Also shown are lobe iterates and preiterates $T^i(L_{1^c,1})$, $i = -2, -1, 0, 1$. Similarly, the lobe dynamics of $L_{2^c,2}$ and iterates for crossing from Ω_{2^c} to Ω_2 are shown. Contrast with the lobe dynamics in Fig. 7.18 for crossing these pseudobarriers in the opposite direction.*

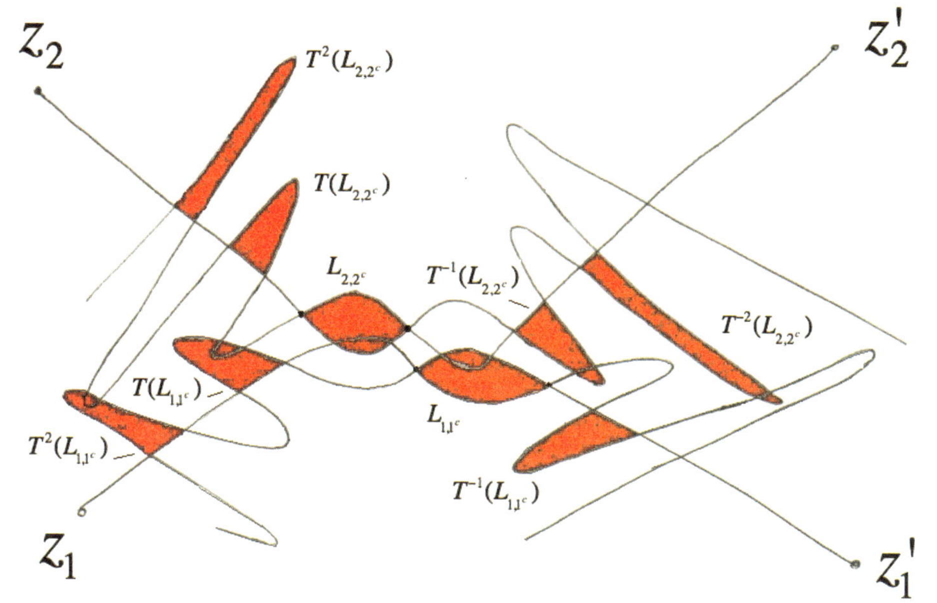

Figure 7.18. *Lobe dynamics continuing the general resonance overlap scenario of Fig.* 7.16, *shown here for transport crossing in opposite directions relative to the transport already shown in Fig.* 7.17. $L_{1,1^c}$ *consists of those points which iterate from* Ω_1 *to* Ω_{1^c}, *and likewise the iterates of this lobe are shown. See Eqs.* (7.34)–(7.36). *Also shown is the lobe dynamics of* $L_{2,2^c}$.

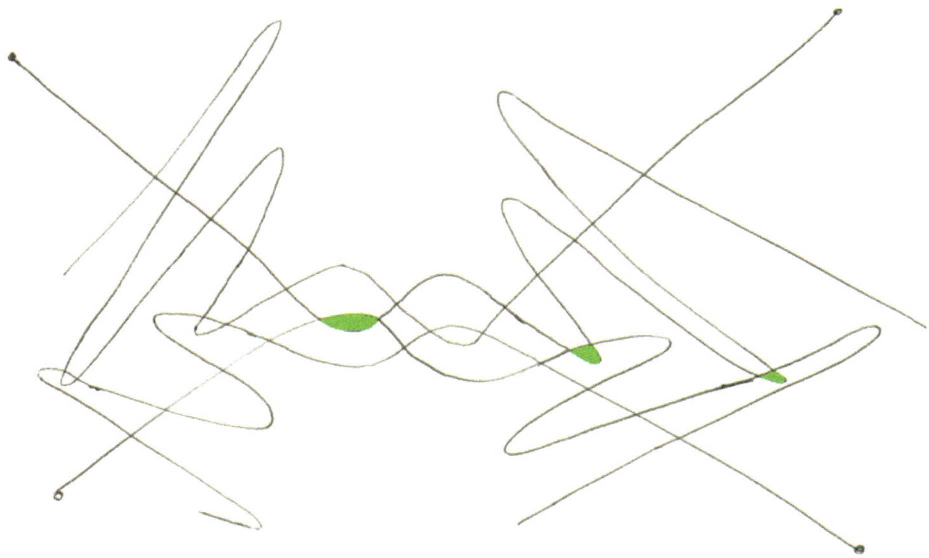

Figure 7.19. *The resonance overlap scenario, marked in green, when lobes intersect,* $T^i(L_{1,1^c}) \cap T^j(L_{2^c,2}) \neq \emptyset$, *for some* i, j, *from Eq.* (7.36).

7.4 Transport Rates

In the previous section we discussed the transport in dynamical systems geometrically in terms of watching the orbits of special sets called lobes and turnstiles consisting of stable and unstable manifold segments. We discussed how stable and unstable manifold segments bound partition elements across which transport can be discussed. Following sets alone is a geometric discussion of transport. On the other hand, questions of flux, expected escape time, and so forth are inherently quantitative questions requiring a measure structure. We suggest in this section that these questions are a natural place to adjust our already useful transfer operator methods. Particularly for area-preserving maps, there has been detailed study of transport rates and flux by careful study of the stable and unstable manifold structures and lobe dynamics [209], for example, and through resulting symbolic dynamics in [270, 318] by the so-called trellis structures of Easton [108, 107]. These structures are descriptive of the topology of the homoclinic tangle, beginning with lobe areas calculated from action integrals such as in [194]. See also the Markov tree model in mixed resonance systems [246], and the interesting discussion of area as a devil's staircase [54]. We will focus here on uniform arbitrary fine partitions in the spirit of the Ulam method by use of the Frobenius–Perron operator methods, which allow us to attack more general systems without the special analytic setup required for Melnikov's method and action-based methods. Questions pertaining to transport rates and flux, and high activity transport barriers, have useful realizations in the underlying Markov chain description implicit in the transfer operator approximated by the graph theory discussion.

7.4.1 Flux across Barriers

When investigating transport, flux across barriers is the natural quantity to calculate. It is natural to ask how flux may vary as system parameters vary; how does a barrier give way to a partial barrier as an invariant set gives way to an almost-invariant set? Our discussion of flux will be in terms of an ensemble density profile ρ. The idea is to measure how much relative quantity of the ensemble is moved from one identified region to another. We will discuss this in terms of transfer operators.

Discussion of transport makes sense only by first assuming a partition relative to which the transport is described. Given a region, say an open set $S = S_i$ to be one of the elements of a topological partition $\{S_i\}$ of the phase space, it is possible to define *mass flux* (or simply *flux*) in and out of the region, across the boundary ∂S of the set being discussed [34]. Let

$$\mathcal{F}_S^{\pm}[\rho] \equiv \text{mass entering/exiting } S \text{ from outside/inside } S \text{ upon one application of the map, due to initial density profile } \rho, \tag{7.37}$$

where $\mathcal{F}_S^{\pm}[\rho]$ can be read as "mass flux into/out of S due to an initial density profile ρ." To calculate $\mathcal{F}_S^{+}[\rho]$, we appropriately restrict the region of integration of the transfer operator which here we have assumed to be in the form of the stochastic Frobenius–Perron operator for the sake of discussion

$$\mathcal{F}_S^{+}[\rho] = \int_S \int_{\tilde{S}} \nu(x - F(y))\rho(y)dydx. \tag{7.38}$$

The symbol $\tilde{S}$ denotes the complement of the set S for discussion of transport away from

S. Observe that the inner integral,

$$\int_{\tilde{S}} \nu(x - F(y))\rho(y)dy,$$

gives the density at x which comes from a point $y \in \tilde{S}$, meaning not in S. The outer integral accounts for the total mass of all such contributions in x. To calculate the flux into S, $\mathcal{F}_S^-[\rho]$, we must simply reverse the regions of integration in Eq. (7.38). The obvious identities follow immediately,

$$\mathcal{F}_S^+[\rho] = \mathcal{F}_{\tilde{S}}^-[\rho] \quad \text{and} \quad \mathcal{F}_S^-[\rho] = \mathcal{F}_{\tilde{S}}^+[\rho], \tag{7.39}$$

due to conservation of mass.

One has the option to calculate $\mathcal{F}_S^{\pm}[\rho]$ by direct (numerical) application of Eq. (7.38). On the other hand, projection onto basis elements corresponding to a fine grid, and corresponding integration quadrature, leads to matrix computations akin to the Ulam method discussed in many other places in this book.

The inner integral becomes

$$\begin{aligned} \int_{\tilde{S}} \nu(x - F(y))\rho(y)dy &\simeq \int_{\tilde{S}} \nu(x - F(y)) \sum_{i=1}^{N} c_i \chi_{B_i}(y)dy \\ &= \sum_{i:B_i \cap \tilde{S} \neq \emptyset} c_i \int_{B_i} \nu(x - F(y))dy. \end{aligned} \tag{7.40}$$

Substitution into Eq. (7.38) gives the approximation; recognize that this last double integral consists of a sum over those entries of a stochastic transition matrix $A_{i,j}$ representing the approximate full Frobenius–Perron operator such that $B_i \in \tilde{S}$ and $B_j \in S$. Hence, we define a flux matrix,

$$\mathcal{A}_S^+ \equiv \left\{ \begin{array}{ccc} A_{i,j} & \text{if} & B_j \in S \text{ and } B_i \in \tilde{S} \\ 0 & \text{otherwise} & \end{array} \right\}, \tag{7.41}$$

which allows us to rewrite Eq. (7.39),

$$\mathcal{F}_S^+[\rho] \simeq \|\mathcal{A}_S^+ \cdot \mathbf{c}\|_1, \tag{7.42}$$

where $\|.\|_1$ is defied by the absolute sum. In this form, we have flux into S terms of $\mathcal{A}_S^+$, which is a masked version of the full transfer matrix A times the coefficient weights vector,

$$\mathbf{c} = (c_1, c_2, \ldots, c_N)^t, \tag{7.43}$$

representing estimation of the density ρ on the grid. One can similarly form and interpret masked transfer matrices $\mathcal{A}_S^-$, $\mathcal{A}_{\tilde{S}}^+$, and $\mathcal{A}_{\tilde{S}}^-$ representing flux out of S, into $\tilde{S}$, and out of $\tilde{S}$, respectively. Correspondingly the conservation statements (7.39) hold.

Taking as a special choice the initial uniform density,

$$\mathbf{c} = \frac{1}{N}\mathbf{1}, \tag{7.44}$$

we can find the stochastic "area flux." In this case, Eq. (7.42) reduces to the absolute sum of matrix entries,

$$\mathcal{F}_S^+[\mathbf{1}] = \sum_{i,j} [\mathcal{A}_S^+]_{i,j}. \tag{7.45}$$

These quantities are highlighted in Example 7.1 and illustrated in Fig. 7.21. The parts of $\mathcal{A}_S^{\pm}$, $\mathcal{A}_{\tilde{S}}^{\pm}$ are visible of each block shown. There also follows a discussion of a stochastic version of completing a heteroclinic connection in Example 7.1.

In the previous section, the discussion of transport mechanism was in terms of lobe dynamics. A lobe defined in terms of segments of stable and unstable manifolds leads to the mechanism corresponding to transport across barriers defined also usually by segments of stable and unstable manifolds. From these barriers we also saw there are partition elements. This section, on the other hand, discusses transport in terms of transfer operators masked according to an a priori choice of partition. Both of these perspectives are complementary. On the one hand we could simply choose S to be the insides of the region defined by the homoclinic orbit labeled green in the Henon example in Fig. 7.13, for example. As such, using an approximate transfer operator approach, we would expect the off-diagonal parts of $\mathcal{A}_S^{\pm}$, $\mathcal{A}_{\tilde{S}}^{\pm}$ to correspond to the lobe dynamics illustrated geometrically in Fig. 7.14 continuing with the example.

7.4.2 Expected Escape Time

Another natural quantity to consider when investigating transport is expected escape time across a barrier of the orbit starting from an initial condition from a point in a region. Again, suppose a region S is an element from a topological partition. Given a point $x \in B_i \subset S$, let $T(x)_i$ be the actual time of escape for a particular sample path of the stochastic dynamical system of a randomly chosen initial condition in the ith box B_i. The expected time of escape is

$$\langle T(x)_i \rangle = \sum_{n=1}^{\infty} n\ P(T(x) = n) = \sum_{i=1}^{\infty} n\ P(F^n(x) \notin S \text{ and } F^m(x) \in S\ \forall m < n). \tag{7.46}$$

The key issue to acknowledge is that we are interested in the mean time of *first* escape. While

$$A_{i,j}^n = P(F^n(x) \in B_j | x \in B_i), \tag{7.47}$$

this probability does not forbid multiple passages or recurrences. In particular, it accounts for orbits which might leave S and return to S multiple times before finally landing in $B_j \subset \tilde{S}$ on the nth iterate.

We define an operator which addresses the probability of first escape from S, again by restricting (masking) the Galerkin matrix A. Let this "escape matrix" be defined[111]

$$[\mathcal{E}_S^-]_{i,j} \equiv \left\{ \begin{array}{ccc} A_{i,j} & \text{if} & B_i \in S \\ 0 & \text{otherwise} & \end{array} \right\}. \tag{7.48}$$

[111] This is an approximation similar to that in Eq. (7.41), except we have less severely restricted the region of integration of the Frobenius–Perron operator: $\int_S \int_M \nu(x - F(y))\rho(y)dydx$.

Since $\mathcal{E}_S^-$ contains zero probability elements of transitions of the type $\tilde{S} \to S$, we now have

$$[\mathcal{E}_S^-]_{i,j}^n = P(F^n(x) \in B_j \subset \tilde{S} \text{ and } F^m(x) \in S \ \forall m < n | x \in B_i \subset S), \tag{7.49}$$

which is exactly the probability of first exit transition that we require to calculate the mean in Eq. (7.46).

Since the events described in the probability in Eq. (7.49) are disjoint events for two different target boxes,

$$P(F^n(x) \in \cup_{j:B_j \subset \tilde{S}} B_j \text{ and } F^m(x) \in S \ \forall m < n | x \in B_i \subset S) = \sum_{j:B_j \subset \tilde{S}} [\mathcal{E}_S^-]_{i,j}^n. \tag{7.50}$$

We now have the notation necessary to state the following theorem.

Theorem 7.2 (see [34]). If the escape matrix defined in Eq. (7.48) is bounded,[112] $\|[\mathcal{E}_S^-]\| < 1$, then the expected mean time of escape from S for an orbit starting in box $B_i \subset S$ to any box $B_j \subset \tilde{S}$ is

$$\langle \mathcal{T}(B_i, B_j) \rangle = \frac{1}{\#\{j : B_j \subset \tilde{S}\}} \sum_{j:B_j \subset \tilde{S}} [[\mathcal{E}_S^-] \cdot (I - [\mathcal{E}_S^-])^{-1} \cdot (I - [\mathcal{E}_S^-])^{-1}]_{i,j}, \tag{7.51}$$

where $\#\{j : B_j \subset \tilde{S}\}$ is the number of boxes $B_j \subset \tilde{S}$, and $\mathcal{T}(B_i, B_j)$ is the time of first escape of a single randomly sampled path starting at B_i, based on paths defined by the graph G_A model of the Frobenius–Perron operator.

Note that this theorem as stated is closely related to the so-called *fundamental matrix* from the topic of *absorbing Markov chains* from the related theory of *finite Markov chains* theory [296, 176]. More on absorbing Markov chains can be found in Section 5.11 about *open systems*.

Proof. By Eq. (7.49),

$$\langle \text{Time of first escape from } B_i \subset S \text{ to } B_j \subset \tilde{S} \rangle = \sum_{n=1}^{\infty} n [\mathcal{E}_S^-]_{i,j}^n. \tag{7.52}$$

By the assumed bound, $\|[\mathcal{E}_S^-]\| < 1$, we have the matrix geometric series [23]

$$(I - [\mathcal{E}_S^-])^{-1} = \sum_{n=0}^{\infty} [\mathcal{E}_S^-]^n, \tag{7.53}$$

from which it is straightforward to derive

$$\sum_{n=1}^{\infty} n [\mathcal{E}_S^-]^n = [\mathcal{E}_S^-] \cdot (I - [\mathcal{E}_S^-])^{-1} \cdot (I - [\mathcal{E}_S^-])^{-1}. \tag{7.54}$$

[112] The notation $\|.\|$ used in this section denotes the matrix natural norm, $\|A\| = \sup_{\|u\|=1} \|A \cdot u\|$, in terms of a vector norm, which we could choose here to be the sup-norm, in which case, $\|A\|_\infty = \max_i \sum_j |A_{i,j}|$ [140], which is the maximum column sum.

Hence, selecting the i,jth entry of this matrix on the right side of Eq. (7.54) gives the mean escape time from S, starting at box B_i, and arriving after n-iterates at box B_j. By independence of the events of arriving at two different boxes B_j and $B_{j'}$, both in $\tilde{S}$, the total mean escape time from box B_i to any box $B_j \subset \tilde{S}$ is the arithmetic mean of the mean escapes times to each individual box, which gives the formula Eq. (7.51). □

The restriction that $\|[\mathcal{E}_S^-]\| < 1$ is not severe, since in all but trivial choices of S (when $\|[\mathcal{E}_S^-]\| = 1$, because the matrix A is stochastic) the inequality should hold. Of course the formula (7.51) gives expectation in terms of the combinatorial model of the Frobenius–Perron operator, formed using a fine grid, and not the full operator. We expect the calculation to be good for a fine grid.

Remark 7.3. By considering the disjointness of the covering by open elements in the grid $\{B_i\}$ and also the topological cover $\{S, \tilde{S}\}$, then the probabilities involved in (7.51) are independent. Therefore, the mean escape times between larger regions, for example, $\langle \mathcal{T}(S,\tilde{S})\rangle$, or any other disjoint set pairing, are easily computed by making the sum over the larger number of boxes therein and likewise counting the larger number of boxes in the denominator.

Example 7.1 (noise-induced basin hopping in stochastically perturbed dynamical system: a model of childhood disease). Consider a model used to describe the dynamics of a childhood disease spread in a large population

$$\begin{aligned} S'(t) &= \mu - \mu S(t) - \beta(t)I(t)S(t), \\ I'(t) &= \left(\frac{\alpha}{\mu+\gamma}\right)\beta(t)I(t)S(t) - (\mu+\alpha)I(t), \\ \beta(t) &= \beta_0(1+\delta\cos 2\pi t). \end{aligned} \tag{7.55}$$

Here $S(t)$ refers to the concentration number of susceptible individuals and $I(t)$ refers to the concentration number of infected individuals. This is a typical rate equation to be found in mathematical epidemiology. These sorts of equations are special cases of those equations found in population dynamics which describe the growth and decay of each population. In this particular model, we see quadratic interaction terms, and the contact rate $\beta(t)$ is taken to be a periodic function to model the concept that childhood diseases among many have a seasonal aspect.

In [20] it was demonstrated that when normally distributed noise $\nu(x)$ with a sufficiently large standard deviation is added, then the behavior of the modified SI model [282, 283] was transformed from regular periodic cycles to something completely different, which we may call stochastic chaos. In the deterministic case with parameter values at $\mu = 0.02$, $\alpha = 1/0.0279$, $\gamma = 100$, $\beta_0 = 1575$, and $\delta = 0.095$, the bifurcation diagram (not seen here but shown in [34]) reveals that there exist two stable periodic orbits and two unstable periodic orbits. Furthermore, the deterministic system exhibits bistability since there are two competing basins of attraction. In Fig. 7.20 we show these two basins in white and pink, and corresponding stable and unstable manifolds from the pair of period-2 stable and then unstable orbits and likewise for the pair of period-3 stable and unstable orbits. The stable manifold of the period-3 orbit naturally outlines the two regions which are the basin of stable period-2 and period-3, respectively.

In [34] we demonstrated that by methods reviewed in this section, there is diffusion between the two basins due to the stochastic effect, a form of noise-induced basin hopping.

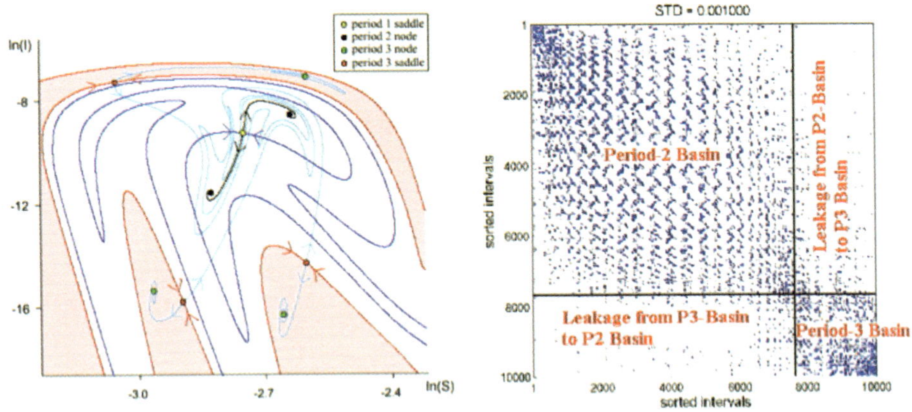

Figure 7.20. *Noise-induced basin hopping in a stochastically perturbed dynamical system—SEIR model with noise. (Left) The phase space depiction from the deterministic model* (7.55) *has two basins corresponding to a stable period-2 point (white region) and a stable period-3 point (pink region). With even a small amount of noise, points on the stable manifold of a period-2 point can jump onto the stable manifold of the period-3 point across the basin boundary formed by the stable manifolds of the saddle-period-3 orbit. (Right) The resulting almost-invariant sets can be found by appropriately permuting the Ulam–Galerkin matrices. The off-diagonal elements carry the transport information.* [34]

There is not yet a heteroclinic tangle, since as it turns out these parameters are not large enough where the global tangency bifurcation occurs to give rise to chaos. Nonetheless, the stable and unstable manifolds serve an important role in noise-induced transport in this subcritical state. The noise facilitates jumping between these manifolds. The tools described herein identify the flux between basins as noise that is added. Multiplying these rates by the probability density function results in a measure of where a trajectory is most likely to escape to another basin. It was found [34] that the highest escape rates as described in Eq. (7.51) occur exactly where we previously conjectured, at the nearby heteroclinic tangencies, thus creating a chaos-like orbit. In Figs. 7.21 and 7.22, we illustrate the effect of increasing noise for this system. In Fig. 7.23 we illustrate the regions which are most active in transport activity. As we see, the reddened area denoting greatest diffusion occurs just where the deterministic version of the system has the unstable manifold of the period-2 orbit most closely approaching to the stable manifold of the period-3 orbit. Effectively it is as if the stochastic effect is completing the heteroclinic tangle. But the analysis that finds this effect is inspection of the transport rates encoded in the masked transfer operators depicted in Fig. 7.20.

Example 7.2 (transport in the Gulf of Mexico). In [36] we studied the Deepwater Horizon oil spill in the Gulf of Mexico, as already referenced at the end of Chapter 1 regarding Figs. 1.16–1.18, 1.20, 1.21. Using the resulting Ulam–Galerkin matrix mentioned in Fig. 4.3, we can discuss transport mechanisms in the Gulf of Mexico. Producing an Ulam–Galerkin matrix of the nonautonomous system progressively as time evolves leads to a set of transfer operators evolving in time. Escape rates by methods discussed in this section

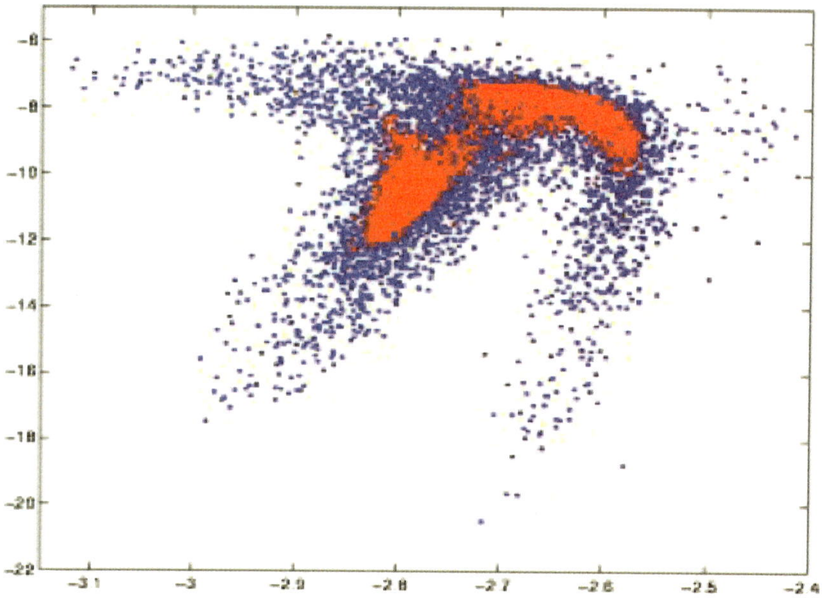

Figure 7.21. *Invariant density (PDF) of the SI model for noise standard deviation $\sigma = 0.03$ due to direct simulation; there is strong mixing between what the deterministic system predicts is separate basins of the bistable system.* [34]

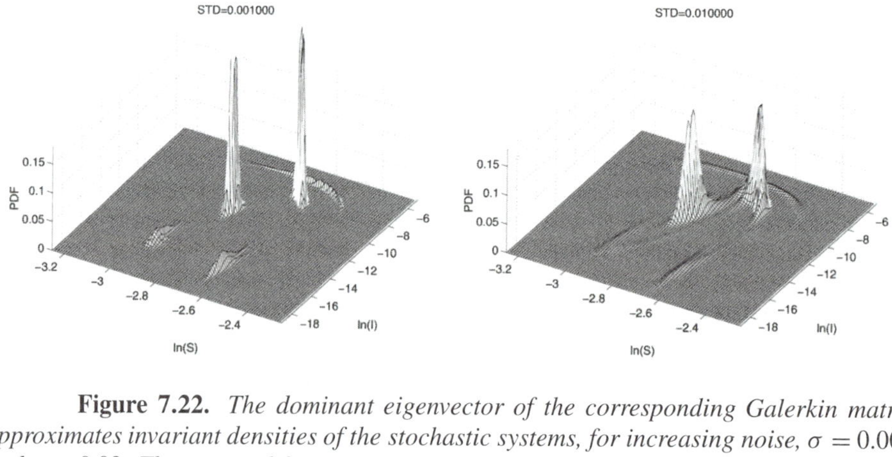

Figure 7.22. *The dominant eigenvector of the corresponding Galerkin matrix approximates invariant densities of the stochastic systems, for increasing noise, $\sigma = 0.001$ and $\sigma = 0.03$. The essential feature is that when $\sigma = 0$, the main density spikes are at the dynamic centers of each respective basin. Initially, as σ increases, the density becomes less diffusely distributed around the stable fixed points, due to the predominantly small stochastic diffusion added to the deterministic dynamics. There persist two stable fixed points and two distinct basins. As we continue to increase σ, a crossover effect occurs near $\sigma = 0.02$, after which the density mass becomes mixed throughout a larger region, and predominantly mixed between the originally separate basins.* [34]

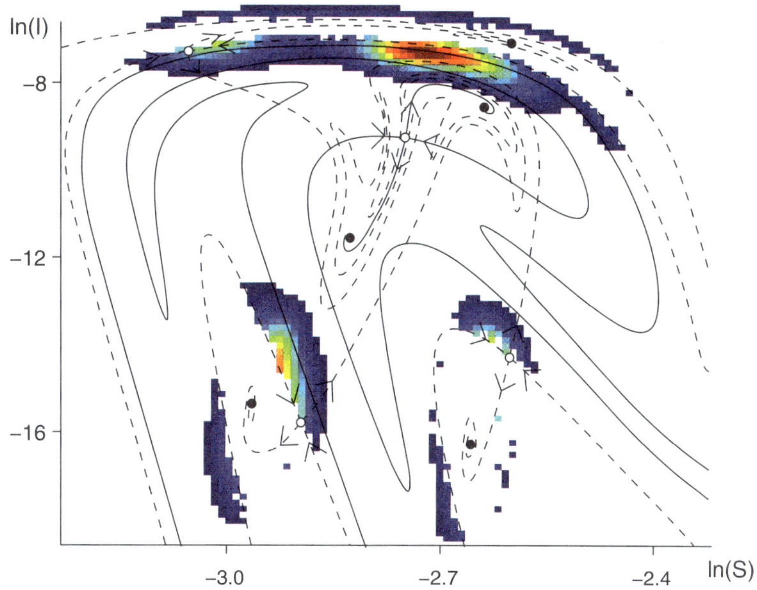

Figure 7.23. *Transport PDF flux. The conditional probability of transition from small amplitudes to large outbreaks. The highest probability regions of transport (dark) point to a bull's-eye monitoring region for control. Overlaid are the stable and unstable manifolds corresponding to the underlying deterministic model. Compare with Fig.* 7.20. [34]

and partitions as discussed in Chapter 5 have been presented in [36]. Further, by method of relative measure and relatively coherent sets, just as we illustrated for the Rossby system in Fig. 5.17 on relative coherence, we produced in [37] a map of relative coherence shown in Fig. 7.24, including the use of the coherence criterion inequality (5.140) of ρ_0 used as a stopping criterion. We choose 20,000,000 points uniformly and randomly in the water region as the initial status; see [36]. The final status is the positions of these points after 6 days. We use 32867 triangles $\{B_i\}_{i=1}^{32867}$ as a partition of X and 32359 triangles $\{C_j\}_{j=1}^{32359}$ as a partition of Y. After applying our subdivision method on these triangles, the results are shown in Fig. 7.24. In this example, we set $\rho_0 = 0.9998$ as the threshold stopping criterion. We find it to be particularly interesting to contrast this operator-based perspective to the FTLE perspective shown in Fig. 7.25, in which we also show tracers. These data are most enlightening as a movie, which can be found at [33].

7.4.3 Escape Time, Long Life, Scattering, and Unstable Chaotic Saddles

Unstable invariant sets are important for understanding the mechanisms behind many dynamically important phenomena such as chaotic transients. These phenomena can be physically relevant in experiments. Take, for example, the so-called edge of chaos scenario whereby a transient turbulence-like phenomenon in a plane Couette flow, or pipe flow, may

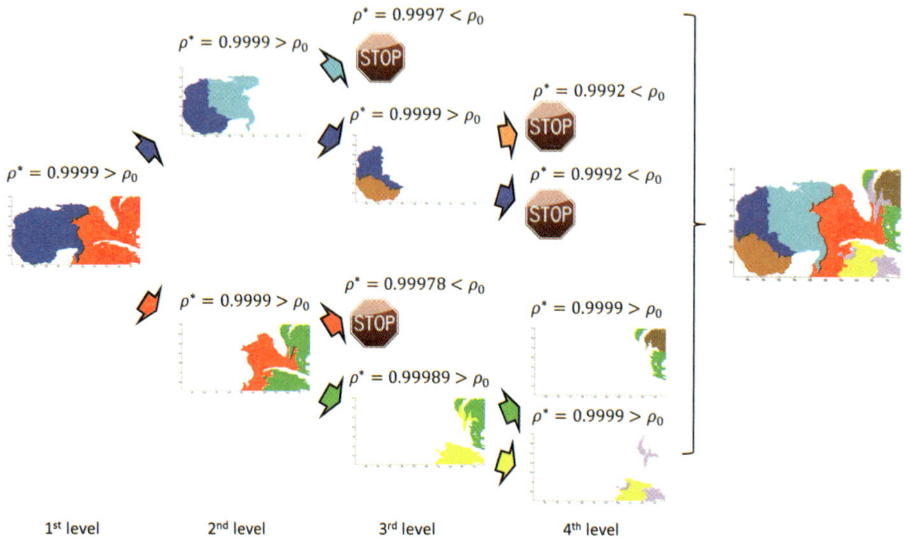

Figure 7.24. *Hierarchical relative coherence in the Gulf of Mexico following the flow according to vector field from the HYCOM model* [173]; *tree structure and relatively coherent pairs coloring as in Fig.* 5.17. *The inequality* (5.140) *of* ρ_0 *is used here as a stopping criterion.*

appear but gives way to linear stability of the laminar flow [280]. This was explained in Skufca–Yorke–Bruno [293] as the presence of an unstable chaotic saddle (Definition 7.6).

As a lead-in to the discussion of unstable chaotic saddles, consider specifically Chapter 6 and refer to Fig. 6.29, wherein a sequence of embedded subshifts was used to approximate the symbolic dynamics representation of the dynamical system. We reprise this discussion, now in the context of unstable chaotic saddles and escape from these sets. We can now describe the embedded subshifts of Chapter 6 and Fig. 6.29 as unstable chaotic saddles.

Unstable chaotic saddles may be Cantor sets embedded in some more regular attractor set [242, 6, 32, 29], as already referenced several times herein. Techniques such as the PIM-triple method [242], simplex method variant [233], and even the step-and-stagger method [305] have been developed to compute long ϵ-chain pseudo-orbits near such sets. The following relates to a long-life function [32] which was designed to both describe the lifetime landscape function and to find points with long-lived orbits before escape.

We use notation for a uniformly continuously differentiable discrete time dynamical system:

$$\mathbf{z}_{n+1} = \mathbf{F}(\mathbf{z}_n) \tag{7.56}$$

The discussion of unstable sets must be for orbits relative to some reference set, which we will call B. We require that the orbit lies in some set B, or at least ask for some time. See Fig. 7.28 for an example of such a set B chosen to be a disc. Then we say that $\{\mathbf{z}_i\}_{i=0}^N$ is a B-*invariant orbit segment* if $\mathbf{z}_i \in B$ for all $i = 0, 1, \ldots, N$, and each $\mathbf{z}_i$ satisfies Eq. (7.56). Exact orbits rarely exist in a finite precision computer, and choosing $\epsilon = 10^{-15}$, slightly bigger than the order of machine precision, is the best that can be constructed. A useful concept which is slightly weaker than the notion of orbits is the following.

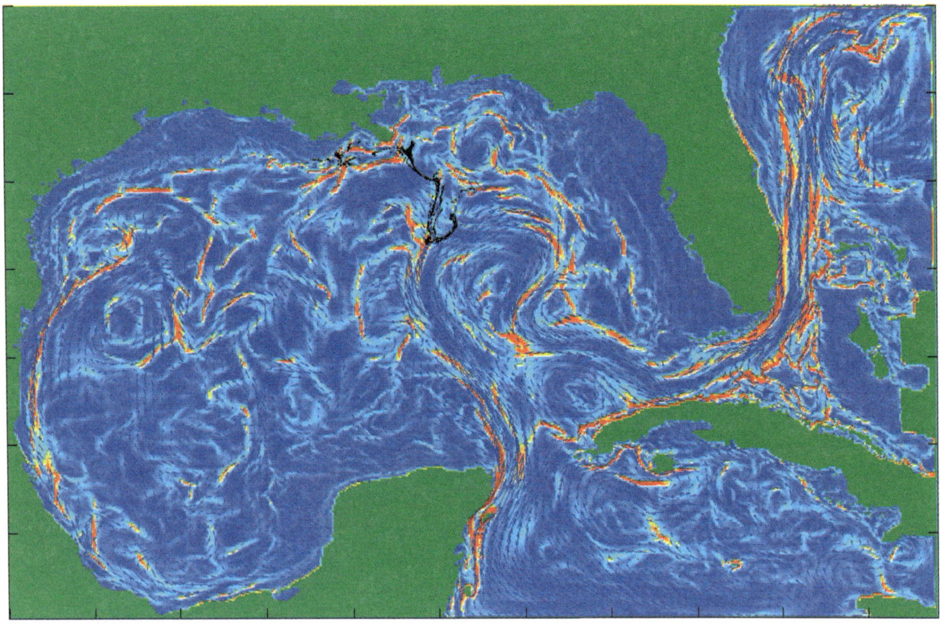

Figure 7.25. *FTLE for the Gulf of Mexico computed from a 3-day range beginning May 24, 2010. Compare to Fig. 7.24. Also shown is the underlying vector field on May 24, 2010, as well as black tracers representing the spread of oil. A month after the initial explosion, the tracer particles have dispersed significantly from the source.* [36]

Definition 7.3 (see [268]). An ϵ-**chain** segment $\{\mathbf{z}_i\}_{i=0}^{N}$, also called a **pseudo-orbit**, satisfies

$$\|\mathbf{z}_{n+1} - \mathbf{F}(\mathbf{z}_n)\| < \epsilon \ \forall i = 0, \ldots, N. \tag{7.57}$$

When discussing unstable invariant sets, it is useful to define the following.

Definition 7.4 (see [32]). A discrete **forward (backward) lifetime escape function**,

$$L_B^{\pm} : B \to Z^{\pm}, \tag{7.58}$$

is as follows:

$$L_B^{\pm}(\mathbf{z}) = |j| \text{ if } \mathbf{F}^{\pm i}(\mathbf{z}) \in B \text{ for } 0 \leq |i| \leq |j|, \text{ but } \mathbf{F}^{\pm(j+1)}(\mathbf{z}) \notin B. \tag{7.59}$$

That is, $\mathbf{F}^{\pm i}$ denotes the ith forward or backward iterate, depending on the sign, and $\mathbf{F}^0$ denotes the identity map, $\mathbf{F}^0(\mathbf{z}) \equiv \mathbf{z}$ for all $\mathbf{z}$.

Example 7.3 (lifetime escape function of a Henon map). In Figs. 7.26 and 7.27, we see lifetime functions plotted over the phase space, for the Henon map (6.55), and where B is a circle of radius 2 centered on the origin. Notice the central role of the chosen set B from which the question of escape is posed. As an exercise in the notation, since the Henon map has an attractor set which we may call $\mathcal{A}$, consider then if we were to choose the escape set B such that $\mathcal{A} \subset B$ and B is in the trapping set. Then it is easy to see that $L_B^{\pm}(\mathbf{z}) = \infty$ for all $z \in B$. Thus the interesting Cantor-like appearance of the invariant set suggested in Fig. 7.28 and the tower-like appearance of the lifetime escape function [32, 230] shown in

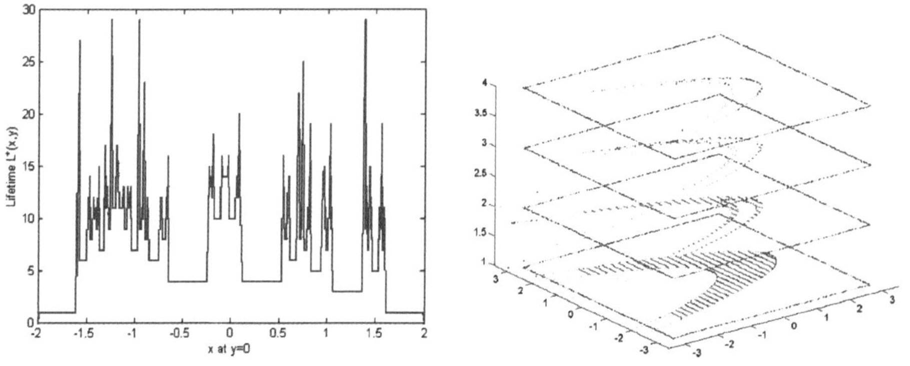

Figure 7.26. *(Left) A cross-section of the forward lifetime function of Henon map* (6.55), *shown in Fig.* 7.27, *where B is a circle of radius* 2 *centered on the origin. Notice the stepping nature of the towers of increasing lifetime. (Right) Layers show points invariant in a box $[-2,2]\times[-2,2]$ for $i=1,2,3,4$ steps successively. The spikes limit on the invariant set $\mathcal{L}^{\pm}(\infty)$, approximated in Fig.* 7.28. [32]

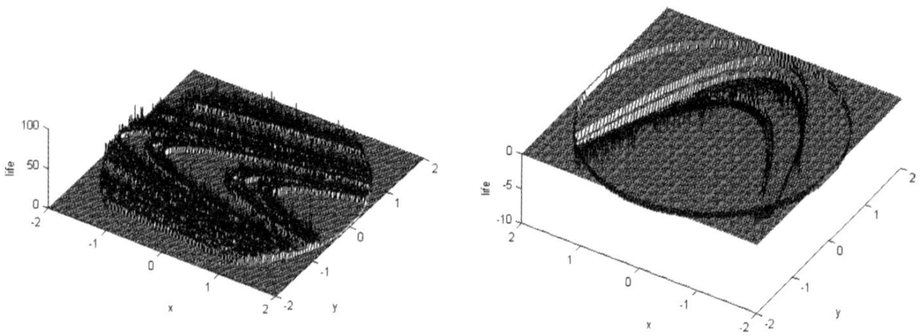

Figure 7.27. *Forward (left) and backward (right) lifetime escape functions* (7.59) *of Henon map* (6.55), *where B is a circle of radius* 2 *centered on the origin.* [32]

Figs. 7.26 and 7.27 is highly dependent on the fact that the chosen disc B intersects the Henon invariant set $\mathcal{A}$.

To elucidate the tower-like appearance of these lifetime escape functions $L_B^{\pm}(\mathbf{z})$ and the relationship to invariant sets, we define the following.

Definition 7.5. Let

$$\mathcal{L}_B(i)^{\pm} = \{\mathbf{z} : L_B^{\pm}(\mathbf{z}) \geq |i|\} \tag{7.60}$$

be the set of points with lifetime of at least i. We name this the ***i*-lifetime escape set,** denoted $\mathcal{L}_B(i)^{\pm}$.

Whereas $L_B^{\pm}(\mathbf{z})$ is a function that measures lifetime from a given point $\mathbf{z}$, a set with specified escape properties is denoted $\mathcal{L}_B(i)^{\pm}$. These definitions help to explain the towering steps of the lifetime escape functions shown in Figs. 7.26 and 7.27. Now we see that

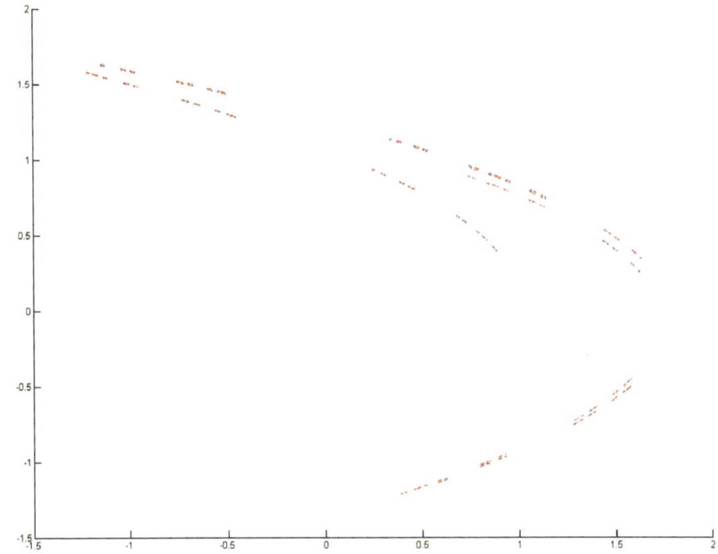

Figure 7.28. $\mathcal{L}^+(60)$ *(almost-)invariant set of the Henon map, using B as the circle of radius 2, approximates the invariant set $\mathcal{I} = \mathcal{L}^+(\infty)$ which is apparently an unstable chaotic saddle; see Definition* 7.6. [32]

the towers follow the natural nesting of steps of increasing heights for continuous maps F

$$\mathcal{L}_B^{\pm}(i \pm 1) \subset \mathcal{L}_B^{\pm}(i) \; \forall i. \tag{7.61}$$

Further, the forward invariant set is

$$\mathcal{I} = \mathcal{L}_B^{+}(\infty). \tag{7.62}$$

As another example, another set B in Fig. 7.29 is shown. Choosing $B = \{(x, y) : y > \epsilon\}$ again intersects the attractor, $\mathcal{A}$. By consequence, any invariant set of $B \cap \mathcal{A}$ must not lie in any preimage of the gap shown, $\overline{B}$, the complement of B.

It is a common scenario that when there is chaos on the attractor set, there is chaos on an invariant set $\mathcal{L}_B^{+}(\infty)$, and furthermore these sets are unstable. For this situation, the following is a useful and commonly discussed concept.

Definition 7.6. An invariant set $\mathcal{I} \subset M$ of a map $F : M \to M$ is an **unstable chaotic saddle** if the following hold:

- $\mathcal{I}$ is an invariant set.
- The map is chaotic even for the map restricted to $\mathcal{I}$, $F|_{\mathcal{I}}$.
- $\mathcal{I}$ is unstable relative to M. That is, no open neighborhood[113] N of $\mathcal{I}$ exists such that *all* of its points remain forever. $F^n(z) \in N$ for all $n > 0$ if $z \in N \cap \overline{\mathcal{I}}$.

[113] N is an open neighborhood of $\mathcal{I}$ means that N is open in the embedding set M and $\mathcal{I} \subset N$. Generally the term neighborhood denotes that not only is N open, but it also "tightly wraps" in that no points in N are far from some points in $\mathcal{I}$.

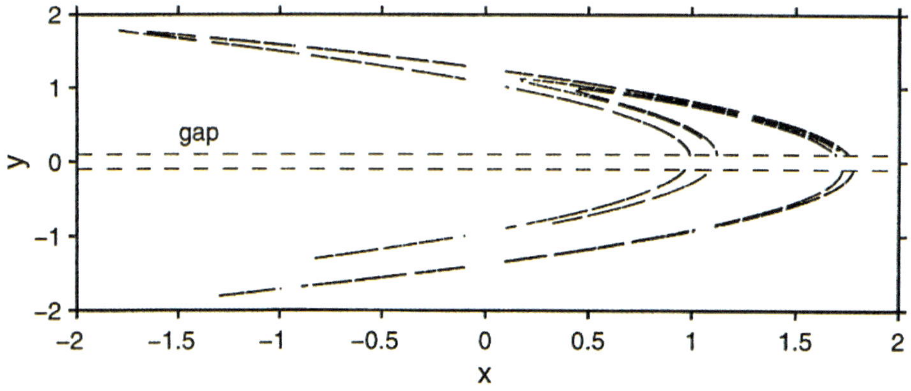

Figure 7.29. *An invariant set unstable chaotic saddle, $\mathcal{I}$, built as Eq.* (7.62) *from the Henon attractor $\mathcal{A}$ with a set B is a horizontal strip removed as Eq.* (7.62). *A few preimages are shown. A succession of holes are removed from the resulting unstable chaotic saddle just as a classic construction of a Cantor set. Compare to Fig.* 3.2. [196]

The term "saddle" associated with unstable chaotic saddle refers to a common scenario where points in $\mathcal{I}$ have both nonempty stable and unstable manifolds, but by practice the sets are still called saddle even if there is no stable direction, as we see in the one-dimensional scenario in Fig. 7.30.

Formally, these chaotic saddles can inherit the symbolic dynamical description as subshifts embedded in a larger shift grammar of $\mathcal{A}$ [28, 29, 196], a statement which is made clear in the context of Section 6.4.4 and as seen in Fig. 6.29. Said symbolically, the words corresponding to the set B cannot be in the subshift of the invariant set $\mathcal{I}$. The analogy to the figures in Section 6.4.4 and symbolic dynamics described by Fig. 6.29 closely relates to the embedded unstable invariant sets by considering a tent map with a hole.

Example 7.4 (lifetime escape function and lifetime escape set of a tent map). See Fig. 7.30, where we choose a full tent map, and thus the invariant set[114] is chosen $\mathcal{A} = [0,1]$. As in [328] we choose

$$B = \left[0, \frac{1}{2} - \epsilon\right] \cup \left[\frac{1}{2} + \epsilon, 1\right], \tag{7.63}$$

which creates a hole $\overline{B} = (\frac{1}{2} - \epsilon, \frac{1}{2} + \epsilon)$ relative to which we can discuss lifetime escape. As shown, $\overline{B} = A_{10}$ is shaded darker gray. The first preimages $A_{20} \cup A_{21} = F^{-1}(\overline{B})$ and second preimages $A_{30} \cup A_{31} \cup A_{32} \cup A_{33} = F^{-2}(\overline{B})$ are shown as lighter gray strips. As a matter of example, according to the lifetime escape function, (7.59), for any $z \in A_{20} \cup A_{21}$ for example, $L^{+}_{\overline{B}}(z) = 1$.

As an aside, we are prepared at this stage to make a remark concerning the interesting devil's staircase function as seen in Eq. (6.30) as studied in [29, 28, 196, 328]. As explained

[114]Notice we are speaking of the invariant set in this example rather than an attractor.

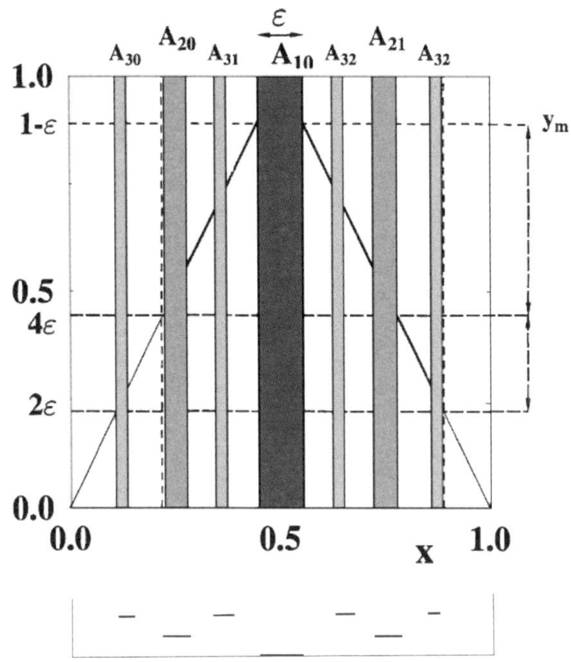

Figure 7.30. *An invariant set unstable chaotic saddle, $\mathcal{I}$, built as Eq.* (7.62) *from the invariant set of a full tent map $\mathcal{A} = [0,1]$ with a set B is an interval near $\frac{1}{2}$ shaded dark gray.* [328]

in Fig. 6.29, increasing the size of the hole removes corresponding words from the grammar of the chaotic saddle. But what causes the flat spots in the entropy function? In Fig. 7.32 we see that as the size of the hole $\overline{B}$ is increased, so is shown the gap $\overline{B}(s) = (\frac{1}{2} - s, \frac{1}{2} + s)$ of this v-shaped region in the x phase space by s parameter space. Also the preimages decrease, and for the simple piecewise linear tent map, they are also v-shaped but narrower. The summary of the story is that when these tongue-like regions intersect, the symbolic word corresponding to an interval disappears from the invariant set for all further larger values of s. The key to the flat spots of the devil's staircase topological entropy function in Eq. (6.30) is that there are open intervals of s when there is no word to eliminate since the next word has already been eliminated. For example, notice the interval of s centered on $s = 0.2$. Likewise, a Cantor set of these tongues[115] of various opening angles emanates from the interval $[0,1] \times \{0\}$ with countable many such intersections and corresponding flat spots. In between these, open intervals of s exist for which the symbolic dynamics of the unstable chaotic saddle does not change. We have even seen similar behavior in multivariate settings [328] such as the Henon map with a hole as a gap of increasing width, as suggested by Fig. 7.30.

[115]Which reminds us of the Arnold tongues from another setting in dynamical systems, but which also gives rise to a devil's staircase function.

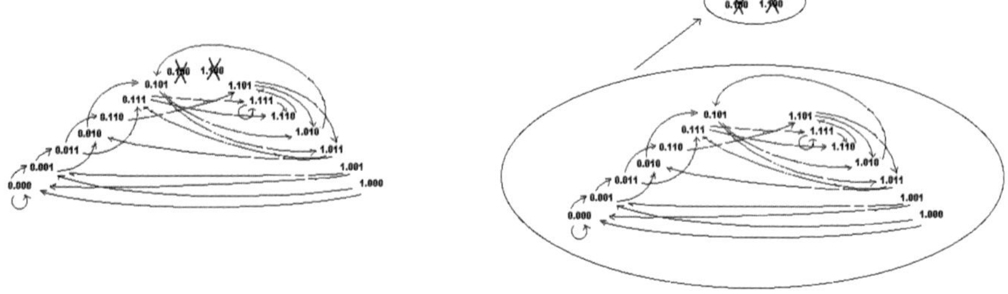

Figure 7.31. *Considering a hole such as in Fig.* 7.30 *that is a dyadic rational allows us to describe the dynamics exactly as a symbolic dynamics whose grammar is a finite directed graph, such as was shown in Fig.* 6.29(b). *Furthermore, here we show a reconfiguring of the embedding of this graph in the plane so as to emphasize the hole-like nature of this dynamics.* [29]

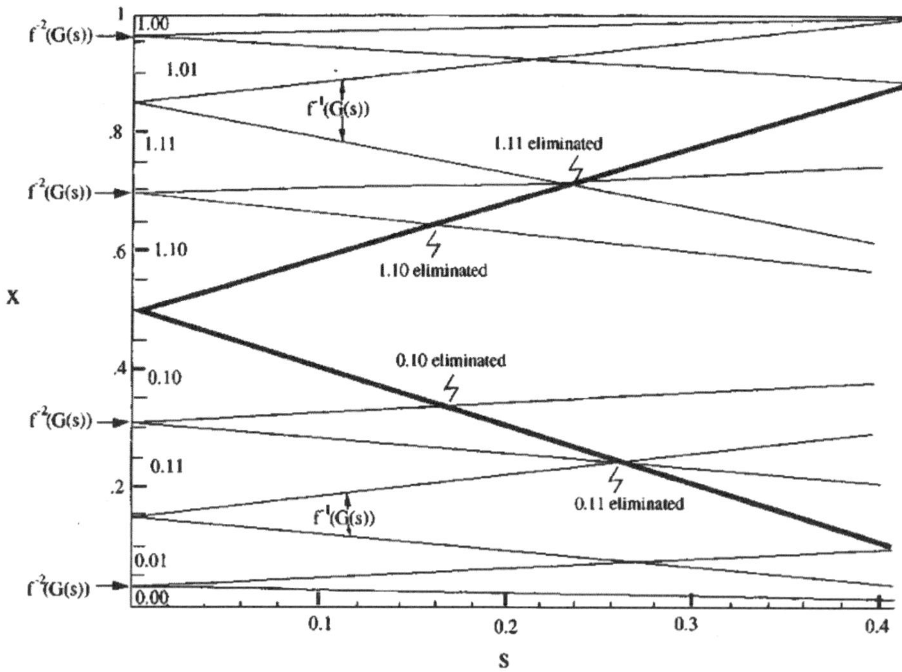

Figure 7.32. *The devil's staircase–like topological entropy function seen in Fig.* 6.29 *is understood here for a tent map by considering a hole parameter s, for $\overline{B}(s) = (\frac{1}{2} - s, \frac{1}{2} + s)$, and all of its preimages. This creates tongue-like structures that overlap at countably many points and open sets of s for which the symbolic dynamics of the unstable chaotic saddle does not change.* [29]

Consider Fig. 6.29(b) which illustrates symbolic dynamics of just such a situation as the hole in the tent map in Figs. 7.30–7.32. If ϵ is a dyadic rational,[116] then the hole $\overline{B}$ has an exact representation by finite symbolic words. The case of Fig. 6.29(b) corresponds to $\epsilon = \frac{1}{16}$. The hole corresponds to the intervals labeled by the 4-bit words 0.100 and 1.100. In this example, we can represent the directed graph to emphasize the hole $\overline{B}$.

[116] $\epsilon = \frac{p}{2^n}$ for some integer n and $0 \leq p \leq 2^n - 1$. In other words it is a fraction whose denominator is a power of 2.

Chapter 8

Finite Time Lyapunov Exponents

8.1 Lyapunov Exponents: One-Dimensional Maps

A somewhat informal introduction to the Lyapunov exponent of one-dimensional differentiable map will be presented in this section. For simplicity, we start with a one-dimensional map. Let X be a compact subset of $\mathbb{R}$ and suppose that T is a piecewise continuously differentiable map on X defined by the rule

$$x \mapsto T(x), \qquad x \in X. \tag{8.1}$$

The iterated map $T^{(n)}(x)$ here will refer to an n-fold composition

$$T^{(n)}(x) = \underbrace{T \circ T \circ \cdots \circ T(x),}_{n \text{ times}} \qquad m \in Z. \tag{8.2}$$

This section is concerned with the determination of the separation rate of the derivative of a map $T^{(n)}$ as n increases. In particular, we would like to study the rate at which the orbits of infinitesimally nearby (initial) points are separating. If we choose $|x-y| < \varepsilon$ for some small $\varepsilon > 0$, we would like to study the quantity

$$\left|T^n(x) - T^n(y)\right| \approx \prod_{i=0}^{n-1} \left|T'(T^i(x))\right| \times |x-y|, \tag{8.3}$$

which is readily derived from the chain rule. It then follows that

$$\frac{1}{n}\log\left|T^n(x) - T^n(y)\right| \approx \frac{1}{n}\sum_{i=0}^{n-1}\log\left|T'(T^i(x))\right|. \tag{8.4}$$

The right-hand side of the above equation represents the exponential growth rate of separation. Observe that if the orbits of x and y converge as n increases, this quantity will be negative; conversely, it will be positive if nearby orbits diverge. In general, as n goes to infinity, the limit of this quantity may not exist; for example, the points may come very close to each other infinitely many times. Nevertheless, it will always be bounded since we assume that X is compact. Therefore, for every (initial) point $x \in X$, we may define the *Lyapunov exponent* of x, $\lambda(x)$, as follows:

$$\lambda(x) := \limsup_{n\to\infty} \frac{1}{n} \log \left| T'(T^i(x)) \right|. \tag{8.5}$$

Example 8.1. Let $T(x) = 2x$. Then $\lambda(x) = \log 2$ for all $x \in X$, since $T' = 2$.

Example 8.2. Let $X = [0,1]$ and consider a tent map defined by

$$T(x) = \begin{cases} 2x & \text{if } 0 \le x \le 0.5, \\ 2(1-x) & \text{if } 0.5 \le x \le 1. \end{cases} \tag{8.6}$$

In this case, $T(x)$ is piecewise continuous with $\left|T'(x)\right| = 2$ for all x except at $x = 0.5$, where $T'(x)$ does not exist. Therefore, $\lambda(x)$ is not defined at $x = 0.5$ and all of x such that $T^{(i)}(x) = 0.5$ for some i. However, it can be shown that a set of those points is countable. Hence, we can say that $\lambda(x) = \log 2$ up to Lebesgue-measure of zero.

Example 8.3. Empirically estimating Lyapunov exponents for the logistic map, $x_{n+1} = 4x_n(1-x_n)$. Rather than as an efficient way to estimate $\lambda(x)$, the following is presented merely as a numerical presentation of the idea behind Eq. (8.4) for the sake of intuition. See Fig. 8.1, where the orbits from two nearby initial conditions and the growth of the error between them are shown. Instead of averaging the derivative along an orbit according to Eq. (8.4), this crude estimate reveals the idea behind the computation. Shown is a graphical description of an average growth rate of error, $\log(|x_n - y_n|)/n$ for small $|x_n - y_n|$.

We close this example with the remark that an averaged quantity along an orbit also corresponds to a spatial average relative to an ergodic measure if it exists, which in this one-dimensional setting is according to the Birkhoff ergodic theorem. However, the Oseledets multiplicative ergodic theorem is required more generally, as discussed in Section 8.2. Comparing the right-hand side of Eq. (8.4), an average logarithmic estimate of the derivative along an orbit is a special case of Eq. (1.5). As such, we can interpret that we are estimating quantity $\log|T'(x_n)|$ as the orbit x_n visits various regions of phase space. Further, the orbit x_n visits various regions of the phase space according to the ergodic measure which we know exists; since this is chosen to be the logistic map, with parameter value 4, we know the ergodic invariant measure in closed form through its corresponding density, $d\mu(x) = \frac{1}{\pi\sqrt{x(1-x)}}$. We could directly compute that $\lambda = 2$.

The discussion in this chapter will lead us not to consider the long time averages as has been traditional practice, as fits a major theme of this book and recent practice; relatively short time estimates of traditional analysis may also give useful information. Here, even if the system is ergodic, it may not mix quickly everywhere. We have already called this concept weakly transitive due to almost-invariant sets as discussed in Chapter 3. In such cases, for short times, estimates of Lyapunov exponents can become highly spatially dependent. This runs counter to the until-recent almost universal folklore that since almost every initial condition gives the same value, then there is no spatial information in the computation. Quite to the contrary, allowing for spatially dependent short time estimates can reveal almost-invariant sets, leading to the finite time Lyapunov exponents (FTLEs) as discussed in Section 8.3.

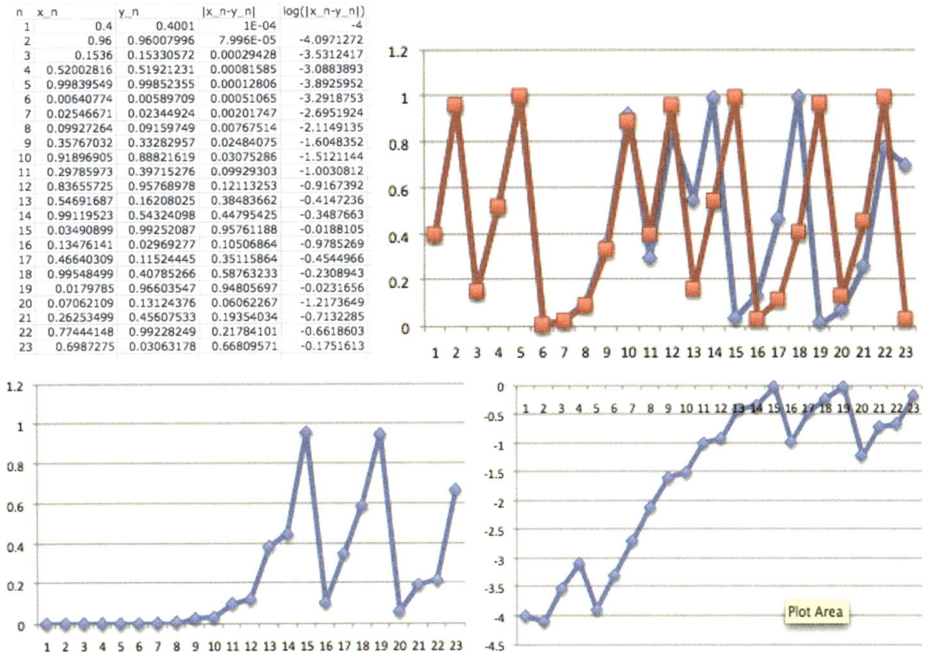

n	x_n	y_n	|x_n-y_n|	log(|x_n-y_n|)
1	0.4	0.4001	1E-04	-4
2	0.96	0.96007996	7.996E-05	-4.0971272
3	0.1536	0.15330572	0.00029428	-3.5312417
4	0.52002816	0.51921231	0.00081585	-3.0883893
5	0.99839549	0.99852355	0.00012806	-3.8925952
6	0.00640774	0.00589709	0.00051065	-3.2918753
7	0.02546671	0.02344924	0.00201747	-2.6951924
8	0.09927264	0.09159749	0.00767514	-2.1149135
9	0.35767032	0.33282957	0.02484075	-1.6048352
10	0.91896905	0.88821619	0.03075286	-1.5121144
11	0.29785973	0.39715276	0.09929303	-1.0030812
12	0.83655725	0.95768978	0.12113253	-0.9167392
13	0.54691687	0.16208025	0.38483662	-0.4147236
14	0.99119523	0.54324098	0.44795425	-0.3487663
15	0.03490899	0.99252087	0.95761188	-0.0188105
16	0.13476141	0.02969277	0.10506864	-0.9785269
17	0.46640309	0.11524445	0.35115864	-0.4544966
18	0.99548499	0.40785266	0.58763233	-0.2308943
19	0.0179785	0.96603547	0.94805697	-0.0231656
20	0.07062109	0.13124376	0.06062267	-1.2173649
21	0.26253499	0.45607533	0.19354034	-0.7132285
22	0.77444148	0.99228249	0.21784101	-0.6618603
23	0.6987275	0.03063178	0.66809571	-0.1751613

Figure 8.1. *Exponential growth of error. (Upper left) An empirical orbit of the logistic map* $x_{n+1} = 4x_n(1-x_n)$ *starting from the initial condition* x_1 *and also from a nearby initial condition* $y_1 = 0.4001$. *Errors* $|x_n - y_n|$ *grow exponentially, thus the* $\log(|x_n - y_n|)$. *(Upper right) A time series of both orbits* x_n *and* y_n *plotted as n vs.* x_n, *and we see errors are initially small but grow quickly. (Lower left) Errors grow exponentially until they saturate at close to the size* $O(1)$ *of the phase space,* $x_n \in [0,1]$, *after which the orbits move chaotically, seemingly randomly relative to each other. As a result, the errors sometimes sporadically become small (recurrence) and then grow again. (Lower right) On average, the logarithmic error growth is linear for small errors until it saturates. The slope of this average initial line estimates the Lyapunov exponent as described by Eqs.* (8.3)–(8.5).

8.2 Lyapunov Exponents: Diffeomorphism and Flow

This section presents a brief introduction of Lyapunov exponents for diffeomorphism and flows in multidimensional cases. To extend the previous treatment on a one-dimensional map onto a multidimensional map, let us consider a manifold $M \in \mathbb{R}^m$ and a diffeomorphism $f : M \to M$. Let $\|\cdot\|$ be the norm on the tangent vectors. Now for a point $\mathbf{x} \in M$, think of a vector emanating from $\mathbf{x}$ on the tangent space, i.e., $\mathbf{v} \in T_x M$. On the tangent space, the evolution is described by the linearized dynamic of f, i.e., $\mathbf{D_x f}$. As in the one-dimensional case, our interest is to determine the exponential growth rate in the direction of $\mathbf{v}$ under the iteration of $\mathbf{D_x f}$. This leads us to the definition of *Lyapunov characteristic exponent* (LCE) in the direction $\mathbf{v}$ at $\mathbf{x}$ as

$$\lambda(\mathbf{x}, \mathbf{v}) := \lim_{k\to\infty} \frac{1}{k} \log \left| \mathbf{D_x f}^k \mathbf{v} \right|, \tag{8.7}$$

if the limit exists. In the multidimensional case, a positive LCE at $\mathbf{x}$ in the direction of $\mathbf{v}$ implies that an infinitesimal line element emanating from $\mathbf{x}$ in the direction of $\mathbf{v}$ will experience exponential expansion along the solution $\mathbf{x}(t;t_0,\mathbf{x}_0)$. Likewise, when LCE is negative, it will see the exponential contraction. It is not difficult to see that $\lambda(\mathbf{x},c\mathbf{v}) = \lambda(\mathbf{x},\mathbf{v})$ for a constant c, so the LCE depends on the initial point $\mathbf{x}$ and the orientation $\mathbf{v}$ but not on the length of $\mathbf{v}$. Therefore, for a given $\mathbf{x}$, it is interesting to ask how many distinct values of $\lambda(\mathbf{x},\mathbf{v})$ there are for all $\mathbf{v} \in T_x M$. The answer to this question was given in the Oseledets multiplicative ergodic theorem (MET) under very general conditions. Precise statements and proof are referred to in [243, 3]. Roughly speaking, the MET states that

1. there is a sequence of subspaces $\{0\} = V_x^{r(\mathbf{x})+1} \subset V_x^{r(\mathbf{x})} \subset \cdots \subset V_x^1 = T_x M$ such that for any $\mathbf{v} \in V_x^j \backslash V_x^{j-1}$, the limit in (8.7) exists and $\lambda(\mathbf{x},\mathbf{v}) = \lambda_j(\mathbf{x})$;
2. the exponents satisfy $-\infty \leq \lambda_{r(\mathbf{x})}(\mathbf{x}) < \cdots < \lambda_2(\mathbf{x}) < \lambda_1(\mathbf{x})$;
3. $\mathbf{D_x f} V_x^j = V_{f(\mathbf{x})^j}$ for all $1 \leq j \leq r(\mathbf{x})$;
4. the functions $r(\mathbf{x})$ and $\lambda_j(\mathbf{x})$ are both measurable and f-invariant, i.e., $r \circ f = r$ and $\lambda_j \circ f = \lambda_j$.

The first two statements imply that there are at most $r(\mathbf{x})$ of $\lambda(\mathbf{x},\mathbf{v})$ depending on which subspace V_x^j, $\mathbf{v}$ belong. Each V_x^j is invariant in a sense of 3. Statement 4 means that for a given $\mathbf{x}$ and $\mathbf{v}$, $\lambda(\mathbf{x},\mathbf{v})$ and $r(\mathbf{x})$ are constant along the orbits of f. These $r(\mathbf{x})$ distinct values $\lambda_j(\mathbf{x})$ are called the *Lyapunov exponents* of $\mathbf{x}$.

We consider a geometric interpretation of the Lyapunov exponents for a multidimensional map. It is clear in the one-dimensional case that there exists only one exponent, say λ_1, and the length of a line will grow as $\exp(\lambda_1)$. In the multidimensional cases, according to the Oseledets theorem there can be many Lyapunov exponents as dimensions, say $\lambda_1 > \lambda_2 > \cdots > \lambda_n$. These exponents also have the same geometric interpretation as the one-dimensional case; an area will grow as $\exp(\lambda_1+\lambda_2)$, a volume as $\exp(\lambda_1+\lambda_2+\lambda_3)$. However, the area and volume will be distorted along the orbit. The direction $\mathbf{v}_1$ of the greatest separation along the orbit of nearby points corresponds to λ_1. Then, choosing among all directions perpendicular to $\mathbf{v}_1$, the direction of greatest separation corresponds to $\mathbf{v}_2$, and so on. Fig. 8.2 shows how an infinitesimal circle is transformed to an ellipse by a three-dimensional map, the sphere is stretched out in the expanding direction while shrunk in the contraction directions. Also, the images ($\mathbf{w}_1$ and $\mathbf{w}_2$) of initially orthogonal vectors $\mathbf{v}_1$ and $\mathbf{v}_2$ remain orthogonal.

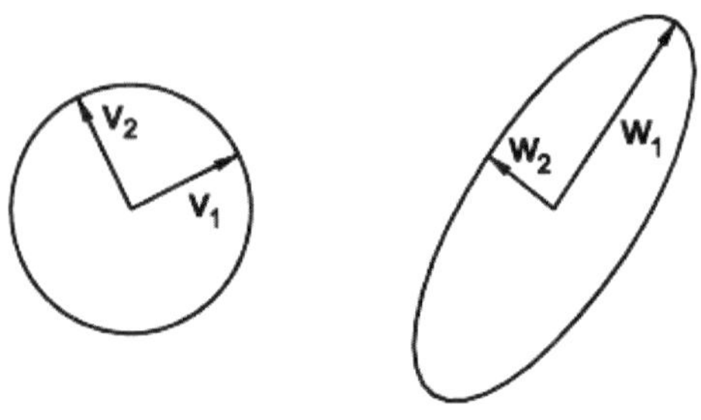

Figure 8.2. *An infinitesimal circle is mapped to an ellipse by a two-dimensional map.*

In fact, some insight into the MET can be obtained by considering this geometric interpretation in view of SVD of the matrix $A_k = \mathbf{D_x f}^k$, say $\mathbf{A} \in \mathbb{R}^{m\times n}$. Then the orthogonal sets obtained from SVD will represent the direction corresponding to the Lyapunov exponent of the dynamic of $\mathbf{A}$. For example, in two dimensions, the vectors $\mathbf{v}_1$ and $\mathbf{v}_2$ must be particularly the preimages of the principle axes $\mathbf{w}_1$ and $\mathbf{w}_2$, respectively (i.e., $\mathbf{D_x f}^k \mathbf{v}_i = \mathbf{w}_i$). Also, the vector $\mathbf{v}_i$ will be expanded (or contracted) by a factor of σ_i, the singular vector of $\mathbf{A}_k$ corresponding to $\mathbf{v}_i$. The MET then says that the choice of $\mathbf{v}_i$ will be independent of k for a sufficiently large k and σ_i yields an approximation of Lyapunov exponent $\lambda_i = \log \sigma_i$ for a large k.

Our interest now turns to the calculation of $\lambda_i(\mathbf{x}, \mathbf{v})$. The MET states the existence the Lyapunov exponents but not how to calculate them. Nevertheless, it suggests a simple calculation of the largest Lyapunov exponent (i.e., $\lambda_1(\mathbf{x})$) for a given $\mathbf{x}$ and arbitrarily chosen $\mathbf{v}$. When choosing arbitrarily, most vectors v will have all components in the direction of the eigenvectors (i.e., most $\mathbf{v}$ will lie in $V_x^1 \backslash V_x^2$), hence

$$\lambda_1(x) = \lim_{k\to\infty} \frac{1}{k} \log\left(|\mathbf{D_x f}^k \mathbf{v}|\right).$$

So, as long as $\mathbf{v}$ does not lie in V_x^2, $\lambda_1(\mathbf{x})$ can be determined from the above limit. Once $\lambda_1(\mathbf{x})$ is known, the calculation of the remaining Lyapunov exponents can be determined based on condition 2, but it is not easy to do. We refer readers to [56] for the details of such calculations.

We now discuss the concept of the Lyapunov exponent for the flow. Consider a vector field

$$\dot{\mathbf{x}} = \mathbf{f}(\mathbf{x}, t), \qquad \mathbf{x} \in \mathbb{R}^n, \tag{8.8}$$

where $\mathbf{f}(\mathbf{x}, t)$ is assumed to be C^r for some $r \geq 1$. At time t, the notation $\mathbf{x}(t; t_0, \mathbf{x}_0)$ is conventionally used to denote the solution of the initial value problem (8.8) to emphasize the explicit dependence on the initial conditions $\mathbf{x}_0$ and t_0. However, we will occasionally abbreviate the notation $\mathbf{x}(t; t_0, \mathbf{x}_0)$ by $\mathbf{x}(t)$ when the initial conditions are understood. The flow map associated with (8.8), which maps the initial position $\mathbf{x}_0$ of a trajectory beginning at time t_0 into its position at time t, can be written in a form of a two-parameter family of transformations (called a flow map) defined as

$$\begin{aligned} \phi_{t_0}^t &: U \to U \\ &: \mathbf{x}(t_0; t_0, \mathbf{x}_0) \mapsto \mathbf{x}(t; t_0, \mathbf{x}_0). \end{aligned} \tag{8.9}$$

This flow map is typically obtained by a numerical integration of the vector field in the first equation of (8.8). Note that in the extended phase space $U \times \mathbb{R}$, we can assume the existence and uniqueness of the solution $\mathbf{x}(t; t_0, \mathbf{x}_0)$, and hence it satisfies the so-called *cocycle property*:

$$\mathbf{x}(t_2; t_1, \mathbf{x}_0) = \mathbf{x}(t_2, t_1, \mathbf{x}(t_1; t_0, x_0)) \quad \text{and} \quad t_0 \leq t_1 \leq t_2. \tag{8.10}$$

The basic concept of Lyapunov exponents for the flows is similar to the diffeomorphism but it requires knowledge of the derivative of the flow map $\phi_{t_0}^t$ with respect to the initial point $\mathbf{x}_0$.

To study the Lyapunov exponent, we then need to describe the dynamics of the solution near $\mathbf{x}(t, t_0, \mathbf{x}_0)$. Denote by $\mathbf{D_x f} \equiv \mathbf{D_x f}_{t_0}^t$ the linearization of f at $\mathbf{x}$, represented by an

$n \times n$ matrix. Note that this matrix varies with time as the point $\mathbf{x}$ evolves along its path line from t_0 to t. The time-dependence of $\mathbf{D_x f}$ arises through the explicit dependence of f on t as well as through the time-dependent solution $\mathbf{x}(t; t_0, \mathbf{x}_0)$. Therefore, even if f is time-independent, $\mathbf{D_x f}$ may be time-dependent.

To find $\mathbf{D_x}\phi_{t_0}^t$, the derivative of the flow map $\phi_{t_0}^t$ for some fixed t with respect to some fiducial point $\mathbf{x}$, we solve the *variational equation*

$$\frac{d}{dt}\mathbf{D_x}\phi = \mathbf{D_x f}\left(\phi_{t_0}^t(\mathbf{x}), t\right) \cdot \mathbf{D_x}\phi, \qquad \mathbf{D_x f}_{t_0}^{t_0} = \mathbf{I}, \tag{8.11}$$

where $\mathbf{D_x}\phi$ abbreviates $\mathbf{D_x}\phi_{t_0}^t$. This is also called the *fundamental matrix solution* of (8.11). Intuitively, the vector $\mathbf{D_x}\phi(\delta\mathbf{x})$ is a linear operator that takes small variations tangent at time t_0 to small variations tangent of the solution of (8.8) at time t:

$$\mathbf{D_x}\phi_{t_0}^t : T_{\mathbf{x}}M \to T_{\phi_{t_0}^t \mathbf{x}}M. \tag{8.12}$$

More precisely, we have

$$\mathbf{D_x}\phi_{t_0}^t(\mathbf{x}) \cdot \mathbf{f}(\mathbf{x}) = \mathbf{f}(\phi_{t_0}^t(\mathbf{x})). \tag{8.13}$$

The LCE emanating from a point $\mathbf{x}_0$ with an orientation $\mathbf{v} \in T_{\mathbf{x}}M$ can then be defined as

$$\lambda(\mathbf{x}_0, \mathbf{v}, t_0) := \limsup_{t\to\infty} \frac{1}{t - t_0} \log\left(\frac{|\mathbf{D_x}\phi_{t_0}^t(\mathbf{x}_0)\mathbf{v}|}{|\mathbf{v}|}\right). \tag{8.14}$$

However, this asymptotic notion of the LCE does not lend itself well to a practical tool for a study of transport and mixing of the flow. Observe that all LCEs are zero for any trajectory that eventually become regular in a sense that $\lim|\lambda(\mathbf{x}_0, \mathbf{v}, t_0)|$ converges regardless of how much strong expansion takes place for a finite time. But such a finite time phenomenon can be practically important to identify transport and mixing regions. In the next section, we will present a direct estimation of the maximal Lyapunov exponent and demonstrate its potential as a tool to identify global time-dependent structures governing the transport and mixing of a flow.

8.3 Finite Time Lyapunov Exponents and Lagrangian Coherent Structure

8.3.1 Setup

It has been demonstrated that the stable and unstable manifolds of hyperbolic fixed points in two-dimensional autonomous systems separate regions of qualitatively different dynamics. In the case of periodic or quasi-periodic systems, the dynamics of lobes between interlacing stable and unstable manifolds governs the transport behavior of the systems. One salient behavior of initial points straddling stable manifolds of hyperbolic fixed points or periodic points is that they will typically experience exponential separation in forward time, which can be deduced from the lambda lemma [146], and likewise for those points straddling unstable manifolds under a time-reversal system. This is illustrated in Fig. 8.3; the expansion of the green neighborhood is a consequence of the lambda lemma. Also, one may recall

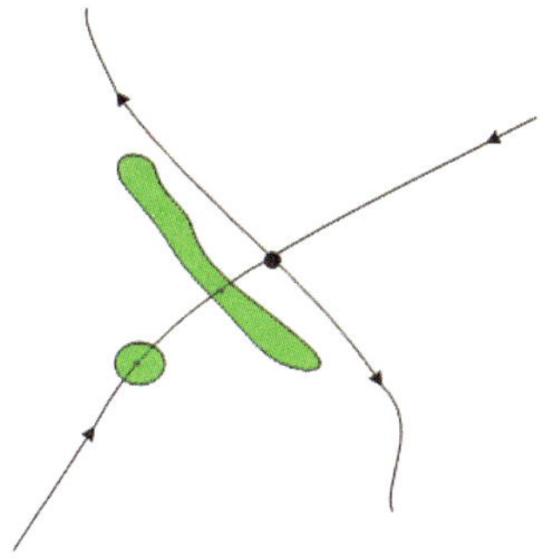

Figure 8.3. *Any small neighborhood of a point q (except the fixed point itself) on a stable manifold will eventually expand exponentially in forward time, which, at least for the homoclinic orbit, follows from the lambda lemma.*

that only those points on the stable manifold can converge to the fixed point. Together with the previous fact, this implies that the separation of points initially straddles the stable manifold. Such characteristics play a pivotal role in approximating stable/unstable manifolds in this chapter.

However, for general time-dependent systems, the stable/unstable manifolds may not even exist with or without a presence of instantaneous hyperbolic fixed points or periodic points. Furthermore, the hyperbolic fixed points may not even be trajectories and may vanish, emerge, or lose their hyperbolic properties in time. So, how do we define the core structures that organize the trajectory pattern around them in a fashion similar to the hyperbolic invariant manifold in the autonomous systems?

In the case of time-dependent systems, we will seek to locate the dynamically evolving structures that form a skeleton of Lagrangian patterns. Such structure is termed Lagrangian coherent structure (LCS), which was proposed in [153]. Therein, LCS is defined to be a material surface, a codimension-one invariant surface in the extended phase space, and exhibit the strongest local attraction or repulsion in the flow. As a material surface, LCS is, therefore, a dynamical structure moving with the underlying flow.

Recall from a discussion in the previous section that certain phenomena in mixing and transport can be identified only over finite time intervals, and hence the ability to locate LCSs assumed to exist over a finite time interval is our goal here. Heuristically, LCS can be captured by the curves with locally relatively large separation rate. Therefore, one would need a way to estimate the separation rate of passive tracers for different initial conditions. A conventional tool to quantify tracer separation and sensitivity to the initial condition of a time-dependent system is the finite time Lyapunov exponent (FTLE). The idea of using FTLE to locate LCS for time-dependent systems can be dated back to the work of Pierrehumbert and Yang [257], in which the spatial distributions of the FTLE fields are used to identify the partial barrier to chaotic mixing on the isentropic surfaces of the Antarctic stratospheric circumpolar region. In [152, 285], it was suggested that a repelling LCS over a finite time interval may be captured through the "ridge" of the FTLE field. Likewise, the attracting LCS should follow the ridge of the backward-time FTLE field. However, the heuristic motivation of using the FTLE ridge to mark the LCSs can be problematic. Some counterexamples to show that LCSs may not be the FTLE ridge and that FTLE ridges may not locate LCSs have been presented in [153]. A discussion of these examples will be deferred until the next section. An introduction of the concept and calculation of FTLE is given below.

Consider again the velocity fields of the form

$$\begin{aligned} \dot{\mathbf{x}}(t;t_0,\mathbf{x}_0) &= \mathbf{v}(\mathbf{x},t), \qquad \mathbf{x} \in U \subset \mathbb{R}^n, \\ \mathbf{x}(t_0) &= \mathbf{x}_0, \end{aligned} \tag{8.15}$$

where $\mathbf{v}(\mathbf{x},t) : \mathbb{R}^n \times \mathbb{R} \to \mathbb{R}^n$ is C^2 in open set $U \in \mathbb{R}^n$ and C^1 in time. In general, the FTLE can also be defined for a higher dimension and can be generalized to Riemannian differentiable manifolds [200].

Recall that we are interested in estimating the maximum stretching rate of trajectories near a point $\mathbf{x}(t_0)$. Let $\mathbf{y}(t) = \mathbf{x}(t) + \delta\mathbf{x}(t)$ for all t. The amount of separation of the infinitesimal perturbation $\delta\mathbf{x}(t_0)$ (with arbitrary orientation) after the time τ is given by

$$\delta\mathbf{x}(t_0+\tau) = \phi_{t_0}^{\tau}(\mathbf{y}(t_0)) - \phi_{t_0}^{\tau}(\mathbf{x}(t_0)) = D\phi_{t_0}^{\tau}(\mathbf{x}(t_0))\delta\mathbf{x}(t_0) + O\left(\|\delta\mathbf{x}(t_0)\|^2\right). \tag{8.16}$$

Therefore, the growth of a small linearized perturbation $\delta\mathbf{x}(t_0)$ in the L_2-norm is given by

$$\begin{aligned} \|\delta\mathbf{x}(t_0+\tau)\| &= \left\|\mathbf{D}\phi_{t_0}^{\tau}(\mathbf{x}(t_0))\delta\mathbf{x}(t_0)\right\| \\ &= \frac{\|\mathbf{D}\phi_{t_0}^{\tau}(\mathbf{x}(t_0))\delta\mathbf{x}(t_0)\|}{\|\delta\mathbf{x}(t_0)\|}\|\delta\mathbf{x}(t_0)\| < \|\mathbf{D}\phi_{t_0}^{\tau}(\mathbf{x}(t_0))\| \; \|\delta\mathbf{x}(t_0)\|. \end{aligned} \tag{8.17}$$

Notice that the separation of trajectories near a point $\mathbf{x}_0$ is controlled by $\|\mathbf{D}\phi_{t_0}^{t_0+\tau}(\mathbf{x}_0)\|$, and it depends not only on position and time but also on the integrating time τ. Depending on the systems, the integrating time τ has to be properly chosen in order to reveal meaningful coherent structure, and this will be demonstrated later on in the chapter.

The FTLE is defined as

$$\sigma_\tau(\mathbf{x}_0) = \frac{1}{|\tau|}\ln\left(\max_{\delta\mathbf{x}_0}\frac{\|\mathbf{D}\phi_{t_0}^{t_0+\tau}(\mathbf{x}_0)\delta\mathbf{x}_0\|}{\|\delta\mathbf{x}_0\|}\right). \tag{8.18}$$

In the above definition, FTLE at $\mathbf{x}$ is taken to be the maximum Lyapunov exponent among all orientations. So, how will we choose the orientation of the infinitesimal perturbation to obtain the maximum stretching? By considering the matrix $\mathbf{D}\phi_{t_0}^{t_0+\tau}$ as a linear operator, the largest infinitesimal stretching along the solution of (8.15) starting at $\mathbf{x}_0$ is, therefore, given by the largest singular value of $\mathbf{D}\phi_{t_0}^{t_0+\tau}$. We introduce a symmetric matrix:

$$\mathbf{C}_{t_0}^{t_0+\tau}(\mathbf{x}_0) = (\mathbf{D}\phi_{t_0}^{t_0+\tau}(\mathbf{x}_0))^T\mathbf{D}\phi_{t_0}^{t_0+\tau}(\mathbf{x}_0). \tag{8.19}$$

Denote by $\xi_1(\mathbf{x}_0,t_0,\tau)\ldots\xi_n(\mathbf{x}_0,t_0,\tau)$ an orthonormal eigenbasis of $\mathbf{C}_{t_0}^{t_0+\tau}(\mathbf{x}_0)$ with the corresponding eigenvalues

$$0 < \lambda_1(\mathbf{x}_0,t_0,\tau) \le \cdots \le \lambda_{n-1}(\mathbf{x}_0,t_0,\tau) \le \lambda_n(\mathbf{x}_0,t_0,\tau). \tag{8.20}$$

Note that in the language of fluid dynamics, the matrix $\mathbf{C}_{t_0}^{t_0+\tau}$ is called the *Cauchy–Green deformation tensor*. The maximum expansion of infinitesimal perturbation at the point $\mathbf{x}_0$

can then be obtained when its orientation is aligned with the eigenvector $\xi_1(\mathbf{x}_0, t_0, \tau)$, i.e.,

$$\max_{\delta\mathbf{x}(t_0)} \|\delta\mathbf{x}(t_0+\tau)\| = \sqrt{\lambda_n(x_0, t_0, \tau)}\|\delta\mathbf{x}(t_0)\|. \tag{8.21}$$

Therefore, we may write the FTLE of the initial point $\mathbf{x}_0$ as

$$\sigma_\tau(\mathbf{x}_0) = \frac{1}{|2\tau|} \ln \lambda_n(\mathbf{x}_0, t_0, \tau). \tag{8.22}$$

We want to stress that the leading singular vector as the solution of the optimization (8.21) has a close relationship to the *true leading Lyapunov vector*, denoted by $\zeta_1(t)$, but they are conceptually different. The latter has to be defined as the vector toward which all perturbations $\delta\mathbf{x}(t_0 - s)$ for any $s > 0$, any perturbation started at a long time s before t_0, must converge. That is,

$$\zeta_1(t_0) = \lim_{s\to\infty} \mathbf{D}\phi_{t_0-s}^{t_0} \delta\mathbf{x}(t_0 - s). \tag{8.23}$$

The definition of the leading Lyapunov vector can be pictorially described as shown in Fig. 8.4.

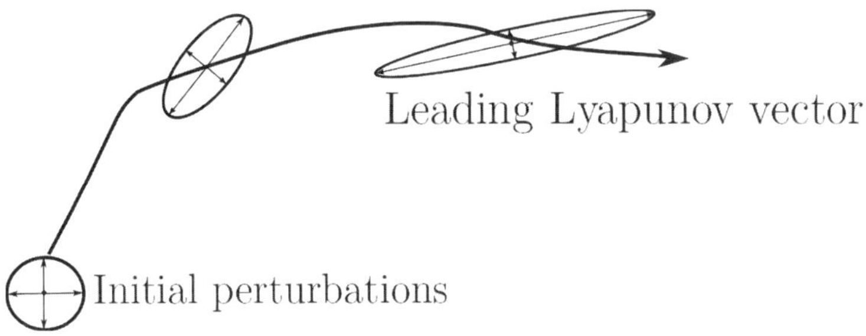

Figure 8.4. *Random perturbations are attracted toward the leading Lyapunov vector as they evolve.*

The leading singular vector of $\mathbf{D}\phi_{t_0}^{t_0+\tau}(\mathbf{x}_0))$, however, is merely a finite time estimate of the leading Lyapunov vector. In fact, any singular vectors, unless lying parallel exactly with another Lyapunov vector, must also approach the leading Lyapunov vector under the evolution by the tangent linear dynamic, but they are initially off the leading Lyapunov vector and could have an initial growth that is higher than the maximum Lyapunov exponent of the system. As an example, consider a two-dimensional linear map

$$\begin{pmatrix} x_1(t+\tau) \\ x_2(t+\tau) \end{pmatrix} = \begin{pmatrix} 2x_1(t) + 16x_2(t) \\ 0.9x_2(t) \end{pmatrix}, \tag{8.24}$$

which has the constant tangent linear map for any t_0 given by a 2×2 matrix:

$$\mathbf{D}\phi_{t_0}^{t_0+\tau} = \begin{pmatrix} 2 & 16 \\ 0 & 0.9 \end{pmatrix}. \tag{8.25}$$

Since the tangent linear map is constant, the leading Lyapunov vector is, by definition, the leading eigenvector of (8.25); thinking of the power method for approximating the largest eigenvalue can give a clue to this claim. Therefore, the largest Lyapunov exponent is log(2). The singular values of (8.25) are $\sqrt{\lambda_1} = 4.02, \sqrt{\lambda_2} = 0.33$; that is, at time $t = t_0 + \tau$ the leading singular vector grows about twice as fast as the leading Lyapunov vector. However, at $t = t_0 + n\tau$ for a positive integer $n > 1$, the singular vectors approach the leading Lyapunov vector, losing its orthogonality, and bear a similar growth rate to that of the Lyapunov vector; see Fig. 8.5.

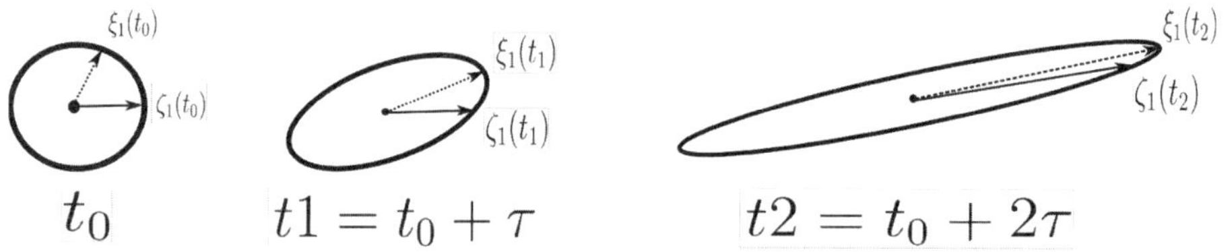

Figure 8.5. *At the initial time t_0, the leading singular vector ξ_1 is far apart from the leading Lyapunov vector ζ_1. At time $t_0 + \tau$, the singular vector grows at a faster rate than the leading Lyapunov vector and the angle between the two vectors becomes smaller. At time $t_0 + 2\tau$, the singular vector approaches the leading Lyapunov vector and their growth rates are now similar.*

Note that the argument in the above example is valid due to the lack of orthogonality among the eigenvectors of the tangent linear map, since if $\mathbf{D}\phi_{t_0}^{t_0+\tau}$ is symmetric, eigenvectors and singular vectors would coincide, and a growth rate higher than the Lyapunov number would not be allowed. We also want to remark that the analogue of the leading Lyapunov vector in the above example for a time-dependent flow is the time-dependent LCS, which will be formally defined later in this chapter by using a concept similar to (8.23).

Last but not least, a possibility of a fast growth rate of the leading singular vector in the above example hints at the fact that even though the dynamic may have a very low growth rate taken as an average over time to infinity, there could be some significant transient, but perhaps long-lived, structures that might be of interest in terms of transport and mixing property of the dynamics, which can be discovered only with the finite time estimate of the leading Lyapunov vector. It is this observation that we try to utilize in order to learn about core structures governing the transport mechanism of a given dynamic. The aim of the FTLE here is not to estimate the leading Lyapunov vector, which may or may not have an important role in transport mechanism.

8.3.2 Algorithm

For simplicity, we present the algorithm for the two-dimensional case in this section. An approximation in a higher dimension is similar; see [285]. Also, the Cartesian grid will be assumed throughout this section. For dynamical systems on a more general coordinate (e.g., on a sphere or S^2), nonstructure grids have been recently developed by Lieken and Ross [200].

Supposed that we want to approximate the FTLE field at time t on a bounded domain $D \in \mathbb{R}^2$ and choose the flow period to be τ. The FTLE can be approximated in the following steps:

1. Initialize Cartesian grid points $(x_{ij}(t_0)), y_{ij}(t_0))$ for $1 \leq i \leq m$ and $1 \leq i \leq n$ on D.

2. Advect these grid points to time $t+\tau$ by some standard integration technique to obtain $(x_{ij}(t_0+\tau), y_{ij}(t_0)+\tau)$.

3. Approximate the spatial gradient of the flow map $\mathbf{D}\phi_\tau(\cdot)$ by a finite-difference technique (e.g., a central difference). In particular, we have

$$\mathbf{D}\phi_t^{t+\tau}(x_{ij}(t), y_{ij}(t)) = \begin{pmatrix} \frac{x_{i+1,j}(t+\tau)-x_{i-1,j}(t+\tau)}{x_{i+1,j}(t)-x_{i-1,j}(t)}, & \frac{x_{i,j+1}(t+\tau)-x_{i,j-1}(t+\tau)}{y_{i,j+1}(t)-y_{i,j-1}(t)} \\ \frac{y_{i+1,j}(t+\tau)-y_{i-1,j}(t+\tau)}{x_{i+1,j}(t)-x_{i-1,j}(t)}, & \frac{y_{i,j+1}(t+\tau)-y_{i,j-1}(t+\tau)}{y_{i,j+1}(t)-y_{i,j-1}(t)} \end{pmatrix}. \tag{8.26}$$

4. Based on the above approximation of the spatial gradient at each grid point, the largest singular value of $\mathbf{D}\phi_\tau(\cdot)$ is calculated and the FTLE can be obtained according to (8.22).

Alternatively, instead of approximating the gradient from the neighboring grid points as above, it is possible to approximate $\mathbf{D}\phi_t^{t+\tau}(\cdot)$ for each grid point by choosing nearby points specifically for each individual point so that one can achieve a better approximation by the finite difference, as shown by the blue dots in Fig. 8.6. However, for the purpose of the LCS visualization, the former implementation will always capture all LCSs in D, whereas the latter may miss some meaningful LCSs. As previously described, the points straddling LCS will exponentially separate forward in time, and so the finite difference of these points will be large. However, the FTLE of LCS can rapidly decrease in the direction perpendicular to the LCS. This implies that if the grid points used in the finite difference do not straddle LCS, the true LCS could be invisible in the FTLE field.

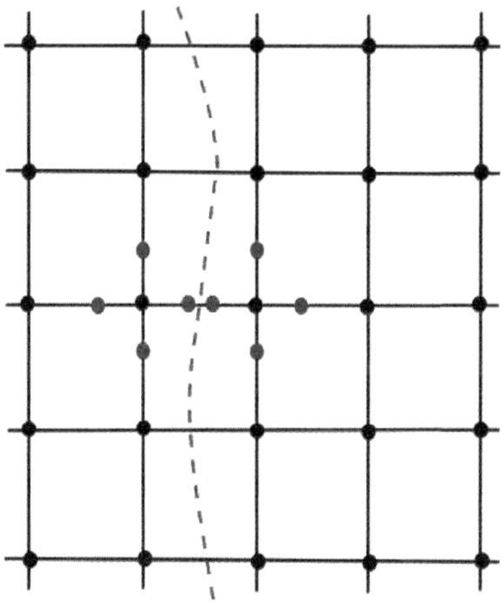

Figure 8.6. *The LCS (shown as the dotted line) could be missed if the off-grid test points (shown in blue) are used to estimate the FTLE. In contrast, the LCS will always be captured when the grid points are used in the FTLE approximate.*

In addition, the approximation of LCS requires a high resolution grid since a coarse grid could underestimate FTLE due to the folding, which is typical in nonlinear systems. Fig. 8.7 depicts this situation. Note that the algorithm outlined above is suitable only for the model flow from which the velocity vectors can be determined all along the trajectory from t to $t+\tau$. However, in the case of a finite data set where the velocity field is limited to a finite domain for some finite period of time, the algorithm is similar in principle but it requires a treatment of an issue, in which the trajectory of a point could invariably leave the domain of data and so integration cannot be continued.

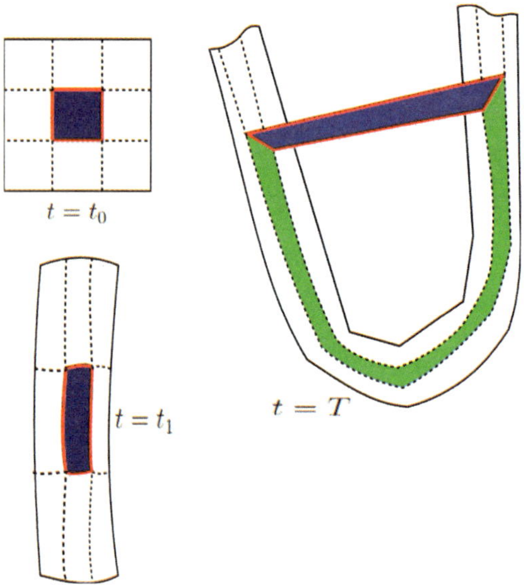

Figure 8.7. *If the grid points are too coarse, there could be an underestimation of the FTLE. The actual stretching (shown in green) could be much larger than the apparent stretching (shown in blue) that is calculated on the finite grid.*

8.3.3 Example 1: Duffing Equation

Consider again the Duffing equation of the form

$$\begin{aligned} \frac{dx}{dt} &= -y, \\ \frac{dy}{dt} &= x - x^3 \end{aligned} \tag{8.27}$$

as our benchmark example. We compute the FTLEs both forward and backward in time using 22500 sample points taken from the domain $[-1,1]\times[-1,1]$ and with the time integration of $|\tau| = 5$. A visualization of FTLEs is shown in Fig. 8.8.

To visualize the LCS, we need to approximate the ridge of the FTLE fields. In general, it may be a difficult task to extract smooth ridges from FTLE fields. However, it is sufficient in this example to visualize the ridges by thresholding the FTLE field; see

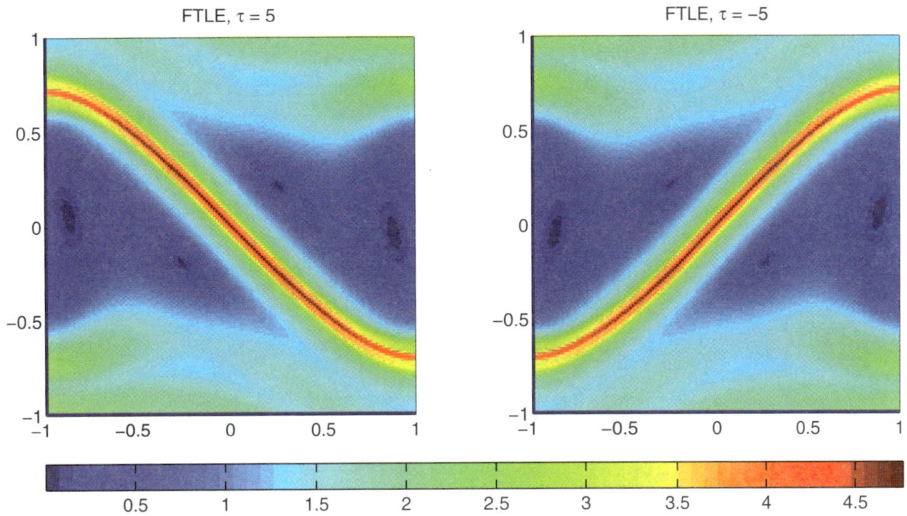

Figure 8.8. *(Left) The FTLE of the Duffing equation with forward-time integration, $\tau = 5$. The skeleton of this FTLE corresponds to the stable manifold of the hyperbolic fixed point at the origin. (Right) The FTLE calculated with $\tau = -5$, which reveals the unstable manifold emerging from the fixed point at the origin.*

Fig. 8.9. Here we plot the attracting LCS computed from $\tau = 5$ in red and the repelling LCS computed from $\tau = -5$ in blue.

8.3.4 Example 2: Periodically Forcing Double Gyre

Consider the autonomous system

$$\begin{aligned} \frac{dx}{dt} &= -\frac{\partial \psi}{\partial x} = A\pi \sin(\pi x)\cos(\pi y), \\ \frac{dy}{dt} &= \frac{\partial \psi}{\partial y} = A\pi \cos(\pi x)\sin(\pi y), \end{aligned} \tag{8.28}$$

where the flow trajectory is bounded in the domain $[0,2] \times [0,1]$. The velocity field consists of two gyres rotating in the opposite direction and divided by a separatrix in the middle $x = 1$; see Fig. 8.10. The system consists of two hyperbolic fixed points, (1,0) and (1,1). A key feature of this system is that the unstable manifold of the fixed point (1,1) coincides with the stable manifold of the fixed point (1,0) and they form the separatrix, which acts as a boundary of the two gyres. In other words, the initial particles on the left of the separatrix will never transverse across the separatrix to the right, and likewise for those on the right.

The forward and backward FTLE is shown in Fig. 8.11 for various integrating time. When τ is too small, the separatrix is not manifested, since most sample points do not experience enough separation. On the other hand, when τ is too large, most sample points become decorrelated and their FTLEs approach their asymptotic value to become indistinguishable.

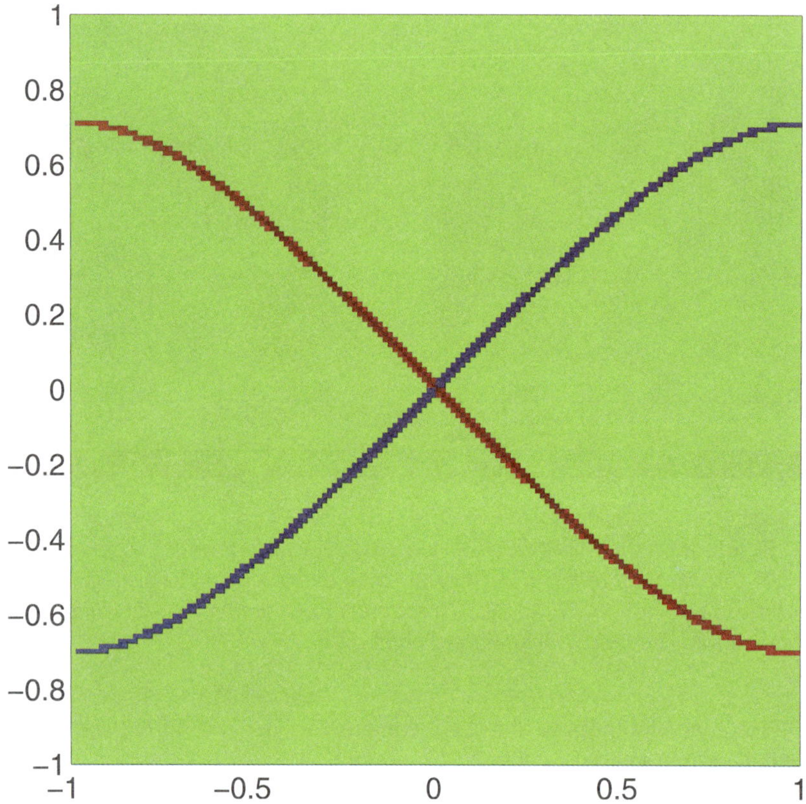

Figure 8.9. *Only the grid points with FTLEs above 85% of the maximum FTLEs are plotted, and the FTLEs of the remaining grid points are suppressed to zero.*

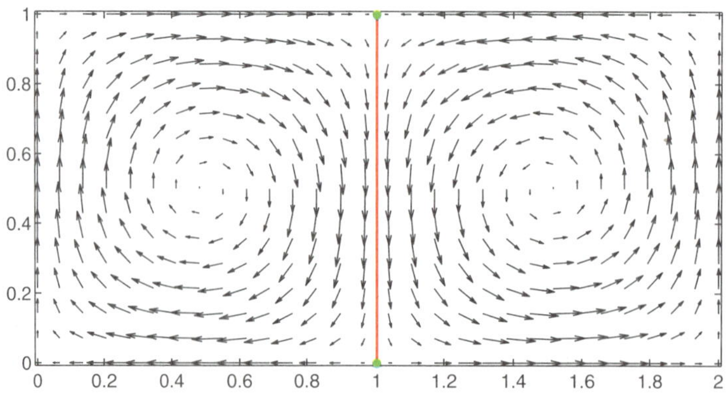

Figure 8.10. *The motion of the autonomous double gyre velocity field. The green points are the hyperbolic fixed points. The unstable manifold of the fixed point at the top boundary coincides with the stable manifold of the fixed point at the lower boundary, with the separatrix (the red line) dividing the two gyres.*

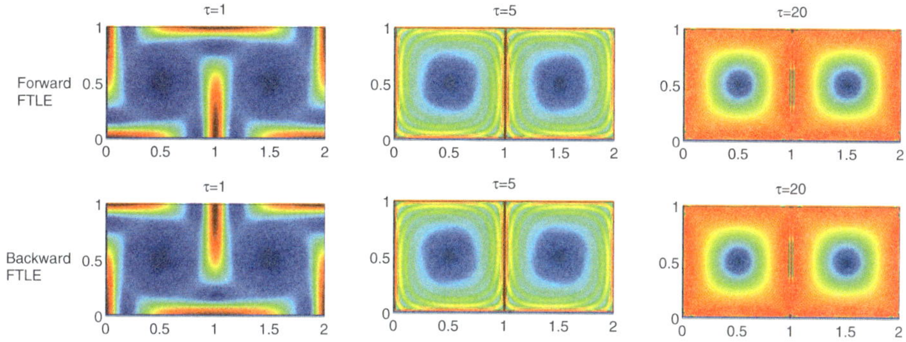

Figure 8.11. *The forward and backward FTLE field for various integrating time.*

8.3.5 Example 3: Periodically Forcing Double Gyre

Although a periodic, Hamiltonian system can be thoroughly understood by some traditional tools such as lobe dynamics, Melenikov methods, or Poincaré maps. For a pedagogical reason, we extend the previous example to demonstrate the validity of the FTLE analysis when applied to a periodic system. The periodically forcing double gyre flow is described by

$$\begin{aligned} \frac{dx}{dt} &= -\frac{\partial \psi}{\partial x} = A\pi \sin(\pi f(x))\cos(\pi y), \\ \frac{dy}{dt} &= \frac{\partial \psi}{\partial y} = A\pi \cos(\pi f(x))\sin(\pi y)\frac{df}{dx}, \end{aligned} \tag{8.29}$$

where the flow trajectory is bounded in the domain $[0,2]\times[0,1]$ and A determines the velocity magnitude. The time-dependent stream function $\psi(x,y,t)$ and the forcing function $f(x,t)$ in the above equation are given by

$$\begin{aligned} \psi(x,y,t) &= A\sin(\pi f(x,t))\sin(\pi y), \\ f(x,t) &= \epsilon \sin(\omega t)x^2 + (1-2\epsilon\sin(\omega t))x. \end{aligned} \tag{8.30}$$

At $t=0$ the separatrix is in the middle of the two gyres and its periodic motion is governed by $f(x,y,t)$, in which the maximum distance of the separatrix away from the middle $x=1$ is approximated by ϵ.

Since the separatrix is connected to the two (Eulerian) periodic points on the boundary $y=0$ and $y=1$ for each time t, we can track the motion of the separatrix via the conditions that $\frac{dx}{dt}=0$ for all t and y, and that at $t=0$ the separatrix, denoted now by $\tilde{x}$, is at $x=1$. These conditions are satisfied when

$$\begin{aligned} \tilde{x} &= 1 + \frac{\sqrt{1+4\epsilon^2\sin^2(\omega t)}-1}{2\epsilon\sin(\omega t)} \\ &\approx \epsilon\sin(\omega t) \qquad \text{for small } \epsilon. \end{aligned} \tag{8.31}$$

In this example, we set $\omega = 2\pi$, $A = 0.25$, and $\epsilon = 0.25$. Fig. 8.12 demonstrates the periodic motion of the separatrix using the above parameters. To visualize the forward and

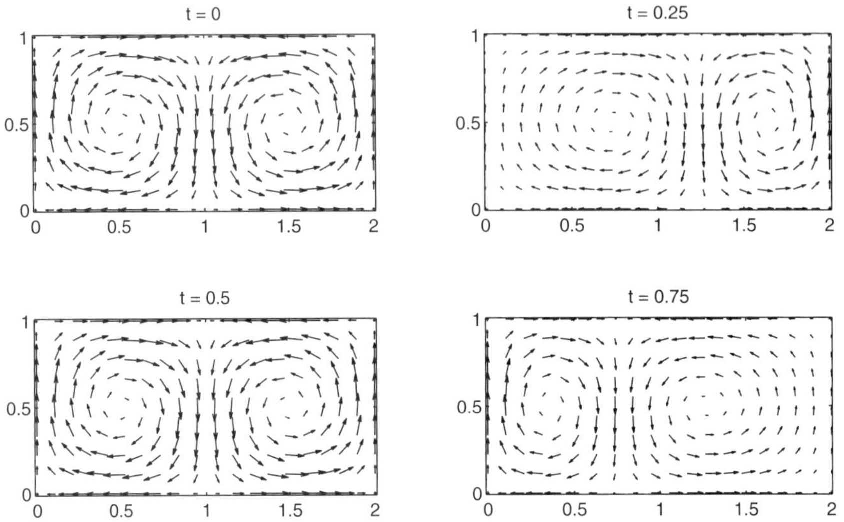

Figure 8.12. *The motion of the double gyre velocity field with the period of* $T = 1$. *The location of the separatrix can be approximated by* (8.31).

backward FTLE fields at time t on the same picture, we superimpose them by plotting a function

$$F(x) = \begin{cases} \sigma_\tau(x(t)) & \text{if } \sigma_\tau(x(t)) > \sigma_{-\tau}(x(t)), \\ -\sigma_{-\tau}(x(t)) & \text{if } \sigma_\tau(x(t)) \le \sigma_{-\tau}(x(t)). \end{cases} \tag{8.32}$$

Approximation of LCS by FTLE ridges may be visualized by plotting $F(x)$ as in (8.32). See Fig. 8.13 for the flow time $\tau = 5$, which is deliberately chosen to be short enough to reveal the lobe dynamics. Increasing the flow time to $\tau = 40$ reveals a different aspect of the dynamics of the double gyre system, namely, the KAM islands, as shown in Fig. 8.14 in a comparison with the stroboscopic map. The result in Fig. 8.14 agrees with the fact that the KAM islands are invariant regions and enclosed by the invariant manifold.

8.3.6 Example 4: ABC Flow

The ABC (Arnold–Beltrami–Childress) flow is given by

$$\begin{aligned} \frac{dx}{dt} &= A\sin(z) + C\cos(y), \\ \frac{dy}{dt} &= B\sin(x) + A\cos(z), \\ \frac{dz}{dt} &= C\sin(y) + B\cos(x), \end{aligned} \tag{8.33}$$

where the domain is defined to be a torus $0 \le x, y, z \le 2\pi$. This flow is known for exhibiting a complicated geometrical structure of stable and unstable invariant manifolds of hyperbolic periodic orbits, which forms a barrier of the invariant sets. The FTLE for this flow has been calculated in [125] for a choice of $A = \sqrt{3}, B = \sqrt{2}, C = 1$ for a comparison

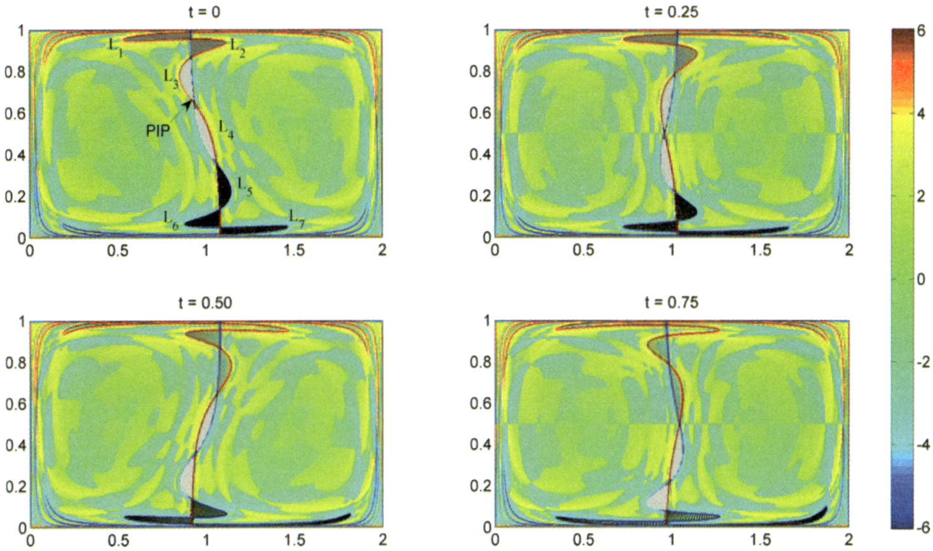

Figure 8.13. *The approximation of the LCS by the FTLE field. The unstable material line (stable manifold) is represented in red and the stable material line (unstable manifold) is represented in blue. Observe that the initial point inside the lobe L_5–L_7 stretches much faster than the others when advecting forward in time. In particular, as the lobe moves toward the (Eulerian) fixed point at the bottom boundary, it will undergo an exponential expanding in a manner similar to those points initially straddling the stable manifold of a fixed point in the time-independent system.*

with the invariant sets; see Fig. 8.15. The result demonstrates that, for this flow, the FTLE field agrees with the barriers of the invariant sets.

In fact, it has been shown that there exist exactly six invariant sets for this ABC flow [52, 145], although it is not clear by observing the three-dimensional FTLE field. All of these six invariant sets can be approximated by using six dominated eigenvectors of the transition matrix; see Fig. 8.16.

8.3.7 Example 5: Stratospheric Coherent Structures in Southern Hemisphere

The "edge" of the Antarctic polar vortex is known to behave as a barrier to the meridional (poleward) transport of ozone during the austral winter [258]. This chemical isolation of the polar vortex from the middle and low latitudes produces an ozone minimum in the vortex region, intensifying the ozone hole relative to what would be produced by photochemical processes alone. Observational determination of the vortex edge remains an active field of research. In this example, we will apply the FTLE technique to European Centre for Medium-Range Weather Forecasts (ECMWF) interim two-dimensional velocity data on the isentropic surface of 850 Kelvin during September 2008 in order to identify the poleward transport of ozone. Traditionally, the contour curve where the potential vorticity (PV) decreases steepest is heuristically used to define the transport barrier. Fig. 8.17(a) shows the

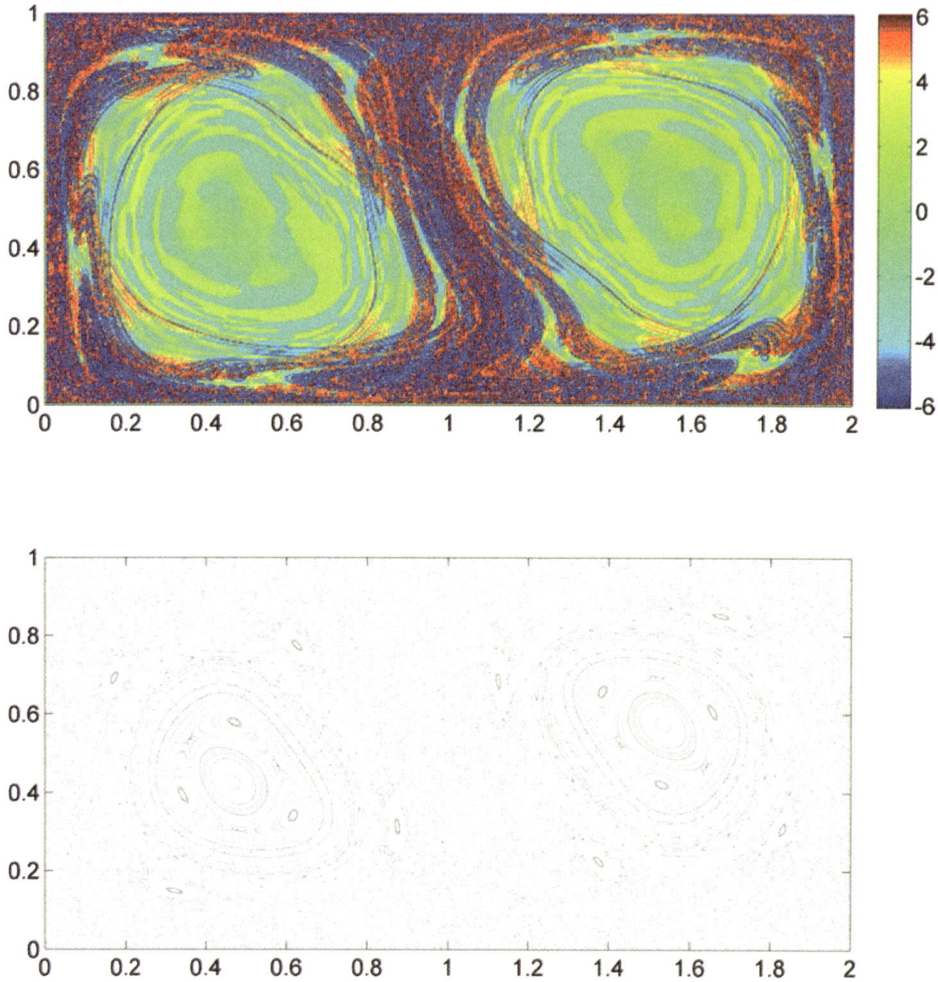

Figure 8.14. *The areas enclosed by the stable/unstable material lines agree very well with the invariant tori, which is manifested by the Poincaré plot.*

PV obtained from ECMWF interim data on the 475 Kelvin isentropic surface on September 14, 2008.

We compute backward and forward FTLEs on September 14, 2008, with the flow duration of 14 days; the results are shown in Fig. 8.17(b), which may be compared with Fig. 8.17(a) to see their similarity. In order to compare the FTLE result with the PV-determined coherent structures (the vortex "boundary"), we use a common approach, developed in [216, 237, 297] to subjectively define the vortex boundary as the PV contour with the highest gradient w.r.t. the *equivalent latitude*. Briefly, the equivalent latitude, ϕ_e, is defined from the area A enclosed by a given PV contour as $A = 2\pi R^2(1 - \sin\phi_e)$ [237]. A transport boundary is then defined to be the location of the highest gradient in PV, as shown in Fig. 8.18. The result of the vortex boundary is plotted in Fig. 8.17(a) as the green curve.

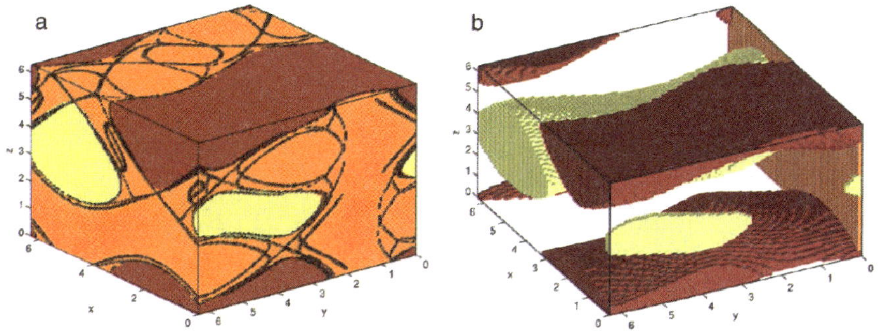

Figure 8.15. *A comparison between the FTLE field, shown as the dark curve in* (a), *and some of the invariant sets in* (b). *A good agreement between the two is evident. Courtesy of* [125].

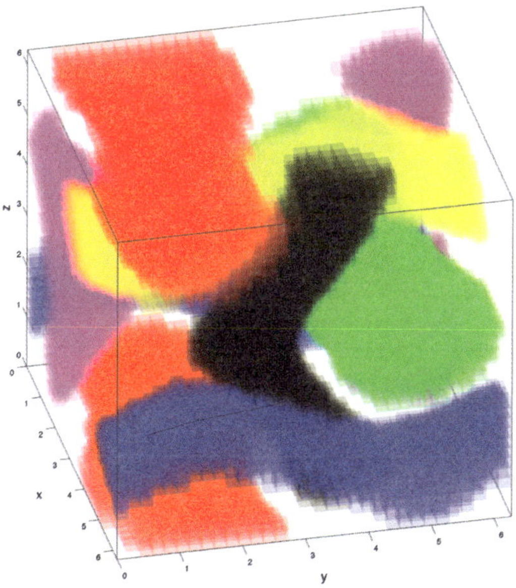

Figure 8.16. *All six invariant sets of the ABC flow.*

Note that the computational cost of the vortex boundary is relatively low, as the PV values have already been provided by the ECMWF interim data.

Fig. 8.17(b) compares the attracting and repelling LCSs with the PV boundary. Interestingly, the LCSs wrap tightly around the PV boundary, and the lobes formed by the interlacing of LCSs occur mainly outside the PV boundary. This suggests that particles (ozone in this case) are completely trapped inside the PV boundary and transport occurs only in the vicinity of the midlatitude region where the LCSs intricately interlace. In the stratosphere literature, such a region is sometimes referred to as a "stochastic" layer [179].

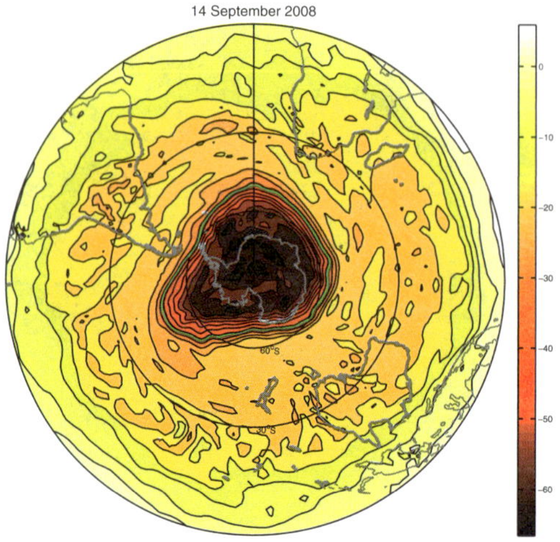

(a) Potential vorticity (PV).

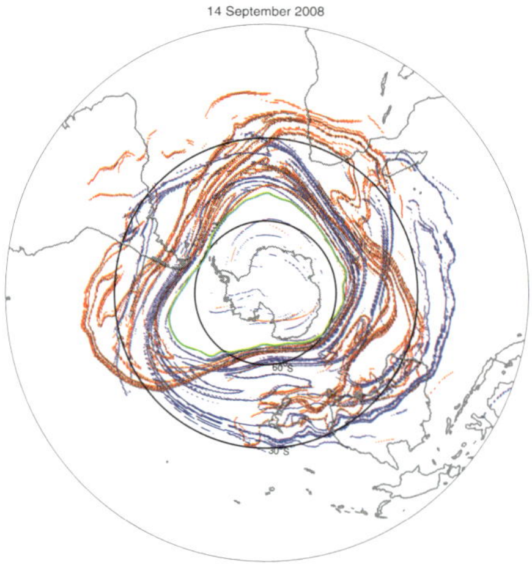

(b) The ridges of FTLEs.

Figure 8.17. *Comparison of PV boundary and the ridge of FTLEs.* (a) *The potential vorticity in the PV units (PVUs)* (1 $PVU = 10^{-6} K m^2 kg^{-1} s^{-1}$) *obtained from ECMWF interim data on the isentropic surface of* 475 *Kelvin on September* 14, 2008. *The "edge" of the PV, shown as the green curve where the gradient is largest, is conventionally used to define the barrier to the poleward transport during the austral winter.* (b) *The ridge of FTLE fields computed from ECMWF interim velocity data on the isentropic surface of* 475 *Kelvin on September* 14, 2008, *with a flow duration of* 14 *days.Potential vorticity (PV)*

8.3.8 Relevance of FTLE as a Tool for LCS Detection

In previous examples, we have heuristically used the FTLE-maximizing curve to detect the LCS. The results were shown to be very reasonable on the physical ground. In this section,

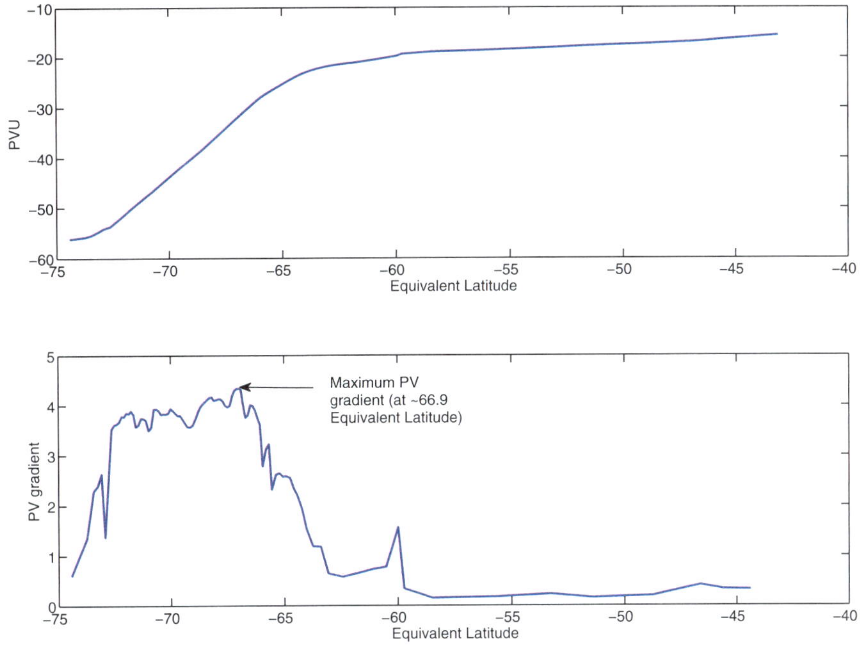

Figure 8.18. *(Top) Potential vorticity plotted in terms of the equivalent latitude in the unit of PVU. (Bottom) PV gradient w.r.t. the equivalent latitude in the unit PVU per degree.*

we will be concerned with their relations based on more rigorous definitions of LCSs and ridges. This formality will raise an awareness of using FTLE ridges to locate LCSs.

We begin with a mathematical definition of the ridge. More formally, a second-derivative ridge of the FTLE field is a codimension-one manifold that maximizes FTLE in the transverse direction along the ridge [285]. The (second-derivative) ridge is defined here as a parameterized curve $c : (a,b) \rightarrow \mathbb{M}$ that satisfies two conditions. First, the derivative of the curve $c'(s)$ has to be parallel to the vector $\nabla\sigma_\tau(c(s))$. This is to force the tangent line of the curve to be oriented in the direction of the largest variation of the FTLE field. Second, the direction of the steepest descent is that of the normal vector to the curve $c(s)$, which can be expressed by

$$n^T H(\sigma_\tau)n = \min_{\|v\|=1} v^T H(\sigma_\tau)v < 0, \tag{8.34}$$

where n is the unit normal vector to the ridge curve $c(s)$, and $H(\sigma_\tau)$ is the Hessian matrix of the FTLE field:

$$H(\sigma_\tau) = \begin{bmatrix} \frac{\partial^2\sigma_\tau}{\partial x^2} & \frac{\partial^2\sigma_\tau}{\partial x \partial y} \\ \frac{\partial^2\sigma_\tau}{\partial y \partial x} & \frac{\partial^2\sigma_\tau}{\partial y^2} \end{bmatrix}. \tag{8.35}$$

When computing in forward time ($\tau > 0$), the ridge of the FTLE field reveals the so-called *repelling Lagrangian coherent structure*, which has a similar feature to the stable material manifold of a hyperbolic trajectory; see [153, 152, 155]. The same is true for the similar feature between attracting LCSs and unstable material manifold.

Roughly speaking, the repelling LCS is a structure that creates stretching in forward time. As proposed in [155], a repelling LCS has two key properties:

1. It is a codimension-one invariant surface (called a *material surface*) in the extended phase space $U \times \mathbb{R}$ that "moves with the flow" and can be observed as evolving Lagrangian patterns; i.e., the LCS at an initial time t_0 must evolve into the LCS at the time $t = t_0 + \tau$ under the flow $\phi_{t_0}^t$. Formally speaking, if $M(t_0) \subset U$ denotes a codimension-one material surface defined at time t_0, which may be viewed as an ensemble of initial conditions at t_0, the codimension-one material surface $M(t)$ at time t must satisfy
$$M(t) = \phi_{t_0}^t(M(t_0)). \tag{8.36}$$
Note that the "bundle" of $M(t)$ generates an invariant manifold M of the ODE (8.15) in the extended phase space $U \times \mathbb{R}$, as schematically illustrated in Fig. 8.19.

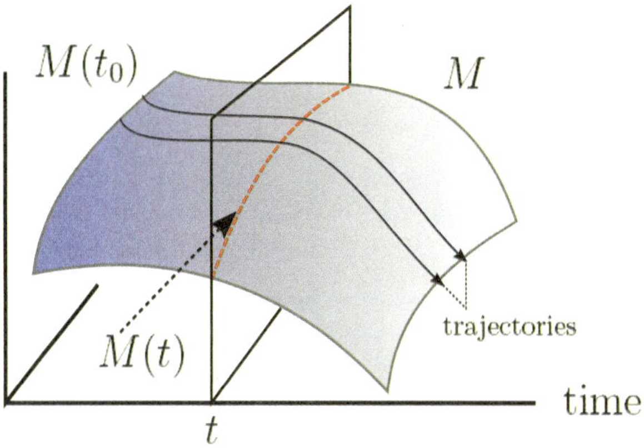

Figure 8.19. *Material surface $M(t) = \phi_{t_0}^t(M(t_0))$ generated in the extended phase space.*

2. A repelling LCS should locally maximize the repulsion rate in the flow in order to be distinguishable from all nearby trajectories that could also be material surfaces with similar repelling behavior due to the continuous dependence on the initial conditions of the flow. If we denote, in the case of two-dimensional flow, the one-dimensional tangent space of $M(t)$ by $T_{\mathbf{x}_t}M(t)$ and the one-dimensional normal space by $N_{\mathbf{x}_t}M(t)$, then the repulsion rate is defined as
$$\rho_{t_0}^t(\mathbf{x}_0, \mathbf{n}_0) = \langle \mathbf{n}_t, \mathbf{D}\phi_{t_0}^t(\mathbf{x}_0)\mathbf{n}_0 \rangle, \tag{8.37}$$
which describes the growth of perturbation in the normal direction $\mathbf{n}_t \in N_{\mathbf{x}_t}M(t)$ of $M(t)$, which is orthogonal to the tangent space $T_{\mathbf{x}_t}M(t)$. As illustrated in Fig. 8.20, the tangent vector $\mathbf{e}_0 \in T_{\mathbf{x}_0}M(t_0)$ based at a point $\mathbf{x}_0$ on $M(t_0)$ is mapped by the flow map $\mathbf{D}\phi_{t_0}^t$ to a tangent vector $\mathbf{D}\phi_{t_0}^t(\mathbf{x}_0)\mathbf{e}_0 \in T_{\mathbf{x}_t}M(t)$ at point $\mathbf{x}_t$ on $M(t)$. In other words, we have that
$$T_{\mathbf{x}_0}M(t_0) = \mathbf{D}\phi_{t_0}^t(\mathbf{x}_0)T_{\mathbf{x}_0}M(t_0). \tag{8.38}$$

However, the normal vector $\mathbf{n}_0 \in N_{\mathbf{x}_0}M(t_0)$ is not necessarily mapped by the flow map to a normal vector $\mathbf{n}_t \in N_{\mathbf{x}_t}M(t)$. The repulsion rate as defined above measures the growth of perturbation via the orthogonal projection of the vector $\mathbf{D}\phi_{t_0}^{t}(\mathbf{x}_0)\mathbf{n}_0$ onto the normal space containing $\mathbf{n}_t$.

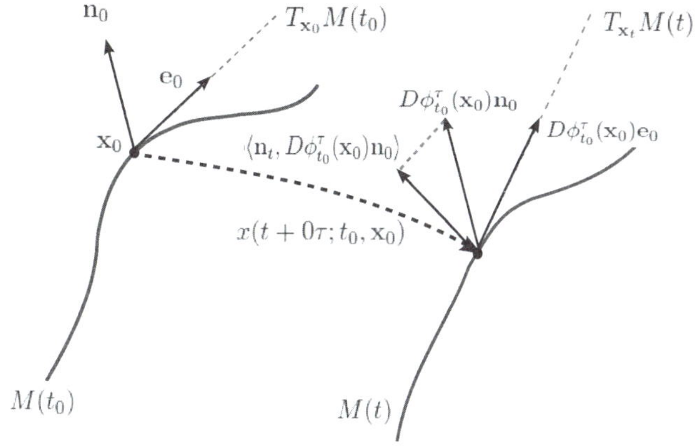

Figure 8.20. *Geometry of the linearized flow map of a two-dimensional flow.*

That is, at any point $\mathbf{x}_0 \in M(t_0)$, the repulsion rate $\rho_{t_0}^{t}(\mathbf{x}_0, \mathbf{n}_0) > \rho_{t_0}^{t}(\hat{\mathbf{x}}_0, \hat{\mathbf{n}}_0)$ for any $t \in [t_0, t_0 + \tau]$, where $\hat{\mathbf{x}}_0$ is a point on a nearby material surface $\hat{M}(t_0)$ intersected by the normal $\mathbf{n}_0$, and $\hat{\mathbf{n}}_0$ is a normal vector associated with $\hat{M}(t_0)$ and based at the point $\hat{\mathbf{x}}_0$; see Fig. 8.21.

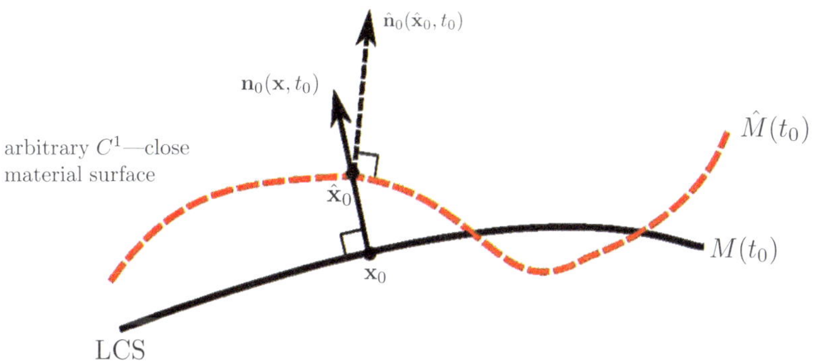

Figure 8.21. *The LCS is locally defined as a material surface that maximizes the repulsion rate in the normal direction.*

A detailed discussion of these properties can be found in Haller [155].

Similarly, the ridge of the FTLE computed with negative integrating time ($\tau < 0$) locates the *attracting Lagrangian coherent structure*, which acts as a core repelling structure in the backward time; hence the previous LCS criterion are still applicable when regarding the attracting LCS advected forward in time as the repelling LCS of backward time [153, 152]. It was analytically and experimentally verified by Shadden [285] that the LCS defined as the second-derivative ridge of the FTLE field is nearly Lagrangian in a sense that the particle flux across the LCS becomes negligible as the integrating time τ increases provided that the true LCS associated with the vector field is hyperbolic for all time. However, the LCS obtained from a finite time data set is likely to be hyperbolic only for a finite time. Therefore, if the integrating time τ increases beyond the hyperbolic time of the trajectory, which is not known a priori,the resulting LCS may not exhibit the Lagrangian property. In general, the ridges may not mark the true LCSs, while many of published works were using the FTLE in a nonrigorous way. The rigorous necessary sufficient conditions for the FTLE ridges to locate the LCSs have only recently been published in Haller [155]. We refer the details of these conditions therein.

We are now concerned with some pathological examples in which the explicit expression of the Cauchy–Green tensors can be explicitly established. It will be seen that the FTLE ridges do not necessarily mark the LCS.

Example 8.4 (LCS exists but there are no FTLE ridges). Consider a decoupled, two-dimensional ODE:

$$\begin{aligned} \dot{x} &= x, \\ \dot{y} &= -y - y^3. \end{aligned} \tag{8.39}$$

The vector field of this flow is shown in Fig. 8.22, along with the stable and unstable

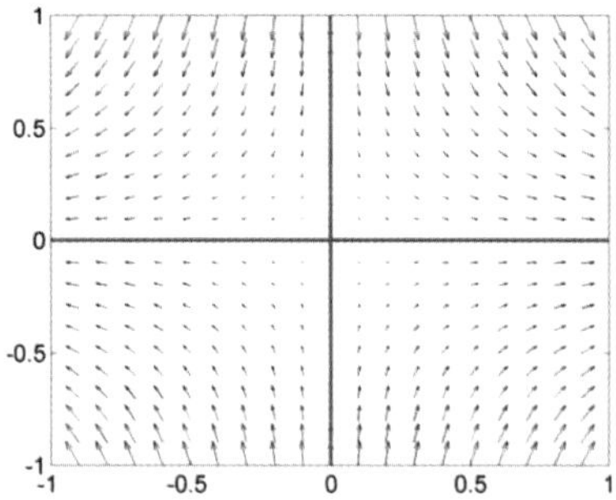

Figure 8.22. *The vector field of the flow* (8.39) *with the attracting LCS at* $y = 0$ *and repelling LCS at* $x = 0$.

manifolds of the fixed point at $(0,0)$. Clearly, the stable manifold ($x = 0$) here represents the repelling LCS, while the unstable manifold ($y = 0$) marks the attracting LCS. However, we will show that the forward FTLE field turns out to be a constant, and hence there are no FTLE ridges. Since the vector field is autonomous and decoupled, we may assume that $t_0 = 0$ and solve for the trajectory for a given initial condition (x_0, y_0) as

$$\begin{aligned} x(t) &= x_0 e^{2t}, \\ y(t) &= \frac{y_0}{\sqrt{(y_0^2+1)e^{2t} - y_0^2}}. \end{aligned} \tag{8.40}$$

For this example, the Cauchy–Green tensor as defined in (8.19) can be written as

$$\Lambda_\tau = \begin{pmatrix} \left[\frac{\partial x(t,0,x_0)}{\partial x_0}\right]^2 & 0 \\ 0 & \left[\frac{\partial y(t,0,y_0)}{\partial y_0}\right]^2 \end{pmatrix} = \begin{pmatrix} e^{2t} & 0 \\ 0 & \frac{e^{4t}}{[(y_0^2+1)e^{2t}-y_0^2]^3} \end{pmatrix}. \tag{8.41}$$

Therefore, for $t > 0$, the maximum eigenvalue is

$$\lambda_{\max}(t) = e^{2t},$$

which, according to (8.22), leads to a constant FTLE field

$$\sigma_t(x(t)) \equiv 1. \tag{8.42}$$

This implies that there are no ridges for all $t > 0$. An illustration of this misconception is shown in Fig. 8.23.

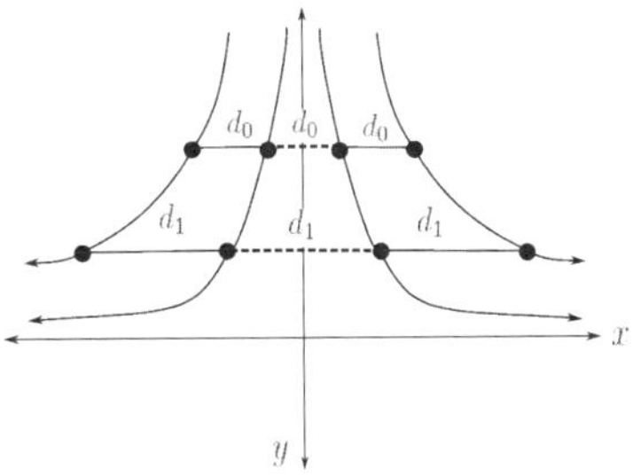

Figure 8.23. *A saddle flow without an FTLE ridge that has a repelling LCS at $x = 0$.*

In contrast, for $t < 0$, the maximum eigenvalue becomes

$$\lambda_{\max}(t) = \frac{e^{4t}}{[(y_0^2+1)e^{2t} - y_0^2]^3},$$

which depends both on the flow time t and the value of y_0. Thus, the backward FTLE field depends only on y. To check the extremum of the backward FTLE we calculate

$$\frac{\partial}{\partial y_0}\lambda_{\max}(y_0,t_0) = -6\left(e^{2t}-1\right)e^{4t}\frac{y_0}{\left(e^{2t}(y_0^2+1)-y_0^2\right)^4},$$

$$\frac{\partial^2}{\partial y_0^2}\lambda_{\max}(y_0,t_0) = -6\left(e^{2t}-1\right)e^{4t}\frac{e^{2t}-7y_0^2e^{2t}+7y_0^2}{\left(e^{2t}(y_0^2+1)-y_0^2\right)^5}.$$

It is now easy to check that for $t < 0$ and $y_0 = 0$, $\frac{\partial}{\partial y_0}\lambda_{\max}(0,t_0) = 0$ and $\frac{\partial^2}{\partial y_0^2}\lambda_{\max}(0,t_0) > 0$. It can then be concluded that the backward FTLE field has a minimum ridge (through) at the x-axis, which coincides with the attracting LCS as demonstrated in Fig. 8.24. Note that in this example we accepted the repelling and attracting LCSs based on our intuition but have not done a formal validation based on those LCSs conditions in [155]. In fact, it was shown that the x-axis is not attracting LCS but only "weakly" LCS as defined therein.

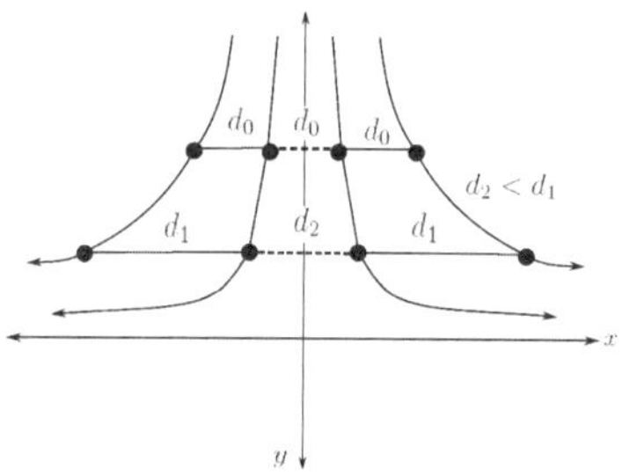

Figure 8.24. *An illustration of the scenario in which the attracting LCS at $y = 0$ appears as the FTLE instead of a ridge.*

Example 8.5 (FTLE ridges that are neither an attracting nor a repelling LCS). Consider the two-dimensional area-preserving system

$$\begin{aligned} \dot{x} &= 2 + \tanh y, \\ \dot{y} &= 0. \end{aligned} \tag{8.43}$$

Since $\dot{y} = 0$, the vector is just a flow parallel to the x-axis as illustrated in Figure 8.24. Next, we will show that the x-axis is indeed a ridge of both forward and backward FTLE in this example. Again, we may assume $t_0 = 0$, and hence the trajectories of (8.43) are given by

$$\begin{aligned} x(t) &= x_0 + t(\tanh y_0 + 2), \\ y(t) &= y_0. \end{aligned} \tag{8.44}$$

The maximum and minimum eigenvalues of the Cauchy–Green tensor can then be calculated as

$$\begin{aligned} \lambda_{\max}(t) &= \frac{1}{2}t^2[\operatorname{sech}(y_0)]^4 + \frac{1}{2}\sqrt{(t^4[\operatorname{sech}(y_0)]^8 + 4)} + 1, \\ \lambda_{\min}(t) &= \frac{1}{2}t^2[\operatorname{sech}(y_0)]^4 - \frac{1}{2}\sqrt{(t^4[\operatorname{sech}(y_0)]^8 + 4)} + 1. \end{aligned} \tag{8.45}$$

The hyperbolicity of the trajectory can be readily seen from the above equation, since we have the relation

$$\log \lambda_{\min}(t) < 0 < \log \lambda_{\max}(t). \tag{8.46}$$

Also, the $\operatorname{sech}(y_0)$ attains a unique maximum at $y_0 = 0$, and so does the quantity $\sigma_t(x(0;t,x_0,y_0))$, independent of x_0. Therefore, for any $t \neq 0$, the maximal ridge of the $\sigma_t(x(0;t,x_0,y_0))$ field is the x-axis with a constant height along the ridge.

However, since the flow in this example is area-preserving, the distance perpendicular to the ridge (x-axis in this example) is constant for all times (which should also be intuitively clear from the geometry of the flow). Therefore, the ridge is neither a repelling nor an attracting LCS.

8.3.9 Conditions for Using FTLE Ridge to Identify LCS

Below we present the sufficient and necessary conditions for a hyperbolic LCS based on the FTLE ridge, which has been proved in [155].

Theorem 8.1. Assume that $M(t_0) \subset U$ is a compact FTLE ridge. Then $M(t) = \phi_{t_0}^t[M(t_0)]$ is a repelling LCS over the time interval $[t_0, t_0+T]$ if and only if at each point $x_0 \in M(t_0)$ the following conditions are satisfied:

1. $u_n(x_0, t_0, T) \perp T_{x_0} M(t_0)$.
2. $\lambda_{n-1} \neq \lambda_n(x_0, t_0, T) > 1$.
3. $\langle \nabla \lambda_n(x_0, t_0, T), u_n(x_0, t_0, T) \rangle = 0$.
4. The matrix $\mathcal{L}(x_0, t_0, T)$ is positive definite, where

$$\mathcal{L} = \begin{pmatrix} \nabla^2 C^{-1}[\xi_n, \xi_n, \xi_n, \xi_n] & 2\frac{\lambda_n - \lambda_1}{\lambda_n \lambda_1}\langle \xi_1, \nabla \xi_n \xi_n \rangle & \cdots & 2\frac{\lambda_n - \lambda_{n-1}}{\lambda_n \lambda_1}\langle \xi_{n-1}, \nabla \xi_n \xi_n \rangle \\ 2\frac{\lambda_n - \lambda_1}{\lambda_n \lambda_1}\langle \xi_1, \nabla \xi_n \xi_n \rangle & 2\frac{\lambda_n - \lambda_1}{\lambda_n \lambda_1} & \cdots & 0 \\ \vdots & \vdots & \ddots & \vdots \\ 2\frac{\lambda_n - \lambda_{n-1}}{\lambda_n \lambda_1}\langle \xi_{n-1}, \nabla \xi_n \xi_n \rangle & 0 & \cdots & 2\frac{\lambda_n - \lambda_1}{\lambda_n \lambda_1} \end{pmatrix}.$$

Proof. See [155]. □

The first diagonal term in $\mathcal{L}$ is the second derivative of the inverse Cauchy–Green tensor, which is equal to

$$\nabla^2 C^{-1}[\xi_n, \xi_n, \xi_n, \xi_n] = -\frac{1}{2}\langle \xi_n, \nabla^2 \lambda_n \xi_n \rangle + 2\sum_{q=1}^{n-1} \frac{\lambda_n - \lambda_q}{\lambda_q \lambda_n}\langle \xi_n, \nabla \xi_n \xi_n \rangle^2. \tag{8.47}$$

The first condition implies that the normal vector $\mathbf{n}_0$ at $\mathbf{x}_0$ should align with the $\xi_n(x_0, t_0, T)$, the eigenvector corresponding to the largest eigenvalue λ_n. The second condition states that the largest eigenvalue must be greater than one and multiplicity one to ensure the dominance of the growth normal over the growth in tangent direction. Together with the second condition, the third condition, which can be interpreted as the directional derivative of the λ_n field in the direction of u_n, completes the stationary requirement for the normal repulsion rate. In the last condition, the required positive definiteness of $\mathcal{L}$, which is not as intuitive as the first two conditions, arises from the variational argument in the proof of this theorem in [155]. Nevertheless, it is shown in [184] that this positive definite condition is equivalent to the condition of $\langle \nabla u_n(x_0), \lambda_n^2(x_0, t_0, T) u_n(x_0, t_0, T) \rangle < 0$. In other words, this condition ensures that the repulsion rate, hence λ_n, has a nondegenerate maximum in the normal direction of the hyperbolic LCS.

Chapter 9

Information Theory in Dynamical Systems

In this chapter, we outline the strong connection between dynamical systems and a symbolic representation through symbolic dynamics. The connection between dynamical systems and its sister topic of ergodic theory can also be emphasized through symbolization by using the language inherent in information theory. Information as described by the Shannon information theory begins with questions regarding code length necessary for a particular message. Whether the message be a poem by Shakespeare, a raster scanned or a multiscaled (wavelet) represented image, or even an initial condition and its trajectory describing the evolution under a given dynamical process, the language of information theory proves to be highly powerful and useful. In the first few sections of this chapter, we will review just enough classical information theory to tie together some strong connections to dynamical systems and ergodic theory in the later sections.

9.1 A Little Shannon Information on Coding by Example

Putting the punchline first, we will roughly state that information is defined to describe a decrease in uncertainty. Further, less frequent outcomes confer more information about a system. The Shannon entropy is a measure of average uncertainty.

Basic questions of data compression begin a story leading directly to Shannon entropy and information theory. For ease of presentation, we shall introduce this story in terms of representation of a simple English phrase. The discussion applies equally to phrases in other human languages, representations of images, encodings of music, computer programs, etc.

The basic idea behind *entropy coding* is the following simple principle:

- We assign short code words to likely frequent source symbols and long code words to rare source symbols.

- Such source codes will therefore tend to be variable length.

- Since long code words will be assigned to those sources which are less likely, they are therefore more surprising, and conversely short codes are less surprising.

Consider a phrase such as the single English word "Chaos"; we choose a single-word phrase only to make our presentation brief. This story would be the same if we were to choose a whole book of words, such as this book in your hands. Encoded in standard ASCII,[117]

$$\text{Chaos} \rightarrow 1000011\ 1101000\ 1100001\ 1101111\ 1110011. \tag{9.1}$$

For convenience and clarity to the reader, a space is indicated between each 7-bit block to denote individual letters. Notice that in this 7-bit version of standard ASCII coding, it takes $5 \times 7 = 35$ bits to encode the 5 letters in the word Chaos, stated including the uppercase beginning letter,

$$\text{"C"} \rightarrow 1000011, \tag{9.2}$$

versus what would be the lowercase in ASCII,

$$\text{"c"} \rightarrow 1100011. \tag{9.3}$$

ASCII is a useful code in that it is used universally on computers around the world. If a phrase is encoded in ASCII, then both the coder and the decoder at the other end will understand how to translate back to standard English, using a standard ASCII table. The problem with ASCII, however, is that it is not very efficient. Consider that in ASCII encoding

$$\text{"a"} \rightarrow 1111010, \quad \text{"z"} \rightarrow 1100001. \tag{9.4}$$

Both the "a" and the "z" have reserved the same 7-bit allocation of space in the encoding, which was designed specifically for English. If it were designed for some language where the "a" and the "z" occurred equally frequently, this would be fine, but in English, it would be better if the more frequently used letters, such as vowels, could be encoded with 1- or 2-bit words, say, and those which are rarely used, like the "z" (or even more so the specialty symbols such as \$, &, etc.), might be reasonably encoded with many bits. On average such an encoding would do well when used for the English language for which it was designed.

Codes designed for specific information streams, or with *assumed prior* knowledge regarding the information streams can be quite efficient. Amazingly, encoding efficiencies of better than 1-bit/letter may even be feasible. Consider the (nonsense) phrase with 20 characters,

$$\text{"chchocpohccchohhchco"}. \tag{9.5}$$

In ASCII it would take $20 \times 7 = 140$ bits for a bit rate of 7 bits/letter. However, in a Huffman code[118] we can do much better. Huffman coding requires a statistical model regarding the expected occurrence rate of each letter. We will take as our model[119]

$$p_1 = P(\text{"c"}) = 0.4,\ p_2 = P(\text{"h"}) = 0.35,\ p_3 = P(\text{"o"}) = 0.2,\ p_4 = P(\text{"p"}) = 0.05, \tag{9.6}$$

[117] American Standard Code for Information Interchange (ASCII) is a character encoding of the English alphabet used commonly in computers. Each character gets the same length of 7 bits even though that some characters are not likely to be used.

[118] The Huffman code is a variable-length code algorithm that is in some sense optimal, as discussed in Section 9.2, especially in Theorem 9.1. This breakthrough was developed by an MIT student, D.A. Huffman, and published in his 1952 paper [169].

[119] The notation $P(A)$ denotes the probability of event A.

which we derive by simply counting occurrences[120] of each letter to be 8, 7, 4, and 1, respectively, and by assuming stationarity.[121] With these probabilities, it is possible the following Huffman code follows,

$$\text{“c”} \to 0, \quad \text{“h”} \to 10, \quad \text{“o”} \to 110, \text{“p”} \to 111, \tag{9.7}$$

from which follows the Huffman encoding,

$$\text{“chchocpohcccchohhchco”} \to$$

$$0\ 10\ 0\ 10\ 110\ 0\ 111\ 110\ 10\ 0\ 0\ 0\ 10\ 110\ 10\ 10\ 0\ 10\ 0\ 110. \tag{9.8}$$

Again, spaces are used here to guide the eye separating bits related to each individual letter. The spaces are not actually part of the code. In this encoding, the 20-letter phrase chchocpohccchohhchco is encoded in 37 bits, for a bit rate of 37 bits/20 letters = 1.85 bits/letter, which is a great deal better than what would have been 7 bits/letter in a 140-bit ASCII encoding.

We have not given the details behind how to form a Huffman code from a given discrete probability distribution as this would be somewhat outside the scope of this book and beyond our needs here. Our point is simply that there are much better ways to encode an information stream by a well-chosen variable-length code as exemplified by the well regarded Huffman code. The dramatic improvement of course comes from this pretty good statistical model used. For example, zero bit resource is allocated for the letter "z", and the rest of the alphabet. Any message that requires those symbols would require a different encoding or the encoding simply is not possible.

How shall the quality of a coding scheme such as the Huffman code be graded? Considering the efficiency of the encoding is a great deal like playing the role of a bookie,[122] hedging bets. Most of the bit resource is allocated for the letter "c" since it is most common, and the least is allocated for the "o" and "p" since they are less common than the others used. When a message is stored or transmitted in agreement with this statistical model, then high efficiency occurs and low bit rates are possible. When a message that is contrary to the model is transmitted with this then ill-fitted model, less efficient coding occurs. Consider a phrase hhhhhhhhhhhhhhhhhhhh ("h" 20 times).[123] Then a 3 bits/letter efficiency occurs, which is a worst-case scenario with this particular Huffman coding. That is still better than ASCII, because the model still assumes only 4 different letters might occur. In other words, it still beats ASCII since at the outset we assumed that there is zero probability for all but those 4 symbols.

---

[120] Without speaking to the quality of the model, note that counting occurrences is surely the simplest way to form a probabilistic model of the likelihood of character occurrences.

[121] Roughly stated, a stochastic process is stationary if the joint probabilities "do not change in time." More will be said precisely in Definition 9.12 in Section 9.5.

[122] A bookie is a person who handles bets and wagers on events, usually sporting events on which gamblers place money in hopes that their favorite team will win and they will win money. Bookies need a good probability model if they expect to win over many bets on average.

[123] Consider another coding method entirely, called run length encoding, in which the "h" is to be repeated some number of times [263]. The longer the repetition, the more efficient such an encoding would be, since the overhead of the annotations to repeat is amortized; the codings of "state 20 h's" and "state 50 h's" are essentially the same, for example. Such codes can only be useful for very special and perhaps trivial phrases with long runs of single characters. This discussion relates to the notion of **Kolmogorov complexity,** which is defined to be the length of the optimal algorithm which reproduces the data. Kolmogorov complexity is not generally computable.

To make the notion of bit rate and efficiency more rigorous, note that the statistical *expectation* of the bit rate in units of bits/letter may be written

$$\begin{aligned}\text{Average bit rate} &= \text{bits/letter}\\ &= \sum_i P(i\text{th letter occurs in message})\ (\text{Length used to encode } i\text{th letter})\\ &= 0.4*1+0.35*2+0.2*3+0.05*3\text{bits/letter}\\ &= 1.85\text{bits/letter}. \end{aligned} \tag{9.9}$$

The perfect coincidence in this example between expectation and actual encoding rate simply reflects the fact that the toy message used matches the probabilities. The optimal bit length of each encoding can be shown to be bounded,

$$(\text{Length used to encode } i\text{th letter}) \leq -\log_2(p_i). \tag{9.10}$$

Consider the relation of this statement to the optimal code rate implicit in Theorem 9.2.

Considering an optimal encoding of a bit stream leads to what is called the Shannon entropy, defined formally in Definition 9.4, specialized for a coding with 2 outcomes (bits):

$$H_2 :\equiv -\sum_i p_i \log_2(p_i). \tag{9.11}$$

Shannon entropy carries the units of bits/letter, or alternatively bits/time if the letters are read at a rate of letters/time. Comparing this to the question of how long the coding is in the previous example, Eq. (9.6),

$$H_2 = -0.4\log_2(0.4) - 0.35\log_2(0.35) - 0.2\log_2(0.2) - 0.05\log_2(0.05) = 1.7394. \tag{9.12}$$

Note that $1.7394 < 1.85$ as the code used was not optimal. The degree to which it was not optimal is the degree to which, in Eq. (9.10),

$$-\left[(\text{Length used to encode } i\text{th letter}) + \log_2(p_i)\right] > 0. \tag{9.13}$$

Specifically, notice that $p_3 = 0.2 > p_4 = 0.05$, but they each are "gambled" by the bit rate bookie who allocates 3 bits each. There are reasons for the suboptimality in this example:

- The probabilities are not d-atic (d-atic with $d = 2$, also spelled dyadic in this case, means $p_i = \frac{r}{2^n}$ for some r, n integers and for every i).
- A longer version of a Huffman coding would be required to differentiate these probabilities. Here only three bits were allowed at maximum. Huffman codes are developed in a tree, and depth is important too.
- Furthermore, the Huffman coding is a nonoverlapping code [169], meaning each letter is encoded before the next letter can be encoded. Huffman code is a special example of so-called entropy coding within the problem of lossless[124] compression.

[124]The object of lossless encoding is to be able to recover the original "message" exactly, as opposed to lossy encoding, in which some distortion is allowed. For example, consider a computer program as the source. A "zipped" file of a computer program must be decompressed *exactly* as the original for the decompressed computer program to work. On the other hand, lossless compression generally borrows from representation theory of functions; a lossy scheme for compressing a digital photograph includes truncating a Fourier expansion.

Huffman codes are optimal within the class of entropy coding methods, beating the original Shannon–Fano code, for example, but it is not as efficient as the Lempel–Ziv (LZ) [325] or arithmetic coding methods [197, 68]. Our purpose in using the Huffman coding here was simply a matter of specificity and simplicity of presentation.

Important aspects of a proper coding scheme are that it must at least be

- One-to-one and therefore invertible[125]: Said otherwise, for each coded word, there must be a way to recover the original letters so as to rebuild the word. Without this requirement, we could quite simply compress every message no matter how long, say the complete works of Shakespeare, to the single bit 0, and not worry that you cannot go back based on that bit alone. The information is lost as such.
- Efficient: A poor coding scheme could make the message length longer than it may have been in the original letter coding. This is a legal feature of entropy encoding, but of course not useful.

The details of these several coding schemes are beyond our scope here, but simply knowing of their existence as related to the general notions of coding theory is leading us to the strong connections of entropy in dynamical systems. Sharpening these statements mathematically a bit further will allow us to discuss the connection.

9.2 A Little More Shannon Information on Coding

The Shannon entropy $H_D(X)$ defined in Definition 9.4 can be discussed in relation to the question of the possibility of an optimal code, as the example leading to Eq. (9.9) in the previous section reveals. To this end, we require the following notation and definitions.

Definition 9.1. An **encoding** $c(x)$ for a random variable X (see Definition 3.3) is a function from the countable set $\{x\}$ of outcomes of the random variable to a string of symbols from a finite alphabet, called a D-ary code.[126]

Remark 9.1. Commonly in digital computer applications which are based on binary bits, representing "0" and "1", or "on" and "off", generally $D = 2$.

Following the above discussion, it is easy to summarize with the following definition.

Definition 9.2. The expectation of the length L of a source encoding $c(x)$ of the random variable X with an associated probability distribution function $p(x)$ is given by

$$L(C) = \sum_{x} p(x) l(c(x)), \tag{9.14}$$

where $l(c(x))$ is the length of the encoding $c(x)$ in units of bits. In Eq. (9.9), we described the units of L to be bits/letter; however, it can be interpreted also simply as length when a fixed positive number C of letters are coded.

[125] A code is called **nonsingular** if for every outcome there is a unique representation by a string from the symbol set ($\{0, 1\}$ if binary); otherwise it is called singular.

[126] D-ary refers to the use of D symbols, which may be taken from the symbol set, $\{0, 1, \ldots, D-1\}$ or $\{0, 1\}$, the usual binary set if $D = 2$. Arithmetic occurs in base-D.

Given an encoding and repeated experiments from a random variable, we can get a code extension which is simply an appending of codes of each individual outcome.

Definition 9.3. Given a code $c(x)$, an **encoding extension** is a mapping from ordered strings of outcomes x_i to an ordered string of symbols from the D-ary alphabet of the code,

$$C = c(x_1 x_2 \ldots x_n) \equiv c(x_1)c(x_2)\ldots c(x_n). \tag{9.15}$$

This is a concatenation of the alphabet representation of each outcome in the sequence $x_1, x_2, \ldots, x_n$.

The formal definition of Shannon entropy can be stated as follows.

Definition 9.4. The **Shannon entropy** of a D-ary code for a random X with probability distribution function $p(x)$ is given by the nonnegative function

$$H_D(X) = -\sum_x p(x) \log_D p(x), \tag{9.16}$$

in terms of the base-D logarithm.

There is a strong connection between the notion of optimal coding and this definition of entropy as revealed by the following classic theorems from information theory, as proven in [68]. Discussion of the existence of optimal codes follows starting from the Kraft inequality.

Theorem 9.1. An instantaneous code[127] C of a random variable X with code word lengths $l(x)$ satisfies the inequality

$$\sum_x D^{-l(x)} \leq 1. \tag{9.17}$$

By converse, the Kraft inequality implies that such an instantaneous code exists. The proof of this second statement is by Lagrange multiplier optimization methods [68]. Furthermore a statement relating Shannon entropy and expected code length $L(C)$ can be summarized as follows.

Theorem 9.2. Given a code C of a random variable X with probability distribution $p(x)$, C is a minimal code if the code word lengths are given by

$$l(x) = -\log_D p(x). \tag{9.18}$$

The following definition of Shannon information may be described as a pointwise entropy in that it describes not only the length of a given optimally coded word, but also the entropy as if we know that the random variable takes on $X = x$, and therefore the corresponding code word is used.

[127] An **instantaneous code**, also called a **prefix code**, completes each code word before the next word begins. No code word is a prefix of any other code word. Therefore, the receiver does not require a prefix before each word to know when to start reading the next word. The converse is that two words share the same coding, and therefore a prefix would be required to distinguish them. Since perhaps the most commonly used prefix code is the Huffman code, favored for its optimality properties, often by habit even other source codes are called Huffman by some.

Definition 9.5. **Shannon information** is defined by the quantity

$$l(x) = -\log_D p(x). \tag{9.19}$$

Shannon information in some sense describes *a degree of surprise we should hold when an unlikely event comes to pass.* A great deal more information is inferred when the unlikely occurs than when the usual, high probability outcomes x occur. Comparing Eqs. (9.14), (9.16), and (9.19), we can describe Shannon entropy $H_D(X)$ as an *information expectation* of the random variable.

Now we can state the relationship of an optimal code C^* to the entropy of the process, which gives further meaning to the notion of Shannon entropy.

Theorem 9.3 (source coding). Let C^* be an optimal code of a random variable X, meaning the expected code length of any other code C is bounded, $L(C) \geq L(C^*)$, and if C^* is an instantaneous D-ary code, then

$$H_D(X) \leq L(C^*) \leq H_D(X)+1. \tag{9.20}$$

Following the example in the previous section, recall that in the ASCII coded version of the 20-character message "chchocpohcchohhchco" in Eq. (9.8), $L(C) = 140$, whereas with the Huffman coded version, $L(C^*) = 37$ bits. Furthermore, with the statement that Huffman is optimal, we know that this is a shortest possible encoding by an instantaneous code. Since Huffman coding embodies an algorithm which provides optimality, this validates existence.

Theorem 9.4. A Huffman code is an optimal instantaneous code.

That Huffman is an optimal code is enough for our discussion here regarding existence of such codes, and the relationship between coding and entropy. The algorithmic details are not necessary for our purposes in this book, and so we skip the details for the sake of brevity. Details on this and other codes, Lempel–Ziv (LZ) [325] or arithmetic coding methods [197, 68] most notably, can be found elsewhere. As it turns out, other nonprefix codes, notably arithmetic coding, can yield even shorter codes, which is a play on the definitions of optimal, code, and encoding extension. In brief, arithmetic coding can be understood as a mapping from letters to base-D representations of real numbers in the unit interval $[0,1]$, distributed according to the probabilities of X, and perhaps in levels by a stochastic process $X_1, X_2, \ldots$ generating letters. In this perspective, a Huffman code is a special case encoding one letter at a time by the same mapping process, whereas arithmetic coding maps the entire message all together. Thus arithmetic coding allows the reading of letters in seemingly overlapping fashion.

9.3 Many Random Variables and Taxonomy of the Entropy Zoo

Given many random variables,

$$\{X_1, X_2, \ldots, X_n\} \tag{9.21}$$

(or two when $n = 2$) in a product probabilities space,

$$\{(\Omega_1, \mathcal{A}_1, P_1) \times (\Omega_2, \mathcal{A}_2, P_2) \times \cdots \times (\Omega_n, \mathcal{A}_n, P_n)\}, \tag{9.22}$$

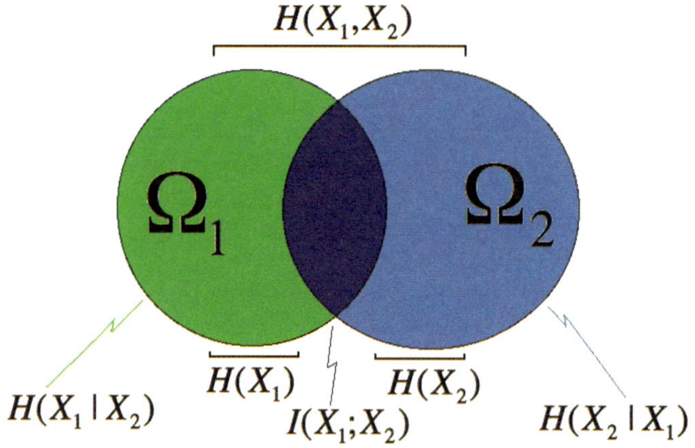

Figure 9.1. *Given a compound event process, here $n = 2$ with Ω_1 and Ω_2 associated with Eqs.* (9.21) *and* (9.22), *we can discuss the various joint, conditional, and individual probabilities, as well as the related entropies as each is shown with their associated outcomes in the Venn diagram.*

one can form probability spaces associated with the many different intersections and unions of outcomes; see Fig. 9.1. Likewise, the associate entropies we will review here give the degree of "averaged surprise" one may infer from such compound events.

Definition 9.6. The **joint entropy** associated with random variables $\{X_1, X_2, \ldots, X_n\}$ is

$$H(X_1, X_2, \ldots, X_n) = -\sum_{x_1, x_2, \ldots, x_n} p(x_1, x_2, \ldots, x_n) \log p(x_1, x_2, \ldots, x_n) \tag{9.23}$$

in terms of the joint probability density function $p(x_1, x_2, \ldots, x_n)$, and with the sum taken over all possible joint outcomes $(x_1, x_2, \ldots, x_n)$.

Joint entropy is sometimes called the *total entropy* of the combined system. See Fig. 9.1, where $H(X_1, X_2)$ is presented as the uncertainty of the total colored regions.

Definition 9.7. The **conditional entropy** associated with two random variables $\{X_1, X_2\}$ is

$$H(X_1|X_2) = -\sum_{x_2} p(x_2) H(X_1|X_2 = x_2) \tag{9.24}$$

in terms of the probability density function $p_2(x_2)$.

Conditional entropy $H(X_1|X_2)$ can be understood as the *remaining entropy* bits in the uncertainty of the random variable X_1 with the information bits already given regarding the intersection events associated with X_2. See the Venn diagram in Fig. 9.1. In other words, measuring $H(X_1|X_2)$ answers the question, "What does X_2 not say about X_1?" An alternative formula for conditional entropy may be derived in terms of the joint probabilities $p(x_1, p_2)$,

$$H(X_1|X_2) = -\sum_{(x_1, x_2)} p(x_1, x_2) \log \frac{p(x_1, x_2)}{p(x_2)}, \tag{9.25}$$

which is easy to see since the term $H(X_1|X_2 = x_2)$ given in Definition 9.7:

$$H(X_1|X_2 = x_2) = \sum_{x_1} p(x_1|x_2) \log p(x_1|x_2) \tag{9.26}$$

Using the relationship for conditional probabilities,

$$p(x_1|x_2) = \frac{p(x_1,x_2)}{p_2(x_2)}, \tag{9.27}$$

substitution into Definition 9.7 yields

$$\begin{aligned} H(X_1|X_2) &= -\sum_{x_2} p_2(x_2) H(X_1|X_2 = x_2) && (9.28)\\ &= -\sum_{x_1,x_2} p_2(x_2) p(x_1|x_2) \log(x_1|x_2) && (9.29)\\ &= -\sum p(x_1,x_2) \log p(x_1|x_2), && (9.30) \end{aligned}$$

the last being a statement of a cross entropy. Finally, again applying Eq. (9.27),

$$\begin{aligned} -\sum p(x_1,x_2) \log p(x_1|x_2) &= -\sum p(x_1,x_2) \log \frac{p(x_1,x_2)}{p_2(x_2)} && (9.31)\\ &= -\sum p(x_1,x_2) \log p(x_1,x_2) && (9.32)\\ &\quad + \sum p(x_1,x_2) \log p_2(x_2). && (9.33) \end{aligned}$$

From this follows the **chain rule** of entropies:

$$H(X_1|X_2) + H(X_2) = H(X_1,X_2). \tag{9.34}$$

A few immediate statements regarding the relationships between these entropies can be made.

Theorem 9.5. $H(X_1|X_2) = 0$ if and only if X_1 is a (deterministic) function of the random variable X_2.

In other words, since X_1 can be determined whenever X_2 is known, the status of X_1 is certain for any given X_2.

Theorem 9.6. $H(X_1|X_2) = H(X_1)$ if and only if X_1 and X_2 are independent random variables.[128]

This can be understood as a statement that knowing X_2 gives no further information regarding X_1 when the two random variables are independent.

Two useful further entropy-like measures comparing uncertainty between random variables are the mutual information and the Kullback–Leibler divergence.

[128] X_1 and X_2 are defined to be independent if $p(x_1,x_2) = p_1(x_1)p_2(x_2)$, or likewise by Eq. (9.27), $p(x_1|x_2) = p_1(x_1)$ and $p(x_2|x_1) = p_2(x_2)$.

Definition 9.8. The **mutual information** associated with two random variables $\{X_1, X_2\}$ is

$$I(X_1;X_2) = \sum_{x_1,x_2} p(x_1,x_2)\log\frac{p(x_1,x_2)}{p_1(x_1)p_2(x_2)}.^{129} \tag{9.35}$$

Alternatively, there follows another useful form of mutual information,

$$I(X_1;X_2) = H(X_1) - H(X_1|X_2). \tag{9.36}$$

Mutual information may be understood as the amount of information that knowing the values of either X_1 or X_2 provides about the other random variable. Stated this way, mutual information should be symmetric, and it is immediate to check that Eq. (9.35) is indeed so. Likewise, inspecting the intersection labeled $I(X_1;X_2)$ in Fig. 9.1 also suggests the symmetric nature of the concept. An example application to the spatiotemporal system pertaining to global climate from [105] is reviewed in Section 9.9.2.

The Kullback–Leibler divergence on the other hand is a distance-like measure between two random variables, which is decidedly asymmetric.

Definition 9.9. The **Kullback–Leibler divergence** between the probability density functions p_1 and p_2 associated with two random variables X_1 and X_2 is

$$D_{KL}(p_1||p_2) = \sum_x p_1(x)\log\frac{p_1(x)}{p_2(x)}. \tag{9.37}$$

The D_{KL} is often described as if it is a metric-like distance between two density functions, but it is not technically a metric since it is not necessarily symmetric; generally,

$$D_{KL}(p_1||p_2) \neq D_{KL}(p_2||p_1). \tag{9.38}$$

Nonetheless, it is always nonnegative, as can be seen from (9.2) by considering $-\log(p_2(x))$ as a length of encoding. Furthermore, $D_{KL}(p_1||p_2)$ can be understood as an entropy-like measure in that it measures the expected number of extra bits which would be required to code samples of X_1 when using the wrong code as designed based on X_2, instead of specifically designed for X_1. This interpretation can be understood by writing

$$\sum_x p_1(x)\log\frac{p_1(x)}{p_2(x)} = \sum_x p_1(x)\log p_1(x) - \sum_x p_1(x)\log p_2(x) \tag{9.39}$$

$$= H_c(X_1,X_2) - H(X_1), \tag{9.40}$$

where $H_c(X_1, X_2)$ is the cross entropy,

Definition 9.10. The **cross entropy** associated with two random variables $\{X_1, X_2\}$ with probability density functions p_1 and p_2,

$$H_c(X_1|X_2) = H(X_1) + D_{KL}(p_1||p_2), \tag{9.41}$$

describes the inefficiency of using the wrong model p_2 to build a code for X_1 relative to a correct model p_1 to build an optimal code whose efficiency would be $H(X_1)$.

[129] It is useful to point out at this stage that $p_1(x_1)$ and $p_2(x_2)$ are the **marginal distributions** of $p(x_1,x_2)$; $p_1(x_1) = \sum_{x_2} p(x_1,x_2)$ and, likewise, $p_2(x_2) = \sum_{x_1} p(x_1,x_2)$.

Thus when $p_1 = p_2$ and, therefore, $D_{KL}(p_1||p_2) = 0$, the coding inefficiency as measured by cross entropy $H_c(X_1|X_2)$ becomes zero. Mutual information can then be written:

$$I(X_1;X_2) = D_{KL}(p(x_1,x_2)||p_1(x_1)p(x_2)). \tag{9.42}$$

A stochastic process allows consideration of entropy rate.

Definition 9.11. Given a stochastic process $\{X_1, X_2, \ldots\}$, the **entropy rate** is defined in terms of a limit of joint entropies,

$$H = \lim_{n\to\infty} \frac{H(X_1,X_2,\ldots,X_n)}{n}, \tag{9.43}$$

if this limit exists.

Assuming the special case that X_i are independent and identically distributed (i.i.d.), and noting that independence[130] gives

$$p(x_1,x_2,\ldots,x_n) = \prod_{i=1}^{n} p(x_i), \tag{9.44}$$

it follows quickly that[131]

$$H(X_1,X_2,\ldots,X_n) = \sum_{i=1}^{n} H(X_i) = nH(X_1). \tag{9.45}$$

The second statement in this equality only requires independence, and the third follows the identically distributed assumption. Now we are in a position to restate the result (9.20) of the source coding Theorem 9.3 in the case of a coding extension. If $L(c(x_1x_2\ldots x_n))$ is the length of the coding extension of n coded words of realizations of the random variables $X_1, X_2, \ldots, X_n$, then Eq. (9.20) generalizes to a statement regarding the minimum code word *length expected per symbol*,

$$H(X_1,X_2,\ldots,X_n) \le nL(C) \le H(X_1,X_2,\ldots,X_n)+1, \tag{9.46}$$

or, in the case of a stationary process,

$$\lim_{n\to\infty} L = H, \tag{9.47}$$

the entropy rate. Notice that this length per symbol is just what was emphasized by example near Eq. (9.9). In the case of an i.i.d. stochastic process, Eq. (9.46) specializes to

$$H(X_1) \le L(C) \le H(X_1) + \frac{1}{n}. \tag{9.48}$$

---

[130] Statistical independence of X_1 from X_2 is defined when given probabilities as follows: $p(x_1,x_2) = p(x_1)$ for each $X_1 = x_1$ and $X_2 = x_2$. And since $p(x_1,x_2) = p(x_1|x_2)p(x_2)$, then independence implies $p(x_1,x_2) = p(x_1)p(x_2)$. Likewise, $P(x_2|x_1) = p(x_2)$ since $p(x_2|x_1) = \frac{p(x_1,x_2)}{p(x_1)} = \frac{p(x_1)p(x_2)}{p(x_1)} = p(x_2)$.

[131] $H(X_1,X_2,\ldots,X_n) = \sum_i p(x_1,x_2,\ldots,x_n)\log p(x_1,x_2,\ldots,x_n) = \sum_i \prod_{i=1}^n p(x_i)\log\prod_{i=1}^n p(x_i) = \sum_i \prod_{i=1}^n p(x_i)[\sum_{i=1}^n \log p(x_i)] = \sum_i \prod_{i=1}^n p(x_i)[\sum_{i=1}^n \log p(x_i)]$.

This expression is symbolically similar to Eq. (9.20), but now we may interpret the entropy rate of the i.i.d. stochastic process to the expected code word length of a coded extension from appending many coded words.

Finally in this section, coming back to our main purpose here to relate information theory to dynamical systems, it will be useful to introduce the notion of channel capacity. We recall that according to the channel coding theorem, which we will discuss below, a chaotic oscillator can be described as such a channel. Channel capacity is the answer to the question, "How much information can be transmitted or processed in a given time?" One may intuit that a communication system may degrade in the sense of increasing error rate as the transmission rate increases. However, this is not at all the case, as the true answer is more nuanced. The part of the channel coding theorem that we are interested in here can be stated as follows.

Theorem 9.7 (channel coding theorem). A "transmission" rate R is achievable with vanishingly small probability if $R < C$, where C is the **information channel capacity**,

$$C = \max_{p(x)} I(X;Y), \tag{9.49}$$

where the maximum is taken over all possible distributions $p(x)$ on the input process, and Y is the output process.

Now to interpret this theorem, a given communication system has a maximum rate of information C known as the channel capacity. If the information rate R is less than C, then one can approach arbitrarily small error probabilities by careful coding techniques, meaning cleverly designed codes, even in the presence of noise. Said alternatively, low error probabilities may require the encoder to work on long data blocks to encode the signal. This results in longer transmission times and higher computational requirements. The usual transmission system as a box diagram is shown in Fig. 9.2, including some description in the caption of standard interpretations of the inputs and outputs X and Y. Fano's converse about error rate relates a lower bound on the error probability of a decoder,

$$H(X|Y) \leq H(e) + p(e)\log(r), \tag{9.50}$$

where

$$H(X|Y) = -\sum_{i,j} p(x_i, y_i)\log p(x_i|y_j) \quad \text{and} \quad p(e) = \sup_i \sum_{i \neq j} p(y_j|x_i). \tag{9.51}$$

To interpret the relevance of the Theorem 9.7 in the context of dynamical systems theory requires only the work of properly casting the roles of X and Y, as is discussed in Section 9.4 and illustrated in Fig. 9.6.

9.4 Information Theory in Dynamical Systems

In Chapter 6 we highlighted the description of a dynamical system as an underlying symbolic dynamics. Here we will further cement this connection by describing a complementary information theoretic perspective of the symbolic dynamics.

Perhaps there is no better way of emphasizing the information theoretic aspects of chaotic dynamical systems with an underlying symbolic dynamics than by explicitly

Figure 9.2. *Channel capacity* (9.52) *box diagram. Random variable X describes the random variable descriptive of the possibilities of the input states message. Coming out of the channel is the output states whose values are according to the random variable Y which one wishes to be closely related (i.e., desires to be identical in many applications). Alternatively, let Y be understood as states of a decoder. Such will be the case if transmission rate is slow enough, $R < C$, according to Theorem* 9.7.

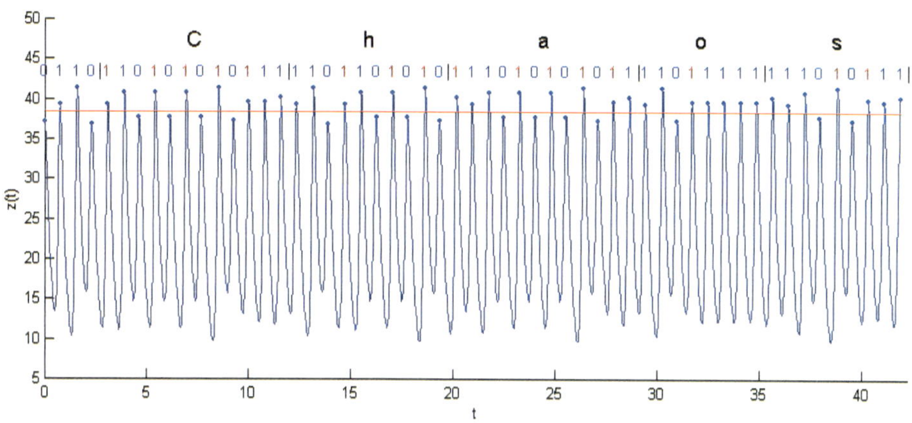

Figure 9.3. *The Lorenz attractor has chaotic orbits and an underlying symbolic dynamics, usually presented according to the successive maxima map of the $z(t)$ time series, which was already discussed in Eq.* (6.4) *according to Eqs.* (6.5)–(6.6), *and the resulting map was already illustrated in Figs.* 6.3 *and* 6.4. *In this case, the property of chaos that there are infinitely many periodic orbits has been leveraged against the symbolic dynamics to choose periodic orbits whose symbolic dynamics encode the word "Chaos" in ASCII form, subject to the non-information-bearing bits shown in red. Compare to the embedding form of the attractor shown in Fig.* 9.4.

demonstrating orbits bearing messages as we wish. In Fig. 6.2, a time series of an orbit segment from a Lorenz equation (6.4) flow is plotted along with its phase space representation of the attractor, in Fig. 6.3. We repeat similar Figs. 9.3 and 9.4. Again, we can read symbols from the $z(t)$ time series by the symbol partition (6.6); zeros and ones can be read by the position of the local maxima of the $z(t)$ coordinate from the differential equation relative to the cusp-maximum of this value z_n in a one-dimensional ansatz $z_{n+1} = f(z_n)$. A 0 bit in the message as indicated in the time series in Fig. 9.4 corresponds to a relatively smaller local maximum (corresponding to the left side of the cusp in Fig. 9.4), meaning local maximum (a red point) occurs on the left "wing" of the attractor. Likewise, 1 bits are encoded in the orbit. Now, however, we choose to read the symbols and interpret them as if they are coded words in ASCII.

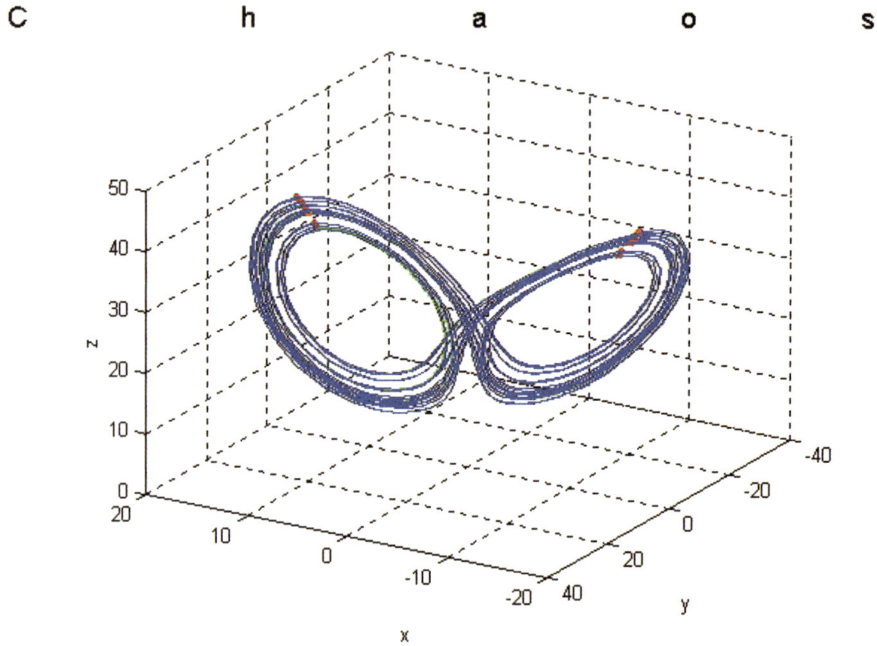

Figure 9.4. *The time series shown in Fig. 9.3 of an orbit segment spelling "Chaos" shown in its phase space presentation of (x,y,z) coordinates on the chaotic attractor. This is a carefully chosen initial condition.*

With the Lorenz parameters set to be the usual famous values as they were in the previous chapter, $(10, 28, 8/3)$, we may observe that the underlying symbolic dynamics grammar never allows two zeros in a row in a symbolic word string, as is described in Fig. 6.26. Therefore, a suitable source code can be built on a prefix coding of the transitions in the graph shown in Fig. 9.5 (right). There are two possible outcomes from a previous symbol 1; either a 0 or a 1 can follow. However, there is only one outcome possible from a symbol 0; simply stated, there is no surprise from the transition, by observing the 1 bit which follows a 0 bit—it was the only possibility. So this transition can be said to be non-information-bearing, or a zero entropy state. To emphasize this, we labeled this transition $*$ in the directed graph since it serves only a role as a pause required to transition back to an information-bearing state—that is, a state where either a 0 *or* a 1 might follow.

Specifically, in terms of this source coding and using the ASCII prefix code, the word "Chaos" has been encoded into the chaotic oscillations. All the non-information-bearing $1/s$ denoting the $*$-transition in the directed graph in Fig. 9.5 are those 1's which are colored red. Counting the ASCII code length to carry the word "Chaos" is $7 \times 5 = 35$ bits, but including the non-information-bearing bits has further required $16 + 35 = 51$ bits. These extra bits represent a nonmaximal channel capacity of the message carrier, which in this case is the dynamical system itself.

The dynamical system encoding the word "Chaos," or any other word, phrase, or arbitrary information stream including sounds or images, etc. [160, 26, 253, 55], has a fundamental associated channel capacity C, (9.52). Thus, by Theorem 9.7, transmission

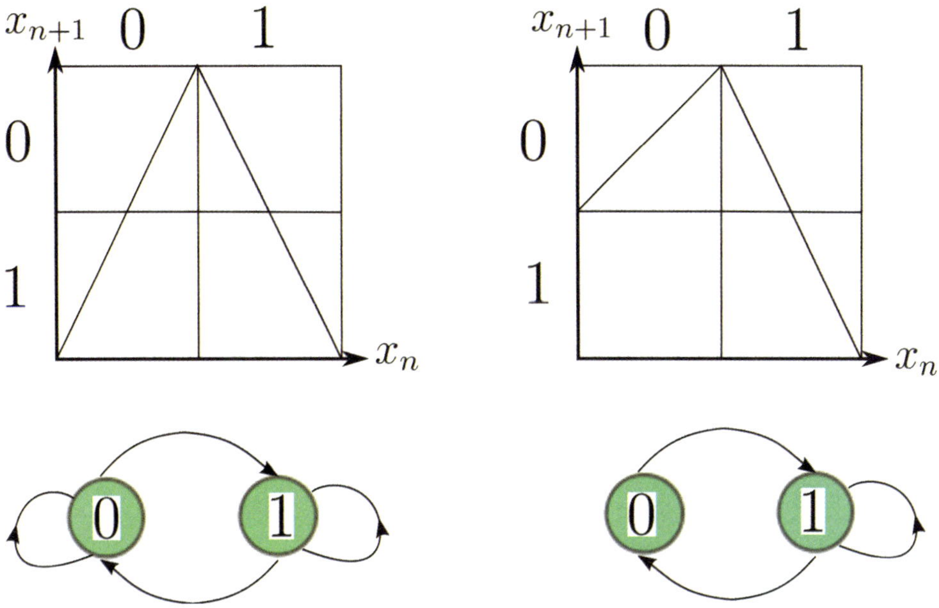

Figure 9.5. *Transitions with no surprise carry no information. (Left) This example full tent map on the Markov partition shown has orbits that at all possible times n, $x_n \in [0,1]$, there exist nearby initial conditions whose evolution will quickly evolve to a state with an opposite symbol 0 or 1 as the case may be; this unrestricted symbolic grammar is generated by the simple (dyadic) two-state directed graph shown below. Each state allows transition to a 0 or 1 outcome, so the surprise of observing the outcome is the information born by random walks through the graph corresponding to iterating the map. (Right) This piecewise linear map drawn on its Markov transition has a symbolic dynamics generated by the graph shown below. This grammar does not allow a 0 symbol to follow a 0 symbol. Thus when at the 0 labeled state ($x < c$), only a 1 can possibly follow; this transition bears no surprise. Said equivalently, it is not information bearing. Thus the required 1 transition serves as nothing other than a delay or pause in the transmission of a message to an information-bearing state—the grammar allows no two 0s in a row. From 1 either a 0 or a 1 may follow. Compare to Fig. 9.3, where such a grammar is used to encode oscillations in a Lorenz attractor to transmit a real message.*

rates R less than C are achievable. This imposes a rate of those non-information-bearing bits we depicted as *red* in Fig. 9.3. See Fig. 9.6, where we illustrate the reinterpretation of the standard box diagram shown in Fig. 9.2 in the setting where the channel is taken to be the dynamical system. In such a case, we reinterpret Eq. (9.52) as

$$C = \max_{p(x)} I(X_n; X_{n+1}). \tag{9.52}$$

The entropy which is most relevant uses the so-called maximum entropy measure (Theorem 9.8) corresponding to the topological entropy $h_{top}(\Sigma')$, which in Eq. (6.74) we

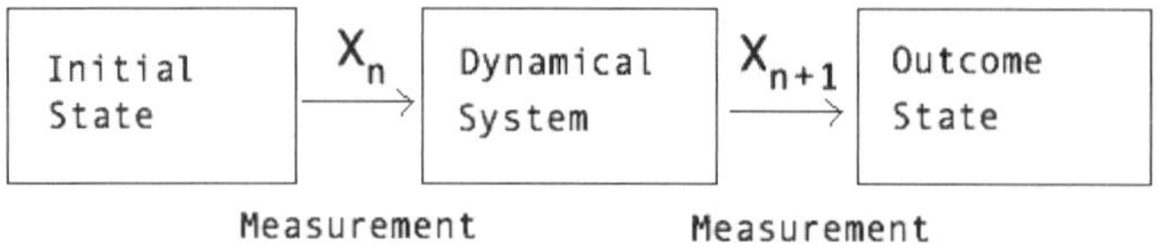

Figure 9.6. *Channel capacity* (9.52) *box diagram interpreted as a dynamical system. Compare to Fig.* 9.2 *and Theorem* 9.7.

see is descriptive of the number of allowable words N_n of a given length n. Perhaps most revealing is the spectral description (6.67) stating that $h_{top}(\Sigma'_k) = \ln \rho(A)$. The trade-off between channel capacity, transmission rates, and noise resistance was revealed in [29]. The corresponding devil's staircase graph is shown in Fig. 6.30 by a computation illustrated by Fig. 6.29 by the spectral method.

Before closing with this example, note that feedback control can be used to stabilize orbits with arbitrary symbolic dynamics [160, 26, 29, 104] starting from control of chaos methods [245, 289]. Specifically, electronic circuitry can be and has been built wherein small control actuations may be used to cause the chaotic oscillator to transmit information by small energy perturbations with simple circuitry to otherwise high powered devices. This has been the research engineering emphasis of [66, 67] toward useful radar devices.

Stated without the emphasis on practical devices, since it can be argued that there is information concerning states in all dynamical systems and a chaotic oscillator could be characterized as a system with positive entropy, then the evolution of the system through these states corresponds to an information-generating system. These systems have been called "information baths" [151].

What does this example tell us in summary?

- Realizing chaotic oscillators as information sources that *forget current measurement states and allow information bits to be inserted at a characteristic rate* as the system evolves (or is forced to evolve by feedback control) into new states summarizes the interpretation of a dynamical system as an information channel; see Fig. 9.6.

- The uncertainty in dynamical systems is in the choice of the initial condition even if the evolution rule may be deterministic.[132] The uncertainty in the symbolic outcome is described as the random variable defining the probability of states, corresponding to symbols. That is realized correspondingly in a deterministic dynamical system as the *unknown precision of state which is amplified upon iteration of an unstable system.* Even though the evolution of the states in the phase space of the dynamical system is deterministic, the exact position in phase space is practically never known exactly. We will make this idea mathematically precise in Section 9.5.

- Small feedback control can be used to steer orbits so that those orbits may bear a symbolic dynamics corresponding to desired information.

[132]This is the difference between a dynamical system (deterministic) and a random dynamical system (nondeterministic). See, for example, the stochastic process in Eq. (3.38) which nonetheless has a deterministic evolution of densities rule by the random dynamical system's Frobenius–Perron operator, (3.43).

9.5 Formally Interpreting a Deterministic Dynamical System in the Language of Information Theory

In the previous section, we illustrated by example that through symbolic dynamics, it is quite natural to think of a dynamical system with such a representation in terms of information theory. Here we will make this analogy more formal, thus describing the connection, which is some foundational aspects of ergodic theory. In this section we will show how to understand a deterministic dynamical system as an information-bearing stochastic process. Here we will describe a fundamental information theoretic quantity called Kolmogorov–Sinai entropy (KS entropy), $h_{KS}(T)$, which gives a concept of information content of orbits in measurable dynamics. By contrast, in the next section we will discuss in greater detail the topological entropy, $h_{top}(T)$.

Assume a dynamical system,

$$T : M \to M, \tag{9.53}$$

on a manifold M, with an invariant measure μ. For the sake of simplicity of presentation, we will assume a symbol space of just two symbols,[133]

$$\Omega = \{0, 1\}. \tag{9.54}$$

As discussed earlier,

$$\begin{aligned} s &: M \to \Omega \\ s(x) &= \chi_{A_0}(x) + \chi_{A_1}(x), \end{aligned} \tag{9.55}$$

but with an arbitrary open topological cover,

$$\overline{A_0 \cup A_1} = M, \text{ but } A_0 \cap A_1 = \emptyset, \tag{9.56}$$

and $\chi_A : M \to [0, 1]$ is the indicator function on sets $A \subset M$ as usual. We further assume probability measure using μ and a corresponding random variable (Definition 3.3),

$$X : \Omega \to \mathbb{R}, \tag{9.57}$$

for any randomly chosen initial condition $x \in M$ and therefore random symbol $s(x)$. Thus with μ, let

$$p_0 = P(X = 0) = \mu(A_0), \qquad p_1 = P(X = 1) = \mu(A_1). \tag{9.58}$$

In this notation, a dynamical system describes a discrete time stochastic process (Definition 4.15) by the sequence of random variables as follows

$$X_k(\omega) = X(s(T^k(x))). \tag{9.59}$$

Now using natural invariant measure μ when it exists, we may write

$$P(X_k = \sigma) = \mu(A_\sigma), \tag{9.60}$$

[133] The symbol partition need not be generating, in which case the resulting symbol dynamics will perhaps be a positive entropy process, but not necessarily fully descriptive of the maximal entropy of the dynamical system, as discussed in Section 6.4.6 and [40].

where

$$\sigma = 0 \text{ or } 1. \tag{9.61}$$

A stochastic process has entropy rate which we describe, a probability space $(P, \mathcal{A}, \Omega)$, and associated stochastic process defined by a sequence of random variables $\{X_1(\omega), X_2(\omega), \dots\}$. The stochastic process is defined to be stationary in terms of the joint probabilities as follows, from [51], and specialized for a discrete outcome space.

Definition 9.12. A stochastic process, $X_1, X_2, \dots$, is stationary if for all $k > 0$ the process $X_{k+1}, X_{k+2}, \dots$ has the same distribution as $X_1, X_2, \dots$. In other words, for every $B \in \mathcal{B}_\infty$,

$$\begin{aligned} &P(X_1 = x_1, X_2 = x_2, \dots) \\ &= P(X_{k+1} = x_1, X_{k+2} = x_2, \dots)\ \forall k > 1, \end{aligned} \tag{9.62}$$

and for each possible experimental outcome $(x_1, x_2, \dots)$ of the random variables.

It follows [51] that the stochastic process $X_1, X_2, \dots$ is stationary if the stochastic process $X_2, X_3, \dots$ has the same distribution as $X_1, X_2, \dots$.

Now the entropy rate of such a stochastic process by Definition 9.11 is

$$H = \lim_{n\to\infty} \frac{1}{n} H^{(n)}(\omega) \tag{9.63}$$

in terms of joint entropies

$$H^{(n)}(\omega) = \sum P(X_1 = x_1, X_2 = x_2, \dots, X_n = x_n) \log P(X_1 = x_1, X_2 = x_2, \dots, X_n = x_n). \tag{9.64}$$

It is straightforward to prove [68] that a sufficient condition for this limit to exist is i.i.d. random variables, in which case,

$$H = \lim_{n\to\infty} \frac{1}{n} H^{(n)}(\omega) = \lim_{n\to\infty} \frac{1}{n} n H(\omega_1) = H(\omega_1). \tag{9.65}$$

The Shannon–McMillan–Breiman theorem [68] states more generally that for a finite-valued *stationary* stochastic process $\{X_n\}$, this limit exists and converges to the entropy rate H.

If the stochastic system is really a dynamical system as described above, one in which a natural invariant measure μ describes the behavior of typical trajectories, then we attain a direct correspondence of the information theoretic description to the dynamical system in terms of its symbolization. We may develop the so-called **metric entropy**, also known as **Kolmogorov–Sinai entropy (KS entropy)**, h_{KS} [185]. Assuming a more general topological partition,

$$\mathcal{P} = \{A_i\}_{i=0}^{k}, \tag{9.66}$$

of $k+1$ components, then the resulting entropy of the stochastic process is

$$H(\mathcal{P}) = -\sum_{i=0}^{k} \mu(A_i) \ln \mu(A_i). \tag{9.67}$$

However, we wish to build the entropy of the stochastic process of the set theoretic **join**[134] of successive refinements that occur by progressively evolving the dynamical system. Let

$$h(\mu, T, \mathcal{P}) = \lim_{n\to\infty} \frac{1}{n} \left(\bigvee_{i=0}^{n} \mathcal{P}^{(n)} \right), \tag{9.68}$$

where we define

$$\mathcal{P}^{(n)} = \bigvee_{i=0}^{n} T^{-n}(\mathcal{P}), \tag{9.69}$$

and T^{-1} denotes the possibly many-branched preimage if T is not invertible. Thus, the join $\bigvee_{i=0}^{n} T^{-n}(\mathcal{P})$ is the set of all set intersections of the form

$$A_{i_1} \cap T^{-1}(A_{i_2}) \cap \cdots \cap T^{-n}(A_{i_{n+1}}), \quad 0 \le i_k \le n. \tag{9.70}$$

Now we should interpret the meaning of these quantities. H for a stochastic process is the limit of the Shannon entropy of the joint distributions. Literally, it is an average time density of the average information in a stochastic process. A related concept of entropy rate is an *average conditional entropy rate,*

$$aH'(X) = \lim_{n\to\infty} H(X_n | X_{n-1}, \dots, X_1). \tag{9.71}$$

Whereas $H(X)$ is an entropy per symbol, $H'(X)$ can be interpreted as the average entropy of seeing the next symbol conditioned on all the previous symbols. There is an important connection which occurs for a stationary process. Under the hypothesis of a stationary stochastic process, there is a theorem [68] that states

$$H(X) = H'(X), \tag{9.72}$$

which further confirms the existence of the limit, Eq. (9.71).

Thus by this connection between entropy rates in dynamical systems we can interpret $h(\mu, T, \mathcal{P}^{(i)})$ in Eq. (9.68) as the information gained per iterate averaged over the limit of long time intervals; call this $h_\mu(T)$ for short. The details of the entropy depend on the **chosen partition** $\mathcal{P}$. As was discussed in Section 6.4.6 and highlighted in Fig. 6.33, the value of entropy measured in a dynamical system depends on the chosen partition. This is most obvious in the extreme case that $\mathcal{P} = P_0$ is defined to be a partition of a single element covering the whole space, in which case all possible symbol sequences of all possible orbits consist of one symbol stream, 0.000.... This would give zero entropy due to zero surprise, and likewise because $p_0 = 1 \implies \log(1) = 0$. It is natural then to ask if there is a fundamental entropy of the dynamical system, rather than having entropy closely associated with the choice of the partition. From this question follows the quantity

$$h_{KS}(T) \equiv h_\mu(T) = \sup_{\mathcal{P}} h(\mu, T, \mathcal{P}^{(i)}). \tag{9.73}$$

[134]The **join** between two partitions $\mathcal{P}_1 = \{P_1^1, P_1^2, \dots, P_1^m\}$ and $\mathcal{P}_2 = \{P_2^1, P_2^2, \dots, P_2^n\}$ is defined $\mathcal{P} = \mathcal{P}_1 \vee \mathcal{P}_2 = (\{P_1^i \cap P_2^j\}_{i=0}^m)_{j=0}^n$, and in general $\mathcal{P} = \bigvee_{i=1}^N \mathcal{P}_i = \mathcal{P}_1 \vee \mathcal{P}_2 \vee \cdots \vee \mathcal{P}_N$ is interpreted successively by repeated application.

This **KS entropy** is the supremum of entropies over all possible partitions. It describes the average bit rate of seeing symbols in terms of all possible partitions and weighted according to natural invariant measure μ. The interpretation of $h_{KS}(T)$ is the description of precision as a degree of surprise of a next prediction with respect to increasing n. When this quantity is positive, in some sense this relates to sensitive dependence on initial conditions.

There is another useful concept of entropy often used in dynamical systems, called **topological entropy** h_{top} [1], which we have already mentioned in Section 6.4.1. We may interpret h_{top} as directly connected to basic information theory and also to h_{KS}. One interpretation of h_{top} is in terms of maximizing entropy by rebalancing the measures so as to make the resulting probabilities extremal. Choosing a simple example of just two states for ease of presentation, we can support this extremal statement by the simple observation that (stating Shannon entropy for two states)

$$(1/2, 1/2) = \underset{p}{\operatorname{argmax}}[-p\log p - (1-p)\log(1-p)]. \tag{9.74}$$

This is easily generalized for finer finite partitions. Expanding the same idea to Eq. (9.73) allows us to better understand Parry's maximal entropy measure when it exists. Stated in its simplest terms for a Markov chain, Parry's measure is a rebalancing of probabilities of transition between states so that the resulting entropy of the invariant measure becomes maximal. See [248] and also [53, 142] and [149] for some discussion of how such maximal entropy measures need not generally exist but do exist at least for irreducible subshifts of finite type. Generally the connection between $h_{top}(T)$ and $h_\mu(T)$ is formal through the following variational theorem.

Theorem 9.8 (variational principle for entropy—connection between measure theoretic entropy and topological entropy [142, 99, 48]). Given a continuous map $f : M \to M$ on a compact metric space M,

$$h_{top}(f) = \sup_{\mu} h_\mu(f), \tag{9.75}$$

where the supremum is taken over those measures μ which are f-invariant Borel probability measures on M.

On the other hand, the direct Definitions 9.13 and 9.14 of topological entropy [248] are in terms of counting numbers of ϵ-separated sets, and how quickly these states of finite precision become separated by iterating the dynamical system; see also [48, 268]. We find the variational principle to be more descriptive than the original definition in terms of understanding the meaning of topological entropy. Further discussion of $h_{top}(T)$ connections with computational methods are made in Section 9.6.

9.6 Computational Estimates of Topological Entropy and Symbolic Dynamics

In Section 9.4, the connection between orbits of a dynamical system and a dynamical system as an entropy process is discussed by example with demonstration of the information present in distinguishing orbits. Further, the connection between measurable dynamics and topological dynamics can be understood in terms of the variational principle for the entropy theorem, Theorem 9.8. In the discussion of symbolic dynamics in Chapter 6, especially in Section 6.4.1, we discussed symbolic dynamics in depth, including formulas describing entropy in terms of cardinality, Eq. (6.74), and also a related spectral formula, Eq. (6.67). In this section, we will reprise this discussion of topological entropy associated with a dynamical system in more detail and in the context of the formal information theory of this chapter.

9.6.1 Review of Topological Entropy Theory

Adler, Konheim, and McAndrew introduced topological entropy in 1965 [1] in terms of counting the growth rate of covers of open sets under the dynamical system. However, the Bowen definition [44, 46] is in terms of ϵ-separated sets.

Definition 9.13 ((n,ϵ)-separated [268]). Given a metric space (M,d) and a dynamical system on this space, $f : M \to M$, a subset $S \subset M$ is (n,ϵ)-separated if

$$d_{n,f}(\mathbf{x},\mathbf{y}) > \epsilon \tag{9.76}$$

for each distinct $\mathbf{x},\mathbf{y} \in S$, $\mathbf{x} \neq \mathbf{y}$, where

$$d_{n,f}(\mathbf{x},\mathbf{y}) = \sup_{0 \le j < n} d(f^j(\mathbf{x}), f^j(\mathbf{y})). \tag{9.77}$$

In terms of the metric topology[135] this can be roughly described as enumerating the coarsening associated with how open sets are maximally cast across open sets. By "cast across" we simply mean that iterates of the set have images intersecting other open sets. Topological entropy is defined in terms of counting the growth rate of the iteration time n of different orbits, where difference is in terms of an ϵ-scale, as time grows, and then in the limit as this ϵ-coarse grain decreases.

Definition 9.14 (topological entropy [44, 268]).

$$h_{top}(f) = \lim_{\epsilon \to 0} \limsup_{n \to \infty} \frac{\log(s(n,\epsilon,f))}{n}, \tag{9.78}$$

where

$$\log(s(n,\epsilon,f)) = \max\{\#(S) : S \subset M \text{ is } (n,\epsilon)\text{- separated by } f\}. \tag{9.79}$$

In this sense, topological entropy is a counting of the rate of new states which develop under the dynamics in terms of states on each coarse scale in the scale limit.

This definition cannot be considered as practical as a computable quantity. Other methods must be deployed. When the dynamical system is a right resolvent sofic shift (see Definition 6.5) we have the simple spectral formula (6.67) to simply and exactly compute topological entropy, which we restate [46, 268]

$$h_{top}(\Sigma'_N) = \ln \rho(A). \tag{9.80}$$

$\rho(A)$ is the spectral radius of the associated transition matrix A, meaning the largest eigenvalue of the matrix of 0's and 1's corresponding to allowed transitions on the generated

[135] A **metric topology** is a topology whose basis of open sets is simply open balls $N_\epsilon(\mathbf{x})$ defined in terms of a metric d on M. $N_\epsilon(\mathbf{x}) = \{\mathbf{y} : d(\mathbf{x},\mathbf{y}) < \epsilon\}$.

directed graph G_A denoting allowed words in the subshift Σ'_N. Recall that following Definition 6.4, there is a finite-sized transition matrix A that generates a graph G_A that presents the grammar of the subshift Σ'_N of a fullshift Σ_N on N symbols with a corresponding Bernoulli mapping $s : \Sigma'_N \to \Sigma'_N$. This formula is practical in general dynamical systems in the scenario of a Markov partition, Definitions 4.1 and 4.2, from which often a finite graph may be associated with transitions of the corresponding dynamical system meant to properly represent transitions of the associated shift map, and using Theorems 9.10 and 9.11 for the equivalence. We have used this computational method for the logistic map in [29, 41] and the Henon map in [196, 41], to name a few. We wish to highlight two perspectives on this approximation by a Markov model.

- A nested sequence of (imbedded chaotic saddle) subsets allows Markov models of the attractor as a whole, as in Example 9.1.
- In terms of the uniform norm, near a given dynamical system, a dynamical system may exist which is exactly Markov, and finite-dimensional computations are exact. Therefore, one way to discuss entropy approximations of a given dynamical system is in terms of sequences of such Markov estimates and understanding the density of the representation.[136]

Another useful description of the topological entropy is the expression, with hypothesis in [47],

$$h_{top}(f) = \limsup_{n\to\infty} \frac{\log(P_n)}{n}, \tag{9.81}$$

where P_n *denotes the number of fixed points of* f^n.[137] Remembering that the popular Devaney definition of chaos (Definition 6.1) includes the requirement that periodic orbits are dense [12, 95], there is some sense in measuring the complexity of the dynamical system in terms of these orbits. A practical perspective is an observation that unstable periodic orbits (**UPO**s) form the "skeleton" of nonlinear dynamics [7, 70]. To this end, [7, 5] have progressed in estimates of thermodynamic properties by using just the short orbits. Numerous theoretical systems were shown to be well described through this approach [5]. Furthermore, recent work using interval arithmetic [132, 131, 130] and computer-assisted proof [134] has allowed for rigorous entropy estimates with this perspective. See also [78] for a discussion of computer-validated proof of symbolic dynamics, and [224, 225, 226], which are based on methods of computational homology [181]. At the forefront and quite impressive, these methods can even be used in infinite-dimensional PDE systems validating the finite-dimensional Galerkin representation as ODEs and the persistence of their periodic orbits despite the finite-dimensional approximation [327, 79, 80].

The use of formula (9.81) is generally to empirically check for convergence for a general mapping, and we emphasize that this formula does not require symbolization as

[136] For example, in the special case of the set of all skew tent maps, a theorem can be found in [19] that states that a given map either is Markov or may be uniformly (sup-norm) estimated to arbitrary accuracy by a map that is Markov; in this sense the family of Markov maps is dense in the uniform topology. See further discussion in Section 4.4 and especially Theorem 4.2.

[137] It is important to count correctly: Given a map $x' = f(x)$, a periodic orbit $\{x_0 x_1, \ldots, x_{n-1}\}$ is n-points which are roots of $f^n(x) = x$. However, not all roots of $f^n(x) = x$ are period n since the period of a point is defined as the *smallest* m such that $f^m(x) = x$. For example, a fixed point ($n = 1$) solves $f(x) = x$, and it also solves $f^2(x) = x$ (and $f^n(x) = x$ for all $n > 0$), but it is called period-1 since 1 is the smallest such n.

shown in Example 9.1 below. However, under the assumption that the mapping is a shift mapping, this formula is similar to a mathematically well-founded statement,

$$h_{top}(\Sigma'_N) = \limsup_{n\to\infty} \frac{\log w_n}{n}, \tag{9.82}$$

where w_n *is the number of words of length n* in Σ'_N. The principled use of both formulas (9.80) and (9.82) are confirmed by the following theorem.

Theorem 9.9 (see [268]). A subshift with the Bernoulli shift map dynamical system, $s : \Sigma'_N \to \Sigma'_N$, has topological entropy that may be computed generally by Eq. (9.80) or Eq. (9.82) when sofic and right resolvent.

Remark 9.2. Understanding the difference between the periodic orbits estimate (9.81) and the word count formula (9.82) may be described in terms of symbolizing orbits with a generating partition. When the periodic orbit estimate (9.81) is used specifically for a system that is already symbolized, we interpret this as counting symbolized periodic orbits. Let

$$u_n = \sigma_0.\sigma_1\sigma_2\ldots\sigma_{n-1}, \quad \sigma_i \in \{0,1,\ldots,N-1\}, \tag{9.83}$$

be a word segment of length n of the N symbols associated with Σ_N. Then by definition w_n *is the number of such blocks of n bits* that appear in points σ in the subshift, $\sigma \in \Sigma'_N$. These word segments may be part of periodic orbit, in which case that word segment is repeated,

$$\sigma = u_n u_n \ldots \equiv \sigma_0.\sigma_1\sigma_2\ldots\sigma_{n-1}\sigma_0.\sigma_1\sigma_2\ldots\sigma_{n-1}\ldots, \tag{9.84}$$

or it would not be repeated if the point is not part of a periodic orbit. So in the symbolized case, the difference between the two formulae is that generally we expect

$$P_n \leq w_n, \tag{9.85}$$

but the hope is that for large n, the difference becomes small.

Remark 9.3. The key to the practical and general use of the periodic orbit formula (9.81) is that symbolization is not necessary. This is a useful transformation of the problem of estimating entropy, since finding a generating partition is generally a difficult problem in its own right for a general dynamical system [71, 144, 75, 41]. Details of the misrepresentation of using a partition that is not generating are discussed in Section 6.4.6 as a review of [41]. Whether or not we symbolize, the number of periodic orbits has changed, which is why the periodic orbit formula is robust in this sense, although it has the alternative difficulty of being confident that *all* of the periodic orbits up to some large period have been found.

Remark 9.4. The key to algorithmically using the periodic orbit formula (9.81) is the possibility of reliably finding all of the periodic orbits of period n for a rather large n. Given that P_n is expected to grow exponentially, it is a daunting problem to solve

$$g(z) = f^n(z) - z = 0 \tag{9.86}$$

for many roots. By saying many roots, we are not exaggerating, since in our experience [75] to reliably estimate the limit ratio in Eq. (9.81), we have worked with on the order of hundreds of thousands of periodic orbits which are apparently complete lists for the relatively large $n \simeq 18$. See Example 9.1 using the Ikeda map. When $n = 1$, the obvious approach would be to use a Newton method or variant. And this works even for $n = 2$ or 3 perhaps, but as n increases and P_n grows exponentially, the seeding of initial conditions for the root finder becomes exponentially more difficult as the space between the basins of attraction of each root becomes small. A great deal of progress has been made toward surprisingly robust methods in this computationally challenging problem of pushing the root finding to produce the many roots corresponding to seemingly complete lists of orbits [279, 18, 98, 76].

Example 9.1 (entropy by periodic orbits of the Ikeda map). Consider the Ikeda map [175, 157]

$$\begin{aligned} I_{keda} &: R^2 \to R^2 \\ (x,y) \mapsto (x',y') &= (a + b[x\cos(\phi) - y\sin(\phi)], b[x\sin(\phi) + y\cos(\phi)]), \end{aligned} \tag{9.87}$$

where

$$\phi = k - \nu/(1 + x^2 + y^2), \tag{9.88}$$

and we choose parameters $a = 1.0, b = 0.9, k = 0.4$, and $\nu = 6.0$. In [74], the authors conjectured a claim construction of all the periodic orbits through period-22, whereas in [75] we used seemingly all of the roughly 373,000 periodic orbits through period-20 to estimate

$$h_{top}(I_{keda}) \simeq 0.602 < \ln 2 \tag{9.89}$$

by Eq. (9.81). In Fig. 9.7 we show those periodic orbits through period-18. Furthermore in [75] we noted the requirement that a generating partition must uniquely differentiate, by the labeling, all of the iterates on all of the periodic orbits; we used this statement to develop a simple construction to successively symbolize (color) each of the periodic orbits in successively longer periods. As these numbers of periodic orbits swell to tens and hundreds of thousands, the attractor begins to fill out, as seen in Fig. 9.7, to become a useful representation of the symbolic dynamics. Notice an interesting white shadow reminiscent of the stable manifolds that are tangent to the unstable manifolds believed to be associated with generating partitions [71, 144].[138] A thorough study of periodic orbits together with a rigorous computer-assisted proof by interval arithmetic [232] is developed in [131] from which we reprise the table of counted orbits, Table 9.1, including comparable estimates of topological entropy commensurate with our own best $h_{top} \simeq 0.602$.

Example 9.2 (entropy of the Henon map). Comparably, for Henon mapping, $h(x,y) = (1 + y - ax^2, bx)$, with $(a,b) = (1.4, 0.3)$ in [131], and finding even more periodic orbits,

$$(n, Q_n, P_n, Q_{\leq n}, P_{\leq n}, h_n) = (30, 37936, 1139275, 109033, 3065317), \tag{9.90}$$

[138] Interestingly, notice that the periodic orbits are expected to distribute roughly according to the invariant measure and thus are more rare at regions of low measure; apparently these "white regions" correspond to just those regions associated by tangencies of stable and unstable manifolds. To see this observation, note that the Sinai–Ruelle–Bowen (SRB) measure is the invariant measure along the closure of all the unstable manifolds of the periodic orbits. A clear shadow of "white" missing periodic orbits (up to the period-18's found) can be seen as transverse curves through the attractor, punctuated at tangency points. This conjecture-observation agrees in principle with the well-accepted conjecture [71, 144] that generating partitions must connect between homoclinic tangencies. See Fig. 6.31 for an explicit construction demonstrating the generating partition for the Henon map constructed directly by this conjecture.

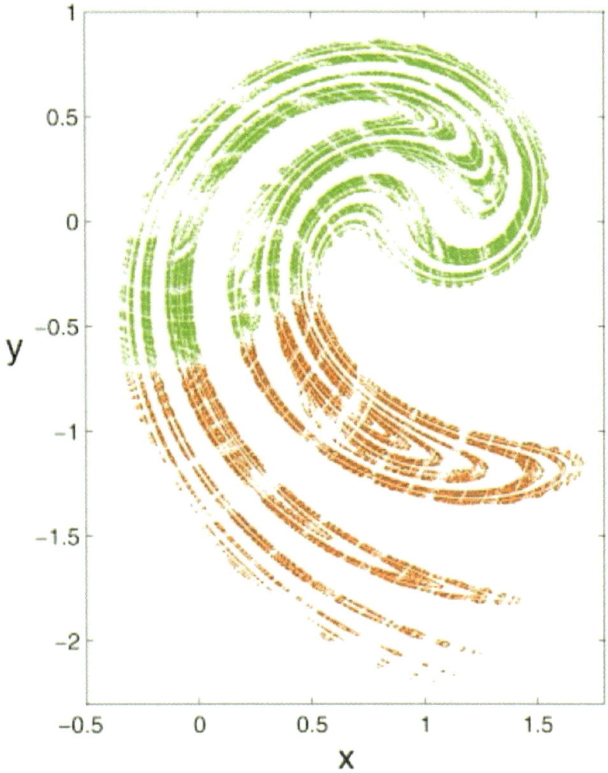

Figure 9.7. *Periodic orbit points up to period-*18 *for the Ikeda–Hammel–Jones–Moloney attractor* (9.87) *from* [75]. *Furthermore, the points on the periodic orbits are colored according to their symbolic representation: green and red dots represent orbit points encoded with symbols* **0** *and* **1**, *respectively. Compare to Table* 9.1 *and Fig.* 9.8. [279]

where rigorous interval arithmetic was used [232]. Compare this to the tabular values for the Ikeda mapping in Table 9.1. Interestingly, this is a symbolic dynamics free computation, but using instead periodic orbits; see Fig. 6.31, where a generating partition for this Henon map is constructed directly by locating tangency points.

Example 9.3 (generating partition and Markov partitions of the Ikeda map). Consider the Ikeda map, Eq. (9.87). In Fig. 9.7, we show a partition from [75] consistent with a generating partition, this having been constructed by requiring uniqueness of representation for each of the hundreds of thousands of periodic orbits through period-18, and the table of periodic orbits, Table 9.1.

In Fig. 9.8, we show two candidate Markov partitions, each using several symbols from [132]. On the left wc see a Markov partition in 4 symbols, and on the right we see a Markov partition in 7 symbols. In [132], a further refined Markov partition in 18 symbols is shown. Generally, a strange attractor may have (infinitely) many embedded Markov partitions representing embedded subshift, where high ordered representations can hope to

Table 9.1. *Periodic orbit counts for the Ikeda–Hammel–Jones–Moloney attractor* (9.87) *from* [75]. *Q_n is the number of periodic orbits of period n. P_n is the number of fixed points of the mapping, f^n. $Q_{\leq n}$ is the number of cycles of period less than or equal to n. $P_{\leq n}$ is the number of fixed points of f^i for $i \leq n$. h_n is the estimate of the topological entropy for n, using Eq.* (9.81). *For comparison, in* [75], *we estimated* $h_{18} \simeq 0.602$. [131]

n	Q_n	P_n	$Q_{\leq n}$	$P_{\leq n}$	h_n
1	2	2	2	2	0.6931
2	1	4	3	4	0.6931
3	2	8	5	10	0.6931
4	3	16	8	22	0.6931
5	4	22	12	42	0.6182
6	7	52	19	84	0.6585
7	10	72	29	154	0.6110
8	14	128	43	266	0.6065
9	26	242	69	500	0.6099
10	46	484	115	960	0.6182
11	76	838	191	1796	0.6119
12	110	1384	301	3116	0.6027
13	194	2524	495	5638	0.6026
14	317	4512	812	10076	0.6010
15	566	8518	1378	18566	0.6033

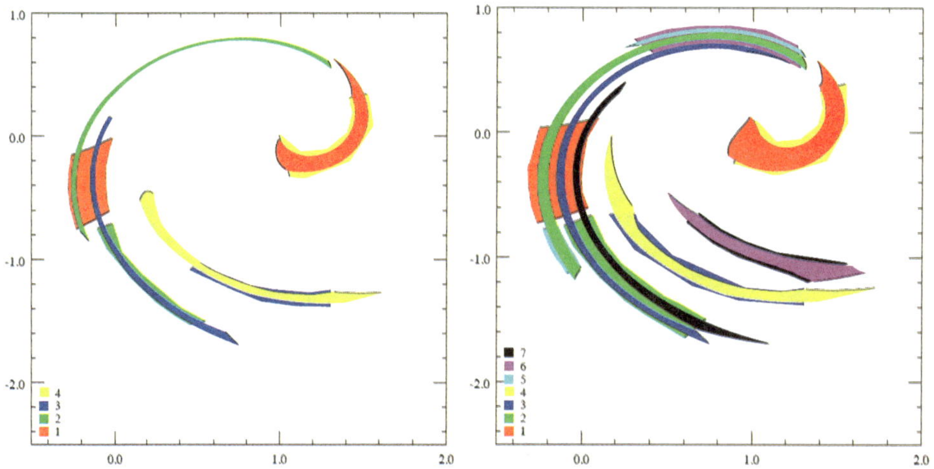

Figure 9.8. *Symbol dynamics of the Ikeda–Hammel–Jones–Moloney attractor* (9.87) *from* [132]. *(Left) Ikeda map, $\alpha = 6$ sets N_i on which the symbolic dynamics on* 4 *symbols exists and their images. (Right) Ikeda map, $\alpha = 6$ sets N_i on which the symbolic dynamics on* 7 *symbols exists and their images. Compare to Definition* 4.4 *and Fig.* 4.5.

represent the symbolic dynamics of a greater and greater subset of the full attractor, such as in [19]; compare to Definition 4.4 and Fig. 4.5. Further discussion of the entropy of this attractor is addressed in Example 9.1.

Example 9.4 (entropy by Markov model of the Ikeda map). In Example 9.3, Fig. 9.8, we recall from [132] two Markov model refinements corresponding to two imbedded subshifts Σ'_4 and Σ'_7 in the Ikeda attractor, using $\alpha = 6$ of Eq. (9.87). See also [131, 130] for techniques of computing enclosures of trajectories, finding and proving the existence of symbolic dynamics, and obtaining rigorous bounds for the topological entropy. From the Ikeda map, the imbedded Markov models yield associated transition matrices:

$$A_4 = \begin{pmatrix} 0 & 0 & 0 & 1 \\ 1 & 0 & 0 & 0 \\ 1 & 1 & 0 & 0 \\ 0 & 0 & 1 & 0 \end{pmatrix}, \qquad A_7 = \begin{pmatrix} 0 & 0 & 0 & 1 & 0 & 0 & 0 \\ 1 & 0 & 0 & 0 & 1 & 1 & 0 \\ 1 & 1 & 0 & 0 & 0 & 0 & 0 \\ 0 & 0 & 1 & 0 & 0 & 0 & 0 \\ 0 & 0 & 0 & 0 & 0 & 1 & 0 \\ 0 & 0 & 0 & 0 & 0 & 0 & 1 \\ 1 & 1 & 0 & 0 & 0 & 0 & 0 \end{pmatrix}. \tag{9.91}$$

Using these Markov representations of the transition matrices of the grammars of the imbedded subshifts, together with Eq. (9.80),

$$h_{top}(\Sigma'_4) = \ln \rho(A_4) = 0.19946 \quad \text{and} \quad h_{top}(\Sigma'_7) = \ln \rho(A_7) = 0.40181, \tag{9.92}$$

and similarly in [132], a further refined 18-symbol Markov model produces

$$h_{top}(\Sigma'_{18}) = \ln \rho(A_{18}) = 0.48585. \tag{9.93}$$

We do not show A_{18} here for the sake of space, and in any case, the principle is to continue to refine and therefore to increase the size of the matrices to build ever-increasing refinements. These estimates are each lower bounds of the full entropy, for example,

$$h_{top}(\mathcal{A}) > h_{top}(\Sigma'_{18}), \tag{9.94}$$

where $h_{top}(\mathcal{A})$ denotes the topological entropy on the chaotic Ikeda attractor $\mathcal{A}$ meaning the entropy of the dynamics, (9.87) on this attractor set. We have often used the phrase "embedded subshift" Σ'_N here, by which we mean that there is a subset $\mathcal{A}'$ of the attractor $\mathcal{A}$ such that the subshift Σ'_N is semiconjugate to the dynamics on that subset $\mathcal{A}'$; "imbedded" is the more accurate term.

Theorem 9.10 (comparing topological entropy of factors; see [188]).[139] Suppose there exist two irreducible subshifts of finite type such that Σ_B is a factor of Σ_A; then

$$h_{top}(\Sigma_B) \leq h_{top}(\Sigma_A). \tag{9.95}$$

[139] A subshift of finite type Σ_B is a **factor** (synonym of **semiconjugate**) of another subshift of finite type Σ_A if there exists a continuous and onto mapping $f : \Sigma_A \to \Sigma_B$ that commutes; $s \circ f(\sigma) = f \circ s(\sigma)$. A conjugacy is a special case where f is a homeomorphism.

From this theorem comes the following related statement.

Lemma 9.1 (topological entropy equivalence). Two conjugate dynamical systems have equivalent topological entropy.

Also related is a slightly weaker than conjugacy condition for equivalence of topological entropy.

Theorem 9.11 (topological entropy compared by finite-one semiconjugacy; see [269]). If $g_1 : X \to X$ and $g_2 : Y \to Y$ are two continuous mappings on compact metric spaces X and Y, and $f : X \to Y$ is a semiconjugacy that is uniformly *finite-one*, then

$$h_{top}(g_1) = h_{top}(g_2). \tag{9.96}$$

It is this third theorem that permits us to compute the topological entropy of a dynamical system in terms of its symbolic dynamics.[140] The topological entropy on the attractor is modeled $h_{top}(\mathcal{A}') = h_{top}(\Sigma'_N)$. The phrase "embedded subshift" is rightly often used. The correct word from topology is *imbedding*.[141]

The first theorem explains that in these examples, it is not a surprise that the successive approximations of this example,

$$\Sigma'_4 \hookrightarrow \Sigma'_7 \hookrightarrow \Sigma'_{18}, \tag{9.97}$$

lead to the estimates as found [132, 75],

$$h_{top}(\Sigma'_4) = 0.19946 \leq h_{top}(\Sigma'_7) = 0.40181 \leq h_{top}(\Sigma'_{18}) = 0.48585 \leq h_{top}(I_{keda}) \simeq 0.602. \tag{9.98}$$

The hooked arrow $\hookrightarrow$ denotes imbedding. While not all imbedded Markov models will be comparable between each other, nor will be their entropies. In the case of these examples the nesting explains the reduced entropies. Further discussion can be found of nested imbeddings of horseshoes in homoclinic tangles [229] and chaotic saddles [28].

9.6.2 A Transfer Operator Method of Computing Topological Entropy

Finally, in this section, we briefly mention the possibility of using a transition matrix version of the Ulam–Galerkin matrix to estimate topological entropy, as was studied in detail in [123] and similar to the computation in [29]. In brief, the idea is to use an outer covering of the attractor by "rectangles" or any other regular topological partition such as the successive refinements of Delaunay triangulations [199]. Such an outer covering of the Henon attractor is shown, for example, in Fig. 9.9. As we will discuss, using the transfer operator seems like an excellent and easy method to produce a transfer matrix–based approach toward entropy estimates for a fine grid. After all this follows the same theme as is done in an Ulam method to compute invariant measure or the spectral partitioning methods highlighted earlier in this writing. However, it turns out the approach is not so simple since care

[140] Further note from the theorem that the entropy of a continuous dynamical system, on a compact set, is equal to the entropy of the map restricted to its nonwandering points [269].

[141] A mapping $f : X \to Y$ is defined as an **imbedding** X into Y if a restriction of the range to Z results in a homeomorphism $f : X \to Z$ [235]. Furthermore, when used in the dynamical systems sense and two mappings, when this restriction is with respect to the two mappings $g_1 : X \to X$ and $g_2 : Y \to Y$, and the domain restriction of Z results in a conjugacy, we still use the term imbedding. Often the word embedding may be used in a context that more accurately denotes imbedding as we have defined it here.

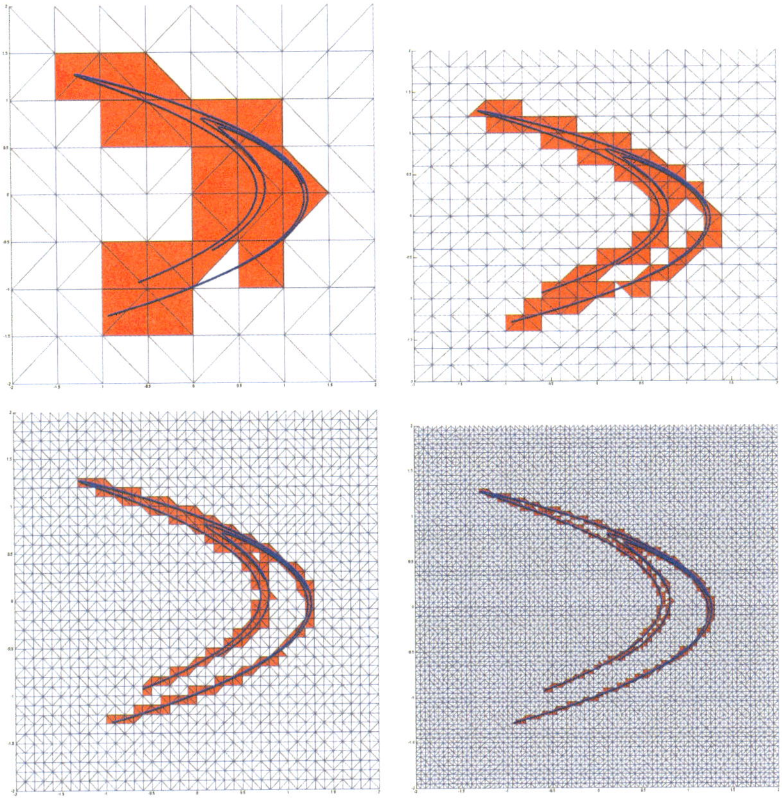

Figure 9.9. *An outer covering (red) of successive partitions by Delaunay triangulations over a Henon attractor. Successively shown are coverings with edge lengths* $h = 0.5, 0.2, 0.1, 0.05$ *resulting in transition matrices* $A_{N\times N}$, $N = 34, 105, 228, 565$, *where* N *is the count of the size of the partition covering the strongly connected component (red) of the full covering.*

must be taken regarding using the generating partition. Since having the generating partition a priori makes any grid method obsolete, we will suggest that transfer matrix–based methods are rarely useful. Nonetheless the theory has some interesting lessons which we will highlight [123, 29].

The idea behind a transfer matrix–based computational method of computing topological entropy hinges on the following refinement theorem relating upper bounds.

Theorem 9.12 (see [123]). Assuming a partition $\mathcal{P}$, let

$$h^*(T,\mathcal{P}) := \lim_{N\to\infty} \frac{\log|w_N(T,\mathcal{P})}{N} = \inf_{N\geq 1} \frac{\log|w_N(T,\mathcal{P})|}{N}; \tag{9.99}$$

then the topological entropy $h_{top}(T)$ of the map T is bounded: If $\mathcal{P}$ is a generating partition (see Definitions 4.5–4.7), then

$$h_{top}(T) \leq \lim_{\text{diam } \mathcal{P}\to 0} \inf h^*(T,\mathcal{P}) \leq h^*(T,\mathcal{P}), \tag{9.100}$$

This theorem provides an upper bound for the topological entropy and suggests a simple constructive algorithm, but one which requires care, as we point out here. We illustrate a direct use of this theorem in Fig. 9.9. In the example in the lower-right frame, $N = 565$ triangular cells cover the attractor. An outer covering of the attractor would result in $h^*(T,\mathcal{P}) \leq \log N$, which we can see is divergent as N is refined. In any case, this is an extreme and not sharp estimate for typical grid refinement.

Example 9.5 (transfer operator for topological entropy for Henon). Direct application of this theorem to the data in Fig. 9.9, building adjacency matrices on the fine Delaunay triangulations,[142]

$$A_{i,j} = \text{if } (1, B_i \to B_j, 0, \text{else}) = ceil(P_{i,j}), \tag{9.101}$$

results in

$$h^* = 1.0123, 1.0271, 1.0641, 1.0245 \tag{9.102}$$

for the specific grid refinements shown,

$$h = 0.5, 0.2, 0.1, 0.05, \tag{9.103}$$

yielding

$$N = 34, 105, 228, 565 \tag{9.104}$$

element coverings, respectively. We see that these are all poor upper bound estimates of a well-regarded $h_{top}(T) \simeq 0.4651$ from [72] derived by methods discussed previously in this section.

So what is wrong? An overestimate is expected, but why does this seemingly obvious and easy approximation method, even with a relatively fine grid, give such large overestimates? The answer lies in the fact that the symbolization is wrong, and not even close. That is, the partition is wrong. $\mathcal{P}$ has 565 elements in our finest cover shown, and we recall that $\log N$ is the upper bound of the entropy of any subshift in Σ_N. While $1.0245 \ll \log 565$, it should not be a surprise that the estimate 1.0245 is not close to 0.4651. Eigenvalues of the transfer matrix are exact if that finite partition happens to be Markov, or close by Eq. (9.100) if the partition of the Markov representation is close to generating.

We recall from footnote 138 and Fig. 6.31 that the generating partition for the Henon map connects primary homoclinic tangencies, which is a zig-zag line that turns out to run roughly near $y = 0$. Therefore, to correctly estimate $h_{top}(T)$, it is necessary to associate each cell B_i with a position relative to the generating partition. Clearly, if the generating partition is

$$\mathcal{P}' = (P'_1, P'_2, \ldots, P'_M), \quad M \ll N, \text{ and cell } B_i \subset P'_j, \tag{9.105}$$

then the new symbol j is associated with each such B_i. Similarly for cells B_i which are not entirely within a single partition element, then a decision must be made, perhaps to choose the largest region of overlap, $P'_j \cap B_i$. In this manner, a projection from the larger symbol space is developed,

$$\Pi : \Sigma_N \to \Sigma_M. \tag{9.106}$$

[142]The adjacency matrix A is easily derived from the stochastic matrix corresponding to the Ulam–Galerkin matrix using the $ceil$ function.

The corresponding projected transition matrix should produce the correct topological entropy, but the arbitrarily presented graph cannot be expected to be right-resolvent (see Definition 6.5). The following theorem guarantees that a right-resolvent presentation exists even if the arbitrary projection may not be right-resolvent.

Theorem 9.13 (Lind and Markus [204]). Every sofic shift has a right-resolving presentation.

The proof of this theorem is constructive by the so-called follower method [204], as used in [41] in a context similar to the discussion here, which is to associate new partitions with transition matrices associated with arbitrary partitions. By this method, a new transfer matrix is a right-resolvent presentation of the grammar is developed. Therefore, its corresponding spectral radius is correctly the topological entropy.

However, proceeding in a manner as described above in two steps (i.e., developing a transition matrix associated with a fine partition and then developing a projection to a right-resolvent presentation by the follower construction) may not be considered to be a useful method, since one still must already know the generating partition to properly associate labels to the fine representation of the transfer matrix. As we have already stated, finding the generating partition is a difficult problem. Further, if we already have the generating partition, then simply counting words associated with long orbit segments is a useful and fast-converging method to estimate entropy without needing to resort to the transition matrix, which skips the computational complexity associated with grid-based methods. Furthermore, an exceedingly fine grid would be needed to properly represent the nuance of the w-shaped generating partition seen in Fig. 6.31. For the sake of simplicity, in [123] the authors chose to associate a nongenerating partition as follows: let

$$\mathcal{P}' = (P_1, P2), \text{ where } P_1 = \{(x,y)|y < 0\}, \text{ and } P_2 = \{(x,y)|y > 0\}. \tag{9.107}$$

Clearly this partition is relatively close to the generating partition. As such, the right-resolvent presentation of the transfer matrix gives an estimate of 0.4628, which we see is less than that from [72],

$$0.4628 < h_{top}(T) \simeq 0.4651. \tag{9.108}$$

That the estimate is close is a reflection of a continuity of the entropy with respect to degree of misplacement discussed in [41, 40]. Furthermore, the theory detailed in [41, 40] shows that using an arbitrary partition risks erratic large errors as emphasized by Figs. 6.32 and 6.33 and Eqs. (6.72)–(6.73) from [41, 40], even if the result is a very interesting devil's staircase–like function describing the consequences of using a nongenerating partition. It could be argued that it is just good luck that $y = 0$, so easily guessed, is in fact close to the generating partition seen in Fig. 6.31 gives a reasonable answer not to be relied upon in general. In any case, the form of the error is not known to be positive or negative, despite the upper bounding statement, Eq. (9.100).

In summary, when estimating topological entropy, the use of transfer operator methods still requires knowledge of the generating partition. Errors will likely be large, as analyzed in [40], if we do not use generating partition information, despite refining grids. However, if we do have the generating partition, then it is perhaps much simpler and more accurate to resort directly to counting words (9.82).

9.7 Lyapunov Exponents, and Metric Entropy and the Ulam Method Connection

In this section we will tighten the connection between metric entropy and how it can be computed in terms of Ulam–Galerkin matrix approximations by considering the Markov action on the corresponding directed graph. This continues in our general theme of connecting concepts from measurable dynamics to computational methods based on transfer operators. Further, we will discuss how this type of computation is exact in the case where a Markov partition is used. Thus, again referring to Section 4.4 and especially Theorem 4.2 concerning density of Markov representations, we can understand a way to analyze the quality of the estimate. Finally, we will review the Pesin identity, which provides a beautiful and deep connection between metric entropy h_{KS} and Lyapunov exponents. We will discuss both estimation and interpretation of these exponents and their information theoretic implications. The main point here is that averaging on single trajectories versus ensemble averages is again the Birkhoff ergodic theorem, Eq. (1.5), which here gives a doubly useful way to compute and understand the same quantities. Compare this section to the introduction with descriptions of two ways of computing Lyapunov exponents discussed in Example 1.4.

9.7.1 Piecewise Linear in an Interval

We start this discussion by specializing to piecewise linear transformations of the interval, specifically to Markov systems that are chaotic; such systems allow the probability density functions to be computed exactly. It is well known that expanded piecewise linear Markov transformations have piecewise constant invariant PDFs, already referred to in Section 4.4.4.

Theorem 9.14 (piecewise constant invariant density; see [50]). Let $\tau : I \to I$ be a piecewise linear Markov transformation of an interval $I = [a,b]$ such that for some $k \geq 1$

$$|(\tau^k)'| > 1,$$

where the derivative exists, which is assumed to be in the interiors of each partition segment. Then τ admits an invariant (probability) density function which is piecewise constant on the partition $\mathcal{P}$ on which τ is Markov.

Using the Frobenius–Perron operator P, the fixed-point function ρ satisfies the definition $P_\tau \rho = \rho$, implying that ρ is the PDF for a measure that is invariant under τ. Since τ is assumed to be a piecewise monotone function, the action of the operator is simply

$$P_\tau \rho(x) = \sum_{z \in \{\tau^{-1}(x)\}} \frac{\rho(z)}{|\tau'(z)|}.$$

The periodic orbit formed by the iteration of $x = a$ forms a partition of the domain $[0,1]$ on which ρ is piecewise constant. On each interval I_i, call the corresponding constant

$$\rho_i = \rho|_{I_i}. \tag{9.109}$$

The probability density function admits an absolutely continuous invariant measure on the Markov partition, the details of which can be found in [50]. For our discussion we note that this measure can be used to find the Lyapunov exponent, and therefore quantify the average rate of expansion or contraction for an interval under iteration. If we have a Markov partition $\mathcal{P} : 0 = c_0 < c_1 < \cdots < c_{n-1} = 1$, then the Lyapunov exponent Λ is exactly computed as

$$\begin{aligned}
\Lambda &= \int_0^1 \ln\left|\tau'(x)\right| \rho(x)\,dx && (9.110)\\
&= \int_{c_0}^{c_1} \ln\left|\tau'(x)\right| \rho_1\,dx + \cdots + \int_{c_{n-2}}^{c_{n-1}} \ln\left|\tau'(x)\right| \rho_{n-1}\,dx\\
&= \ln[\tau'(c_{\frac{1}{2}})] \int_{c_0}^{c_1} \rho_1\,dx + \cdots + [\tau'(c_{n-\frac{1}{2}})] \int_{c_{n-2}}^{c_{n-1}} \ln \rho_{n-1}\,dx\\
&= \sum_{i=1}^{(n-1)} \ln|\tau'(c_{i-\frac{1}{2}})|(c_i - c_{i-1})\rho_i .
\end{aligned}$$

9.7.2 Nonlinear in an Interval, as a Limit of Piecewise Linear

Given a general transformation of the interval $\tau : I \to I$, which is not assumed to be either Markov or piecewise linear, we may estimate Lyapunov and other measurable and ergodic quantities by refinement in terms of sequences of Markov transformations $\{\tau_n\}$ which uniformly estimate τ. Recall that non-Markov transformations can be written as a weak limit of Markov transformations using Theorem 4.7 of [19], at least in the scenario proved for skew tent maps, as discussed elsewhere in this text.

9.7.3 Pesin's Identity Connects Lyapunov Exponents and Metric Entropy

The famous Pesin entropy identity [252, 322, 193],

$$h_\mu(T) = \sum_{i:\Lambda_i>0} \Lambda_i, \tag{9.111}$$

provides a profound connection between entropy h_{KS} and the (positive) Lyapunov exponents Λ_i, under the hypothesis of ergodicity. In fact, a theorem of Ruelle [273] established

$$h_\mu(T) \leq \sum_{i:\Lambda_i>0} \Lambda_i \tag{9.112}$$

under the hypothesis that T is differentiable and μ is an ergodic invariant measure on a finite-dimensional manifold with compact support. In [109], Eckmann and Ruelle assert that this inequality holds as equality often, but not always for natural measures. However, Pesin proved the equality holds at least if μ is a Lebesgue absolutely continuous invariant measure for a diffeomorphism T [252]. Since then a great deal of work has proceeded in various settings including considerations of natural measure, of the infinite-dimensional setting, and perhaps most interesting in the case of the nonhyperbolic setting of the presence of zero Lyapunov exponents. See [322, 193] for further discussion.

A geometric interpretation of Pesin's entropy formula may be stated as follows. On the one side of the formula, metric entropy describes growth rate of information states with respect to evolution of the dynamics through partitions, as stated directly in Eq. (9.73). However, Lyapunov exponents describe an "average" growth rate of perturbations in characteristic directions of orthogonal successively maximal growth rate directions. Thus we can understand the formula as stating that initial conditions with a given initial precision corresponding to initial hypercubes grow according to the positive exponents in time, thus spreading the initial states across elements of the partition, implying new information generated at a rate descriptive of these exponents. Considering this as an information production process infinitesimally for small initial variations suggests the Pesin formula. Considering further that Lyapunov exponents may be computed in two ways by the Birkhoff formula, either by averaging in time the differential information along "typical" (μ-almost every) initial conditions or by averaging among ensembles of initial conditions but weighting by the ergodic measure μ when it exists, this statement of the Birkhoff ergodic theorem provides two ways of computing and understanding metric entropy. See Eq. (1.5) and Example 1.4. Furthermore, often sampling along a test orbit may provide the simplest means to estimate Lyapunov exponents [319] and hence entropy h_μ according to the Pesin formula. Alternatively, computing Lyapunov exponents by Ulam's method provides another direct method for estimating h_μ through Pesin's identity.

9.8 Information Flow and Transfer Entropy

A natural question in measurable dynamical systems is to ask which parts of a partitioned dynamical system influence other parts of the system. Detecting dependencies between variables is a general statistical question, and in a dynamical systems context this relates to questions of causality. There are many ways one may interpret and computationally address dependency. For example, familiar linear methods such as correlation have some relevance to infer coupling from output signals from parts of a dynamical system, and these methods are very popular especially for their simplicity of application [182]. A popular method is to compute mutual information, $I(X_1;X_2)$ in Eq. (9.35), as a method to consider dynamical influence such as used in [105] in the context of global weather events, as we review in Section 9.9.2. However, both correlation and mutual information more so address overlap of states rather than information flow. Therefore, time dependencies are also missed.

The transfer entropy $T_{J\rightarrow I}$ was recently developed by Schreiber [281] to be a statistical measure of information flow, with respect to time, between states of a partitioned phase space in a dynamical system to other states in the dynamical system. Unlike other methods that simply consider common histories, transfer entropy explicitly computes information exchange in a dynamical signal. Here we will review the ideas behind transfer entropy as a measurement of causality in a time evolving system. We present here our work in [31] on this subject. Then we will show how this quantity can be computed using estimates of the Frobenius–Perron transfer operator by carefully masking the resulting matrices.

9.8.1 Definition and Interpretations of Transfer Entropy

To discuss transfer entropy in the setting of dynamical systems, suppose that we have a partitioned dynamical systems on a skew product space $X \times Y$,

$$T : X \times Y \rightarrow X \times Y. \tag{9.113}$$

This notation of a single dynamical system with phase space written as a skew product space allows a broad application, as we will highlight in the examples, and helps to clarify the transfer of entropy between the X and Y states. For now, we will further write this system as if it is two coupled dynamical systems having x and y parts describing the action on each component and perhaps with coupling between components.

$$T(x,y) = (T_x(x,y), T_y(x,y)), \tag{9.114}$$

where

$$\begin{aligned} T_x : X \times Y &\to X \\ x_n &\mapsto x_{n+1} = T_x(x_n, y_n), \end{aligned} \tag{9.115}$$

and likewise

$$\begin{aligned} T_y : X \times Y &\to Y \\ y_n &\mapsto y_{n+1} = T_y(x_n, y_n). \end{aligned} \tag{9.116}$$

This notation allows that $x \in X$ and $y \in Y$ may each be vector (multivariate) quantities and even of different dimensions from each other. See the caricature of this arrangement in Fig. 9.10.

Let

$$x_n^{(k)} = (x_n, x_{n-1}, x_{n-2}, \ldots, x_{n-k+1}) \tag{9.117}$$

be the measurements of a dynamical system T_x, at times

$$t^{(k)} = (t_n, t_{n-1}, t_{n-2}, \ldots, t_{n-k+1}), \tag{9.118}$$

sequentially. In this notation, the space X is partitioned into states $\{x\}$, and hence x_n denotes the measured state at time t_n. Note that we have chosen here not to index in any way the partition $\{x\}$, which may be some numerical grid as shown in Fig. 9.10, since subindices are already being used to denote time, and superindices denote time-depth of the sequence discussed. So an index to denote space would be a bit of notation overload. We may denote simply x, x', and x'' to distinguish states where needed. Likewise, $y_n^{(k)}$ denotes sequential measurements of y at times $t^{(k)}$, and Y may be partitioned into states $\{y\}$ as seen in Fig. 9.10.

The main idea leading to transfer entropy will be to measure the **deviation from the Markov property,** which would presume

$$p(x_{n+1}|x_n^{(k)}) = p(x_{n+1}|x_n^{(k)}, y_n^{(l)}), \tag{9.119}$$

that the state $(x_{n+1}|x_n^{(k)})$ does not include dependency on $y_n^{(l)}$. When there is a departure from this Markovian assumption, the suggestion is that there is no information flow as conditional dependency in time from y to x. The measurement of this deviation between these two distributions will be by a conditional Kullback–Leibler divergence, which we will build toward in the following.

The joint entropy[143] of a sequence of measurements written in the notation of Eqs. (9.117)–(9.118) is

$$H(x_n^{(k)}) = -\sum_{x_n^{(k)}} p(x_n^{(k)}) \log p(x_n^{(k)}). \tag{9.120}$$

[143] Definition 9.6.

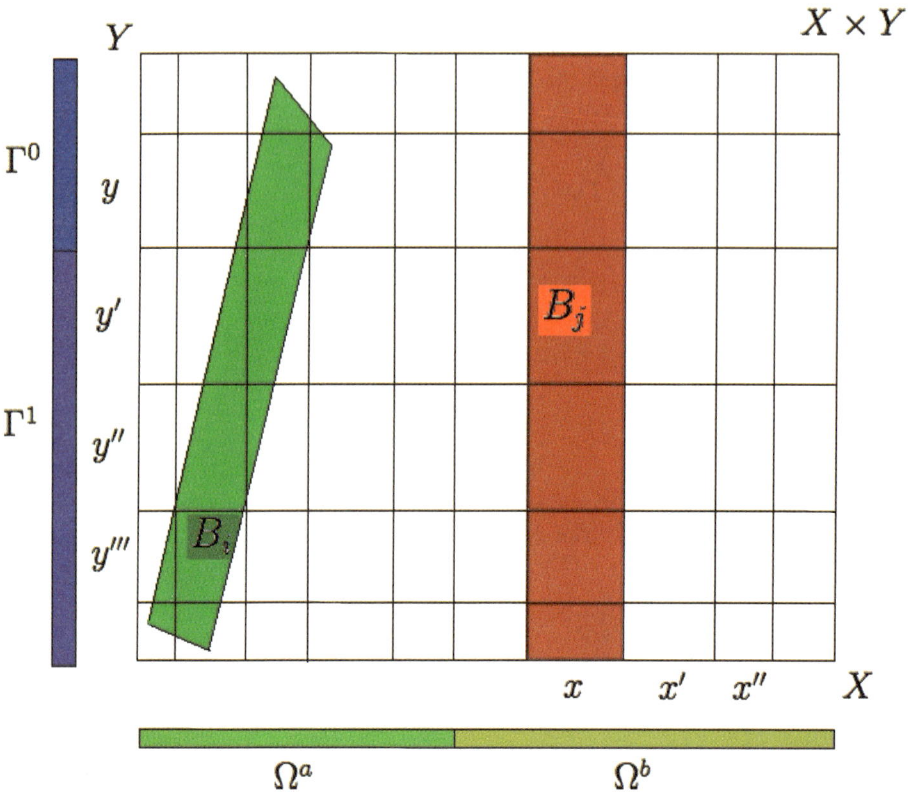

Figure 9.10. *In a skew product space $X \times Y$, to discuss transfer entropy between states $\{x\}$ a partition of X and states $\{y\}$ of Y, some of which are illustrated as x, x', x'' and y, y', y'', y'''. A coarser partition $\{\Omega^a, \Omega^b\}$ of X in symbols a and b and likewise $\{\Gamma^0, \Gamma^1\}$ of Y in symbols 0 and 1 are also illustrated.* [31]

A conditional entropy,[144]

$$\begin{aligned} H(x_{n+1}|x_n^{(k)}) &= -\sum p(x_{n+1}, x_n^{(k)}) \log p(x_{n+1}|x_n^{(k)}) \\ &= H(x_{n+1}, x_n^{(k)}) - H(x_n^{(k)}) \\ &= H(x_{n+1}^{(k+1)}) - H(x_n^{(k)}), \end{aligned} \tag{9.121}$$

is approximately an *entropy rate*,[145] which as it is written quantifies the amount of new information that a new measurement of x_{n+1} allows following the k-prior measurements, $x_n^{(k)}$. Note that the second equality follows the probability chain rule,

$$p(x_{n+1}|x_n^{(k)}) = \frac{p(x_{n+1}^{(k+1)})}{p(x_n^{(k)})}, \tag{9.122}$$

[144]Definition 9.7.

[145]This is an entropy rate in the limit $k \to \infty$ according to Definition 9.11.

and the last equality from the notational convention for writing the states is

$$(x_{n+1}, x_n^{(k)}) = (x_{n+1}, x_n, x_{n-1}, \ldots, x_{n-k+1}) = (x_{n+1}^{(k+1)}). \tag{9.123}$$

Transfer entropy is defined in terms of a Kullback–Leibler divergence, $D_{KL}(p_1||p_2)$, from Definition 9.9 but adapted for the conditional probabilities:[146]

$$D_{KL}(p_1(A|B)||p_2(A|B)) = \sum_{a,b} p_1(a,b) \log \frac{p_1(a|b)}{p_2(a|b)}. \tag{9.124}$$

The states are specifically designed to highlight transfer of entropy between the states X to Y (or vice versa Y to X) of a dynamical system written as a skew product, Eq. (9.113) Define [281],

$$T_{x \to y} = \sum p(x_{n+1}, x_n^{(k)}, y_n^{(l)}) \log \frac{p(x_{n+1}|x_n^{(k)}, y_n^{(l)})}{p(x_{n+1}|x_n^{(k)})}, \tag{9.125}$$

which we see may be equivalently written as a difference of entropy rates, like conditional entropies:[147]

$$T_{y \to x} = H(x_{n+1}|x_n^{(l)}) - H(x_{n+1}|x_n^{(l)}, y_n^{(k)}). \tag{9.126}$$

The key to computation is joint probabilities and conditional probabilities as they appear in Eqs. (9.126) and (9.129). There are two major ways we may make estimates of these probabilities, but both involve coarse-graining the states. A direct application of formulae (9.120) and (9.121), and likewise for the joint conditional entropy, to Eq. (9.125) allows

$$T_{y \to x} = [H(x_{n+1}, x_n) - H(x_n)] - [H(x_{n+1}, x_n, y_n) - H(x_n, y_n)], \tag{9.127}$$

which serves as a useful method of direct computation.

This may be a most useful form for computation, but for interpretation, a useful form is in terms of a conditional Kullback–Leibler divergence,

$$T_{y \to x} = D_{KL}(p(x_{n+1}|x_n^{(k)}, y_n^{(l)})||p(x_{n+1}|x_n^{(k)})), \tag{9.128}$$

found by putting together Eqs. (9.124) and (9.125). In this form, as already noted in Eq. (9.119), transfer entropy has the interpretation as a measurement of the deviation from the Markov property, which would be the truth of Eq. (9.119). That the state $(x_{n+1}|x_n^{(k)})$ does not include dependency on $y_n^{(l)}$ suggests that there is no information flow as a conditional dependency in time from y to x causing an influence on transition probabilities of x. In this sense, the conditional Kullback–Leibler divergence (9.128) describes the deviation of the information content from the Markovian assumption. In this sense, $T_{y \to x}$ describes an information flow from the marginal subsystem y to marginal subsystem x. Likewise, and asymmetrically,

$$T_{x \to y} = H(y_{n+1}|y_n^{(l)}) - H(y_{n+1}|x_n^{(l)}, y_n^{(k)}), \tag{9.129}$$

[146] Recall that the Kullback–Leibler of a single random variable A with probability distribution is an error-like quantity describing the entropy difference between the true entropy using the correct coding model $\log p_1(A)$ versus a coding model $\log p_2(A)$ with a model distribution $p_2(A)$ of A. Thus, conditional Kullback–Leibler is a direct application for conditional probability $p_1(A|B)$ with a model $p_2(A|B)$.

[147] Again, these become entropy rates as $k, l \to \infty$, as already discussed in Eq. (9.121).

and it is immediate to note that generally

$$T_{x\to y} \neq T_{y\to x}. \tag{9.130}$$

This is not a surprise both on the grounds that it has already been stated that Kullback–Leibler divergence is not symmetric; also, there is no prior expectation that influences should be directionally equal.

A partition $\{z\}$ serves as a symbolization which in projection by Π_x and Π_y is the grid $\{x\}$ and $\{y\}$, respectively. It may be more useful to consider information transfer in terms of a coarser statement of states. For example, see Fig. 9.10, where we represent a partition Ω and Γ of X and Y, respectively. For convenience of presentation we represent two states in each partition:

$$\Omega = \{\Omega^a, \Omega^b\} \quad \text{and} \quad \Gamma = \{\Gamma^0, \Gamma^1\}. \tag{9.131}$$

In this case, the estimates of all of the several probabilities can be summed in a manner just discussed above. Then the transfer entropy $T_{x\to y}$ becomes in terms of the states of the coarse partitions. The question of how a coarse partition may represent the transfer entropy of a system relative to what would be computed with a finer partition has been discussed in [151], with the surprising result that the direction of information flow can be effectively measured as not just a poor estimate by the coarse partition, but possibly even of the wrong sign. [31]

9.9 Examples of Transfer Entropy and Mutual Information in Dynamical Systems

9.9.1 An Example of Transfer Entropy: Information Flow in Synchrony

In our recent paper [31], we chose the perspective that synchronization of oscillators is a process of transferring information between them. The phenomenon of synchronization has been found in various aspects of nature and science [300]. It was initially perhaps a surprise when it was discovered that two, or many, oscillators can oscillate chaotically, but if coupled appropriately, they may synchronize and then oscillate identically; this is a description of the simplest form of synchronization which is of identical oscillators. See Fig. 9.11. Applications have ranged widely from biology [302, 141] to mathematical epidemiology [161], chaotic oscillators [250], communication devices in engineering [69], etc. Generally, the analysis of chaotic synchronization has followed a discussion of stability of the synchronization manifold, which is taken to be the identity function when identical oscillators [249], or some perturbation thereof for nonidentical oscillators [304], often by some form of master stability function analysis.

Considering as we have reviewed in this text that chaotic oscillators have a corresponding symbolic dynamics description, then coupling must correspond to some form of exchange of this information. Here we describe our perspective in [31] of coupled oscillators as sharing information, and then the process of synchronization is one where the shared information is an entrainment of the entropy production. In this perspective, when oscillators synchronize, it can be understood that they must be sharing symbols in order that they may each express the same symbolic dynamics. Furthermore, depending on the degree of cocoupling, or master-slave coupling or somewhere in between, the direction-

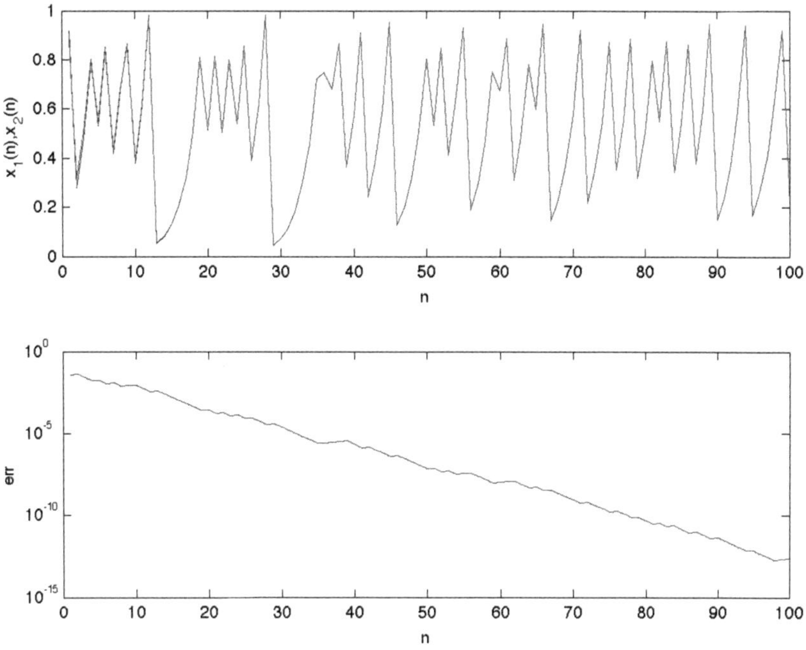

Figure 9.11. *In a nonlinearly coupled skew tent map system, Eq.* (9.133), *of identical oscillators,* $a_1 = a_2 = 0.63$, *and master-slave configuration,* $\delta = 0.6$, $\epsilon = 0.0$ *(parameters as in* [159]*). Note (above) how the signals entrain and (below) the error,* $error(n) = |x(n) - y(n)|$, *decreases exponentially.* [159]

ality of the information flow can be described by the transfer entropy. A study in [151] shows that anticipating synchronization with a transfer entropy perspective emphasizes the importance of the scale necessary to infer directionality.

Consider the following skew tent map system as an example coupling element to highlight our discussion [159], which is a full folding form [19], meaning two-one:

$$f_a(x) = \left\{ \begin{array}{lll} \frac{x}{a} & \text{if} & 0 \le x \le a \\ \frac{1-x}{1-a} & \text{if} & a \le x \le 1 \end{array} \right\}. \tag{9.132}$$

Let us couple these in the following nonlinear manner [159]:

$$\begin{pmatrix} x_{n+1} \\ y_{n+1} \end{pmatrix} = \mathbf{G} \begin{pmatrix} x_n \\ y_n \end{pmatrix} = \begin{pmatrix} f_{a_1}(x_n) + \delta(y_n - x_n) \\ f_{a_2}(y_n) + \epsilon(x_n - y_n) \end{pmatrix}. \tag{9.133}$$

Note that written in this form, if $a_1 = a_2$ and $\epsilon = 0$ but $\delta > 0$, we have a master-slave system of identical systems, as illustrated in Fig. 9.11, where we see a stable synchronized identity manifold with error decreasing exponentially to zero. On the other hand, if $\epsilon = \delta$ but $a_1 \neq a_2$, we can study symmetrically coupled but nonidentical systems in Fig. 9.12. There, the identity manifold is not exponentially stable but is apparently a Lyapunov-stable manifold, since the error, $error(n) = |x(n) - y(n)|$, remains small for both scenarios shown in the figures, $a_1 = 0.63$ but $a_2 = 0.65$ and $a_2 = 0.7$, respectively, with progressively larger

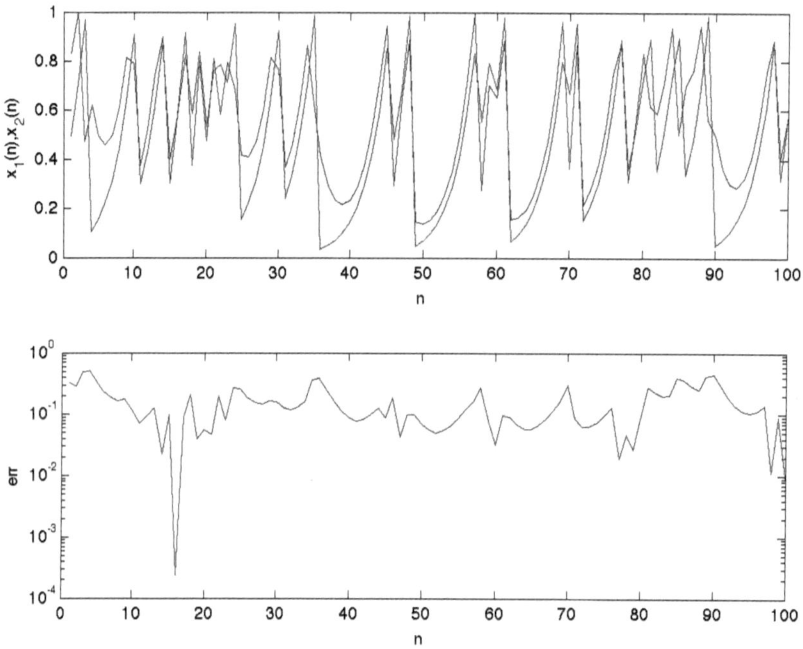

Figure 9.12. *A nonlinearly coupled skew tent map system, Eq.* (9.133), *of nonidentical oscillators,* $a_1 = 0.63, a_2 = 0.65$, *and master-slave configuration,* $\delta = 0.6$, $\epsilon = 0.0$. *Note (above) how the signals approximately entrain and (below) the error,* $error(n) = |x(n) - y(n)|$, *decreases close to zero, where it remains close to an identity manifold,* $x = y$, *where it is stable in a Lyapunov stability sense.* [31]

but stable errors. Our presentation here is designed to introduce the perspective of transfer entropy to understand the process of synchronization in terms of information flow, and from this perspective to gain not only an idea of when oscillators synchronize but perhaps if one or the other is acting as a master or a slave. Furthermore, the perspective is distinct from a master stability formalism.

With coupling resulting in various identical and nonidentical synchronization scenarios as illustrated in Fig. 9.12, we will analyze the information transfer across a study of both parameter matches and mismatches and across various coupling strengths and directionalities. In Figs. 9.13 and 9.14, we see the results of transfer entropies, $T_{x\to y}$ and $T_{y\to x}$, respectively, in the scenario of identical oscillators $a_1 = a_2 = 0.63$ for coupling parameters being swept $0 \le \delta \le 0.8$ and $0 \le \epsilon \le 0.8$. We see that due to the symmetry of the form of the coupled systems, Eq. (9.133), the mode of synchronization is opposite as expected. When $T_{x\to y}$ is relatively larger than $T_{y\to x}$, then the interpretation is that relatively more information is flowing from the x system to the y system, and vice versa. This source of communication is due to coupling the formulation of synchronization. Large changes in this quantity signal the sharing of information leading to synchronization.

In the asymmetric case, $0.55 \le a_1, a_2 \le 0.65$, we show a master-slave coupling $\epsilon = 0$, $\delta = 0.6$ in Fig. 9.15 and compare to Figs. 9.11 and 9.12. In the master-slave scenario chosen, the x-oscillator is driving the y-oscillator. As such, the x-oscillator is sending its

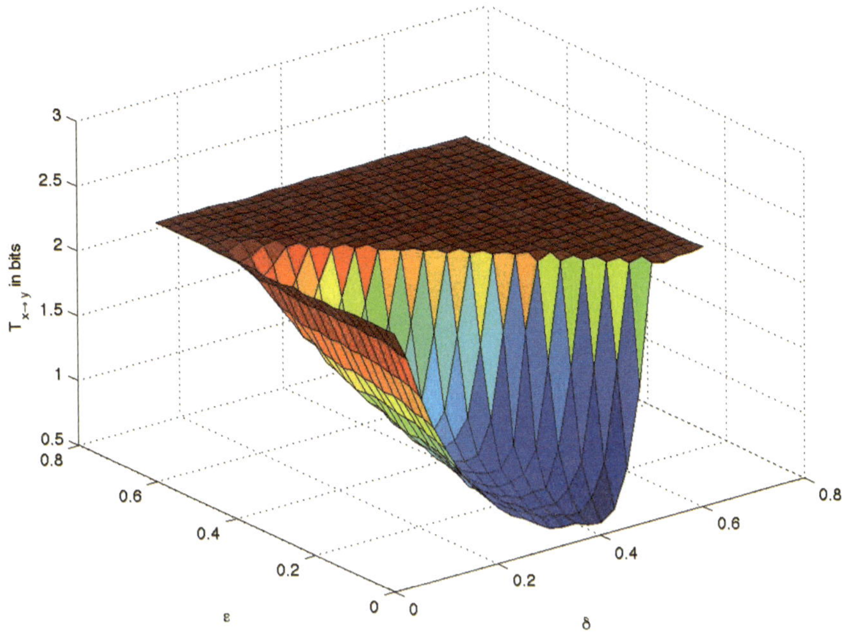

Figure 9.13. *Transfer entropy, $T_{x \to y}$ measured in bits, of the system* (9.133), *in the identical parameter scenario, $a_1 = a_2 = 0.63$, which often results in synchronization depending on the coupling parameters swept, $0 \le \delta \le 0.8$ and $0 \le \epsilon \le 0.8$ as shown. Contrast to $T_{y \to x}$ shown in Fig.* 9.14, *where the transfer entropy clearly has an opposite phase relative to the coupling parameters, (ϵ, δ).* [31]

states in the form of bits to the y-oscillator as should be measured that $T_{x \to y} > T_{y \to x}$ when synchronizing and more so when a great deal of information "effort" is required to maintain synchronization. This we interpret as what is seen in Fig. 9.15 in that when the oscillators are identical, $a_1 = a_2$ shown on the diagonal, the transfer entropy difference $T_{x \to y} > T_{y \to x}$ is smallest since the synchronization requires the smallest exchange of information once started. In contrast, $T_{x \to y} > T_{y \to x}$ is largest when the oscillators are most dissimilar, and we see in Fig. 9.13 how "strained" the synchronization can be, since the error cannot go to zero as the oscillators are only loosely bound.

9.9.2 An Example of Mutual Information: Information Sharing in a Spatiotemporal Dynamical System

Whereas transfer entropy is designed to determine direction of information flow, mutual information $I(X_1; X_2)$ in Eq. (9.35) is well suited to decide the simpler question as to whether there is simply a coupling in a large and complex dynamical system. The advantage of using the simpler but perhaps less informative measure, as it does not give directionality, is that it may require less data.

A recent and exciting application of mutual information comes from an important spatiotemporal dynamical system from analysis of global climate [105], as seen in Fig. 9.16.

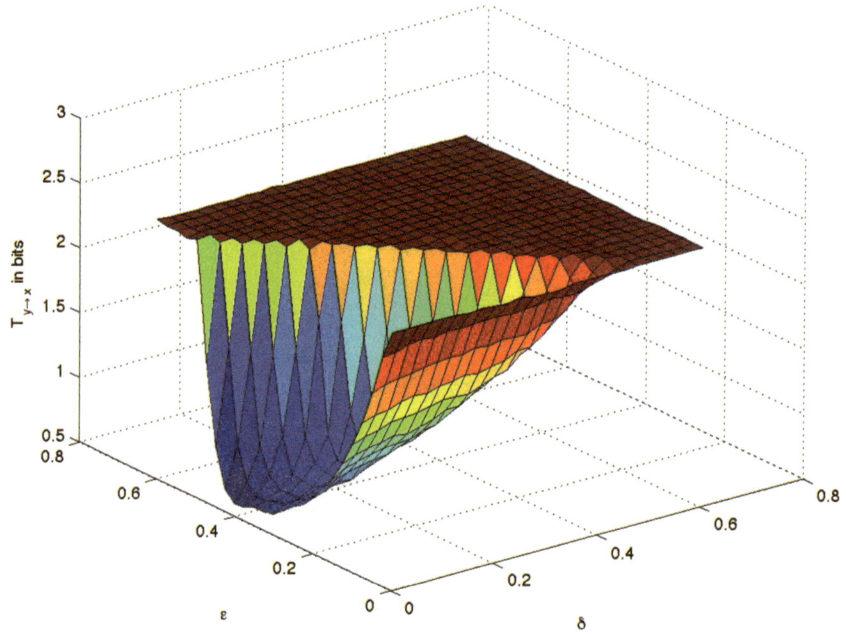

Figure 9.14. *Transfer entropy, $T_{y\to x}$ measured in bits, of the system* (9.133) *in the identical parameter scenario, $a_1 = a_2 = 0.63$, which often results in synchronization depending on the coupling parameters swept, $0 \le \delta \le 0.8$ and $0 \le \epsilon \le 0.8$ as shown. Compare to $T_{x\to y}$ shown in Fig.* 9.13. [31]

The study in [105] used a monthly averaged global SAT field to capture the complex dynamics in the interface between ocean and atmosphere due to heat exchange and other local processes. This allowed the study of atmospheric and oceanic dynamics using the same climate network. They used data provided by the National Center for Environmental Prediction/National Center for Atmospheric Research (NCEP/NCAR) and model output from the World Climate Research Programmes (WCRPs) Coupled Model Intercomparison Project phase 3 (CMIP3) multimodel data set.

These spatiotemporal data can be understood as a time series, $x_i(t)$, at each spatial site i modeled on the globe. Pairs of sites, i, j, can be compared for the mutual information in the measured values for states in the data $x_i(t)$ and $x_j(t)$ leading to $I(X_i;X_j)$. Following a thresholding decision leads to a matrix of couplings $A_{i,j}$ descriptive of mutual information between sites on the globe. The interpretation is that the climate at sites with large values recorded in $A_{i,j}$ are somehow dynamically linked. In Fig. 9.16, we illustrate what was shown in [105], which is a representation of the prominence of each site i on the globe colored according to that prominence. The measure of prominence shown is the vertex betweenness centrality, labeled BC_i. Betweenness centrality is defined as the total number of shortest paths[148] in the corresponding undirected graph which pass through the

[148] In a graph, $G = (V, E)$ consisting of the set of vertices and edges which are simply vertex pairs, a path between i and j of steps along edges of the graph which connect a pair of vertices. A shortest path between i and j is a path which is no longer than any other path between i and j.

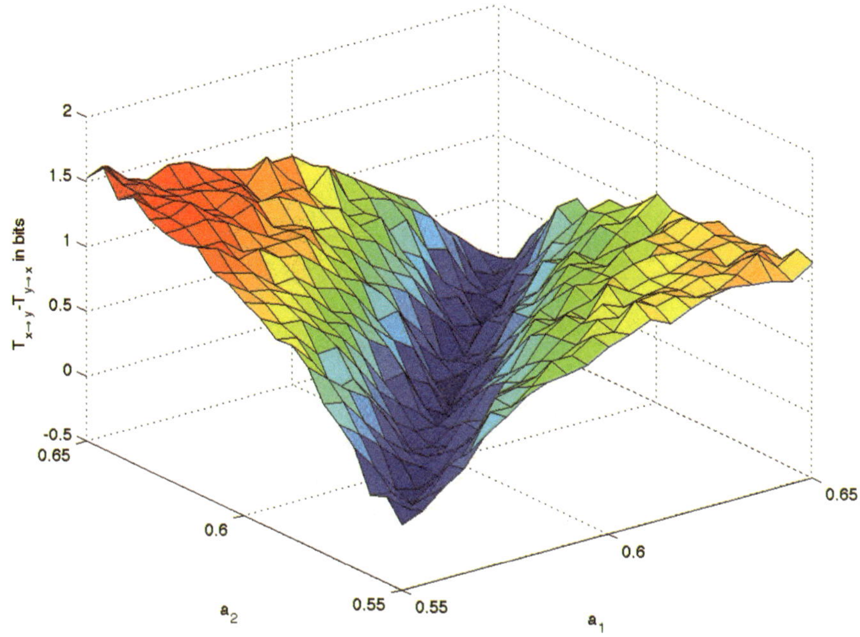

Figure 9.15. *Transfer entropy difference, $T_{y\to x} - T_{y\to x}$, measured in bits, of the system* (9.133) *in the nonidentical parameter scenario sweep,* $0.55 \le a_1, a_2 \le 0.65$, *and master-slave coupling* $\epsilon = 0$, $\delta = 0.6$. *Compare to $T_{x\to y}$ shown in Fig.* 9.13. *Contrast to $T_{y\to x}$ shown in Fig.* 9.14, *where the transfer entropy clearly has an opposite phase relative to the coupling parameters,* (ϵ, δ). *Also compare to Figs.* 9.11 *and* 9.12. [31]

vertex labeled i. BC_i can be considered as descriptive as to how important the vertex i is to any process running on the graph. Since the graph is built by considering mutual information, it may be inferred that a site i with high BC_i is descriptive of a dynamically important site in the spatiotemporal process, in this case the process being global climate. It is striking how this information theoretic quantification of global climate agrees with known oceanographic processes as shown in Fig. 9.16.

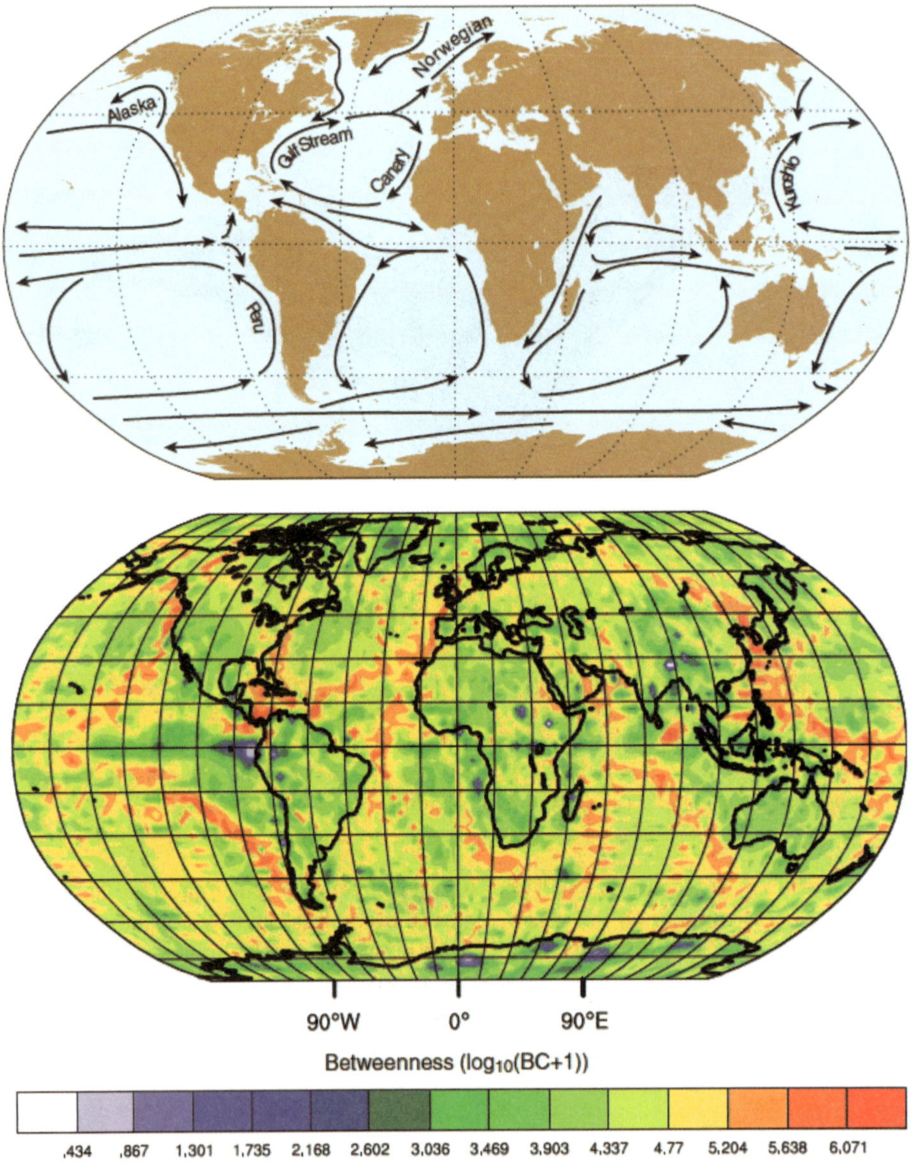

Figure 9.16. *Mutual information mapping of a global climate network from* [105]. *(Top) Underlying and known global oceanographic circulations. (Bottom) Betweenness centrality BC_i from a network derived from mutual information $I(X_i;X_j)$ between the global climate data from spatial sites i, j across the globe. Note that the mutual information theoretic quantification of the global climate is agreeable with the known underlying oceanographic processes.*

Appendix A

Computation, Codes, and Computational Complexity

A.1 MATLAB Codes and Implementations of the Ulam–Galerkin Matrix and Ulam's Method

In Chapter 4, we discussed several different ways of developing the Ulam–Galerkin matrix $P_{i,j}$ approximation of the Frobenius–Perron matrix. We presented the theoretical construction (4.4) in terms of projection onto a finite subspace of basis functions. In Eq. (4.6) and in Remark 4.1 we discussed the description of mapping sets across sets, which is more closely related to the original Ulam work [307], and closely related to the exact descriptions using Markov partitions discussed in Section 4.2. Here we will highlight the sampling method using a test orbit and Eq. (4.7), which we repeat:

$$P_{i,j} \approx \frac{\#(\{x_k | x_k \in B_i \text{ and } F(x_k) \in B_j\})}{\#(\{x_k \in B_i\})}. \tag{A.1}$$

In the code which follows in Section A.1.1, we choose $\{B_i\}$ to be a grid as a tessellation of triangles to cover the domain of interest, which should contain the entire test orbit $\{x_k\}_{k=1}^N$. Theoretical discussion will be presented in Section A.5.2 of how the grid size, the test orbit length N, and system regularity parameters are related for good approximations of the Frobenius–Perron operator.

A.1.1 Ulam–Galerkin Code by Delaunay Triangulation

The first MATLAB code presented here assumes a test orbit $\{x_k\}_{k=1}^N \subset \mathbb{R}^2$ and uses a Delaunay triangulation as the grid to cover the region of input interest uniformly by triangles. The output of the code is primarily the stochastic matrix, A, and the arrays concerning the grid. We present triangles which for technical computer science indexing reasons are easy to work with. Delaunay triangulations are particularly popular in the numerical PDE finite element literature for this reason. For our purposes here, the popularity of such methods allows us to save many steps toward writing compact and simple MATLAB code regarding indexing, identifying occupancy, plotting, and so forth, since there are already many Delaunay triangulation subroutines built into MATLAB. See MATLAB help pages concerning "Tessellation and Interpolation of Scattered Data."

While our code below is specialized for a two-dimensional dynamical system, generalizing it for a three-dimensional system is straightforward in part due to the strength of tessellations which are much easier to work with in general domains than other popular grid geometries, such as rectangles. Also, the MATLAB built-in routines really shine in this application. While we have assumed a uniform grid, if a nonuniform grid is needed, such as for refining regions of high observed invariant density, and perhaps large Lipschitz constants as discussed in Section A.5.2, then the tessellation structures are particularly well suited for refinement.

```
%%%%%%%%%%%%%%%
%% by Erik Bollt
%%%Build an Ulam-Galerkin Matrix Based on a Test Orbit Visiting Triangles
%%%of a Tessellation
%%%%%%%%%%%%%%%
% Input:
%       z - Test orbit is n x 2 for an n-iterate orbit sample
%       h - is the side length of the triangle edges which share a right
%       angle
%       ax, bx, ay, by - the low and high ends of a box bounding data
% Output:
%       dt - a DelaunayTri
%%

function [dt,ll,A,zz]=TransitionMatrix(z,h,ax,bx,ay,by)

%%

%low=-2; high=2; low=0; high=1;
[X1,X2] = ndgrid(ax:h:bx, ay:h:by);
[m,n]=size(X1);
x1=reshape(X1,m*n,1); x2=reshape(X2,m*n,1);

%Formulate Delaunay Triangulation of region
dt= DelaunayTri([x1 x2]);   %See Matlab Subroutine DelaunayTri for
                            %input/output information
                            %dt is the triangulation class
                            %
                            %   where
                            %
                            %dt.dt.Triangulation is an m1 by 3 array of
                            %integers labeling the vertex corner numbers
                            %of the triangles
                            %
                            %   and
                            %
                            %dt.X is an m2 by 2 array of real numbers
                            %defining positions

%triplot(dt) %Optional plot command of this triangulation

%%


%Count number of orbit points in z which cause a triangle to be counted as
%occupied (and otherwise a triangle is not counted as it is empty until
%observed occupied)

nottrue=0;
while(nottrue<1)
    nottrue=1;
    SI = pointLocation(dt,z); %Matlab command, Locate the simplex element in
                              %dt containing the specified locations of each
```

```
                                    %of the elements of the array z of orbit
                                    %samples
                                    %
    l=unique(SI);  %Matlab subrouting: Count number of unique instances in SI
    k=1;
    while(isnan(l(k))<1&&k<length(l)) %Matlab subroutine True if Not-a-Number
        [ii,j]=find(SI==l(k)); %Collect locations corresponding to each unique l.
         cnt(k)=sum(j);
        k=k+1;
    end
    k=k-1;    ll=l(1:k); cnt=reshape(cnt,size(ll));
end
%%

% Plot those  simplex elements of the dt which are occupied by an orbit
% iterate of z - dt(ll,:) are those occupied simplex elements
patch('faces',dt(ll,:), 'vertices', dt.X, 'FaceColor','r');
N=length(ll);

%%

%Translate from ll back to phase space positions z by using the center
%positions with Matlab subroutine "mean"
zz=zeros(N,2);
for i=1:N
    zz(i,:)=mean([dt.X(dt.Triangulation(ll(i),1),:)
    dt.X(dt.Triangulation(ll(i),2),:)
    dt.X(dt.Triangulation(ll(i),3),:)]);
end

%%Now build the transition matrix A
%
%So that A(i,j)>0 iff there is an element in simplex element i such that
% there is an iterate z(k,:) and that
% the next  iterate, z(k+1,:), transitions to simplex element j
A=zeros(N,N);
for i=1:(length(SI)-1)
    ii= find(ll==SI(i));
    jj=find(ll==SI(i+1));
    %[ii jj size(A)]
    A(ii,jj)=A(ii,jj)+1;
end
%Now make A into a stochastic matrix by row normalizing
for i=1:N
    q=length(find(SI==ll(i)));
    A(i,:)=A(i,:)./q;
    A(i,:)=A(i,:)./sum(A(i,:));
end
```

A.1.2 Routine to Test Run Subroutine TransitionMatrix.m

We give a sample test code to run the Ulam–Galerkin code TransitionMatrix.m above, with a test orbit produced using the standard map, from standard.m below; we use parameter $k = 1.2$, which is greater than the "magic" value $k = 0.97\ldots$ such that the last torus disappears, corresponding to the golden mean, thus allowing orbits to have unbounded conjugate-momentum [219]. But still the phase space is highly compartmentalized into almost-invariant sets as this system is only weakly transitive. See Figs. A.1–A.4 and comments therein describing the code and linking to the algorithms that they illustrate. The figure references are at the appropriate locations in the test MATLAB code.[149]

```
1  close all; clear;
2  nPlot = 100000;    % nPlot-iterate test orbit many points after the transient.
3  M=250; diam=10; a=0; b=1;
4  h=0.02
5
6  %%
7  %%%%%%%Make a Sample Orbit of the Standard Map
8  k=1.2; %k is chosen >0.97... for the magic breakup of the "golden mean" torus
9  z=[];
10 % Set some variables
11 transientSteps = 100;   %length of initial transient of test orbit to ignore
12 initialX = rand(2,1);
13 x = initialX; hold on;
14
15 %%
16 % Throw away transients:
17 for i=1:transientSteps
18     x = standard(x,k); %Use as a test example, the standard map.
19 end
20
21 %%
22 % Plot the next nPlot points visited:
23 for i=1:nPlot
24     x = standard(x,k);
25     z=[z;x'];
26 end
27 z=mod(z,1);
28 plot(z(:,1),z(:,2),'b.','markersize',10); hold on;
29 %
30 %%
31
32
33 %%%%%%%%%%%%%%%
34 %%%Build an Ulam-Galerkin Matrix Based on a Test Orbit Visiting Triangles
35 %%%of a Tessellation
36 [dt,ll,A,zz]=TransitionMatrix(z,h,a,b,a,b);
37 triplot(dt) %Matlab routine to draw the triangulation simplex
38 drawnow;
39
40 %%
41 [v,d]=eigs(A',2); w=abs(v(:,1)); %Compute 1st and send eigenvalues/vectors of
42                                   %Ulam-Galerkin Matrix A
43 %
44 figure; stem3(zz(:,1),zz(:,2),w(:),'fill'); %Show the dominant eigenvector
45                                             %(d(1)=1) meant to roughly
46                                             %estimate invariant density
47 %
48 w2=v(:,1); [i,ww]=find(w2>0);
```

[149] MATLAB becomes faster and more efficient as a computing platform if we leverage its strength using array arithmetic. Without giving great details as to the programming intricacies of this language, we state simply that loops, and multiply nested loops especially, can often be replaced with array arithmetic. For example, lines 52 through 56 above can be replaced with a single line, $PI = v(:,1); Phat = \mathrm{diag}(1./PI) * P' * \mathrm{diag}(PI)$, which is both briefer and easier to read (once we are used to the techniques) and many times faster for the computer. An even more efficient method would be to leverage sparse versions of matrix manipulations (especially useful for huge matrices where creating a full diagonal matrix is prohibitive) by using, for example, $PI = v(:,1); Phat = spdiags((1./PI),0,length(PI),length(PI)) * P' * spdiags(PI,0,length(PI),length(PI))$. We will not point out any further specific lines of efficiencies to be gained by eliminating multiple loops. We emphasize that we have chosen the multiple loop style of programming here for pedagogical reasons, as this is likely the most common easily read format for most readers.

```
figure; stem3(zz(i,1),zz(i,2),w2(i),'r', 'fill')

%% Produce a reversible Markov Chain R
P=A;
[v,d]=eigs(A',1);
N=size(A,1);
for i=1:N
PI= v(:,1); Phat=
spdiags((1./PI),0,length(PI),length(PI))*P'*spdiags(PI,0,length(PI),length(PI)
);
    for j=1:N
        Phat(i,j)=v(j,1)*P(j,i)/v(i,1);
    end
end
R=(P+Phat)/2;

%%
%Partition!
[w,l]=eigs(R,4);
figure; plot(w(:,2))
c=0; eps=0.005;
[i,ww]=find(w(:,2)>c);
[ii,ww]=find(w(:,2)<=c);
[iii,ww]=find(abs(w(:,2))<eps);

figure; hold on;
patch('faces',dt(ll(i),:), 'vertices', dt.X, 'FaceColor','r');
patch('faces',dt(ll(ii),:), 'vertices', dt.X, 'FaceColor','b');
patch('faces',dt(ll(iii),:), 'vertices', dt.X, 'FaceColor','k');
```

Below is a subroutine for the standard map function:

```
function xvecNew = standard(xvec,k)
% Matlab coded in vectorized form of standard
%Input:
%        xvec - vector of initial conditions
%        k    - standard map parameter
%
%Output:
%        xvecNew - vector of their images
%
xvecNew = zeros(size(xvec));
xvecNew(1,:) = xvec(1,:)+xvec(2,:)-k*sin(2*pi*xvec(1,:))/(2*pi);
xvecNew(2,:) = xvec(2,:)-k*sin(2*pi*xvec(1,:))/(2*pi);
```

A.2 Ulam–Galerkin Code by Rectangles

In the previous section we gave subroutines to build a stochastic matrix model of the Frobenius–Perron operator following covering the region by a Delaunay triangulation. Although we gave a two-dimensional and uniform example, we argue that such a code is more robust and flexible regarding difficulties of domain irregularities and dimensionality complexity such as dimensionality greater than two, and it is more easily developed into an adaptive refinement method for nonuniform grid sizes to compensate for nonhomogeneity of the dynamical system on the domain. These are presented in subsequent sections. These strengths are all significantly difficult to implement as a practical code if the basic grid elements are rectangles. Rectangles are nonetheless a very popular basic element often used

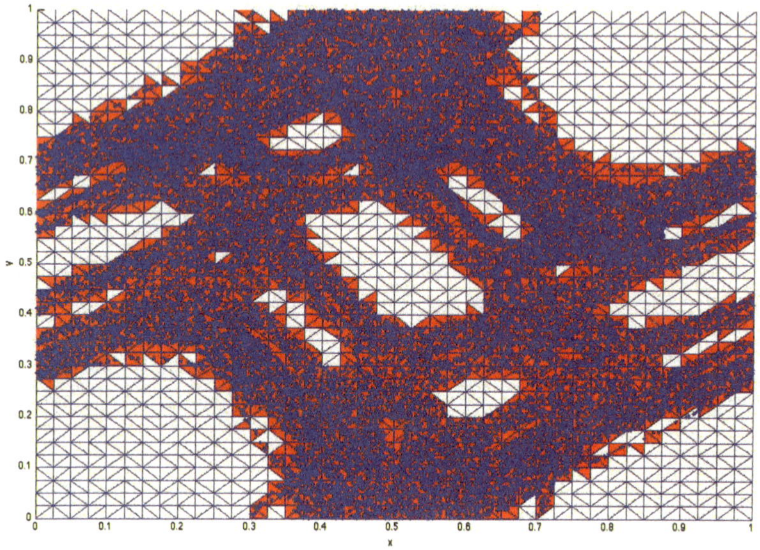

Figure A.1. *A Delaunay triangulation in the domain* $[0,1] \times [0,1]$ *for a test orbit* $\{x_i\}_{i=1}^N$ *from a standard map with* $k = 1.2$. *Using the code TransitionMatrix.m, the initial triangulation is pruned to a smaller collection of triangles to include only those triangles which are visited by the given test orbit. The test orbit is shown as blue dots, and visited triangles are shown as filled red. This smaller number of triangles,* M, *defines the dimensions of the* $M \times M$ *stochastic matrix* A *which is developed in the code according to Eq.* (4.7) *as an Ulam–Galerkin estimate of the Frobenius–Perron operator.*

when studying a dynamical system in terms of cell-mapping methods [166]. For this reason, we also give here a simple rectangle-based code to develop an adjacency matrix, and to demonstrate a few standard codes as an efficient coding example. An extremely robust code called GAIO, which is based on rectangles, has been quite popular among several groups, as it has been used in both three-dimensional problems and problems with grid refinement [82], but we will continue to suggest the strength of tessellations in each of these scenarios.

The code follows, together with Fig. A.5, which shows the resulting adjacency matrix of the operator estimate. Recall that the stochastic matrix is developed according to Eq. (A.1) by relative occupancy in the grid elements of the sample orbit. In the code that follows, we develop instead the adjacency matrix for the sake of discussion and presentation. Recall that an adjacency matrix B is a matrix of zeros and ones, and $B_{i,j} = 1$ if $P_{i,j} > 0$ and $B_{i,j} = 0$ otherwise in terms of the stochastic matrix A from Eq. (A.1); it simply identifies transitions without weights. The code below is easily adjusted to produce the stochastic matrix if we wish.

Another useful memory structure to store a sparse matrix, since Ulam–Galerkin matrices are often quite sparse, is the link list. If A is a sparse stochastic matrix of size $N \times N$, with n nonzero values $P_{i,j} > 0$, then let L be $n \times 3$. If $P_{i,j}$ is the kth nonzero element of

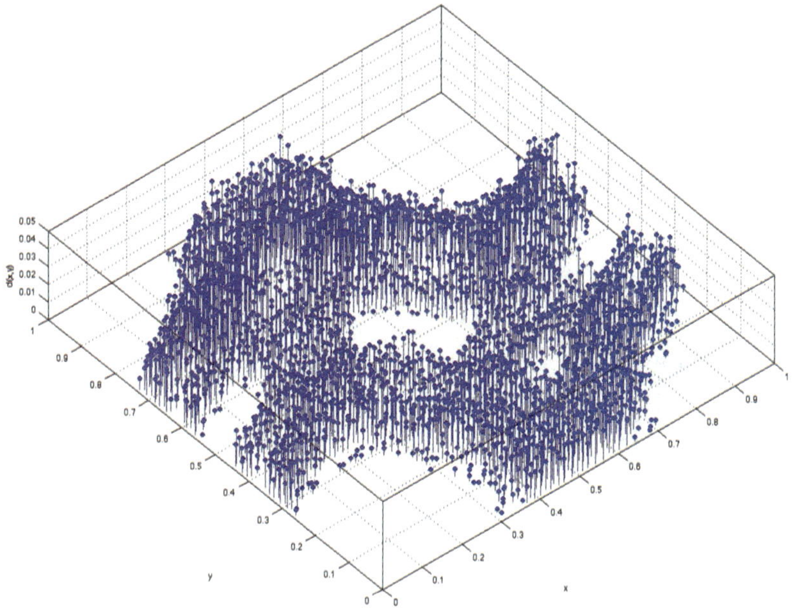

Figure A.2. *The dominant eigenvector of the stochastic matrix Ulam–Galerkin estimate of the Frobenius–Perron operator A is often used as an Ulam method estimator of the invariant density. Here, using the same* $\{x_i\}_{i=1}^N$ *test orbit from a standard map with* $k = 1.2$ *yields the eigenvector shown, which is reshaped to agree with the original domain. We emphasize with this example that while the Ulam method is popular for estimating invariant density, the standard map satisfies none of the few theorems that guarantee the success of the Ulam method. Furthermore, it is not even known if the standard map even has an invariant measure. Nonetheless, this illustration demonstrates the use of the code, and at least we see the support on the tessellation of the test orbit. Even with an incorrect computation of invariant density, some success can still be found regarding estimates of almost-invariant sets shown in Figs.* A.3–A.4.

the matrix (the specifics of this order are not important), then

$$\begin{aligned} L(k,1) &= i, \\ L(k,2) &= j, \\ L(k,3) &= P_{i,j}. \end{aligned} \tag{A.2}$$

Such sparse representations matrices can be huge memory savers for large matrices. N scales as $(1/h)^d$ for a domain of dimension d (here $d = 2$) and a rectangle size h. A scales as N^2 since it is a square matrix, and L is of size

$$card(L) = 3n, \text{ where } 0 \leq n \leq N^2, \tag{A.3}$$

and thus potentially $3n$ could be larger than the matrix size N^2, $3n > N^2$. In practice n is *much* smaller. If each rectangle stretches across several rectangles under the action of the

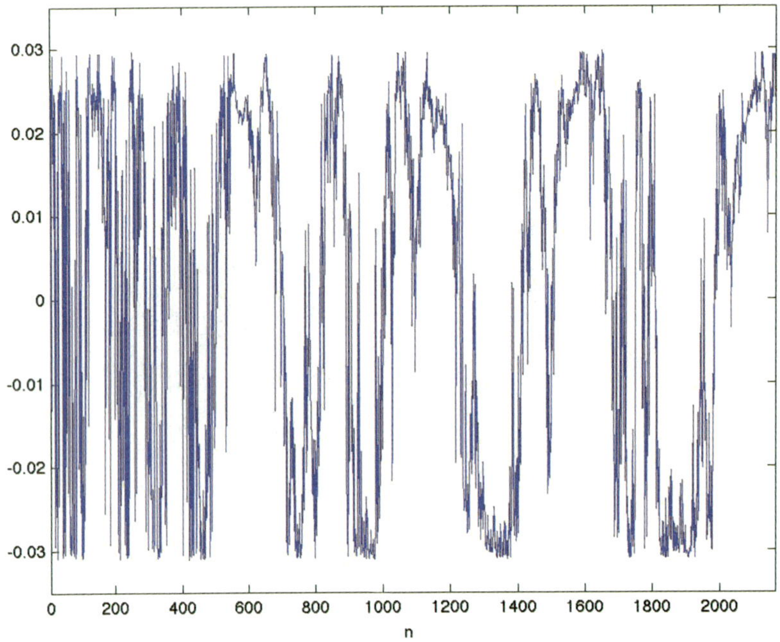

Figure A.3. *Second eigenvector v_2 of the reversible Markov chain, as defined by* (5.57) *in Section* 5.4.

map (see Fig. 4.1), roughly according to the Lyapunov number l on average, then roughly we expect the scaling

$$n \sim lN \quad \text{and} \quad lN \ll N^2. \tag{A.4}$$

```
clear; close all;
%Initial value x0, which is taken from 0 to 1
x0=[0;0];
%Set the grid size, ie. number of boxes are M by M
M=100;
%Set the size of the sampling points
N=100000;

%% Envolve the initial value x0 for N times, and record each point
x=zeros(N+1,2);
x(1,:)=x0;
a=1.2;
b=0.4;
% Iterate the Henon Map
for i=1:N
    x(i+1,:) = henon(x(i,:),a,b);
end

%% Now using the test orbit array x,
%% find which unique boxes the iterates land in
I0=FindBox(x,M); %Which boxes are hit - using FindBox Below
I=Reorder(I0);   %A sorting command - using subroutine below

%Develop directed graph as a Link List structure
L1=I(1:N);  % from node
```

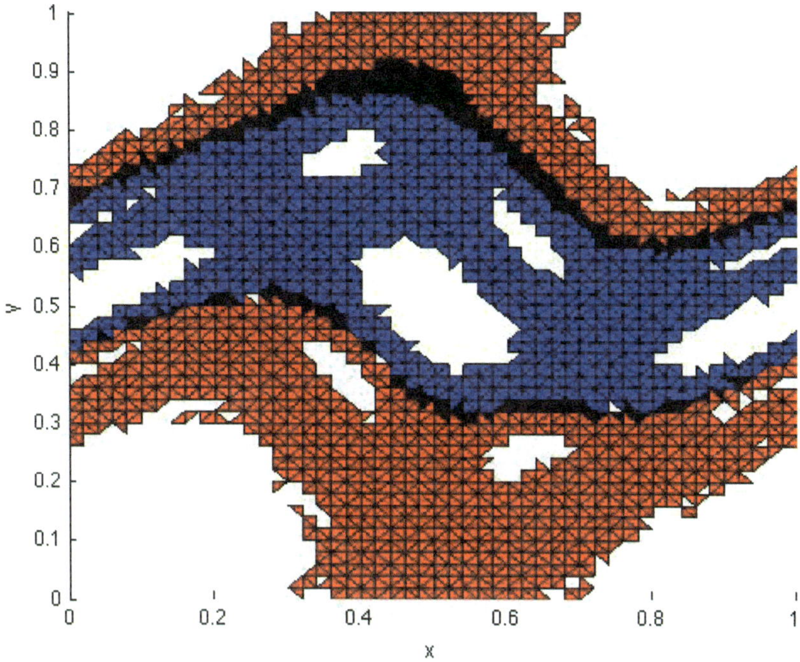

Figure A.4. *Observed almost-invariant sets of the standard map test orbit* $\{x_i\}_{i=1}^N$, *on the pruned tessellation shown in Fig.* A.1. *Following the methods in Section* 5.4 *from the Courant–Fischer theorem (see Theorem* 5.2*) for almost-invariant sets in reversible Markov chain developed from the stochastic matrix, as defined in* (5.57) [124], *the second eigenvector* v_2 *shown in Fig.* A.3 *can be thresholded to reveal those tessellation elements corresponding to weakly transitive components, colored red and blue, with those triangles on the boundary, colored black. Note that these components agree with the expectation that the cantori remnants of the golden mean frequency make for slow transport between the colored regions.* [219]

```
L2=I(2:(N+1)); % to node
%%

%% Convert Link List to Adjacency Matrix
A=zeros(max(I),max(I));
% A(L1,L2)=1; this is wrong
L_ind=sub2ind(size(A),L1,L2);
A(L_ind)=1;

%View the adjacency matrix by spy-plot, built in Matlab plotting program
spy(A);title('Adjacent Matrix of Henon Map'); %See Fig. A.5

Asp=sparse(A); %turn Matrix A into Sparse Matrix

%%%%%%%%%%%
%%
%This function determines which points belong to which box
%input x is a vector
function y=FindBox(x,BoxSize)

```

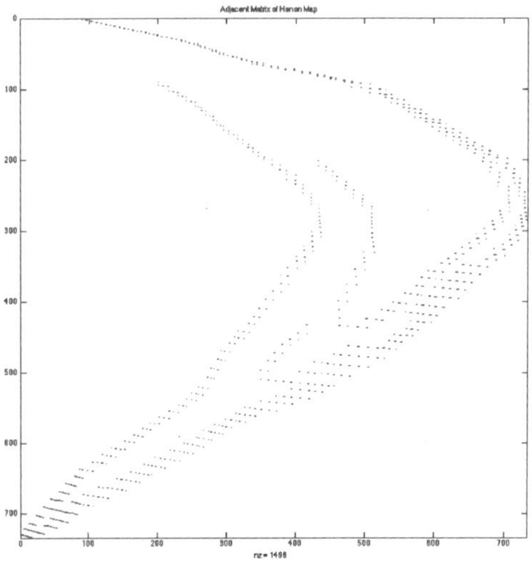

Figure A.5. *Spy plot of adjacency matrix from orbit from the Henon map.*

```
for nb = BoxSize
  % First convert coords to integers:
  ix = scaleToIntegers(x(:,1),nb);
  iy = scaleToIntegers(x(:,2),nb);
  ixy = ix + (iy-1)*nb;          % Each box now has a unique integer address.
end
y=ixy;


%%%%%%%%%%%
%%
% Scaling function
function ix = scaleToIntegers(x,n)
% Return x scaled into 1:n and rounded to integers.
ix = 1+round((n-1)*(x-min(x))/(max(x)-min(x)));


%%%%%%%%%%%
%%
% re-order vector I
% for example if I=[2001 4010 1 44 1], then Reorder(I)=[4 5 1 3 1]
function y=Reorder(vector)

[b,m,n]=unique(vector);
    %From Matlab help: for the array A returns the same values as in A
    %but no repetitions.
y=n;
```

A.3 Delaunay Triangulation in a Three-Dimensional Phase Space

A Delaunay tessellation really shines when we cover an attractor in a three-dimensional phase space. The code in this section should be compared to the code TransitionMatrix.m

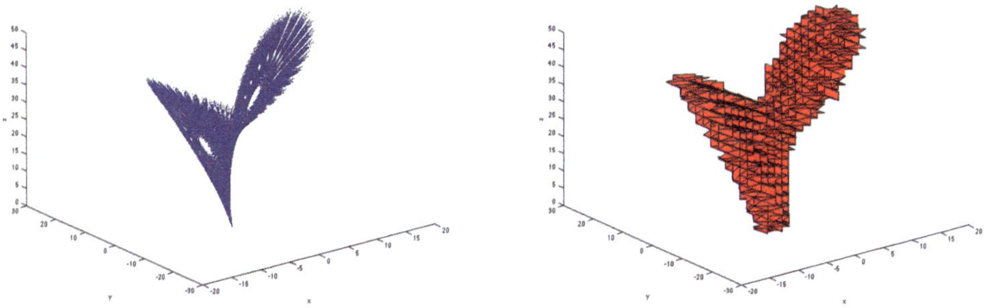

Figure A.6. *(Left) Test points of a sampled orbit of the Lorenz equations on (near) the Lorenz attractor. (Right) A tessellation by Delaunay triangulation out-covering the Lorenz attractor.*

in Section A.1.1, which produces the analogous outer cover by h-sized triangles, but now in terms of h-sized tessellations. Note that a key difference in the array that stores the tessellations was dt.Triangulation, that is, an m1 $\times$ 3 array of integers labeling the vertex corner numbers in a two-dimensional phase space, but m1 $\times$ 4 in a three-dimensional tessellation, as expected by the dimensionality of the corresponding objects. See Fig. A.6.

Note that in using a Delaunay triangulation, the heavy lifting of the covering, indexing, and accounting of triangles is handled by the Delaunay MATLAB subroutines DelaunayTri(.), and supporting routines such as pointLocation(.) for deciding in which tessellation element a given point is located. These are highly refined subroutines that are included in the MATLAB PDE toolbox suite. Delaunay tessellations are commonly used in the PDE community, especially finite element methods, for their power in grid generation in complex domains, and for their use in grid refinement. We highlight both these aspects in this section and the next section, respectively.

This code may be contrasted to a popular code suite called GAIO [82], which performs a covering by rectangles rather than a tessellation, which we offer as having geometric advantages.

Here we do not include the code to produce a stochastic matrix approximation of the Frobenius–Perron operator, as once covered with indexed triangles, the same formula (A.1) applies using code as in the previous sections for the two-dimensional mappings.

```
clear; close all;
%This should be any n x 3 array of real value numbers
%as a test orbit in a 3D phase space
%in this case the test orbit is from a Lorenz attractor
load 'LorenzDat.mat'
z=X;

%Produce a grid that covers the range n each coordinate
xlow=floor(min(X));  xhigh=ceil(max(X));  h=2.5; %And choose a step size of h
[X1,X2,X3]=ndgrid(xlow(1):h:xhigh(1),xlow(2):h:xhigh(2),xlow(3):h:xhigh(3));
m=size(X1);
x1=reshape(X1,m(1)*m(2)*m(3),1);
x2=reshape(X2,m(1)*m(2)*m(3),1);
x3=reshape(X3,m(1)*m(2)*m(3),1);

%Formulate Delaunay Triangulation of region
dt= DelaunayTri([x1 x2, x3]);
```

```
                                    %See Matlab Subroutine DelaunayTri for
                                    %input/output information
                                    %dt is the triangulation class
                                    %
                                    %dt.dt.Triangulation is an m1 by 4 array of
                                    %integers labeling the vertex corner numbers
                                    %of the triangles
                                    %
                                    %dt.X is an m2 by 3 array of real numbers
                                    %defining positions

%Count number of orbit points in z which cause a triangle to be counted as
%occupied (and otherwise a triangle is not counted as it is empty until
%observed occupied
nottrue=0;
while(nottrue<1)
    nottrue=1;
    SI = pointLocation(dt,z); %Matlab command, Locate the simplex element in dt
                              %containing the specified locations of each
                              %of the elements of the array z of orbit
                              %samples
                              %
    l=unique(SI);  %Matlab subrouting: Count number of unique instances in SI
    k=1;
    while(isnan(l(k))<1&&k<length(l)) %Matlab subroutine True for Not-a-Number
        [ii,j]=find(SI==l(k)); %Collect locations corresponding to unique l.
        cnt(k)=sum(j);
        k=k+1;
    end
    k=k-1;   ll=l(1:k); cnt=reshape(cnt,size(ll));
end
%%

% Plot those  simplex elements of the dt which are occupied by an orbit
% iterate of z - dt(ll,:) are those occupied simplex elements
figure; subplot(1,2,1); plot3(X(:,1),X(:,2),X(:,3),'.');
xlabel('x'); ylabel('y'); zlabel('z');
subplot(1,2,2); plot3(0,0,0);
patch('faces',dt(ll,:), 'vertices', dt.X, 'FaceColor','r');
xlabel('x'); ylabel('y'); zlabel('z');
```

A.4 Delaunay Triangulation and Refinement

A major strength of the Delaunay tessellations is the flexibility for grid refinement, especially in several variables. The codes here can be contrasted to the popular GAIO codes [82], which have also demonstrated adaptive refinement, but for rectangular elements.

The adaptive strategy used in this section in our MATLAB codes implements the simplest of possible strategies:

1. Starting with a coarse grid, of size h triangles (on the triangles' short side), then progressively split triangles in half (half with respect to area) if there is more orbit data sampled in each triangle than a minimum threshold. See captions of Figs. A.7–A.8 for a discussion regarding the refinements.

2. Furthermore, any empty triangles are removed from consideration.

The main point is that relatively few elements (triangles in our code) are needed to cover the attractor, and furthermore a fine grid improves accuracy. This second issue can be

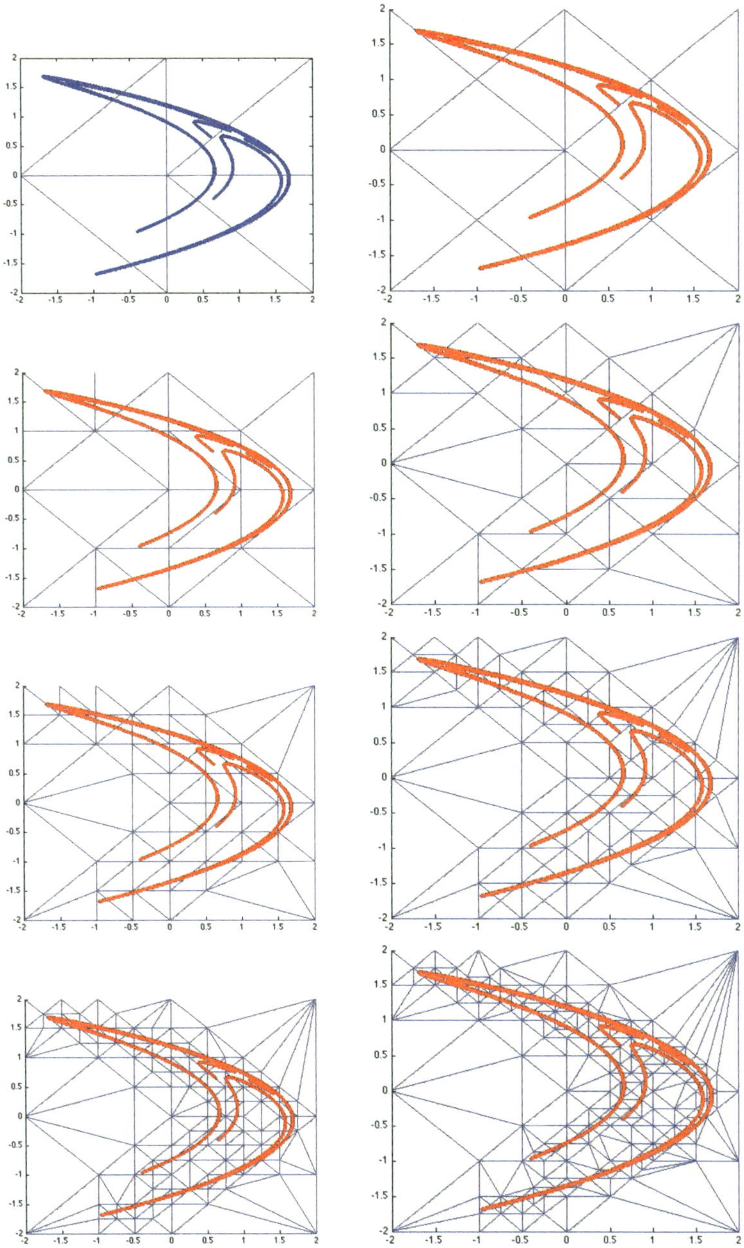

Figure A.7. *Successive refinements of an initially coarse* $h = 2.0$ *Delaunay tessellation grid covering* 10,000 *sample orbit points from the Henon map. In each finer grid shown, triangles have been split in half if there are more than* $thresh = 10$ *sample orbit elements in that triangle. From the coarsest initial grid in the upper left to the finest grid shown after* 8 *steps in the lower right, refinement proceeds until the stopping criterion* $thresh = 10$ *is attained after* 27 *refinement steps, as shown in Fig.* A.8.

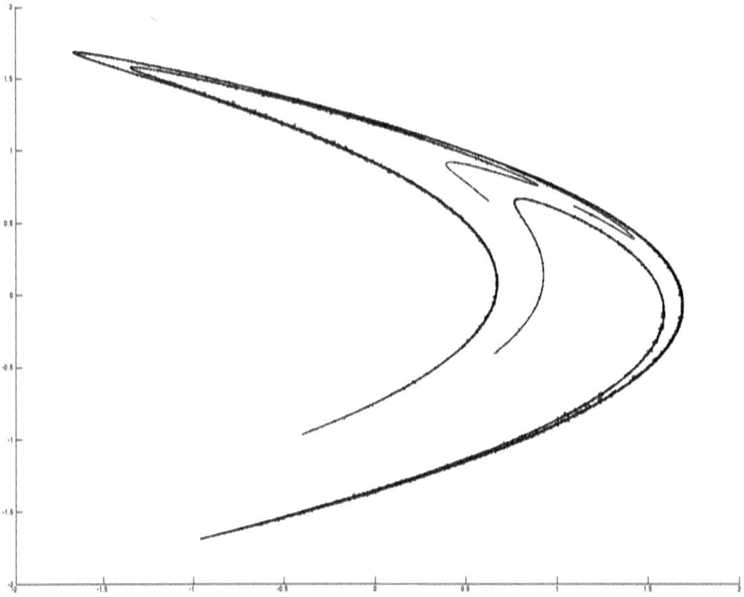

Figure A.8. *From* $h = 2.0$ *shown in Fig.* A.7, 27 *refinement steps leave an adaptive Delaunay tessellation grid as shown that meets a stopping criterion that states that there are no more than* $thresh = 10$ *orbit points in each triangle. The full grid covering the domain, but those triangles covering the orbit are small of edge size* $h \sim 1.7263 * 10.^{-4}$. *Thus we have a fine cover with relatively fewer triangles than the uniform grid of tiny triangles using the codes of the previous section.*

argued analytically by discussing Lipschitz constants as in Section A.5.2 and associated literature, but we give a novel discussion of both of these operations in terms of sharpening the variance of Monte Carlo estimators of the integrals, as we argue in Section A.5.1.

Further, removing empty elements represents a particularly useful computational savings when the box-counting fractal dimension d_b of the attractor [244] is relatively small compared to the embedding dimension d. Then the number of triangles for a total covering grows exponentially with respect to the "dimension," but not as quickly if the box dimension is used instead of the embedding dimension where $d_b < d$,

$$h^{-d_B} \sim\ll h^{-d}. \tag{A.5}$$

For example, taking the Henon mapping shown, an accepted value [143] is $d_b = 1.28 \pm 0.01 < d = 2$. In Figs. A.7–A.8, there results $n = 36439$ triangles total covering the attractor with a fine grid to the same level of refinement as a fine uniform triangulation of $N = 1.0738 \cdot 10^9$ triangles. Since developing a stochastic matrix of sizes $n \times n$ versus $N \times N$, we see that substantial quadratic memory and computational savings are made. See discussion in the next sections concerning estimation precision of the transfer operator and covering precision.

```
close all; clear;
h=2; %Initial Grid size
```

```
[X1,X2] = ndgrid(-2:h:2, -2:h:2); %Create uniform grid of triangles sized h;
[m,n]=size(X1);
x1=reshape(X1,m*n,1); x2=reshape(X2,m*n,1);

%Initialize
a=1.2; global b; b=0.4; %Henon map parameters
z=[]; %collected orbit, initially empty
transientSteps = 10;
initialX = [0;0];
nPlot = 100000;  % Plot this many points after the transient.

x = initialX;
for i=1:transientSteps % Throw away transients:
    x = henon(x,a);
end
% Plot the next nPlot points visited:
for i=1:nPlot
    x = henon(x,a); %Iterate the Henon map
    z=[z;x'];
end
plot(z(:,1),z(:,2),'b.','markersize',10); hold on;

dt= DelaunayTri([x1 x2]); %dt= DelaunayTri(zsamp);
triplot(dt)


thresh=10;  %The threshold number of samples we require in each triangle,
            %else split the triangle
P=[1 2; 1 3; 2 3]; %Labels for triangle corners
nottrue=0; qq=0; %Triggers used in deciding if refinement is done
while(nottrue<1)
    nottrue=1;
    SI = pointLocation(dt,z);
    ln=unique(SI);
    n=hist(SI,ln);

    if(max(n)>thresh) %keep refining the grid as long as there are triangles
                      %with more than thresh points filling
        nottrue=0;
    end

    xmk=[]; %Start with empty list of triangle corners which will be appended
    for i=1:length(ln)
        if(n(i)>thresh) %population in ln(i) triangle greater than thresh,
                        %then split it
            r=ln(i);
            for l=1:3 %Choose the longest edge of triangle
                      %to divide that triangle -
                      %Note use of P - corners matrix
                d(l)=norm(dt.X(dt.Triangulation(r,P(l,1)),:)
                    -dt.X(dt.Triangulation(r,P(l,2)),:));
            end
            [Y,I]=max(d); %Choose longest edge of triangle
            xm=mean(dt.X(dt.Triangulation(r,P(I(1),:)),:));
                     %Place a point in the middle of that longest edge
            xmk=[xmk; xm]; %Keep this dividing point amongst collected list
                           %of these midpoints to add to Delaunay triangulation
        end
    end
    pause

    %If there was some triangle splitting, then append to the triangulation
    if(length(xmk)>0)
        x1=[x1; xmk(:,1)]; x2=[x2; xmk(:,2)]; dt= DelaunayTri([x1 x2]);
    end
```

```
    figure; hold on; triplot(dt); plot(z(:,1),z(:,2),'r.','markersize',10);
        %plot(x1,x2,'.r'); %plot(xmk(:,1),xmk(:,2),'.r')
end
%Plot only those occupied triangles as filled triangle elements
figure; hold on;  ll = pointLocation(dt,z); triplot(dt);
patch('faces',dt(ll,:), 'vertices', dt.X, 'FaceColor','r');

function xvecNew = henon(xvec,a)
% Vectorized Henon, i.e. return images of lots of points.
% xvec is a 2 by m matrix, representing m points.  Returns a 2 by m matrix.
global b

xvecNew = zeros(size(xvec));
xvecNew(1,:) = a - xvec(1,:).^2 + b*xvec(2,:);
xvecNew(2,:) = xvec(1,:);
```

A.5 Analysis of Refinement

Grid refinement is generally performed when computing integrals for the sake of improving accuracy, and our Ulam's matrix estimation problem suffers some of the same computational issues as any such problem. Generally, finer grids should give better accuracy, but finer grids require more sample points, which in turn requires more computational time. In this section we give three complementary discussions of the issues herein, one statistical based on Monte Carlo's method and the usual random sampling method of Eq. (A.1), one analytical based on Lipschitz continuity and stretching properties of the particular dynamical system; finally we discuss the ability to resolve fine-scale structures when underlying sets of interest tend to be Cantor sets in so many interesting dynamical systems.

A.5.1 Why Refine the Grid? A Monte Carlo Discussion

Some theory of approximation in terms of coverings and Lipschitz constants leads to a theory for refinement, and even adaptive refinement can be found in Section A.5.2. Here we will give a complementary description in terms of Monte Carlo integration. As the grid refines, that (4.6) produces a matrix whose dominant eigenvector converges in measure to the invariant measure is the statement of Ulam's method [307]. Beyond that, better estimates of the quantities can be understood for any fixed grid size in terms of integration theory. The finite size h discussion is most relevant for the many finite time questions, such as almost-invariance and coherence, and many other issues discussed in this book.

See [170] for a theory of Monte Carlo integration in the context of Ulam's method with uniform grids and refinement, and then see [100] for the improvement to the (pseudo)random Monte Carlo integration by quasi–Monte Carlo integration (using carefully designed uniform sampling, rather than random uniform sampling) and the improvements therein. We assert here that adaptive refinement can be nicely understood in terms of importance sampling from Monte Carlo theory.

The basic computed quantity for the action of the Frobenius–Perron operator on a box (or triangle) labeled B_i is to estimate how much of B_i maps across box (triangle) B_j. Rewrite $P_{i,j}$ in terms of Lebesgue measure as an integration [84],

$$P_{i,j} = \frac{m(B_i \cap F^{-1}(B_j))}{m(B_i)} = \frac{1}{area(B_i)} \int_{B_i} \chi_{F^{-1}(B_j)}(x)dx := \langle\langle \chi_{F^{-1}(B_j)} \rangle\rangle, \qquad \text{(A.6)}$$

where

$$\chi_B(x) = \text{if}(1, x \in B, 0, \text{else}) \tag{A.7}$$

is the indicator function. Note the standard assumption for Monte Carlo integration specializes here,

$$x \sim U(B_i), \tag{A.8}$$

drawing from the uniform distribution on the grid element B_i. Following [264], we have used the notation $\langle\langle \chi_{F^{-1}(B_j)} \rangle\rangle$ to denote the true average of the function $\chi_{F^{-1}(B_j)}$ on the region B_i.

Refining the grid of $\{B_i\}$ has two useful outcomes.

1. Improves Ulam's method [307], convergence [201]. This is an infinite time problem, and many discussions and analyses have focused on convergence of Ulam's method including refinement as well as adaptive refinement [83, 180, 147, 85]. See notably an analysis of the rate of convergence in [43].

2. Monte Carlo estimate of the integrals on a finite-sized grid improves. In other words, how well do we estimate the integrals in Eq. (A.6)?

Note that finite time analysis using the Ulam–Galerkin matrix such as almost-invariance rely on point 2 more than point 1.

The integral presentation of Ulam's method problems may be estimated by Monte Carlo methods [170, 100] by random samples for (A.6). In the adaptive grid application, we should not directly use the slightly easier approximation, Eq. (A.1), since equal-sized units (triangles) are implicit in the form of the equation.

First we review the basic Monte Carlo method, but we specialize notation for a given grid element B_i and to the discussion of our integrand, $\chi_{F^{-1}(B_j)}(x)$ in Eq. (A.6), but otherwise this is the usual beginning discussion. A collection of *uniformly* sampled points $\{x_k\}_{k=1}^m \subset B_i$ leads to the standard Monte Carlo integration formula [267, 264],

$$\int_{B_i} \chi_{F^{-1}(B_j)}(x)dx \approx area(B_i)\left(\langle \chi_{F^{-1}(B_j)} \rangle \pm \sqrt{\frac{\langle [\chi_{F^{-1}(B_j)}]^2 \rangle - \langle \chi_{F^{-1}(B_j)} \rangle^2}{m}}\right), \tag{A.9}$$

where $\langle \cdot \rangle$ denotes the sample mean, and likewise used for the second moment,

$$\langle \chi_{F^{-1}(B_j)} \rangle = \frac{1}{m}\sum_{k=1}^{m} \chi_{F^{-1}(B_j)}(x_k), \quad \langle [\chi_{F^{-1}(B_j)}]^2 \rangle = \frac{1}{m}\sum_{k=1}^{m} [\chi_{F^{-1}(B_j)}(x_k)]^2. \tag{A.10}$$

Also, the relationship between the true function variance $Var(\chi_{F^{-1}(B_j)})$ and its estimator $\text{Var}(\langle \chi_{F^{-1}(B_j)} \rangle)$ scales according to

$$\text{Var}(\langle \chi_{F^{-1}(B_j)} \rangle) = \frac{\text{Var}(\chi_{F^{-1}(B_j)})}{m}. \tag{A.11}$$

Variance refinements are designed to improve this denominator. Let $P_{i,j}$ denote the true value of the integration in Eq. (A.6), and $\langle \chi_{F^{-1}(B_j)} \rangle$ is the estimator. We write the set of points $\{x_k\}_{k=1}^m \subset B_i$, and the subindex does not denote time; it is simply a list of uniformly randomly sampled points in the triangle element B_i.

A common scenario of sampling the behavior of the dynamical system is a large sample of initial conditions, perhaps a uniform sample, that we may advect. On the other hand, if we have a single orbit sample instead of a uniform sample, then we write $\{x_{i_k}\}_{k=1}^m \subset B_i$ as the set of those points from a long orbit $\{x_i\}_{i=1}^N$ that fall in B_i, but do not emphasize that these are not likely consecutive in time. This sample is not expected to be uniform, but rather to fall roughly according to the invariant density (if it exists). Our discussion here proceeds as if this invariant density is absolutely continuous, which we find serves well for computationally relevant problems where there is a bit of noise present. Great improvements by reduction of variance methods leverage variable distributions by adaptive strategies.

Stratified sampling is a powerful method of variance refinement for nonuniformly distributed samples, which further can be used recursively. We specialize the presentation in [264], but otherwise we closely adopt the notation. If we split B_i into equal disjoint volumes $area(B_i)/2$ each, we label them a and b. If furthermore $m/2$, half of the sampled points $\{x_{i_k}\}_{k=1}^m \subset B_i$ land in each of these subtriangles. Now we have a new Monte Carlo estimator of $P_{i,j}$, which we denote $\langle \chi_{F^{-1}(B_j)} \rangle'$:

$$\langle \chi_{F^{-1}(B_j)} \rangle' = \frac{\langle \chi_{F^{-1}(B_j)} \rangle_a + \langle \chi_{F^{-1}(B_j)} \rangle_b}{2}. \tag{A.12}$$

The power of the stratified sampling method is the new estimator of variance:

$$\mathrm{Var}(\langle \chi_{F^{-1}(B_j)} \rangle') = \frac{\mathrm{Var}(\langle \chi_{F^{-1}(B_j)} \rangle_a) + \mathrm{Var}(\langle \chi_{F^{-1}(B_j)} \rangle_b)}{4}. \tag{A.13}$$

Substituting each of the estimators on the right of this equation in Eq. (A.11) yields

$$\mathrm{Var}(\langle \chi_{F^{-1}(B_j)} \rangle') = \frac{\frac{\mathrm{Var}_a(\chi_{F^{-1}(B_j)})}{m/2} + \frac{\mathrm{Var}_b(\chi_{F^{-1}(B_j)})}{m/2}}{4} = \frac{\mathrm{Var}_a(\chi_{F^{-1}(B_j)}) + \mathrm{Var}_b(\chi_{F^{-1}(B_j)})}{2m}. \tag{A.14}$$

Comparing Eqs. (A.11) and (A.14), together with the Huygens–Steiner parallel axis theorem,

$$\begin{aligned} &\mathrm{Var}(\chi_{F^{-1}(B_j)}) \\ &= \frac{1}{2}[\mathrm{Var}_a(\chi_{F^{-1}(B_j)}) + \mathrm{Var}_b(\chi_{F^{-1}(B_j)})] + \frac{1}{4}(\langle\langle \chi_{F^{-1}(B_j)} \rangle\rangle_a - \langle\langle \chi_{F^{-1}(B_j)} \rangle\rangle_b)^2, \end{aligned} \tag{A.15}$$

one can conclude that variance never increases when splitting the region equally and with equal samples, and the variance decreases when $\mathrm{Var}_a(\chi_{F^{-1}(B_j)})$ and $\mathrm{Var}_b(\chi_{F^{-1}(B_j)})$ are different.

In reality, if the triangles are split into equal areas, it is not expected that there will be equal numbers of points in each half. Suppose in each half triangle a and b there are unequal numbers of points, m_a and m_b, and $m_a + m_b = m$. Following the discussion as above, it follows that

$$\mathrm{Var}(\langle \chi_{F^{-1}(B_j)} \rangle') = \frac{1}{4}\left[\frac{\mathrm{Var}_a(\chi_{F^{-1}(B_j)})}{m_a} + \frac{\mathrm{Var}_b(\chi_{F^{-1}(B_j)})}{m_b}\right]. \tag{A.16}$$

This formula can be used to conclude that there will be an improvement in the stratified Monte Carlo estimator due to reduced variance as long as we are not close to the scenario that $\mathrm{Var}(\chi_{F^{-1}(B_j)}) = \mathrm{Var}_a(\chi_{F^{-1}(B_j)}) = \mathrm{Var}_b(\chi_{F^{-1}(B_j)})$. Since the estimate does not get worse, we can repeat the splitting as long as there are points in each split.

A.5.2 A Lipschitz Continuity Discussion of Computational Complexity

The phrase "computational complexity" as used here denotes the question of how much computer work is expected to fulfill the required task, and how does this computational work scale with changes of the parameters such as when developing the Ulam–Galerkin matrix. Generally, a fine grid is desirable for more accurate representation of the dynamical system (unless a Markov partition is used), but a finer grid element such as a small triangle, or rectangle, requires integration of the integral in Eq. (A.6), and to do this requires more sample orbit points as the grid is refined. Also, more points are needed for dynamical systems that have stronger stretching properties so that the covered regions of $F^{-1}(B_j)$ are properly sampled for grid elements B_j. Here we discuss these trade-offs in terms of Lipschitz constants of the mapping, and the Gronwall inequality [251] assuming a continuous time system with corresponding regularity. A related discussion of grid refinement in terms of Lipschitz continuity can be found in [86, 88, 148, 180]. The discussion here is drawn closely from [37].

Assume square boxes of length q. A triangulation can be handled similarly, but rectangles will simplify this discussion even if triangles allow for the powerful Delaunay triangulation algorithms in practice. We will give arguments as to how many points are required so as to not "miss" any dynamics. We present these arguments as a list. Let $f : X \times [t_0, t] \to X$ be a flow and let X be compact.

1. To find an ε such that for all $x_1, x_2 \in X, |x_1(t_0) - x_2(t_0)| < \varepsilon$ we have

   $$|x_1(t) - x_2(t)| < q, \tag{A.17}$$

   where $x_1(t_0)$ and $x_2(t_0)$ are the initial status of x_1, x_2 and $x_1(t), x_2(t)$ are the final status of x_1, x_2. The purpose of this step is to avoid the distance d between the final positions of two closed initial points under the flow changing dramatically. Thus we require $d \geq q$. See Fig. A.9 (left). So, we must control ε such that the case shown in Fig. A.9 (right) holds for all initial points.

   Recall Gronwall's inequality,

   $$|x_1(t) - x_2(t)| \leq |x_1(t_0) - x_2(t_0)| e^{M|t-t_0|}, \tag{A.18}$$

   where M is the Lipschitz constant. Assuming f is uniformly continuously differentiable,

   $$M = \max_{(x,t) \in X \times [t_0, t]} \left| \frac{\partial f}{\partial x}(x,t) \right|. \tag{A.19}$$

   We must control the distance between $x_1(t)$ and $x_2(t)$ through controlling the distance between $x_1(t_0)$ and $x_2(t_0)$. Let

   $$\varepsilon = \frac{q}{e^{M|t-t_0|}}. \tag{A.20}$$

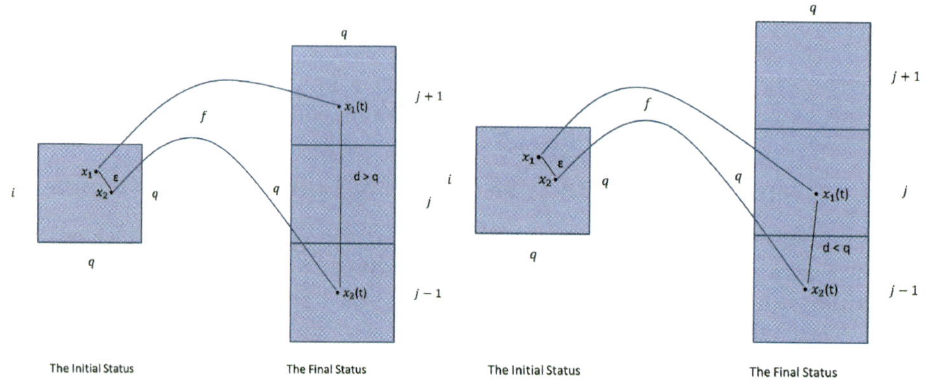

Figure A.9. *(Left) A failure of sampling. Two points in box i evolve to boxes $j+1$ and $j-1$, and without further samples the resulting Ulam–Galerkin matrix would miss a transition $i \to j$ which is expected in a continuous transformation f. $d > q$. (Right) The successful sampling scenario. Points x_1 and x_2 are close enough that the f never casts them further than across successive rectangles $j-1$ and j. Recall that a continuous f will map a connected set to a connected set, and this should be reflected in the images of the coarse representation of boxes cast across boxes. $d < q$.*

Then we have

$$|x_1(t) - x(t)| \leq |x_1(t_0) - x(t_0)|e^{M|t-t_0|} < \varepsilon e^{M|t-t_0|} = q \tag{A.21}$$

for all $x \in X$ satisfying $|x_1(t_0) - x(t_0)| < \varepsilon$. That is, for any $x(t_0)$ in an ε-ball of $x_1(t_0)$, $x(t_0)$ will keep its distance with $x_1(t_0)$ less than q through time $t - t_0$.

2. Next, we consider the more general case of the ε-ball of $x_1(t_0)$; consider ε-balls centered on each sample data x which must cover the whole X. Since X is compact, the ε-balls of a finite subset $\{x_k\}_{k=1}^P$ cover X. Let $\{x_k\}_{k=1}^P$ be initial conditions of the flow, with images under the flow $\{x_k(t)\}_{k=1}^P$ in a finite time $t - t_0$. Define

$$l = \min_{\{x_k(t)\}_{k=1}^P} \{\text{distance between each of four boundaries of some } j \text{ box and } x_k(t)\},$$

where the j box is as defined above, and is the image of X, which we denote $X(t)$. See Fig. A.10.

Here, again, we use Gronwall's inequality as follows:

$$|x(t) - x_k(t)| \leq |x(t_0) - x_k(t_0)|e^{M|t-t_0|} \leq l,$$

where $k = 1, 2, \ldots, P$. Let

$$\varepsilon' = \frac{l}{e^{M|t-t_0|}}, \tag{A.22}$$

and we have

$$|x_k(t) - x(t)| \leq |x_k(t_0) - x(t_0)|e^{M|t-t_0|} < \varepsilon e^{M|t-t_0|} = l, \tag{A.23}$$

which is to say, all points x in an ε'-ball of x_k in X for some k will map into the l-ball of $x_k(t)$ in $X(t)$ for the same k.

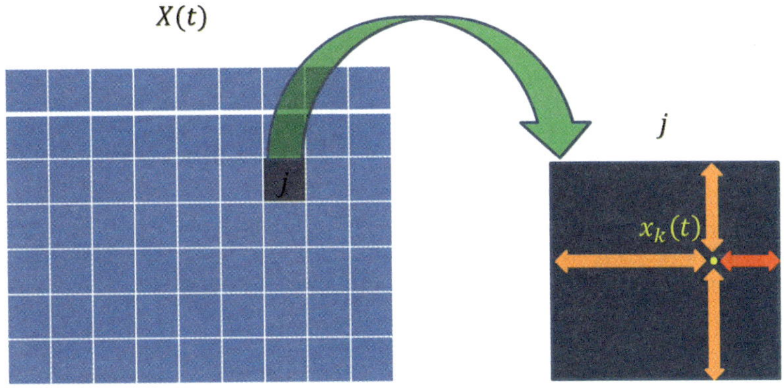

Figure A.10. *The red arrow indicates the minimum distance between a point $x_k(t)$ for some k and four boundaries of a box j for some j that contains $x_k(t)$. We must check all x_k in $\{x_k(t)\}_{k=1}^P$ with all the boxes of the partition of $X(t)$ to develop l.*

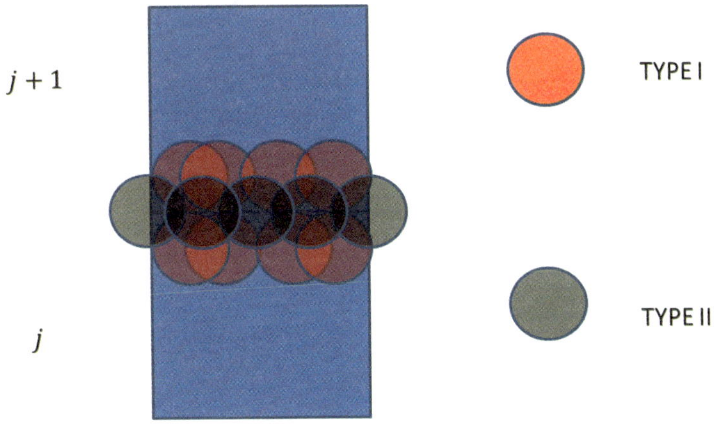

Figure A.11. *Covering geometry scenarios.*

3. In this step, we build a new cover for $X(t)$ with l-balls by the following. There are two types of balls of the cover. Type I covers are those l-balls that never touch the boundaries of the grid; Type II covers are those l-balls centered at boundaries and their centers never in the l-balls of Type I; see Fig. A.11.

   $X(t)$ is compact, so we have a finite subset of this new cover still covering $X(t)$. Let the centers of these balls be $\{x_q(t)\}_{q=1}^Q$ for some integer Q. Since the flow is assumed continuous, as is its inverse the preimage of these $(x_q(t), l)$-balls is still a cover of the preimage X of $X(t)$, which is also finite. By the above discussion, more specifically, the finite cover of X consists of ε'-balls of some of $\{x_q(t)\}_{q=1}^Q$. That is, the preimage of Type I l-balls in $X(t)$ and Type II l-balls in $X(t)$ vary as shown. Thus, in X, we have the following case; see Fig. A.12.

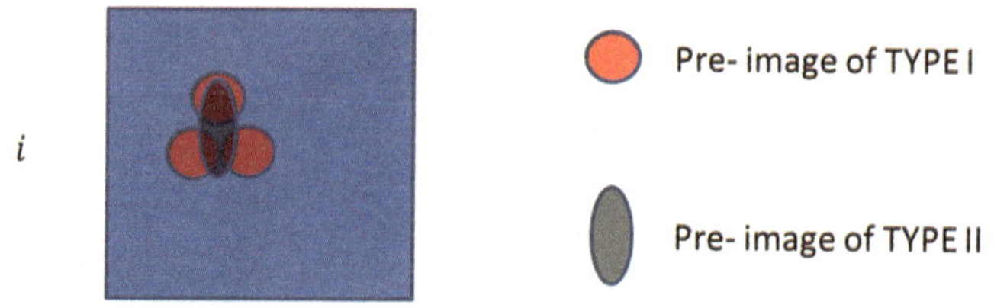

Figure A.12. *Preimage of covers in two different scenarios. The ellipse denotes the preimage of some Type* II *l-ball, but the preimage may be any shape which depends on the inverse of the flow.*

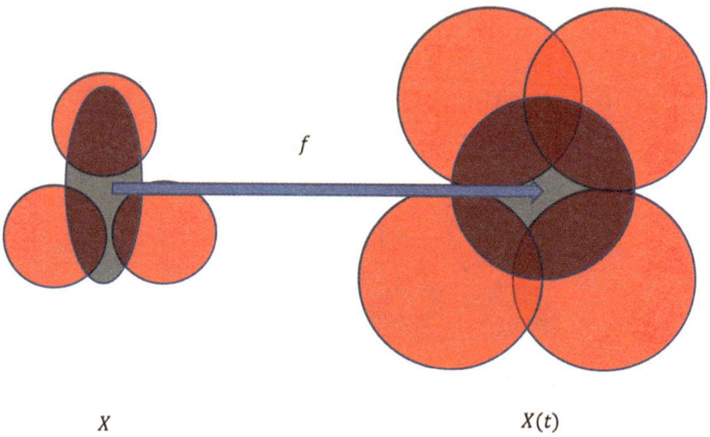

Figure A.13. *Geometry between nonoverlapping areas sample.*

Consider the areas of the balls of both X and $X(t)$ which are not overlapping. The flow is a bijection between these two areas; (see Fig. A.13), which is to say, the points in red ε'-balls in X will never go out of the l-balls in $X(t)$.

Notice the important connection between the nonoverlapping areas of X and $X(t)$; we can control the nonoverlapping area in X by shrinking the size of l-balls in $X(t)$. In other words, if l become smaller, so does the nonoverlapping area in X, that is, the nonoverlapping area in X can be ignored for some small l. Then we can guarantee a sufficient sampling for a given grid with length q if we choose a mesh grid of points with distance ε' or smaller. The number of points we need in each box of the partition of the initial status X should be more than $\frac{q}{\varepsilon'}$. To obtain an ε', we just choose a small l which depends on the Lipschitz constant of a specific problem. Note that this description is in terms of a uniform ε' grid. However, a uniformly random cover where this ε' condition is satisfied is also sufficient.

A.5.3 On Cantor Structures When Developing Ulam–Galerkin Matrices

A common scenario in chaotic dynamical systems and turbulent flow is that fine-scale structures develop by some sort of Cantor set type of construction. The following example highlights this propensity with a simple example and discusses the implications to computational complexity when developing an Ulam–Galerkin matrix.

Example A.1 (a Baker map and discussion of computational complexity [276]). Consider a Baker map as an example to describe the trade-off between the number of box coverings and the Lyapunov exponent. A Baker map on the unit square $[0,1]\times[0,1]$ may be written

$$x_{n+1} = \begin{cases} \lambda_1 x_n & \text{if } y_n < \alpha, \\ (1-\lambda_2)+\lambda_2 x_n & \text{if } y_n > \alpha, \end{cases}$$
$$y_{n+1} = \begin{cases} y_n/\alpha & \text{if } y_n < \alpha, \\ (y_n-\alpha)/(1-\alpha) & \text{if } y_n > \alpha, \end{cases} \tag{A.24}$$

where $\lambda_1+\lambda_2<1$ and $0<\alpha<1$. The forward-time invariant set of this map is a Cantor set of parallel vertical lines. Consider learning these structures with box covering methods, and what can be discovered by such finite computations. Suppose that we want to cover all parallel stripes left in the unit box *after n iterations* by square boxes of size ϵ so that two stripes are not contained in the same box; see Fig. A.14. The size of the square box required to do this is

$$\epsilon(n) = \min\{\lambda_1^n, \lambda_2^n\}. \tag{A.25}$$

Therefore, the required number of boxes to reveal the fractal structure of these parallel stripes is

$$N(\epsilon) \geq \epsilon(n)^{-2}, \tag{A.26}$$

and $d=2$ is the dimension of the embedding phase space. Using more boxes than the minimum requirement (A.26) would increase required computer storage and computational

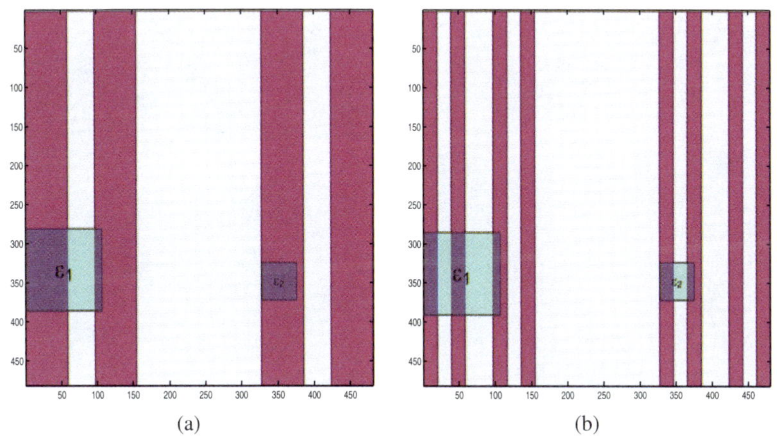

Figure A.14. *Two iterations of the Baker map with $\lambda_1=\lambda_2$.* (a) *All stripes after $n=2$. A square box of size ϵ_1 cannot resolve a structure finer than ϵ_1. But we see that a smaller square box of size ϵ_2 is able to resolve a stripe of width up to ϵ_2.* (b) *Here, the box of size ϵ_2 cannot resolve the stripe left in the unit box after $n=3$.*

work but would not give us a better knowledge of the stripes left in the unit box. One lesson we see with this is that there is a trade-off between Lyapunov exponents (Chapter 8) , λ_i, time, n, box size, ϵ, the size of the memory requirements, and the size of the resulting digraph, $N(\epsilon)$, and the dimension of the underlying space, d.

Bibliography

[1] R.L. Adler, A.G. Konheim, and M.H. McAndrew. Topological entropy. *Transactions of the American Mathematical Society*, 114(2):309–319, 1965. (Cited on pp. 284, 285)

[2] K.T. Alligood, T. Sauer, and J.A. Yorke. *Chaos: An Introduction to Dynamical Systems*. Springer, 1996. (Cited on pp. xii, 8, 62, 168, 180, 183)

[3] L. Arnold. *Random Dynamical Systems*. Monographs in Mathematics. Springer, 1998. (Cited on p. 240)

[4] L. Arnold. *Random Dynamical Systems*, 2nd ed. Springer, New York, 2003. (Cited on p. xii)

[5] R. Artuso, E. Aurell, and P. Cvitanovic. Recycling of strange sets: I. Cycle expansions. *Nonlinearity*, 3:325–359, 1990. (Cited on p. 286)

[6] P. Ashwin, J. Buescu, and I. Stewart. From attractor to chaotic saddle: A tale of transverse instability. *Nonlinearity*, 9:703–737, 1996. (Cited on p. 228)

[7] D. Auerbach, P. Cvitanović, J.P. Eckmann, G. Gunaratne, and I. Procaccia. Exploring chaotic motion through periodic orbits. *Physical Review Letters*, 58(23):2387–2389, 1987. (Cited on p. 286)

[8] L.W. Bagget. *Functional Analysis: A Primer.* Volume 157 of Pure and Applied Mathematics. Marcel Dekker, 1992. (Cited on pp. 67, 94)

[9] J.P. Bagrow and E. M. Bollt. A local method for detecting communities. *Phys. Rev. E*, 72(046108), 2005. (Cited on pp. 84, 132)

[10] J.P. Bagrow and E.M. Bollt. Local method for detecting communities. *Physical Review E*, 72(4):46108, 2005. (Cited on pp. 132, 133, 134, 136)

[11] V. Baladi. *Positive Transfer Operators and Decay of Correlations*. Volume 16 of Advanced Series in Nonlinear Dynamics. World Scientific, River Edge, NJ, 2000. (Cited on p. xii)

[12] J. Banks, J. Brooks, G. Cairns, G. Davis, and P. Stacey. On Devaney's definition of chaos. *The American Mathematical Monthly*, 99(4):332–334, 1992. (Cited on pp. 8, 55, 155, 167, 286)

[13] R.B. Bapat and T.E.S. Raghavan. *Nonnegative Matrices and Applications*. Volume 64 of Encyclopedia of Mathematics and Its Applications. Cambridge University Press, Cambridge, UK, 1997. (Cited on pp. 89, 90)

[14] S. Bassein. The dynamics of a family of one-dimensional maps. *The American Mathematical Monthly*, 105(2):118–130, 1998. (Cited on pp. 95, 96)

[15] H. van den Bedem and N. Chernov. Expanding maps of an interval with holes. *Ergodic Theory and Dynamical Systems*, 12:637–654, 2002. (Cited on p. 141)

[16] M. Benedicks and L. Carleson. The dynamics of the Hénon map. *The Annals of Mathematics*, 133(1):73–169, 1991. (Cited on p. 171)

[17] N. Berglund. *Perturbation Theory of Dynamical Systems*. Arxiv preprint math/0111178, 2001. (Cited on pp. 56, 58)

[18] O. Biham and W. Wenzel. Unstable periodic orbits and the symbolic dynamics of the complex Hénon map. *Physical Review A*, 42(8):4639–4646, 1990. (Cited on p. 288)

[19] L. Billings and E.M. Bollt. Probability density functions of some skew tent maps. *Chaos, Solitons and Fractals*, 12(2):365–376, 2001. (Cited on pp. 93, 94, 96, 97, 98, 99, 286, 291, 297, 303)

[20] L. Billings and I.B. Schwartz. Exciting chaos with noise: Unexpected dynamics in epidemic outbreaks. *Journal of Mathematical Biology*, 44(1):31–48, 2002. (Cited on p. 224)

[21] G.D. Birkhoff. *Dynamical Systems*. AMS, Providence, RI, 1966. (Cited on p. 60)

[22] B. Bollobás. *Modern Graph Theory*. Springer, New York, 1998. (Cited on pp. 82, 133)

[23] N. Boccara. *Functional Analysis: An Introduction for Physicists*. Academic Press, Boston, MA, 1990. (Cited on pp. 94, 223)

[24] E.M. Bollt. Communicating with chaos using high-dimensional symbolic dynamics. *Physics Letters A*, 255(1-2):75–81, 1999. (Not cited)

[25] E.M. Bollt. Controlling chaos and the inverse Frobenius-Perron problem: Global stabilization of arbitrary invariant measures. *International Journal of Bifurcation and Chaos*, 10(5):1033–1050, 2000. (Cited on pp. 53, 54)

[26] E.M. Bollt and E.J. Kostelich. Review of chaos communication by feedback control of symbolic dynamics. *International Journal of Bifurcation and Chaos*, 13(2):269–186, 2003. (Cited on pp. 150, 153, 154, 181, 186, 189, 191, 192, 278, 280)

[27] E.M. Bollt and E.J. Kostelich. Optimal targeting of chaos. *Physics Letters A*, 245(5):399–406, 1998. (Cited on p. 3)

[28] E. Bollt, Y.C. Lai, and C. Grebogi. Coding, channel capacity, and noise resistance in communicating with chaos. *Physical Review Letters*, 79(19):3787–3790, 1997. (Cited on pp. 60, 141, 153, 154, 232, 292)

[29] E. Bollt and Y.C. Lai. Dynamics of coding in communicating with chaos. *Physical Review E*, 58(2):1724–1736, 1998. (Cited on pp. 103, 141, 183, 185, 186, 228, 232, 234, 280, 286, 292, 293)

[30] E. Bollt. Combinatorial control of global dynamics in a chaotic differential equation. *International Journal of Bifurcations and Chaos*, 11(8):2145–2162, 2001. (Cited on p. 19)

[31] E. Bollt. Synchronization as a process of sharing and transferring information. *International Journal of Bifurcations and Chaos*, 22:1250261, 2012. (Cited on pp. 298, 300, 302, 304, 305, 306, 307)

[32] E.M. Bollt. The path towards a longer life: On invariant sets and the escape time landscape. *International Journal of Bifurcations and Chaos in Applied Sciences and Engineering*, 15(5):1615–1624, 2005. (Cited on pp. 228, 229, 230, 231)

[33] E.M. Bollt. Gulf movie. http://people.clarkson.edu/~ebollt/Papers/GulfFTLE.mov. (Cited on p. 227)

[34] E. Bollt, L. Billings, and I. Schwartz. A manifold independent approach to understanding transport in stochastic dynamical systems. *Physica D*, 173:153–177, 2002. (Cited on pp. 69, 220, 223, 224, 225, 226, 227)

[35] E.M. Bollt and A. Klebanoff. A new and simple chaos toy. *The International Journal of Bifurcations and Chaos*, 12:1843–1857, 2002. (Cited on p. 178)

[36] E.M. Bollt, A. Luttman, S. Kramer, and R. Basnayake. Measurable dynamics analysis of transport in the Gulf of Mexico during the oil spill. *The International Journal of Bifurcations and Chaos*, 22(03):1230012. (Cited on pp. 22, 23, 27, 28, 225, 227, 229)

[37] E. Bollt and T. Ma. Relatively coherent sets as a hierarchical partition method. *International Journal of Bifurcation and Chaos*, 23:1330026, 2013. (Cited on pp. 145, 146, 147, 227, 327)

[38] E. Bollt and J. Skufca. Markov Partitions. *Encyclopedia of Nonlinear Science*. Editor: A. Scot, Routledge, New York, 2005. (Cited on pp. 73, 74, 75)

[39] E.M. Bollt, T. Stanford, and Y.C. Lai. What symbolic dynamics do we get with a misplaced partition? On the validity of threshold crossings analysis of chaotic time-series. *Physica D*, 154(3-4):259–286, 2001. (Cited on p. 74)

[40] E.M. Bollt, T. Stanford, Y.C. Lai, and K. Życzkowski. Validity of threshold-crossing analysis of symbolic dynamics from chaotic time series. *Physical Review Letters*, 85(16):3524–3527, 2000. (Cited on pp. 183, 189, 192, 281, 295)

[41] E.M. Bollt, T. Stanford, Y.C. Lai, and K. Życzkowski. What symbolic dynamics do we get with a misplaced partition? On the validity of threshold crossings analysis of chaotic time-series. *Physica D*, 154(3-4):259–286, 2001. (Cited on pp. 152, 177, 183, 189, 192, 286, 287, 295)

[42] L. Boltzmann and J.T. Blackmore. *Ludwig Boltzmann: His Later Life and Philosophy, 1900–1906. A Documentary History*. Springer, 1995. (Cited on p. 60)

[43] C.J. Bose, and R. Murray. The exact rate of approximation in Ulam's method. Discrete and continuous dynamical systems. *Discrete and Continuous Dynamical Systems*, 7(1):219–235, 2001. (Cited on p. 325)

[44] R. Bowen. Markov partitions for Axiom A diffeomorphisms. *American Journal of Mathematics*, 92(3):725–747, 1970. (Cited on pp. 74, 285)

[45] R. Bowen. ω-limit sets for axiom A diffeomorphisms. *J. Differential Equations*, 2:333–339, 1975. (Cited on p. 81)

[46] R. Bowen. *Equilibrium States and the Ergodic Theory of Anosov Diffeomorphisms*. Volume 470 of Lecture Notes in Mathematics. Springer, Berlin, New York, 1975. (Cited on pp. 74, 81, 163, 285)

[47] R. Bowen. Periodic points and measures for Axiom A diffeomorphisms. *Transactions of the American Mathematical Society*, 154:377–397, 1971. (Cited on p. 286)

[48] R. Bowen. Topological entropy for noncompact sets. *Transactions of the American Mathematical Society*, 184:125–136, 1973. (Cited on p. 284)

[49] A. Boyarsky and Y.-S. Lou. Approximating measure invariant under higher-dimensional chaotic transformation. *Journal of Approximation Theory*, 65(2):231–244, 1991. (Cited on pp. 68, 69)

[50] A. Boyarsky and P. Gora. *Laws of Chaos: Invariant Measures and Dynamical Systems in One Dimension*. Probability and Its Applications. Birkhäuser Boston, Boston, MA, 1997. (Cited on pp. xii, 6, 69, 73, 94, 95, 98, 99, 296)

[51] L. Breiman. *Probability*. SIAM, Philadelphia, PA, 1992. (Cited on p. 282)

[52] M. Budišić and I. Mezić. Geometry of the ergodic quotient reveals coherent structures in flows. *Physica D*, 15:1255–1269, 2012. (Cited on p. 253)

[53] H. Buljan and V. Paar. Parry measure and the topological entropy of chaotic repellers embedded within chaotic attractors. *Physica D*, 172(1-4):111–123, 2002. (Cited on p. 284)

[54] Q. Chen. Area as a devil's staircase in twist maps. *Physics Letters A*, 123:444–450, 1987. (Cited on p. 220)

[55] G. Chen, Y. Mao, and C.K. Chui. A symmetric image encryption scheme based on 3D chaotic cat maps. *Chaos, Solitons and Fractals*, 21(3):749–761, 2004. (Cited on p. 278)

[56] F. Christiansen and H.H. Rugh. Computing Lyapunov spectra with continuous Gram–Schmidt orthonormalization. *Nonlinearity*, 10(5):1063–1072, 1997. (Cited on p. 241)

[57] C. Chui, Q. Du, and T. Li. Error estimates of the Markov finite approximation of the Frobenius Perron operator. *Nonlinear Analysis*, 19(4):291–308, 1992. (Cited on p. 69)

[58] F. Chung. Laplacians and the Cheeger inequality for directed graphs. *Annals of Combinatorics*, 9:1–19, 2005. (Cited on pp. 118, 127)

[59] F. Chung. *Spectral Graph Theory*. AMS, Providence, RI, 1997. (Cited on p. 138)

[60] A. Clauset, M.E.J. Newman, and C. Moore. Finding community structure in very large networks. *Physical Review E*, 70:066111, 2004. (Cited on p. 137)

[61] P. Collet and J.P. Eckmann. *Iterated Maps on the Interval as Dynamical Systems*. Birkhäuser Boston, Boston, MA, 2009. (Cited on p. 96)

[62] P. Collet, S. Martínez, and B. Schmitt. On the enhancement of diffusion by chaos, escape rates and stochastic stability. *Transactions of the American Mathematical Society*, 351(7):2875–2897, 1999. (Cited on p. 103)

[63] P. Collins. Symbolic dynamics from homoclinic tangles. *International Journal of Bifurcation and Chaos in Applied Sciences and Engineering*, 12(3):605–618, 2002. (Cited on p. 210)

[64] P. Constantin. *Integral Manifolds and Inertial Manifolds for Dissipative Partial Differential Equations*. Springer, 1989. (Cited on p. 31)

[65] N.J. Corron and S.D. Pethel. Control of long-period orbits and arbitrary trajectories in chaotic systems using dynamic limiting. *Chaos*, 12:1, 2002. (Cited on p. 150)

[66] N.J. Corron, S.D. Pethel, and B.A. Hopper. Controlling chaos with simple limiters. *Physical Review Letters*, 84(17):3835–3838, 2000. (Cited on p. 280)

[67] N.J. Corron, S.D. Pethel, and K. Myneni. Synchronizing the information content of a chaotic map and flow via symbolic dynamics. *Physical Review E*, 66(3):36204, 2002. (Cited on pp. 154, 280)

[68] T.M. Cover and J.A. Thomas. *Elements of Information Theory*. John Wiley, 2006. (Cited on pp. 269, 270, 271, 282, 283)

[69] K.M. Cuomo and A.V. Oppenheim. Circuit implementation of synchronized chaos with applications to communications. *Physical Review Letters*, 71(1):65–68, 1993. (Cited on p. 302)

[70] P. Cvitanović. Invariant measurement of strange sets in terms of cycles. *Physical Review Letters*, 61:2729–2732, 1988. (Cited on p. 286)

[71] P. Cvitanović, G.H. Gunaratne, and I. Procaccia. Topological and metric properties of Hénon-type strange attractors. *Physical Review A*, 38(3):1503–1520, 1988. (Cited on pp. 76, 174, 187, 189, 287, 288)

[72] G. D'Alessandro, P. Grassberger, S. Isola, and A. Politi. On the topology of the Hénon Map. *Journal of Physics A: Mathematical and General*, 23:5285, 1990. (Cited on pp. 294, 295)

[73] L. Dannon, A. Diaz-Guilera, J. Duch, and A. Arenas. *Comparing Community Structure Identification*. Preprint cond-mat/0505245, 2005. (Cited on pp. 84, 132)

[74] R.L. Davidchack and Y.C. Lai. Efficient algorithm for detecting unstable periodic orbits in chaotic systems. *Physical Review E*, 60(5):6172–6175, 1999. (Cited on p. 288)

[75] R.L. Davidchack, Y.C. Lai, E.M. Bollt, and M. Dhamala. Estimating generating partitions of chaotic systems by unstable periodic orbits. *Physical Review E*, 61(2):1353–1356, 2000. (Cited on pp. 287, 288, 289, 290, 292)

[76] R.L. Davidchack, Y.C. Lai, A. Klebanoff, and E.M. Bollt. Towards complete detection of unstable periodic orbits in chaotic systems. *Physics Letters A*, 287(1-2):99–104, 2001. (Cited on p. 288)

[77] C.S. Daw, C.E.A. Finney, and E.R. Tracy. A review of symbolic analysis of experimental data. *Review of Scientific Instruments*, 74:915–930, 2003. (Cited on p. 189)

[78] S. Day, R. Frongillo, and R. Treviño. Algorithms for rigorous entropy bounds and symbolic dynamics. *SIAM Journal on Applied Dynamical Systems*, 7:1477–1506, 2008. (Cited on p. 286)

[79] S. Day, O. Junge, and K. Mischaikow. A rigorous numerical method for the global analysis of infinite-dimensional discrete dynamical systems. *SIAM Journal on Applied Dynamical Systems*, 3(2):117–160, 2004. (Cited on p. 286)

[80] S. Day, J.-P. Lessard, and K. Mischaikow. Validated continuation for equilibria of PDEs. *SIAM Journal on Numerical Analysis*, 45(4):1398–1424, 2007. (Cited on p. 286)

[81] A. de Carvalho and T. Hall. How to prune a horseshoe. *Nonlinearity*, 15(3):19–68, 2002. (Cited on pp. 174, 177, 189)

[82] M. Dellnitz, G. Froyland, and O. Junge. The algorithms behind GAIO-set oriented numerical methods for dynamical systems. In *Ergodic Theory, Analysis, and Efficient Simulation of Dynamical Systems*, Springer, Berlin, 2001, pages 145–174. (Cited on pp. xii, 314, 319, 320)

[83] M. Dellnitz, R. Guder, and E. Kreuzer. An adaptive method for the approximation of the generalized cell mapping. *Chaos, Solitons and Fractals*, 8(4):525–534, 1997. (Cited on p. 325)

[84] M. Dellnitz, A. Hohmann, O. Junge, and M. Rumpf. Exploring invariant sets and invariant measures. *Chaos*, 7(2):221–228, 1997. (Cited on p. 324)

[85] M. Dellnitz and O. Junge. An adaptive subdivision technique for the approximation of attractors and invariant measures. *Computing and Visualization in Science*, 1(2):63–68, 1998. (Cited on p. 325)

[86] M. Dellnitz and O. Junge. An adaptive subdivision technique for the approximation of attractors and invariant measures. *Computing and Visualization in Science*, 1(2):63–68, 1998. (Cited on p. 327)

[87] M. Dellnitz and O. Junge. On the approximation of complicated dynamical behavior. *SIAM Journal on Numerical Analysis*, 36(2):491–515, 1999. (Cited on pp. xii, 62, 64)

[88] M. Dellnitz, R. Guder, and E. Kreuzer. An adaptive method for the approximation of the generalized cell mapping. *Chaos, Solitons and Fractals*, 8(4):525–534, 1997. (Cited on p. 327)

[89] W. De Melo and S. van Strien. *One-Dimensional Dynamics*. Springer, Berlin, Heidelberg, 1993. (Cited on pp. xii, 154, 156)

[90] M. Demers and L.-S. Young. Escape rates and conditionally invariant measures. *Nonlinearity*, 19:377–397, 2006. (Cited on pp. 103, 145)

[91] P. Deuflhard, W. Huisinga, A. Fischer, and C. Schütte. Identification of almost invariant aggregates in reversible nearly uncoupled Markov chains. *Linear Algebra and Its Applications*, 315(1-3):39–59, 2000. (Cited on pp. 107, 108, 110, 113)

[92] P. Deuflhard and F. Bornemann. *Scientific Computing with Ordinary Differential Equations*. Springer, 2002. (Cited on p. 58)

[93] P. Deuflhard, M. Dellnitz, O. Junge, and C. Schütte. Computation of essential molecular dynamics by subdivision techniques. In *Computational Molecular Dynamics: Challenges, Methods, Ideas*. Springer, Berlin, 1999, pages 98–115. (Cited on pp. 107, 108)

[94] P. Deuflhard and M. Weber. Robust Perron cluster analysis in conformation dynamics. *Linear Algebra and Its Applications*, 398:161–184, 2005. (Cited on pp. 107, 110, 113)

[95] R.L. Devaney. *An Introduction to Chaotic Dynamical Systems*. Westview Press, Boulder, CO, 2003. (Cited on pp. xii, 8, 52, 55, 96, 167, 180, 183, 286)

[96] R. Devaney and Z. Nitecki. Shift automorphisms in the Henon mapping. *Communications in Mathematical Physics*, 67(2):137–146, 1979. (Cited on pp. 157, 172)

[97] P. Diaconis, S. Holmes, and R. Montgomery. Dynamical bias in the coin toss. *SIAM Review*, 49(2):211–235, 2007. (Cited on pp. 169, 170)

[98] F.K. Diakonos, P. Schmelcher, and O. Biham. Systematic computation of the least unstable periodic orbits in chaotic attractors. *Physical Review Letters*, 81(20):4349–4352, 1998. (Cited on p. 288)

[99] E.I. Dinaburg. On the relations among various entropy characteristics of dynamical systems. *Izvestiya: Mathematics*, 5(2):337–378, 1971. (Cited on p. 284)

[100] J. Ding and Z. Wang. Parallel computation of invariant measures. *Annals of Operations Research*, 103(1-4):283–290, 2001. (Cited on pp. 324, 325)

[101] J. Ding and A.H. Zhou. The projection method for computing multidimensional absolutely continuous invariant measures. *Journal of Statistical Physics*, 77(3-4):899–908, 1994. (Cited on pp. 68, 69)

[102] J. Ding and A.H. Zhou. Piecewise linear Markov approximations of Frobenius–Perron operators associated with multi-dimensional transformations, *Nonlinear Analysis*, 25(4):399–408, 1995. (Cited on p. 68)

[103] J. Ding and A. Zhou. Finite approximations of Frobenius-Perron operators. A solution of Ulam's conjecture to multi-dimensional transformations. *Physica D*, 92(1-2):61–68, 1996. (Cited on p. 54)

[104] M. Dolnik and E.M. Bollt. Communication with chemical chaos in the presence of noise. *Chaos*, 8:702, 1998. (Cited on p. 280)

[105] J.F. Donges, Y. Zou, N. Marwan, and J. Kurths. The backbone of the climate network. *Europhysics Letters*, 87:48007, 2009. (Cited on pp. 274, 298, 305, 306, 308)

[106] R.W. Easton. *Geometric Methods for Discrete Dynamical Systems*. Oxford University Press, New York, 1998. (Cited on p. 210)

[107] R.W. Easton. Transport through chaos. *Nonlinearity*, 4:583–590, 1991. (Cited on p. 220)

[108] R.W. Easton. Trellises formed by stable and unstable manifolds in the plane. *Transactions of the American Mathematical Society*, 294(2):719–732, 1986. (Cited on pp. 204, 220)

[109] J.P. Eckmann and D. Ruelle. Ergodic theory of chaos and strange attractors. *Reviews of Modern Physics*, 57(3):617–656, 1985. (Cited on p. 297)

[110] C.H. Edwards, D.E. Penney, and D. Calvis. *Elementary Differential Equations*. Prentice–Hall, 2008. (Cited on p. 57)

[111] L.C. Evans. *Partial Differential Equations*. Volume 19 of Graduate Studies in Mathematics. AMS, Providence, RI, 2010. (Cited on p. 47)

[112] B. Fiedler. *Ergodic Theory, Analysis, and Efficient Simulation of Dynamical Systems*. Springer, 2001. (Cited on p. xii)

[113] M. Fiedler. Algebraic connectivity of graphs. *Czechoslovak Mathematical Journal*, 23(2):298–305, 1973. (Cited on p. 138)

[114] G.W. Flake, S. Lawrence, C.L. Giles, and F.M. Coetzee. Self-organization of the Web and identification of communities. *Communities*, 35(3):66–71, 2002. (Cited on p. 132)

[115] A.L. Fradkov and R.J. Evans. Control of chaos: Methods and applications in engineering. *Annual Reviews in Control*, 29(1):33–56, 2005. (Cited on p. 150)

[116] D. Fritzsche, V. Mehrmann, D.B. Szyld, and E. Virnik. An SVD approach to identifying metastable states of Markov chains. *Electronic Transactions on Numerical Analysis*, 29:46–69, 2008. (Cited on p. 113)

[117] G. Froyland. Finite approximation of Sinai-Bowen-Ruelle measures for Anosov systems in two dimensions. *Random Comput. Dynamics*, 3(4):251–263, 1995. (Cited on pp. 68, 69, 123)

[118] G. Froyland. *Estimating Physical Invariant Measures and Space Averages of Dynamical Systems Indicators*. Ph.D. thesis, University of Western Australia, 1996. (Cited on p. 69)

[119] G. Froyland. Computing physical invariant measures. In *Proceedings of the* 1997 *International Symposium on Nonlinear Theory and Its Applications*, Volume 2, 1997, pages 1129–1132. (Cited on p. 68)

[120] G. Froyland. Approximating physical invariant measures of mixing dynamical systems in higher dimensions. *Nonlinear Analysis*, 32(7):831–860, 1998. (Cited on pp. 68, 123)

[121] G. Froyland. Extracting dynamical behaviour via Markov models. In *Nonlinear Dynamics and Statistics*, A. Mees, ed., Birkhäuser Boston, Boston, MA, 2001, pages 283–324. (Cited on pp. xii, 63)

[122] G. Froyland. Statistically optimal almost-invariant sets. *Physica D*, 200(3-4):205–219, 2005. (Cited on pp. 64, 101, 113, 116)

[123] G. Froyland, O. Junge, and G. Ochs. Rigorous computation of topological entropy with respect to a finite partition. *Physica D*, 154(1-2):68–84, 2001. (Cited on pp. 292, 293, 295)

[124] G. Froyland, S. Lloyd, and N. Santitissadeekorn. Coherent sets for nonautonomous dynamical systems. *Physica D*, 239:1527–1541, 2010. (Cited on pp. 127, 317)

[125] G. Froyland and K. Padberg. Almost-invariant sets and invariant manifolds–Connecting probabilistic and geometric descriptions of coherent structures in flows. *Physica D*, 238(16):1507–1523, 2009. (Cited on pp. 101, 107, 113, 252, 255)

[126] G. Froyland, P.K. Pollett, and R. M Stuart. A closing scheme for finding almost-invariant sets in open dynamical systems. *Journal of Computational Dynamics*, to appear. (Cited on p. 141)

[127] G. Froyland, N. Santitissadeekorn, and A. Monahan. Normalized cut and image segmentation. *IEEE Transactions on Pattern Analysis and Machine Intelligence*, 22(8):888–905, 2000. (Cited on pp. 124, 125)

[128] G. Froyland, N. Santitissadeekorn, and A. Monahan. Transport in time-dependent dynamical systems: Finite-time coherent sets. *Chaos*, 20:043116, 2010. (Cited on pp. 116, 120, 126, 127, 128, 129, 130, 131, 145, 146, 147)

[129] G. Froyland and O. Stancevic. Open and closed dynamical systems. *Discrete and Continuous Dynamical Systems Series B*, 14:457–472, 2010. (Cited on pp. 141, 145)

[130] Z. Galias. Computational methods for rigorous analysis of chaotic systems. In *Intelligent Computing Based on Chaos*, Springer, Berlin, 2009, pages 25–51. (Cited on pp. 286, 291)

[131] Z. Galias. Interval methods for rigorous investigations of periodic orbits. *International Journal of Bifurcation and Chaos*, 11(9):2427–2450, 2001. (Cited on pp. 182, 286, 288, 290, 291)

[132] Z. Galias. Rigorous investigation of the Ikeda map by means of interval arithmetic. *Nonlinearity*, 15:1759, 2002. (Cited on pp. 286, 289, 290, 291, 292)

[133] Z. Galias. Rigorous numerical studies of the existence of periodic orbits for the Hénon map. *Journal of Universal Computer Science*, 4(2):114–124, 1998. (Cited on p. 182)

[134] Z. Galias and W. Tucker. Rigorous study of short periodic orbits for the Lorenz system. In *IEEE International Symposium on Circuits and Systems (ISCAS* 2008*)*. IEEE, 2008, pages 764–767. (Cited on p. 286)

[135] Z. Galias and W. Tucker. Validated study of the existence of short cycles for chaotic systems using symbolic dynamics and interval tools. *International Journal of Bifurcation and Chaos*, 21(2):551–563, 2011. (Cited on p. 182)

[136] A. Gervois and M.L. Mehta. Broken linear transformations. *Journal of Mathematical Physics*, 18:1476, 1977. (Cited on p. 96)

[137] J. Gleick, J. Glazier, and G. Gunaratne. Chaos: Making a new science. *Physics Today*, 41:79, 1988. (Cited on p. 166)

[138] A.V. Goldberg and R.E. Tarjan. A new approach to the maximum flow problem. *Journal of the ACM*, 35:921–940, 1988. (Cited on p. 138)

[139] H. Goldstein, C. Poole, J. Safko, and S.R. Addison. Classical mechanics. *American Journal of Physics*, 70:782, 2002. (Cited on p. 150)

[140] G.H. Golub and C.F. Van Loan. *Matrix Computations*, 3rd ed. Johns Hopkins University Press, Baltimore, MD, 1996. (Cited on pp. 104, 108, 109, 115, 116, 223)

[141] M. Golubitsky, I. Stewart, P.L. Buono, and J.J. Collins. Symmetry in locomotor central pattern generators and animal gaits. *Nature*, 401(6754):693–695, 1999. (Cited on p. 302)

[142] T.N.T. Goodman. Relating topological entropy and measure entropy. *Bulletin of the London Mathematical Society*, 3(2):176, 1971. (Cited on p. 284)

[143] P. Grassberger. On the fractal dimension of the Henon attractor. *Physics Letters A*, 6(5):224–226, 1983. (Cited on p. 322)

[144] P. Grassberger, H. Kantz, and U. Moenig. On the symbolic dynamics of the Hénon map. *Journal of Physics A: Mathematical and General*, 22:5217–5230, 1989. (Cited on pp. 76, 174, 187, 189, 287, 288)

[145] J.M. Greene, M. Henon, A. Mehr, A.M. Soward, T. Dombre, and U. Frisch. Chaotic streamlines in the ABC flows. *Journal of Fluid Mechanics*, 167:353–391, 1986. (Cited on p. 253)

[146] J. Guckenheimer and P. Holmes. *Nonlinear Oscillators, Dynamical Systems, and Bifurcations of Vector Fields*. Volume 42 of Applied Mathematical Sciences. Springer, 1983. (Cited on pp. xii, 29, 32, 36, 169, 178, 201, 242)

[147] R. Guder and E. Kreuzer. Control of an adaptive refinement technique of generalized cell mapping by system dynamics. *Nonlinear Dynamics*, 20(1):21–32, 1999. (Cited on p. 325)

[148] R. Guder and E. Kreuzer. Control of an adaptive refinement technique of generalized cell mapping by system dynamics. *Nonlinear Dynamics*, 20(1):21–32, 1999. (Cited on p. 327)

[149] B.M. Gurevich and S.V. Savchenko. Thermodynamic formalism for countable symbolic Markov chains. *Russian Mathematical Surveys*, 53:245, 1998. (Cited on p. 284)

[150] L. Hagen and A.B. Kahng. New spectral methods for ratio cut partitioning and clustering. *IEEE Transactions on Computer-Aided Design of Integrated Circuits and Systems*, 11(9):1074–1085, 1992. (Cited on p. 138)

[151] D.W. Hahs and S.D. Pethel. Distinguishing anticipation from causality: Anticipatory bias in the estimation of information flow. *Physical Review Letters*, 107(12):128701, 2011. (Cited on pp. 280, 302, 303)

[152] G. Haller. Distinguished material surfaces and coherent structures in three-dimensional fluid flows. *Physica D*, 149:248–277, 2001. (Cited on pp. 243, 258, 260)

[153] G. Haller. Finding finite-time invariant manifolds in two-dimensional velocity fields. *Chaos*, 10:99–108, 2000. (Cited on pp. 41, 128, 243, 244, 258, 260)

[154] G. Haller. Lagrangian coherent structures from approximate velocity data. *Phys. Fluids A*, 14:1851–1861, 2002. (Cited on p. 41)

[155] G. Haller. A variational theory of hyperbolic Lagrangian coherent structures. *Physica D*, 240(7):574–598, 2011. (Cited on pp. 258, 259, 260, 261, 262, 263)

[156] G. Haller and A. C. Poje. Finite-time transport in aperiodic flows. *Physica D*, 119:352–380, 1998. (Cited on p. 41)

[157] S.M. Hammel, C. Jones, and J.V. Moloney. Global dynamical behavior of the optical field in a ring cavity. *Optical Society of America, Journal B: Optical Physics*, 2:552–564, 1985. (Cited on p. 288)

[158] T.E. Harris. The existence of stationary measures for certain Markov processes. In *Proceedings of the Third Berkeley Symposium on Mathematical Statistics and Probability: Held at the Statistical Laboratory, University of California*, 1954–1955, Volume 1. University of California Press, Berkeley, Los Angeles, 1956, page 113. (Cited on p. 93)

[159] M. Hasler and Y.L. Maistrenko. An introduction to the synchronization of chaotic systems: Coupled skew tent maps. *IEEE Transactions on Circuits and Systems* I: Fundamental Theory and Applications, 44(10):856–866, 1997. (Cited on p. 303)

[160] S. Hayes, C. Grebogi, and E. Ott. Communicating with chaos. *Physical Review Letters*, 70(20):3031–3034, 1993. (Cited on pp. 154, 278, 280)

[161] D. He and L. Stone. Spatio-temporal synchronization of recurrent epidemics. *Proceedings of the Royal Society of London. Series B: Biological Sciences*, 270(1523):1519, 2003. (Cited on p. 302)

[162] M. Hénon. Numerical study of quadratic area-preserving mappings. *Quarterly of Applied Mathematics*, 27:291–312, 1969. (Cited on p. 210)

[163] M. Hénon. A two-dimensional mapping with a strange attractor. *Communications in Mathematical Physics*, 50(1):69–77, 1976. (Cited on p. 171)

[164] M.W. Hirchh and S. Smale. *Differential Equations, Dynamical Systems, and Linear Algebra*. Academic Press, Orlando, FL, 1974. (Cited on p. 19)

[165] B.K.P. Horn and B.G. Schunck. Determining optical flow. *Artificial intelligence*, 17:185–203, 1981. (Cited on p. 125)

[166] C.S. Hsu. A theory of cell-to-cell mapping dynamical systems. *Journal of Applied Mechanics*, 47:931, 1980. (Cited on pp. xii, 314)

[167] C.S. Hsu. *Cell-to-Cell Mapping: A Method of Global Analysis for Nonlinear Systems*. Springer, 1987. (Cited on pp. xi, xii)

[168] U. Huebner, N.B. Abraham, and C.O. Weiss. Dimensions and entropies of chaotic intensity pulsations in a single-mode far-infrared NH_3 laser. *Physical Review A*, 40(11):6354–6365, 1989. (Cited on p. 152)

[169] D.A. Huffman. A method for the construction of minimum-redundancy codes. *Proceedings of the IRE*, 40(9):1098–1101, 1952. (Cited on pp. 266, 268)

[170] E.Y. Hunt. A Monte Carlo approach to the approximation of invariant measures. *Random Comput. Dynamics*, 2(1):111–133, 1994. (Cited on pp. 324, 325)

[171] B.R. Hunt, J.A. Kennedy, T.Y. Li, and H.E. Nusse. SLYRB measures: Natural invariant measures for chaotic systems. *Physica D*, 170(1):50–71, 2002. (Cited on pp. 6, 62)

[172] F. Hunt. Approximating the invariant measures of randomly perturbed dissipative maps. *Journal of Mathematical Analysis and Applications*, 198(2):534–551, 1996. (Cited on p. 69)

[173] HYCOM. Hybrid Coordinate Ocean Model (HYCOM). http://www.hycom.org/, 2010. (Cited on pp. 23, 25, 228)

[174] K. Ide, D. Small, and S. Wiggins. Distinguished hyperbolic trajectories in time-dependent fluid flows: Analytical and computational approach for velocity fields defined as data sets. *Nonlinear Processes in Geophysics*, (9):237–263, 2002. (Cited on p. 41)

[175] K. Ikeda, H. Daido, and O. Akimoto. Optical turbulence: Chaotic behavior of transmitted light from a ring cavity. *Physical Review Letters*, 45(9):709–712, 1980. (Cited on p. 288)

[176] M. Iosifescu. *Finite Markov Processes and Their Applications*. Dover, 1980. (Cited on pp. 143, 223)

[177] M. Jordan. Spectral clustering: Analysis and an algorithm. *Machine Learning Journal*, 5, 2001. (Cited on p. 138)

[178] M.I. Jordan, A.Y. Ng, and Y. Weiss. On spectral clustering: Analysis and an algorithm. In *Proceedings of the conference on Neural Information Processing Systems (NIPS)*, 2001. (Cited on p. 138)

[179] B. Joseph and B. Legras. On the relation between kinematic boundaries, stirring, and barriers for the antarctic polar vortex. *Journal of the Atmospheric Sciences*, 59:1198–1212, 2002. (Cited on pp. 128, 255)

[180] O. Junge. An adaptive subdivision technique for the approximation of attractors and invariant measures: Proof of convergence. *Dynamical Systems. An International Journal*, 16(3):213–222, 2001. (Cited on pp. 325, 327)

[181] T. Kaczyński, K.M. Mischaikow, and M. Mrozek. *Computational Homology*. Springer, 2004. (Cited on p. 286)

[182] H. Kantz, T. Schreiber, and R.S. Mackay. *Nonlinear Time Series Analysis*. Cambridge University Press, Cambridge, UK, 1997. (Cited on p. 298)

[183] D.R. Karger and C. Stein. A new approach to the minimum cut problem. *Journal of the ACM*, 43(4):601–640, 1996. (Cited on p. 138)

[184] D. Karrasch. Comment on "A variational theory of hyperbolic Lagrangian coherent structures, Physica D, 240 (2011), 574–598." *Physica D*, 241(17):1470–1473, 2011. (Cited on p. 263)

[185] A. Katok. Fifty years of entropy in dynamics: 1958–2007. *Journal of Modern Dynamics*, 1(4):689–718, 2007. (Cited on p. 282)

[186] M.B. Kennel and M. Buhl. Estimating good discrete partitions from observed data: Symbolic false nearest neighbors. *Physical Review Letters*, 91(8):84102, 2003. (Cited on p. 194)

[187] B.W. Kernighan and S. Lin. An efficient heuristic procedure for partitioning graphs. *Bell Systems Technical Journal*, 49:291–307, 1970. (Cited on p. 138)

[188] B. Kitchens. *Symbolic Dynamics: One-Sided, Two-Sided, and Countable State Markov Shifts*. Springer, 1998. (Cited on pp. 180, 181, 182, 183, 291)

[189] B.P. Koch and R.W. Leven. Subharmonic and homoclinic bifurcations in a parametrically forced pendulum. *Physica D*, 16(1):1–13, 1985. (Cited on p. 179)

[190] A.N. Kolmogorov and S.V. Fomin. *Introductory Real Analysis*. Dover, New York, 1975. (Cited on pp. 6, 14, 31, 167)

[191] A.N. Kolmogorov and S.V. Fomin. *Elements of the Theory of Functions and Functional Analysis*. Dover, 1999. (Cited on pp. 67, 94)

[192] W.S. Koon, M.W. Lo, J.E. Marsden, and S.D. Ross. Dynamical systems, the three-body problem and space mission design. In *International Conference on Differential Equations (Berlin, 1999)*, Volumes 1 and 2. World Scientific, River Edge, NJ, 2000, pages 1167–1181. (Cited on p. 218)

[193] N. Korabel and E. Barkai. *Pesin-Type Identity for Weak Chaos*. Arxiv preprint arXiv:0808.1398, 2008. (Cited on p. 297)

[194] G. Kovačič. Lobe area via action formalism in a class of Hamiltonian systems. *Physica D*, 51(1-3):226–233, 1991. (Cited on p. 220)

[195] V. Kumaran. Josiah Willard Gibbs. *Resonance*, 12(7):4–11, 2007. (Cited on p. 60)

[196] Y.C. Lai, E. Bollt, and C. Grebogi. Communicating with chaos using two-dimensional symbolic dynamics. *Physics Letters A*, 255(1-2):75–81, 1999. (Cited on pp. 232, 286)

[197] G.G. Langdon Jr. and J.J. Rissanen. *Method and Means for Arithmetic Coding Utilizing a Reduced Number of Operations*, August 25, 1981. U.S. Patent 4,286,256. (Cited on pp. 269, 271)

[198] A. Lasota and M. C. Mackey. *Chaos, Fractals, and Noise: Stochastic Aspects of Dynamics*, 2nd ed. Springer, New York, 1994. (Cited on pp. xii, 43, 44, 46, 47, 48, 52, 96, 193)

[199] D.T. Lee and B.J. Schachter. Two algorithms for constructing a Delaunay triangulation. *International Journal of Parallel Programming*, 9(3):219–242, 1980. (Cited on p. 292)

[200] F. Lekien and S.D. Ross. The computation of finite-time Lyapunov exponents on unstructured meshes and for non-Euclidean manifolds. *Chaos*, 20(1):1–20, 2010. (Cited on pp. 244, 246)

[201] T.Y. Li. Finite approximation for the Frobenius-Perron operator. A solution to Ulam's conjecture. *Journal of Approximation Theory*, 17(2):177–186, 1976. (Cited on pp. 67, 68, 325)

[202] A.J. Lichtenberg and M.A. Lieberman. *Regular and Stochastic Motion*. Volume 38 of Applied Mathematical Sciences. Springer, New York, 1983. (Cited on p. 178)

[203] D.R. Lide. *CRC Handbook of Chemistry and Physics*. CRC, 2003. (Cited on p. 86)

[204] D.A. Lind and B. Marcus. *An Introduction to Symbolic Dynamics and Coding*. Cambridge University Press, Cambridge, UK, 1995. (Cited on pp. 156, 180, 181, 182, 183, 295)

[205] D. Lippolis and P. Cvitanović. How well can one resolve the state space of a chaotic map? *Physical Review Letters*, 104(1):14101, 2010. (Cited on pp. 193, 194)

[206] C. Liverani and V. Maume-Deschamps. Lasota-Yorke maps with holes: Conditionally invariant probability measures and invariant probability measures on the survivor set. *Annales de l'Institut Henri Poincaré. Probabilités et Statistiques*, 39:385–412, 2003. (Cited on p. 141)

[207] E.N. Lorenz. Deterministic nonperiodic flow. *Journal of the Atmospheric Sciences*, 20:130–141, 1963. (Cited on p. 152)

[208] R.S. MacKay. *Renormalisation in Area-Preserving Maps*. World Scientific, River Edge, NJ, 1993. (Cited on pp. 209, 211)

[209] R.S. Mackey, J.D. Meiss, and I.C. Percival. Transport in Hamiltonian systems. *Physica D*, 13(1-2):55–81, 1984. (Cited on pp. 204, 206, 215, 220)

[210] A.M. Mancho, D. Small, S. Wiggins, and K. Ide. Computation of stable and unstable manifolds of hyperbolic trajectories in two-dimensional, aperiodically time-dependent vectors fields. *Physica D*, 182:188–222, 2003. (Cited on p. 41)

[211] A.M. Mancho, D. Small, and S. Wiggins. Computation of hyperbolic trajectories and their stable and unstable manifolds for oceanographic flows represented as data sets. *Nonlinear Processes in Geophysics*, 11:17–33, 2004. (Cited on p. 41)

[212] A.M. Mancho, D. Small, and S. Wiggins. A tutorial on dynamical systems concepts applied to Lagrangian transport in oceanic flows defined as finite time data sets: Theoretical and computational issues. *Physics Reports*, 437(3-4):55–124, 2006. (Cited on p. 41)

[213] M. Martinelli, M. Dang, and T. Seph. Defining chaos. *Mathematics Magazine*, 71:112–122, 1998. (Cited on p. 8)

[214] S. Martínez, P. Collet, and J. San Martín. Asymptotic laws for one-dimensional diffusions conditioned to nonabsorption. *The Annals of Probability*, 23:1300–1314, 1995. (Cited on p. 145)

[215] J.L. McCauley. *Chaos, Dynamics and Fractals: An Algorithmic Approach to Deterministic Chaos*. Cambridge Nonlinear Science Series. Cambridge University Press, Cambridge, UK, 1993. (Cited on pp. xii, 12)

[216] M.E. McIntyre and T.N. Palmer. The "surf zone" in the stratosphere. *Journal of Atmospheric and Terrestrial Physics*, 46:825–849, 1983. (Cited on pp. 132, 254)

[217] F.A. McRobie and J.M.T. Thompson. Lobe dynamics and the escape from a potential well. *Proceedings of the Royal Society. London. Series A: Mathematical, Physical and Engineering Sciences*, 435(1895):659–672, 1991. (Cited on p. 210)

[218] J.D. Meiss. *Differential Dynamical Systems*. Volume 14 of Mathematical Modeling and Computation. SIAM, Philadelphia, PA, 2007. (Cited on pp. xii, 29, 30)

[219] J.D. Meiss. Symplectic maps, variational principles, and transport. *Reviews of Modern Physics*, 64(3):795–848, 1992. (Cited on pp. 206, 210, 212, 215, 311, 317)

[220] I. Mezić and F. Sotiropoulos. Ergodic theory and experimental visualization of chaos. *Physics of Fluids*, 14:2235–2243, 2002. (Cited on p. 66)

[221] I. Mezić and S. Wiggins. A method for visualization of invariant sets of dynamical systems based on ergodic partition. *Chaos*, 9:213–218, 1999. (Cited on p. 66)

[222] W.M. Miller. Stability and approximation of invariant measures for a class of nonexpanding transformations. *Nonlinear Analysis*, 23(8):1013–1015, 1994. (Cited on p. 63)

[223] J. Milnor and W. Thurston. On iterated maps of the interval. In *Dynamical Systems*. Volume 1342 of Lecture Notes in Mathematics. Springer, Berlin, 1988, pages 465–563. (Cited on pp. 96, 154, 156, 183, 184, 185, 189)

[224] K. Mischaikow and M. Mrozek. Chaos in the Lorenz equations: A computer assisted proof. *American Mathematical Society. Bulletin. New Series*, 32(1):66–72, 1995. (Cited on pp. 182, 286)

[225] K. Mischaikow and M. Mrozek. Chaos in the Lorenz equations: A computer assisted proof. Part II: Details. *Mathematics of Computation*, 67(223):1023–1046, 1998. (Cited on p. 286)

[226] K. Mischaikow, M. Mrozek, and A. Szymczak. Chaos in the Lorenz equations: A computer assisted proof. III. Classical parameter values. *Journal of Differential Equations*, 169(1):17–56, 2001. (Cited on p. 286)

[227] M. Misiurewicz. Possible jumps of entropy for interval maps. *Qualitative Theory of Dynamical Systems*, 2(2):289–306, 2001. (Cited on pp. 96, 99)

[228] M. Misiurewicz and E. Visinescu. Kneading sequences of skew tent maps. *Annales de l'Institut Henri Poincaré. Probabilités et Statistiques*, 27:125–140, 1991. (Cited on pp. 96, 97, 98, 99)

[229] K.A. Mitchell. The topology of nested homoclinic and heteroclinic tangles. *Physica D*, 238(7):737–763, 2009. (Cited on pp. 210, 216, 218, 292)

[230] K.A. Mitchell and J.B. Delos. A new topological technique for characterizing homoclinic tangles. *Physica D*, 221(2):170–187, 2006. (Cited on p. 229)

[231] K.A. Mitchell, J.P. Handley, J.B. Delos, and S.K. Knudson. Geometry and topology of escape. II. Homotopic lobe dynamics. *Chaos*, 13:892–902, 2003. (Cited on pp. 216, 218)

[232] R.E. Moore. *Methods and Applications of Interval Analysis*. SIAM, Philadelphia, PA, 1979. (Cited on pp. 288, 289)

[233] P. Moresco and S. Ponce Dawson. The PIM-simplex method: An extension of the PIM-triple method to saddles with an arbitrary number of expanding directions. *Physica D*, 126(1-2):38–48, 1999. (Cited on p. 228)

[234] J. Moser. On Quadratic Symplectic Mappings. *Mathematische Zeitschrift*, 216:417–430, 1994. (Cited on pp. 157, 211)

[235] J.R. Munkres. *Topology*, 2nd ed. Prentice–Hall, Upper Saddle River, NJ, 2000. (Cited on pp. 73, 94, 292)

[236] R. Murray. Ulam's method for some non-uniformly expanding maps. *Discrete and Continuous Dynamical Systems*, 26(3):1007–1018, 2010. (Cited on p. 69)

[237] E.R. Nash, P.A. Newman, J.E. Rosenfield, and M.R. Schoeberl. An objective determination of the polar vortex using Ertel's potential vorticity. *Journal of Geophysical Research: Atmospheres*, 101:9471–9476, 1996. (Cited on p. 254)

[238] M.E.J. Newman. Fast algorithm for detecting community structure in networks. *Physical Review E*, 69:066133, 2004. (Cited on pp. 133, 134)

[239] M.E.J. Newman. Finding community structure in networks using the eigenvectors of matrices. *Physical Review E*, 74:036104, 2006. (Cited on p. 138)

[240] M.E.J. Newman and M. Girvan. *Statistical Mechanics of Complex Networks*. Springer, Berlin, 2004. (Cited on pp. 80, 133)

[241] M.E.J. Newman and M. Girvan. Finding and evaluating community structure in networks. *Phys. Rev. E*, 69(2):026113, 2004. (Cited on pp. 80, 84, 132, 133, 135)

[242] H.E. Nusse and J.A. Yorke. A procedure for finding numerical trajectories on chaotic saddles. *Physica D*, 36(1-2):137–156, 1989. (Cited on pp. 60, 228)

[243] V.I. Oseledets. A multiplicative ergodic theorem. Lyapunov characteristic numbers for dynamical systems. *Transactions of the Moscow Mathematical Society*, 19:197–231, 1968. (Cited on p. 240)

[244] E. Ott. *Chaos in Dynamical Systems*, 1st ed. Cambridge University Press, Cambridge, UK, 1993. (Cited on pp. xii, 322)

[245] E. Ott, C. Grebogi, and J.A. Yorke. Controlling chaos. *Physical Review Letters*, 64(11):1196–1199, 1990. (Cited on pp. 150, 280)

[246] E. Ott and J.D. Meiss. Markov-tree model of intrinsic transport in Hamiltonian systems. *Physical Review Letters*, 55:2741–2744, 1985. (Cited on p. 220)

[247] K. Padberg. *Numerical Analysis of Transport in Dynamical Systems*. Ph.D. thesis, Universität Paderborn, Paderborn, 2005. (Cited on p. xii)

[248] W. Parry. Symbolic dynamics and transformations of the unit interval. *Transactions of the American Mathematical Society*, 122(2):368–378, 1966. (Cited on p. 284)

[249] L.M. Pecora and T.L. Carroll. Master stability functions for synchronized coupled systems. *Physical Review Letters*, 80(10):2109–2112, 1998. (Cited on p. 302)

[250] L.M. Pecora and T.L. Carroll. Synchronization in chaotic systems. *Physical Review Letters*, 64(8):821–824, 1990. (Cited on p. 302)

[251] L. Perko. *Differential Equations and Dynamical Systems*. Springer, 2001. (Cited on pp. xii, 19, 30, 31, 32, 178, 201, 327)

[252] Y.B. Pesin. Characteristic Lyapunov exponents and smooth ergodic theory. *Russian Mathematical Surveys*, 32(4):55–114, 1977. (Cited on p. 297)

[253] S.D. Pethel, N.J. Corron, and E. Bollt. Deconstructing spatiotemporal chaos using local symbolic dynamics. *Physical Review Letters*, 99(21):214101, 2007. (Cited on p. 278)

[254] S.D. Pethel, N.J. Corron, and E. Bollt. Symbolic dynamics of coupled map lattices. *Physical Review Letters*, 96(3):34105, 2006. (Cited on p. 150)

[255] S.D. Pethel, N.J. Corron, Q.R. Underwood, and K. Myneni. Information flow in chaos synchronization: Fundamental tradeoffs in precision, delay, and anticipation. *Physical Review Letters*, 90(25):254101, 2003. (Cited on p. 150)

[256] G. Pianigiani and J. A. Yorke. Expanding maps on sets which are almost invariant. Decay and chaos. *Transactions of the American Mathematical Society*, 252:351–366, 1979. (Cited on pp. 141, 145)

[257] R.T. Pierrehumbert and H. Yang. Global chaotic mixing on isentropic surfaces. *Journal of the Atmospheric Sciences*, 50:2462–2480, 1993. (Cited on pp. 183, 243)

[258] R.A. Plumb. Stratospheric transport. *Journal of the Meteorological Society of Japan*, 80:793–809, 2002. (Cited on p. 253)

[259] H. Poincaré and R. Magini. Les méthodes nouvelles de la mécanique céleste. *Il Nuovo Cimento* (1895–1900), 10(1):128–130, 1899. (Cited on p. 215)

[260] M. Pollicott and M. Yuri. *Dynamical Systems and Ergodic Theory*. Volume 40 of London Mathematical Society Student Texts. Cambridge University Press, Cambridge, UK, 1998. (Cited on p. 35)

[261] M.A. Porter, J.P. Onnela, and P.J. Mucha. Communities in networks. *Notices of the American Mathematical Society*, 56(9):1082–1097, 2009. (Cited on p. 132)

[262] A. Pothen, H.D. Simon, and K.-P. Liou. Partitioning sparse matrices with eigenvectors of graphs. *SIAM Journal on Matrix Analysis and Applications*, 11(3):430–452, 1990. (Cited on p. 138)

[263] D. Pountain. Run-length encoding. *Byte*, 12(6):317–319, 1987. (Cited on p. 267)

[264] W.H. Press, S.A. Teukolsky, W.T. Vetterling, and B.P. Flannery. *Numerical Recipes in C: The Art of Scientific Computing*, 2nd ed. Cambridge University Press, Cambridge, UK, 1992. (Cited on pp. 325, 326)

[265] A. Pressley. *Elementary Differential Geometry*. Springer, 2010. (Cited on pp. 33, 34)

[266] F. Radicchi, C. Castellano, F. Cecconi, V. Loreto, and D. Parisi. Defining and identifying communities in networks. *Proceedings of the National Academy of Sciences*, 101(9):2658, 2004. (Cited on p. 132)

[267] C.P. Robert and G. Casella. *Monte Carlo Statistical Methods*. Springer, New York, 1998. (Cited on p. 325)

[268] C. Robinson. *Dynamical Systems: Stability, Symbolic Dynamics, and Chaos*, 2nd ed. CRC Press, Boca Raton, FL, 1999. (Cited on pp. xii, 29, 36, 37, 38, 40, 75, 149, 167, 172, 181, 182, 183, 190, 201, 206, 209, 229, 284, 285, 287)

[269] J.C. Robinson. *Infinite-Dimensional Dynamical Systems: An Introduction to Dissipative Parabolic PDEs and the Theory of Global Attractors*. Cambridge University Press, Cambridge, UK, 2001. (Cited on pp. 31, 47, 292)

[270] V. Rom-Kedar. Transport rates of a class of two-dimensional maps and flows. *Physica D*, 43:229–268, 1990. (Cited on p. 220)

[271] H.L. Royden and P. Fitzpatrick. *Real Analysis*, Volume 2. Macmillan, New York, 1968. (Cited on pp. 14, 94)

[272] D.J. Rudolph. *Fundamentals of Measurable Dynamics: Ergodic Theory on Lebesgue Spaces*. Clarendon Press, Oxford University Press, New York, 1990. (Cited on p. 74)

[273] D. Ruelle. An inequality for the entropy of differentiable maps. *Bulletin of the Brazilian Mathematical Society*, 9(1):83–87, 1978. (Cited on p. 297)

[274] I.I. Rypina, M.G. Brown, F.J. Beron-Vera, H. Kocak, M.J. Olascoaga, and I.A. Udovydchenkov. On the Lagrangian dynamics of atmospheric zonal jets and the permeability of the stratospheric polar vortex. *Journal of the Atmospheric Sciences*, 64:3595–3610, 2007. (Cited on p. 127)

[275] N. Santitissadeekorn and E. Bollt. The infinitesimal operator for the semigroup of the Frobenius–Perron operator from image sequence data: Vector fields and computational measurable dynamics from movies. *Chaos*, 17:023126, 2007. (Cited on p. 48)

[276] N. Santitissadeekorn and E.M. Bollt. Identifying stochastic basin hopping and mechanism by partitioning with graph modularity. *Physica D*, (231):95–107, 2007. (Cited on pp. 51, 331)

[277] N. Santitissadeekorn, G. Froyland, and A. Monahan. Optimally coherent sets in geophysical flows: A new approach to delimiting the stratospheric polar vortex. *Physical Review E*, 82(056311), 2010. (Cited on p. 126)

[278] O.M. Šarkovs'kiĭ. Co-existence of cycles of a continuous mapping of the line into itself. *Ukrainskiĭ Matematičeskiĭ Žurnal*, 16:61–71, 1964. (Cited on p. 183)

[279] P. Schmelcher and F.K. Diakonos. Detecting unstable periodic orbits of chaotic dynamical systems. *Physical Review Letters*, 78(25):4733–4736, 1997. (Cited on pp. 288, 289)

[280] A. Schmiegel and B. Eckhardt. Fractal stability border in plane Couette flow. *Physical Review Letters*, 79:5250–5253, 1997. (Cited on p. 228)

[281] T. Schreiber. Measuring information transfer. *Physical Review Letters*, 85(2):461–464, 2000. (Cited on pp. 298, 301)

[282] I.B. Schwartz. Multiple stable recurrent outbreaks and predictability in seasonally forced nonlinear epidemic models. *Journal of Mathematical Biology*, 21(3):347–361, 1985. (Cited on p. 224)

[283] I.B. Schwartz. Small amplitude, long period outbreaks in seasonally driven epidemics. *Journal of Mathematical Biology*, 30(5):473–491, 1992. (Cited on p. 224)

[284] I. B. Schwartz, L. Billings, and E. Bollt. Dynamical epidemic suppression using stochastic prediction and control. *Physical Review E*, 70:046220, 2004. (Not cited)

[285] S.C. Shadden. *A Dynamical Systems Approach to Unsteady Systems*. Dissertation, California Institute of Technology, Pasadena, CA, 2006. (Cited on pp. 243, 246, 257, 260)

[286] S.C. Shadden, J.O. Dabiriand, and J.E. Marsden. Lagrangian analysis of fluid transport in empirical vortex ring flows. *Physics of Fluids*, 18(4):271–304, 2006. (Cited on p. 42)

[287] S.C. Shadden, F. Lekien, and J.E. Marsden. Definition and properties of Lagrangian coherent structures from finite-time Lyapunov exponents in two-dimensional aperiodic flows. *Physica D*, 212:271–304, 2005. (Cited on p. 42)

[288] T. Shinbrot. Progress in the control of chaos. *Advances in Physics*, 44(2):73–111, 1995. (Cited on p. 150)

[289] T. Shinbrot, C. Grebogi, E. Ott, and J.A. Yorke. Using small perturbations to control chaos. *Nature*, 363:411–417, 1993. (Cited on p. 280)

[290] E. Simiu. *Chaotic Transition in Deterministic and Stochastic Dynamical Systems*. Princeton University Press, Princeton, NJ, 2002. (Cited on p. 179)

[291] Y.G. Sinai, editor. *Dynamical Systems* II. *Ergodic Theory with Applications to Dynamical Systems and Statistical Mechanics.* Volume 2 of Encyclopaedia of Mathematical Sciences. Springer, Berlin, New York, 1989. (Cited on p. 59)

[292] J.D. Skufca and E.M. Bollt. A concept of homeomorphic defect for defining mostly conjugate dynamical systems. *Chaos*, 18:013118, 2008. (Cited on pp. 13, 14)

[293] J.D. Skufca, J.A. Yorke, and B. Eckhardt. Edge of chaos in a parallel shear flow. *Physical Review Letters*, 96:174101, 2006. (Cited on p. 228)

[294] S. Smale. Differentiable dynamical systems. *Bulletin of the American Mathematical Society*, 73:747–817, 1967. (Cited on pp. 152, 156, 176, 207)

[295] S. Smale. Finding a horseshoe on the beaches of Rio. *The Mathematical Intelligencer*, 20(1):39–44, 1998. (Cited on pp. 152, 157)

[296] J.L. Snell and J.G. Kemeny. *Finite Markov Chains: With a New Appendix "Generalization of a Fundamental Matrix."* Undergraduate Texts in Mathematics. Springer, 1976. (Cited on pp. 143, 223)

[297] A.H. Sobel, R.A. Plumb, and D.W. Waugh. On methods of calculating transport across the polar vortex edge. *Journal of the Atmospheric Sciences*, 54:2241–2260, 1997. (Cited on pp. 132, 254)

[298] J. Stoer and R. Bulirsch. *Introduction to Numerical Analysis*. Springer, 2002. (Cited on pp. 85, 86, 87)

[299] M. Stoer and F. Wagner. A simple min-cut algorithm. In *Algorithms—ESA94*. Volume 855 of Lecture Notes in Computer Science. Springer, Berlin, 1994, pages 141–147. (Cited on p. 138)

[300] S.H. Strogatz. *Nonlinear Dynamics and Chaos: With Applications to Physics, Biology, Chemistry, and Engineering*. Westview Press, Boulder, CO, 1994. (Cited on p. 302)

[301] S.H. Strogatz. *Nonlinear Dynamics and Chaos: With Applications to Physics, Biology, Chemistry, and Engineering*. Westview Press, Boulder, CO, 2000. (Cited on pp. xii, 30)

[302] S.H. Strogatz and I. Stewart. Coupled oscillators and biological synchronization. *Scientific American*, 269(6):102–109, 1993. (Cited on p. 302)

[303] R. Sturman, J.M. Ottino, and S. Wiggins. *The Mathematical Foundations of Mixing. The Linked Twist Map as a Paradigm in Applications: Micro to Macro, Fluids to Solids*. Cambridge University Press, Cambridge, UK, 2006. (Cited on p. 63)

[304] J. Sun, E.M. Bollt, and T. Nishikawa. Master stability functions for coupled nearly identical dynamical systems. *Europhysics Letters*, 85:60011, 2009. (Cited on p. 302)

[305] D. Sweet, H.E. Nusse, and J.A. Yorke. Stagger-and-step method: Detecting and computing chaotic saddles in higher dimensions. *Physical Review Letters*, 86(11):2261–2264, 2001. (Cited on p. 228)

[306] A.N. Tikhonov. Systems of differential equations containing small parameters in the derivatives. *Matematicheskii Sbornik*, 73(3):575–586, 1952. (Cited on p. 58)

[307] S. M. Ulam. *Problems in Modern Mathematics*. Science Editions. John Wiley, New York, 1970. (Cited on pp. xii, 15, 54, 63, 68, 69, 122, 309, 324, 325)

[308] N. G. Van Kampen. *Stochastic Processes in Physics and Chemistry*, 4th ed. North–Holland, 2003. (Cited on pp. 82, 143)

[309] A.B. Vasil'eva and V.M. Volosov. The work of Tikhonov and his pupils on ordinary differential equations containing a small parameter. *Russian Mathematical Surveys*, 22(2):124–142, 1967. (Cited on p. 58)

[310] J. Von Neumann. Various techniques used in connection with random digits. *Applied Mathematics Series*, 12(36-38):1, 1951. (Cited on p. 8)

[311] S. Wasserman and K. Faust. *Social Network Analysis: Methods and Applications*. Cambridge University Press, Cambridge, UK, 1994. (Cited on p. 132)

[312] C.O. Weiss and J. Brock. Evidence for Lorenz-type chaos in a laser. *Physical Review Letters*, 57(22):2804–2806, 1986. (Cited on p. 152)

[313] C.O. Weiss, W. Klische, N.B. Abraham, and U. Hübner. Comparison of NH_3 laser dynamics with the extended Lorenz model. *Applied Physics B: Lasers and Optics*, 49(3):211–215, 1989. (Cited on p. 152)

[314] C. Werndl. What are the new implications of chaos for unpredictability? *The British Journal for the Philosophy of Science*, 60(1):195, 2009. (Cited on p. 8)

[315] S. Wiggins. *Chaotic Transport in Dynamical Systems*. Springer, 1992. (Cited on pp. 38, 149, 169, 176, 178, 204, 207, 210, 215)

[316] S. Wiggins. *Introduction to Applied Nonlinear Dynamical Systems and Chaos*, 2nd ed. Springer, 2003. (Cited on pp. xii, 29, 35, 36, 37, 38, 39, 40, 169, 178)

[317] S. Wiggins and J.M. Ottino. Foundations of chaotic mixing. *Philosophical Transactions of the Royal Society of London. Series A: Mathematical, Physical and Engineering Sciences*, 362(1818):937, 2004. (Cited on p. 210)

[318] S. Wiggins and V. Rom-Kedar. Transport in two-dimensional maps. *Archive for Rational Mechanics and Analysis*, 109(3):239–298, 1990. (Cited on p. 220)

[319] A. Wolf, J.B. Swift, H.L. Swinney, and J.A. Vastano. Determining Lyapunov exponents from a time series. *Physica D*, 16(3):285–317, 1985. (Cited on p. 298)

[320] K. Yagasaki. The method of Melnikov for perturbations of multi-degree-of-freedom Hamiltonian systems. *Nonlinearity*, 12:799–822, 1999. (Cited on p. 179)

[321] Z.P. You, E.J. Kostelich, and J.A. Yorke. Calculating stable and unstable manifolds. *International Journal of Bifurcation and Chaos*, 1(3):605–623, 1991. (Cited on p. 41)

[322] L.S. Young. Dimension, entropy and Lyapunov exponents. *Ergodic Theory and Dynamical Systems*, 2(01):109–124, 1982. (Cited on p. 297)

[323] L.S. Young. What are SRB measures, and which dynamical systems have them? *Journal of Statistical Physics*, 108(5):733–754, 2002. (Cited on pp. 21, 62)

[324] A. Zhou and J. Ding. *Statistical Properties of Deterministic Systems*. Springer, 2008. (Cited on p. xii)

[325] J. Ziv and A. Lempel. A universal algorithm for sequential data compression. *IEEE Transactions on Information Theory*, 23(3):337–343, 1977. (Cited on pp. 269, 271)

[326] P. Zgliczynski. Computer assisted proof of chaos in the Rössler equations and in the Hénon map. *Nonlinearity*, 10:243–252, 1997. (Cited on pp. 171, 182)

[327] P. Zgliczynski. Rigorous numerics for dissipative partial differential equations II. Periodic orbit for the Kuramoto–Sivashinsky PDE: A computer-assisted proof. *Foundations of Computational Mathematics*, 4(2):157–185, 2004. (Cited on p. 286)

[328] K. Życzkowski and E.M. Bollt. On the entropy devil's staircase in a family of gap-tent maps. *Physica D*, 132(3):392–410, 1999. (Cited on pp. 14, 232, 233)

Index